Discrete Mathematics and Theoretical Computer Science

Springer
London
Berlin
Heidelberg
New York
Barcelona
Hong Kong
Milan
Paris
Santa Clara
Singapore
Tokyo

C. Ding, T. Helleseth and H. Niederreiter (Eds)

Sequences and their Applications

Proceedings of SETA '98

Springer

C. Ding
Department of Computer Science, School of Computing, National University of Singapore, Lower Kent Ridge Road, Singapore 119260

T. Helleseth
Department of Informatics, University of Bergen, N-5020 Bergen, Norway

H. Niederreiter
Institute of Information Processing, Austrian Academy of Sciences, Sonnenfelsgasse 19, A01010 Vienna, Austria

ISBN 1-85233-196-8 Springer-Verlag Berlin Heidelberg New York

British Library Cataloguing in Publication Data
Sequences and their applications : proceedings of SETA '98.
- (Discrete mathematics and theoretical computer science)
1.Sequences (Mathematics) - Congresses 2.Sequences (Mathematics) - Data processing - Congresses
I.Ding, C. (Cunsheng) II.Helleseth, Tor III.Niederreiter, Harald, 1944-
515.2'4
ISBN 1852331968

Library of Congress Cataloging-in-Publication Data
A catalog record for this book is available from the Library of Congress

Printed in Great Britain

Typesetting: Camera ready by contributors
Printed and bound at the Athenæum Press Ltd., Gateshead, Tyne & Wear
34/3830-543210 Printed on acid-free paper SPIN 10739473

Preface

This volume contains the refereed proceedings of the International Conference on Sequences and Their Applications which was held at the River View Hotel in Singapore during December 14-17, 1998. The program of this conference was arranged by a committee consisting of Claude Carlet (University of Caen), Agnes Chan (Northeastern University), Cunsheng Ding (National University of Singapore, co-chair), Dieter Gollmann (Microsoft Research), Tor Helleseth (University of Bergen, co-chair), Kyoki Imamura (Kyushu Institute of Technology), Andrew Klapper (University of Kentucky), Vijay Kumar (University of Southern California), Siu Lun Ma (National University of Singapore), Harald Niederreiter (Austrian Academy of Sciences, co-chair), Dilip Sarwate (University of Illinois at Urbana-Champaign), Hans Schotten (Aachen University of Technology), Jeffrey Shallit (University of Waterloo), Neil Sloane (AT&T Shannon Lab), and Aimo Tietäväinen (University of Turku). The local organization was in the hands of Cunsheng Ding, Kwok Yan Lam (chair), Sjauntele Lau, and Sew Kiok Toh, all of the National University of Singapore.

The idea for the conference grew out of the recognition that sequences in discrete structures like the ring of integers, residue class rings of the integers, and finite fields have found many important applications in modern information and communication technologies. Among these applications we mention cryptographic schemes, ranging systems, spread spectrum communication systems, multi-terminal system identification, code-division multiple-access communication systems, global positioning systems, software testing, circuit testing, and computer simulation. There are also connections between sequences in discrete structures and error-correcting codes. In view of these exciting applications of sequences in discrete structures, it seemed worthwhile and timely to have a conference focusing on the properties of such sequences. The excellent response we received to this idea encouraged us to organize the conference, and we hope that these proceedings will be a testimony to the success of the conference.

We are grateful to the members of the Program Committee for screening abstracts and refereeing full papers. We also thank Jean-Paul Allouche, Eric Bach, Simon Blackburn, Anna Frid, Iiro Honkala, Tohru Kohda, Jyrki Lahtonen, Tero Laihonen, Halvard Martinsen, Alfred Menezes, David Pointcheval, Ari Renvall, Jean-Pierre Tillich, Jinzhong Xu, and Muxiang Zhang for serving as additional referees. All these colleagues have contributed enormously to the quality of the conference presentations and to guaranteeing high standards of these proceedings.

We appreciate the financial support for the conference received from the Centre for Systems Security of the National University of Singapore and the Lee Foundation. The conference would not have been successful without the essential organizational tasks performed in a skillful manner by Sjauntele Lau and Sew Kiok Toh. We express our gratitude to Springer-Verlag for publishing this volume, and especially to Ian Shelley and Caroline Ching for the encouragement and support provided so generously. We thank Professor Cristian Calude for his support of publishing this volume in the Springer DMTCS book series.

May 1999

Cunsheng Ding   Tor Helleseth   Harald Niederreiter

CONTENTS

The Ubiquitous Prouhet-Thue-Morse Sequence

Jean-Paul Allouche[1] and Jeffrey Shallit[2]

[1] CNRS, LRI, Université Paris-Sud, Bâtiment 490
F-91405 Orsay Cedex, France
Email: allouche@lri.fr

[2] Department of Computer Science, University of Waterloo
Waterloo, Ontario N2L 3G1, Canada
Email: shallit@uwaterloo.ca

Abstract. We discuss a well-known binary sequence called the Thue-Morse sequence, or the Prouhet-Thue-Morse sequence. This sequence was introduced by Thue in 1906 and rediscovered by Morse in 1921. However, it was already implicit in an 1851 paper of Prouhet. The Prouhet-Thue-Morse sequence appears to be somewhat ubiquitous, and we describe many of its apparently unrelated occurrences.

1 Introduction

In his seminal 1906 and 1912 papers [65, 66], the Norwegian mathematician Axel Thue (1863–1922) noted that any binary sequence of length ≥ 4 *must* contain a square, i.e., two consecutive identical blocks (the easy proof is left to the reader). He then asked whether it is possible to find an infinite sequence on three letters without squares, i.e., without two consecutive identical blocks. He also asked whether it is possible to find an infinite binary sequence that contains no cube, i.e., no three consecutive identical blocks, or even no overlap, i.e., no sub-block of the form $awawa$, where $a \in \{0, 1\}$ and w is a binary block. The answer to all three questions is positive. Thue used a sequence $\mathbf{t}$ whose construction is given in the next section,

$$\mathbf{t} = 0\,1\,1\,0\,1\,0\,0\,1\,1\,0\,0\,1\,0\,1\,1\,0\,1\,0\,0\,1\,0\,1\,1\,0 \cdots$$

It happens that this sequence $\mathbf{t}$ is really ubiquitous in the literature. In this paper we survey a few of its occurrences in combinatorics on words, differential geometry, number theory, iteration of continuous functions, and mathematical physics. Note that we do not give *all* properties of the sequence $\mathbf{t}$, but rather show how it occurred as a "natural" answer to various apparently unrelated questions.

2 Definition

We first give a formal definition of the Prouhet-Thue-Morse sequence.

Definition 1. *We denote by* $\mathbf{t} = (t_n)_{n \geq 0}$ *the* Prouhet-Thue-Morse *sequence over* $\{0, 1\}$*, defined recursively by* $t_0 = 0$ *and* $t_{2n} = t_n$*,* $t_{2n+1} = \overline{t_n}$ *for all* $n \geq 0$*, where, for* $x \in \{0, 1\}$*, we define* $\overline{x} = 1 - x$*.*

Denote by $s_k(n)$ the sum of the digits in the base-k representation of the integer n. Since we clearly have $s_2(2n) = s_2(n)$ and $s_2(2n+1) = s_2(n) + 1$ for every $n \geq 0$, we easily obtain the following equivalent definition:

Proposition 1. *The Prouhet-Thue-Morse sequence* $\mathbf{t}$ *is equal to the sequence* $(s_2(n) \bmod 2)_{n \geq 0}$.

Yet another definition, easily seen to be equivalent to the previous two, is the following:

Proposition 2. *Let* X *be an indeterminate. Then we have*

$$\begin{aligned}\prod_{i \geq 0}(1 - X^{2^i}) &= (1 - X)(1 - X^2)(1 - X^4) \cdots \\ &= 1 - X - X^2 + X^3 + \cdots \\ &= \sum_{j \geq 0} (-1)^{t_j} X^j.\end{aligned}$$

3 Combinatorics on words

3.1 The pioneering work of Thue

Our first theorem is the one we mentioned in the introduction. It is due to Thue [65, 66].

Theorem 1 (Thue). *The Prouhet-Thue-Morse sequence* $\mathbf{t}$ *is overlap-free.*

For $n \geq 1$ let v_n be the number of 1's between the n'th and $(n+1)$'st occurrence of 0 in the sequence $\mathbf{t}$. Let $\mathbf{v} = (v_n)_{n \geq 1}$. Hence

$$\mathbf{v} = 2\,1\,0\,2\,0\,1\,2\,1\,0\,1\,2 \cdots$$

Thue proved, as a corollary of Theorem 1 above, the following:

Corollary 1. *The sequence* $\mathbf{v} = (v_n)_{n \geq 1}$ *is square-free.*

This work of Thue (see also Berstel [20] and references therein) was the starting point of an important branch of combinatorics, now called *combinatorics on words*. It is worth noting that Thue explained he had no particular application in mind, but he thought the problem was interesting enough in itself to deserve attention.

Thue's papers appeared in an obscure Norwegian journal, and for a long time were not widely known or appreciated. His original results were rediscovered by several different authors, including Marston Morse; see [38, 19].

Although there are uncountably many overlap-free sequences on two symbols, the Prouhet-Thue-Morse sequence is, roughly speaking, the "canonical" example. For example, if in addition to being overlap-free we add some extra requirement, then we often find that the only such sequence is the Prouhet-Thue-Morse sequence or a simple variant. For example, consider the following theorem of Berstel [20].

Theorem 2. *The lexicographically largest overlap-free binary sequence beginning with* 0 *is* $\mathbf{t}$, *the Prouhet-Thue-Morse sequence.*

Recently, together with J. Currie [10], we generalized this theorem, proving in particular the following result:

Theorem 3. *The lexicographically largest overlap-free binary sequence is the sequence* $110110\mathbf{t}$.

3.2 The problem of infinite play in chess

In a little-known 1929 paper, the Dutch chess grandmaster Machgielis (Max) Euwe (1901–1981, world champion 1935–1937) independently discovered the Prouhet-Thue-Morse sequence $\mathbf{t}$ and applied it to an interesting problem in chess [35].

The so-called German rule states that a draw occurs if the same sequence of moves occurs three times in succession. Euwe proved, using the cube-free property of $\mathbf{t}$, that under such a rule infinite games of chess are possible.

For example, one can take the Prouhet-Thue-Morse sequence $\mathbf{t} = t_0 t_1 t_2 \cdots$ and map each 0 to the sequence of four moves (Ng1-f3, Ng8-f6, Nf3-g1, Nf6-g8) and each 1 to the sequence of four moves (Nb1-c3, Nb8-c6, Nc3-b1, Nc6-b8). The resulting sequence of moves represents a legal infinite game of chess, and no draw occurs under the German rule.

Later, Morse rediscovered the same technique [49, 51].

3.3 Morphisms of the free monoid

Let A be an *alphabet*, i.e., a finite set of symbols. The set of *words* over A (i.e., blocks — or strings — of symbols of A) equipped with the operation of *concatenation* is denoted by A^*: this is the *free monoid* generated by A.

Definition 2. *Let A be an alphabet. Let A^* be the free monoid generated by A. A map $\sigma : A^* \to A^*$ is called a* morphism *if for all words x and y in A^* we have $\sigma(xy) = \sigma(x)\sigma(y)$. Such a map is called a* uniform morphism *if all the images by σ of elements of the set A have the same number of letters.*

Note that a morphism is defined by its values on the elements of A. Note also that the morphism σ can be extended to infinite sequences by *continuity* (the set of finite and infinite sequences being equipped with the topology of simple convergence). This means that, for an infinite sequence $\mathbf{z} = (z_n)_{n \geq 0}$, the sequence $\sigma(\mathbf{z})$ is defined by $\sigma(\mathbf{z}) = \sigma(z_0)\sigma(z_1)\sigma(z_2)\cdots$

Proposition 3. *Define the morphism μ on the alphabet $\{0,1\}$ by $\mu(0) = 01$, $\mu(1) = 10$. Then the Prouhet-Thue-Morse sequence* $\mathbf{t}$ *is the unique fixed point of μ that begins with* 0.

Proof. We first note that if an infinite sequence is a fixed point of μ, and begins with 0, it must begin with $\mu(0)$. Since $\mu(0) = 01$, the sequence must begin with $\mu(01) = \mu(0)\mu(1)$, hence with $\mu^2(0)$. Iterating, this means that the sequence must begin with $\mu^k(0)$ for every $k \geq 0$. This proves uniqueness. Since $\mu(0)$ begins with 0, we have that $\mu^{k+1}(0)$ begins with $\mu^k(0)$ for every k. Hence the sequence of words $(\mu^k(0))_{k\geq 0}$ converges towards an infinite sequence, say $\mathbf{z} = (z_n)_{n\geq 0}$, that clearly is a fixed point of μ.

Now, for $x \in \{0,1\}$, we have $\mu(x) = x\overline{x}$, where, as previously, $\overline{x} = 1 - x$. Since $\mathbf{z}$ is a fixed point of μ, we thus have for every $n \geq 0$, that $z_{2n} = z_n$ and $z_{2n+1} = \overline{z_n}$. Hence the sequence $\mathbf{z}$ is equal to the Prouhet-Thue-Morse sequence. □

Is it possible to build another binary sequence that is both overlap-free and generated by a morphism? The next theorem, due to Séébold [62], answers this question negatively. Another proof of this result was given by Berstel and Séébold in [21].

Theorem 4 (Séébold). *If an overlap-free binary sequence is a fixed point of a non-trivial morphism, then it is either equal to* $\mathbf{t}$*, the Prouhet-Thue-Morse sequence, or its complement* $\overline{\mathbf{t}} = (\overline{t_n})_{n\geq 0} = 1001011001101001\cdots$.

The Thue-Morse sequence is the prototype of a class of sequences called 2-*automatic sequences.* Roughly speaking, a sequence is k-automatic if its n'th term is generated by a finite-state machine which takes as input the base-k expansion of n. For more about this class of sequences, see, for example, [29, 33, 4]. For the general subject of combinatorics on words, see [43].

4 Differential geometry

The Prouhet-Thue-Morse sequence has the nice property that it exhibits regularity without being ultimately periodic. Morse rediscovered the sequence $\mathbf{t}$ in 1921 in connection with differential geometry [48]. He proved the following:

Theorem 5 (Morse). *On a surface of negative curvature, having at least two different normal segments, there exists a set of geodesics that are recurrent without being periodic, and this set has the power of the continuum.*

To prove this result, one of the steps was the following proposition, given as a lemma in [48, p. 95]. We say a sequence $\mathbf{a} = a_0a_1a_2\cdots$ is *uniformly recurrent* if for each finite block of symbols w occurring in $\mathbf{a}$ there exists an integer n such that for all i, the sub-block $a_{i+1}\cdots a_{i+n}$ contains an occurrence of w. We say a sequence $\mathbf{a} = a_0a_1a_2\cdots$ is *ultimately periodic* if there exist integers $p \geq 1$, $N \geq 0$ such that $a_i = a_{i+p}$ for all $i \geq N$.

Proposition 4. *There exists an infinite sequence over* $\{1, 2\}$ *which is uniformly recurrent but not ultimately periodic.*

The sequence that Morse gives is exactly $\mathbf{t}$, where 0's have been replaced by 1's and 1's by 2's.

5 Number theory

5.1 The Prouhet-Tarry-Escott problem

As already noted by Adler and Li [2], the sequence $\mathbf{t}$ appears implicitly in an 1851 paper of Prouhet [55]. Prouhet was interested in a problem that was also studied more than fifty years later by Tarry and Escott, and which is now known as the "Prouhet-Tarry-Escott" or "multigrades" problem.

Prouhet addressed the following question: is it possible to find a partition of the set $\{0, 1, 2, \ldots, 2^N - 1\}$ into two disjoint sets I and J, such that $\sum_{i \in I} i^k = \sum_{j \in J} j^k$ for $k = 0, 1, 2, \ldots, t$? Of course we take $0^0 = 1$, so that in particular the case $k = 0$ shows that I and J must have the same number of elements. Prouhet proved that such a partition is possible if $N = t + 1$.

Theorem 6 (Prouhet). *The Prouhet-Thue-Morse sequence* $\mathbf{t} = (t_n)_{n \geq 0}$ *has the following property. Define*

$$\begin{aligned} I &= \{i \in \{0, 1, 2, 3, \ldots, 2^N - 1\} : \ t_i = 0\}, \\ J &= \{j \in \{0, 1, 2, 3, \ldots, 2^N - 1\} : \ t_j = 1\}. \end{aligned}$$

Then for $0 \leq k \leq N - 1$ *we have*

$$\sum_{i \in I} i^k = \sum_{j \in J} j^k.$$

For example, we have

$$0^k + 3^k + 5^k + 6^k + 9^k + 10^k + 12^k + 15^k = 1^k + 2^k + 4^k + 7^k + 8^k + 11^k + 13^k + 14^k$$

for $k = 0, 1, 2, 3$.

Prouhet actually studied the more general problem of finding a partition of $[0, q^N - 1]$ into q sets $I_1, I_2, \ldots, I_q$ such that the q sums $\sum_{i \in I_j} i^k$ (with $j = 1, 2, \ldots, q$ and $k = 0, 1, \ldots, N - 1$) do not depend on j. He gave the following solution (for a proof see e.g., [42]): for each $q \geq 2$ define the sequence $\mathbf{T}_q = (T_q(n))_{n \geq 0}$ by $T_q(n) = s_q(n) \bmod q$. Then let $I_j = \{0 \leq i \leq q^N - 1 \ : \ T_q(i) = j\}$.

For an occurrence of related sequences, see [59]. A very nice relationship between magic cubes, Prouhet sequences, and the Prouhet-Tarry-Escott problem was given by Adler and Li [2]. For the state of art about the Prouhet-Tarry-Escott problem the reader can look at one of the surveys [24, 60].

5.2 Curious infinite products

Woods asked [68], what is the limit of the sequence

$$1/2, \quad \frac{1}{2}\Big/\frac{3}{4}, \quad \frac{\frac{1}{2}}{\frac{3}{4}}\Big/\frac{\frac{5}{6}}{\frac{7}{8}}, \quad \ldots \quad ?$$

Robbins [58] proved that this limit is $\frac{\sqrt{2}}{2}$. More precisely, we have the following:

Proposition 5. *Let $\varepsilon_n = (-1)^{t_n}$, where $(t_n)_{n\geq 0}$ is the Prouhet-Thue-Morse sequence. Then*

$$\left(\frac{1}{2}\right)^{\varepsilon_0}\left(\frac{3}{4}\right)^{\varepsilon_1}\left(\frac{5}{6}\right)^{\varepsilon_2}\cdots = \prod_{n=0}^{\infty}\left(\frac{2n+1}{2n+2}\right)^{\varepsilon_n} = \frac{\sqrt{2}}{2}. \tag{1}$$

Proof. We give a simple proof, discovered by the first author in 1987. Let P and Q be the infinite products defined by

$$P = \prod_{n=0}^{\infty}\left(\frac{2n+1}{2n+2}\right)^{\varepsilon_n}, \quad Q = \prod_{n=1}^{\infty}\left(\frac{2n}{2n+1}\right)^{\varepsilon_n}.$$

Then

$$PQ = \frac{1}{2}\prod_{n=1}^{\infty}\left(\frac{n}{n+1}\right)^{\varepsilon_n} = \frac{1}{2}\prod_{n=1}^{\infty}\left(\frac{2n}{2n+1}\right)^{\varepsilon_{2n}}\prod_{n=0}^{\infty}\left(\frac{2n+1}{2n+2}\right)^{\varepsilon_{2n+1}}.$$

Of course all products are convergent by Abel's theorem. Now, since $\varepsilon_{2n} = \varepsilon_n$, and $\varepsilon_{2n+1} = -\varepsilon_n$, we get

$$PQ = \frac{1}{2}\prod_{n=1}^{\infty}\left(\frac{2n}{2n+1}\right)^{\varepsilon_n}\left(\prod_{n=0}^{\infty}\left(\frac{2n+1}{2n+2}\right)^{\varepsilon_n}\right)^{-1} = \frac{1}{2}\frac{Q}{P}.$$

Since $Q \neq 0$, this gives $P^2 = 1/2$, and the result follows since P is positive. □

Note that the mysterious number Q does not appear in the final result! No expression for Q in terms of known constants is currently known, nor is it known if Q is transcendental or even irrational. This number first occurred in a paper of Flajolet and Martin [36], who studied a class of probabilistic counting algorithms for estimating the number of distinct elements in a large collection of data. Their asymptotic analysis involves the constant $\varphi = .77351\cdots$ given by

$$\varphi = 2^{-1/2}e^{\gamma}\frac{2}{3}\prod_{n=1}^{\infty}\left(\frac{(4n+1)(4n+2)}{4n(4n+3)}\right)^{(-1)^{t_n}}$$

where γ is Euler's constant [36, Theorem 3.A]. It is clear that

$$Q = 2^{-1/2}e^{\gamma}\varphi^{-1}.$$

It is precisely while he was trying to compute Q (and hence φ) that the first author came across the proof above.

The infinite product (1) suggests trying to obtain the expansion

$$\left(\frac{1}{2}\right)^{\pm 1}\left(\frac{3}{4}\right)^{\pm 1}\left(\frac{5}{6}\right)^{\pm 1}\cdots$$

of a number β (for instance, $\frac{\sqrt{2}}{2}$) by a greedy algorithm, where the signs in the exponent are iteratively chosen so that the product thus far approximates β as closely as possible at each step. The following conjecture of the second author [63] was proved by the first author and H. Cohen [6].

Theorem 7 (Allouche, Cohen). *Define a sequence of signs $(\alpha_n)_{n\geq 0}$ by $\alpha_0 = 1$, and, if $\alpha_0, \alpha_1, \ldots, \alpha_n$ are known, define α_{n+1} by:*

$$\alpha_{n+1} = \begin{cases} +1 \text{ if } \left(\frac{1}{2}\right)^{\alpha_0}\left(\frac{3}{4}\right)^{\alpha_1}\cdots\left(\frac{2n+1}{2n+2}\right)^{\alpha_n} > \frac{\sqrt{2}}{2}, \\ -1 \text{ if } \left(\frac{1}{2}\right)^{\alpha_0}\left(\frac{3}{4}\right)^{\alpha_1}\cdots\left(\frac{2n+1}{2n+2}\right)^{\alpha_n} < \frac{\sqrt{2}}{2}. \end{cases}$$

Then the sequence $(\alpha_n)_{n\geq 0}$ is equal to the Prouhet-Thue-Morse sequence on the alphabet $\{-1, +1\}$, i.e., $\alpha_n = (-1)^{t_n}$ for all $n \geq 0$.

For generalizations of these results, see [63, 6, 13, 7].

5.3 Partitions of the set of integers

Let A be the (lexicographically) smallest set of integers such that 0 and 1 belong to A, and for each $n \geq 1$ that belongs to A, the number $2n$ does not belong to A. Hence the first few elements of A are

$$0 \;\; 1 \;\; 3 \;\; 4 \;\; 5 \;\; 7 \;\; 9 \;\; 11 \;\; 12 \;\; 13 \;\; 15 \;\; 16 \;\; 17 \;\; 19 \;\; 20 \;\; 21 \;\; 23 \;\; \cdots$$

It is not hard to see that A and $2A = \{2x \; : \; x \in A\}$ form a partition of the set of non-negative integers; see [26, 5, 64]. An unexpected connection with the Thue-Morse sequence, proved in [5], is given below.

Theorem 8 (Allouche, Arnold, Berstel, Brlek, Jockusch, Plouffe, Sagan). *Let $(a_n)_{n\geq 0} = 0\; 1\; 3\; 4\; 5\; 7\; 9\; 11\; 12\; 13\; 15\; 16\; 17\; 19\; 20\; 21\; 23\; \cdots$ be the increasing sequence of the elements of A. Define the sequence $\mathbf{z} = (z_n)_{n\geq 0}$ by*

$$\mathbf{z} = 0^{a_1 - a_0} \;\; 1^{a_2 - a_1} \;\; 0^{a_3 - a_2} \;\; \cdots \;\; 0^{a_{2n+1} - a_{2n}} \;\; 1^{a_{2n+2} - a_{2n+1}} \cdots$$

where by c^j for $c \in \{0, 1\}$, we mean the string $\overbrace{cc\cdots c}^{j}$. In other words, the sequence of runs *of* $\mathbf{z}$ *is the first difference of the sequence $(a_n)_{n\geq 0}$. Then* $\mathbf{z}$ *is equal to the Prouhet-Thue-Morse sequence.*

5.4 Algebraicity of formal power series in positive characteristic

The Prouhet-Thue-Morse real number $\sum_{n\geq 0} t_n 2^{-n}$ was proved transcendental by Mahler [44, p. 363]; also see Dekking [32]. What can be said about the formal power series $\sum_{n\geq 0}^{\infty} t_n X^{-n}$? This series is transcendental over $\mathbb{Q}(X)$, as noted, for example, in [32]. But, considering this series modulo 2, we have the following proposition:

Proposition 6. *Let $F(X) = \sum_{n\geq 0} t_n X^{-n}$. Then F, considered as an element of $\mathbb{F}_2[[X^{-1}]]$, is quadratic over $\mathbb{F}_2(X)$. More precisely, we have*

$$(1+X)^3F^2 + X(1+X)^2F + X^2 = 0. \tag{2}$$

Proof. This is an easy consequence of the recurrence relations satisfied by the sequence $\mathbf{t}$. Namely, all computations being done modulo 2, we have

$$\begin{aligned}
F &= \sum_{n\geq 0} t_n X^{-n} = \sum_{n\geq 0} t_{2n} X^{-2n} + \sum_{n\geq 0} t_{2n+1} X^{-2n-1} \\
&= \sum_{n\geq 0} t_n X^{-2n} + X^{-1}\sum_{n\geq 0}(1+t_n)X^{-2n} \\
&= F^2 + X^{-1}\left(\frac{X^2}{1+X^2} + F^2\right) \\
&= \left(\frac{1+X}{X}\right)F^2 + \frac{X}{1+X^2} = \left(\frac{1+X}{X}\right)F^2 + \frac{X}{(1+X)^2}.
\end{aligned}$$

Hence, multiplying through by $X(1+X)^2$, we obtain Eq. (2). The fact that F is not a rational function is an easy consequence of the overlap-free property of the sequence $\mathbf{t}$.

More generally, a formal power series with coefficients in $\mathbb{F}_p$, where p is a prime number, is algebraic over the field $\mathbb{F}_p(X)$ if and only if the sequence of its coefficients is p-automatic. This theorem was proved by Christol [27], and more details are given in the paper of Christol, Kamae, Mendès France and Rauzy [28].

There is a theory of continued fractions for Laurent series with coefficients in a finite field [16]. In particular the continued fraction expansion of a quadratic series is ultimately periodic (see [47]; see [61] for a careful study when the ground field is not finite). The continued fraction expansion of $\sum_{n\geq 0} t_n X^{-n}$ is ultimately periodic with a pleasantly short period. It is given by

$$\sum_{n\geq 0} t_n X^{-n} = [0,\ X+1,\ \overline{X,\ X,\ X^3+X,\ X}\ ]$$

where, as usual, the vinculum denotes the repeating portion of the ultimately periodic continued fraction.

5.5 β-Expansions

Representing real numbers in non-integer bases goes back to Rényi [56] and Parry [54]. These expansions — sometimes called β-expansions — differ in some respects from the usual base-k expansions where k is an integer. For example, some numbers may have multiple representations. However, Komornik and Loreti recently proved the following theorem [41]:

Theorem 9 (Komornik, Loreti). *There exists a smallest real number β in the interval $(1,2)$, for which there exists a unique β-expansion of 1 in the form $1 = \sum_{n=1}^{\infty} \delta_n \beta^{-n}$, with $\delta_n \in \{0,1\}$. Furthermore, for this smallest β, the sequence of "digits" $(\delta_n)_{n\geq 1}$ satisfies $\delta_n = t_n$ for $n \geq 1$, where $\mathbf{t} = t_0 t_1 t_2 \cdots$ is the Prouhet-Thue-Morse sequence. The number β is the unique positive root of the equation $1 = \sum_{n=1}^{\infty} t_n \beta^{-n}$, and we have $\beta \doteq 1.787231650$.*

Komornik and Loreti proved [41] that the above result is a consequence of the following proposition:

Proposition 7. *The lexicographically least binary sequence $(w_n)_{n\geq 1}$ such that*

$$\begin{array}{ll} w_{n+1}\ w_{n+2} \cdots < w_1\ w_2 \cdots & \text{if } w_n = 0, \\ \overline{w_{n+1}}\ \overline{w_{n+2}} \cdots < w_1\ w_2 \cdots & \text{if } w_n = 1, \end{array}$$

(where the order is lexicographical order, and $\overline{0} = 1$, $\overline{1} = 0$) satisfies $w_n = t_n$ for $n \geq 1$.

The second author observed that this last result was previously stated in a slightly different form by the first author and M. Cosnard in [8]. See [9] and Section 7.1 below.

6 Semigroup and group theory

The Prouhet-Thue-Morse sequence $\mathbf{t}$ (or one of its variants) occurs in the solution of the Burnside problem for groups: *Is every group with a finite number of generators and satisfying the identity $x^n = 1$ finite?* The answer is yes (and well-known) if $n = 2$, since the group must be abelian in this case. But the answer is no for large odd n, since, as Novikov and Adian showed [52], an infinite group $\Gamma(m,n)$ on m generators and satisfying $x^n = 1$ for all $x \in \Gamma(m,n)$ exists for all $m > 1$, and for all odd n with $n \geq 4381$. Adian's book [1] gives more details about the result and its history, and improves the constant 4381 to 665. One of the steps in the proof consists of finding a cube-free binary sequence (see [1, p. 5]). Actually the cube-free binary sequence given there is *not* the Prouhet-Thue-Morse sequence, since it is not overlap-free. The author uses a result of Arshon [15] in order to construct a cube-free binary sequence, but in that paper Arshon actually gave a cube-free binary sequence that *is equal* to the Prouhet-Thue-Morse sequence on the alphabet $\{1,2\}$ (see [15, p. 779]).

One may also consider the Burnside problem for semigroups. As remarked by Morse and Hedlund [50, 57], with the aid of **v**, the square-free sequence over $\{0,1,2\}$ given above, one can construct an infinite semigroup S on three generators such that $x^3 = x^2$ for all $x \in S$. Indeed, this is accomplished by letting $S = \{0,1,2\}^* \cup \{z\}$, where z is the zero element (i.e., $wz = zw = z$ for all $w \in S$) and subject to the relation $w^2 = z$ for all $w \in S$. Related questions were discussed by Brzozowski, Culik, and Gabrielian [25].

There is another occurrence of the Thue-Morse sequence in group theory, [22], as follows:

Theorem 10 (Boffa, Point). *Define the* Thue-Morse *group identities* I_n *by:* $I_0(x,y)$ *if* $x = y$; $I_{n+1}(x,y)$ *is* $I_n(xy,yx)$. *Then a finite group satisfies a Thue-Morse identity if and only if it is an extension of a nilpotent group by a 2-group.*

The reader will have noticed that

$$\begin{array}{l} I_0(x,y) \text{ if and only if } x = y \\ I_1(x,y) \text{ if and only if } xy = yx \\ I_2(x,y) \text{ if and only if } xyyx = yxxy \\ I_3(x,y) \text{ if and only if } xyyxyxxy = yxxyxyyx \\ \qquad\vdots \end{array}$$

and understood the terminology "Thue-Morse identities". See [23] for a generalization.

7 Real analysis

7.1 Iteration of continuous functions

Iterating a unimodal continuous function from $[0,1]$ into $[0,1]$ yields various behaviors going from convergent orbits to chaos; see the general reference [30]. M. Cosnard and the first author proved the following theorem [8] (also see [3, 31, 39]). Let f be a continuous map from $[0,1]$ into $[0,1]$, that is unimodal, i.e., increasing for x between 0 and some c, then decreasing for x between c and 1. Suppose that $f(1) = 0$. With the orbit $(f^{(n)}(1))_{n\geq 0}$ we associate the *itinerary* $(a_n)_{n\geq 0}$, defined by $a_n = 0$ if $0 \leq f^{(n)}(1) < c$ and $a_n = 1$ if $c < f^{(n)}(1) \leq 1$. With the sequence $\mathbf{a} = (a_n)_{n\geq 0}$ we associate the sequence $\widetilde{\mathbf{a}} = (\widetilde{a_n})_{n\geq 1}$ defined by

$$\widetilde{a_n} = \left(\sum_{j=0}^{n-1} a_j\right) \bmod 2.$$

Note that the sequence **a** is the first difference, taken modulo 2, of the sequence $\widetilde{\mathbf{a}}$.

Theorem 11 (Allouche, Cosnard). *The set of binary sequences $\widetilde{\mathbf{a}} = (\widetilde{a_n})_{n\geq 1}$ corresponding to unimodal continuous functions is exactly the set of binary sequences Γ defined by*

$$\Gamma = \{\mathbf{b} = (b_n)_{n\geq 1} : \ b_1 = b_2 = 1; \ \forall k, \ \overline{\mathbf{b}} \leq T^k(\mathbf{b}) \leq \mathbf{b}\},$$

where the order is lexicographical order, where $\overline{\mathbf{b}} = (\overline{b_n})_{n\geq 0} = (1 - b_n)_{n\geq 0}$, and where the map T is the shift defined by $T((w_n)_{n\geq 1}) = ((w_{n+1})_{n\geq 1})$. The least non-periodic element of Γ (which is also the least accumulation point of Γ) is $(t_n)_{n\geq 1}$ where $\mathbf{t} = t_0 t_1 t_2 \cdots$ is the Prouhet-Thue-Morse sequence.

Note that the classical approach does not use the transformation $\mathbf{a} \to \widetilde{\mathbf{a}}$; only the itineraries $\mathbf{a}$ are considered. The order between sequences is somewhat more complicated, although it boils down to the lexicographical order after applying the transformation $\mathbf{a} \to \widetilde{\mathbf{a}}$. The sequence $\mathbf{a}$ such that $\widetilde{\mathbf{a}} = (t_n)_{n\geq 1}$, i.e., the first difference of the Prouhet-Thue-Morse sequence, is

$$\mathbf{a} = 1\ 0\ 1\ 1\ 1\ 0\ 1\ 0\ \cdots$$

This sequence is called the *period-doubling sequence.* It is not hard to show that $\mathbf{a}$ is a fixed point of the morphism θ defined by $\theta(1) = 10$, $\theta(0) = 11$. For connections with Gray code, see [37].

The link between Theorem 11 and Proposition 7 above is easy [9]. Note also that Theorem 11 can be reformulated in number-theoretical terms:

Corollary 2. *Let Γ' the set of real numbers defined by*

$$\Gamma' = \{x \in [0,1] : \ \forall k \geq 0, \ 1 - x \leq \{2^k x\} \leq x\},$$

where $\{y\}$ denotes the fractional part of the real number y. Then the least irrational element of Γ' (which is also the least accumulation point of Γ') is the number $\tau = \sum_{n\geq 1} t_n 2^{-n} \doteq 0.824908$, where $\mathbf{t} = (t_n)_{n\geq 0}$ is the Prouhet-Thue-Morse sequence.

The number τ appears in other contexts. For example, let p be the probability that a randomly-chosen language L over $\{0,1\}$ has the property that there is at least one word of each possible length. (We flip a fair coin for each word w to decide if it is in L.) Then, as the second author has observed,

$$p = \prod_{i\geq 0}(1 - \frac{1}{2^{2^i}}) = \sum_{j\geq 0} \frac{(-1)^{t_j}}{2^j} = 2 - 2\tau.$$

7.2 The Knopp function

The Knopp function (see the introduction of [34]) is defined, for $a \in (0,1)$ and $b \in \mathbb{N} \setminus \{0\}$, by

$$K_{a,b}(x) = \sum_{n=0}^{\infty} a^n \|b^n x\|$$

where $\|y\|$ is the distance from y to the nearest integer. In 1990, S. Dubuc and A. Elqortobi [34] came across the Prouhet-Thue-Morse sequence $\mathbf{t} = (t_n)_{n\geq 0}$ while studying the maximum of the Knopp function. They proved the following.

Theorem 12 (Dubuc, Elqortobi). *Let $a \in (0,1)$ and let b be an even integer ≥ 2. Let $X^*(a,b)$ be the set of points where the function $K_{a,b}$ takes its maximum. Then the limit of the set $X^*(a,b)$, as $a \to (1/b)_-$, is the set $\{x, 1-x\}$, where $x = ((b^2-b)/2)\sum_{n=0}^{\infty} t_n/b^{n+1}$.*

8 Physics

Since the Prouhet-Thue-Morse sequence is both "easy to generate" and "non-trivial", it permits to generate a kind of *controlled disorder*. In particular this sequence has analogies (but also differences) with one-dimensional quasi-crystals: actually a typical one-dimensional quasicrystal is the Fibonacci sequence, i.e., the fixed point of the Fibonacci morphism $0 \to 01$, $1 \to 0$. Hence a large number of papers in physics study the Prouhet-Thue-Morse sequence. We only mention [17, 18], and the papers given in the bibliography of [12].

9 Generalizations

The alternative definitions of the Prouhet-Thue-Morse sequence given in Section 2 each suggest possible approaches to generalize the sequence.

For example, Proposition 1 suggests studying the generalized Prouhet-Thue-Morse sequence $\mathbf{t}_{k,m} = (s_k(n) \bmod m)_{n\geq 0}$ for integers $k \geq 2$ and $m \geq 1$. For example, we have

$$\mathbf{t}_{3,4} = 0\ 1\ 2\ 1\ 2\ 3\ 2\ 3\ 0\ 1\ 2\ 3\ \cdots.$$

Note that the sequence $\mathbf{t}_{2,m}$ has been studied by J. Tromp and the second author in [67], and that the sequence $\mathbf{t}_{q,q}$ is the sequence $\mathbf{T}_q$ of Section 5.1. Very recently the authors proved the following theorem [14], which generalizes the work of Thue:

Theorem 13. *Let $k \geq 2$, $m \geq 1$ be integers. The generalized Prouhet-Thue-Morse sequence $\mathbf{t}_{k,m}$ is overlap-free if and only if $m \geq k$.*

Several other generalizations of the Prouhet-Thue-Morse sequence have been studied, see for example [40, 45, 46, 53, 69].

10 Conclusion

The Prouhet-Thue-Morse sequence occurs in various fields, so that many apparently unrelated *definitions* of this sequence are equivalent. For example, Proposition 1, Theorem 2, Proposition 3, Theorem 4, Theorem 7, Theorem 8, Theorem 9, Proposition 7, Theorem 11, and even Theorem 10 or Corollary 2 can be turned into definitions. Automatic sequences, of which the Prouhet-Thue-Morse sequence is a simple example, are also useful because they are both "simple to generate" and "non-trivial": in physics as mentioned above, but also in other fields, such as music (see for example [11]). Searching for the many occurrences of the Prouhet-Thue-Morse sequence in the literature can be used as a pretext to take a delightful stroll through many fascinating areas of mathematics.

Acknowledgments

The first author wants to thank Professor Cunsheng Ding and Professor Kwok Yan Lam for having made it possible for him to come to Singapore for SETA98.

References

1. S. I. Adian, "The Burnside problem and identities in groups," Ergebnisse der Mathematik und ihrer Grenzgebiete **95**, Springer Verlag, 1979.
2. A. Adler and S.-Y. R. Li, Magic cubes and Prouhet sequences, *Amer. Math. Monthly* **84** (1977), 618–627.
3. J.-P. Allouche, "Théorie des Nombres et Automates," Thèse d'État, Université Bordeaux I, 1983.
4. J.-P. Allouche, Automates finis en théorie des nombres, *Exposition. Math.* **5** (1987), 239–266.
5. J.-P. Allouche, A. Arnold, J. Berstel, S. Brlek, W. Jockusch, S. Plouffe, and B. E. Sagan, A relative of the Thue-Morse sequence, *Discrete Math.* **139** (1995), 455–461.
6. J.-P. Allouche and H. Cohen, Dirichlet series and curious infinite products, *Bull. Lond. Math. Soc.* **17** (1985), 531–538.
7. J.-P. Allouche, H. Cohen, J. Shallit, and M. Mendès France, De nouveaux curieux produits infinis, *Acta Arith.* **49** (1987), 141–153.
8. J.-P. Allouche and M. Cosnard, Itérations de fonctions unimodales et suites engendrées par automates, *C. R. Acad. Sci. Paris Sér. I* **296** (1983), 159–162.
9. J.-P. Allouche and M. Cosnard, Non-integer bases, iteration of continuous real maps, and an arithmetic self-similar set, Preprint (submitted), 1998.
10. J.-P. Allouche, J. Currie, and J. Shallit, Extremal infinite overlap-free binary words, *Electronic J. Combinatorics* **5** (1) (1998), #R27.
11. J.-P. Allouche and T. Johnson, Finite automata and morphisms in assisted musical composition, *Journal of New Music Research* **24** (1995), 97–108.
12. J.-P. Allouche and M. Mendès France, Automata and automatic sequences, in "Beyond quasicrystals," F. Axel and D. Gratias, eds., Springer/Les Éditions de Physique, 1995, pp. 293–367.
13. J.-P. Allouche and J. Shallit, Infinite products associated with counting blocks in binary strings, *J. London Math. Soc.* **39** (1989), 193–204.
14. J.-P. Allouche and J. Shallit, Sums of digits and overlap-free words, in preparation.
15. C. E. Arshon, Proof of the existence of asymmetric infinite sequences, *Mat. Sb.* **44** (1937), 769–779.
16. E. Artin, Quadratische Körper im Gebiet der höheren Kongruenzen I, II, *Math. Zeitschrift* **19** (1924), 163–187. Reprinted in "Collected papers," pp. 1–104.
17. F. Axel, J.-P. Allouche, M. Kleman, M. Mendès France, and J. Peyrière, Vibrational modes in a one dimensional "quasi-alloy", *J. Physique*, Colloque C3, Supplement to No. 7 **47** (1986), C3-181–C3-186.

18. F. Axel and J. Peyrière, Spectrum and extended states in a harmonic chain with controlled disorder: effects of the Thue-Morse symmetry, *J. Statist. Phys.* **57** (1989), 1013–1047.
19. J. Berstel, Some recent results on squarefree words, in: "STACS84", M. Fontet and K. Mehlhorn, eds., Lecture Notes in Computer Science **166**, Springer Verlag, 1984, pp. 14–25.
20. J. Berstel, "Axel Thue's Papers on Repetitions in Words: a Translation," Publications du Laboratoire de Combinatoire et d'Informatique Mathématique **20**, Université du Québec à Montréal, 1995.
21. J. Berstel and P. Séébold, A characterization of overlap-free morphisms, *Disc. Appl. Math.* **46** (1993), 275–281.
22. M. Boffa and F. Point, Identités de Thue-Morse dans les groupes, *C. R. Acad. Sci. Paris Sér. I* **312** (1991), 667–670.
23. M. Boffa and F. Point, m-Identities, *C. R. Acad. Sci. Paris Sér. I* **314** (1991), 879–880.
24. P. Borwein and C. Ingalls, The Prouhet-Tarry-Escott problem revisited, *Enseign. Math.* **40** (1994), 3–27.
25. J. A. Brzozowski, K. Culik II, and A. Gabrielian, Classification of noncounting events, *J. Comput. System Sci.* **5** (1971), 41–53.
26. L. Carlitz, R. Scorville, and V. E. Hoggatt jr., Representations for a special sequence, *Fibonacci Quart.* **10** (1972), 499–518, 550.
27. G. Christol, Ensembles presque périodiques k-reconnaissables, *Theoret. Comput. Sci.* **9** (1979), 141–145.
28. G. Christol, T. Kamae, M. Mendès France, and G. Rauzy, Suites algébriques, automates et substitutions, *Bull. Soc. Math. France* **108** (1980), 401–419.
29. A. Cobham, Uniform tag sequences, *Math. Systems Theory* **6** (1972), 164–192.
30. P. Collet, J.-P. Eckmann, "Iterated maps on the interval as dynamical systems," Progress in Physics, Birkhäuser, 1980.
31. M. Cosnard, Étude de la classification topologique des fonctions unimodales, *Ann. Inst. Fourier* **35** (1985) 59–77.
32. M. Dekking, Transcendance du nombre de Thue-Morse, *C. R. Acad. Sci. Paris Sér. I* **285** (1977), 157–160.
33. F. M. Dekking, M. Mendès France, and A. van der Poorten, Folds!, *Math. Intelligencer* **4** (1982), 130–138, 173–181, 190–195.
34. S. Dubuc, A. Elqortobi, Le maximum de la fonction de Knopp, *Information Systems and Operational Research* **28** (1990), 311–323.
35. M. Euwe, Mengentheoretische Betrachtungen über das Schachspiel. *Proc. Konin. Akad. Wetenschappen, Amsterdam* **32** (1929), 633–642.
36. P. Flajolet and G. Nigel Martin, Probabilistic counting algorithms for data base applications, *J. Comput. Syst. Sci.* **31** (1985), 182–209.
37. H. Fredricksen, Gray codes and the Thue-Morse-Hedlund sequence, *J. Combin. Math. Combin. Comput.* **11** (1992), 3–11.
38. G. A. Hedlund, Remarks on the work of Axel Thue on sequences, *Nordisk Mat. Tidskrift* **15** (1967), 148–150.
39. L. Jonker, Periodic orbits and kneading invariants, *Proc. London Math. Soc.* **3** (1979), 428–450.
40. M. Keane, Generalized Morse sequences, *Z. Wahrscheinlichkeitstheorie Verw. Geb.* **10** (1968), 335–353.
41. V. Komornik and P. Loreti, Unique developments in non-integer bases, *Amer. Math. Monthly* **105** (1998), 636–639.

42. D. H. Lehmer, The Tarry-Escott problem, *Scripta Math.* **13** (1947), 37–41.
43. M. Lothaire, "Combinatorics on words," 2nd ed., Encyclopedia of Mathematics and its Applications **17**, 1997.
44. K. Mahler, Arithmetische Eigenschaften der Lösungen einer Klasse von Funktionalgleichungen, *Math. Annalen* **101** (1929), 342–266. Corrigendum, **103** (1930), 532.
45. J. C. Martin, Generalized Morse sequences on n symbols, *Proc. Amer. Math. Soc.* **54** (1976), 379–383.
46. J. C. Martin, The structure of generalized Morse minimal sets on n symbols, *Trans. Amer. Math. Soc.* **232** (1977), 343–355.
47. B. de Mathan, Approximations diophantiennes dans un corps local, *Bull. Soc. Math. France, Suppl. Mém.* **21** (1970).
48. M. Morse, Recurrent geodesics on a surface of negative curvature, *Trans. Amer. Math. Soc.* **22** (1921), 84–100.
49. M. Morse, Abstract 360: a solution of the problem of infinite play in chess, *Bull. Amer. Math. Soc.* **44** (1938), 632.
50. M. Morse and G. A. Hedlund, Symbolic dynamics, *Amer. J. Math.* **60** (1938), 815–866.
51. M. Morse and G. A. Hedlund, Unending chess, symbolic dynamics, and a problem in semigroups, *Duke Math. J.* **11** (1944), 1–7.
52. P. S. Novikov and S. I. Adian, Infinite periodic groups, I, II, III, *Izv. Akad. Nauk. SSSR Ser. Mat.* **32** (1968), 212–244, 251–524, 709–731.
53. R. Nürnberg, All generalized Morse-sequences are loosely Bernoulli, *Math. Zeitschrift* **182** (1983), 403–407.
54. W. Parry, On the β-expansions of real numbers, *Acta Math. Acad. Sci. Hung.* **11** (1960), 401–416.
55. E. Prouhet, Mémoire sur quelques relations entre les puissances des nombres, *C. R. Acad. Sci. Paris Sér. I* **33** (1851), 225.
56. A. Rényi, Representations for real numbers and their ergodic properties, *Acta Math. Acad. Sci. Hung.* **8** (1957), 477–493.
57. A. Restivo and C. Reutenauer, Rational languages and the Burside problem, *Theoret. Comput. Sci.* **40** (1985), 13–30.
58. D. Robbins, Solution to problem E 2692, *Amer. Math. Monthly* **86** (1979), 394–395.
59. J. B. Roberts, A curious sequence of signs, *Amer. Math. Monthly* **64** (1957), 317–322.
60. O. Salon, Le problème de Prouhet-Tarry-Escott, *Prétirage du LMD, Marseille* **94-23** (1994).
61. W. M. Schmidt, On continued fractions and Diophantine approximation in power series fields, Preprint, 1998.
62. P. Séébold, Sequences generated by infinitely iterated morphisms, *Disc. Appl. Math.* **11** (1985), 255–264.
63. J. Shallit, On infinite products associated with sums of digits, *J. Number Theory* **21** (1985), 128–134.
64. J.-i. Tamura, Partitions of the set of positive integers, nonperiodic sequences, and transcendence, in: "Analytic Number Theory," Kyoto, 1995, Sūrikaisekikenkyūsho Kōkyūroku, No. 961 (1996), pp. 161–182.
65. A. Thue, Über unendliche Zeichenreihen, *Norske vid. Selsk. Skr. Mat. Nat. Kl.* **7** (1906), 1–22. Reprinted in "Selected mathematical papers of Axel Thue," T. Nagell, ed., Universitetsforlaget, Oslo, 1977, pp. 139–158.

66. A. Thue, Über die gegenseitige Lage gleicher Teile gewisser Zeichenreihen, *Norske vid. Selsk. Skr. Mat. Nat. Kl.* **1** (1912), 1–67. Reprinted in "Selected mathematical papers of Axel Thue," T. Nagell, ed., Universitetsforlaget, Oslo, 1977, pp. 413–478.
67. J. Tromp and J. Shallit, Subword complexity of a generalized Thue-Morse word, *Info. Proc. Letters* **54** (1995), 313–316.
68. D. R. Woods, Elementary problem proposal E 2692, *Amer. Math. Monthly* **85** (1978), 48.
69. J.-Y. Yao, Généralisations de la suite de Thue-Morse, *Ann. Sci. Math. Québec* **21** (1997), 177–189.

On Ideal Autocorrelation Sequences Arising from Hyperovals

Anchung Chang[1], Solomon W. Golomb[1], Guang Gong[2], and P. Vijay Kumar[1]

[1] University of Southern California
EE-Systems, Los Angeles, CA 90089-2565
[2] University of Waterloo
Department of C & O
Waterloo, Ontario N2L 3G1
CANADA

Abstract. Recently, Maschietti showed how one can derive cyclic difference sets from monomial hyperovals. The incidence functions of these difference sets give rise to binary sequences with ideal autocorrelation function. An overview of this result is provided as well as a derivation of the trace expansion and linear span of sequences relating to the Segre hyperoval.

1 INTRODUCTION

This paper is concerned primarily with binary, i.e., $\{0,1\}$, sequences of period 2^n-1 having ideal autocorrelation function, i.e., sequences $\{s(t)\}$ of period 2^n-1 satisfying

$$\sum_{t=0}^{2^n-2}(-1)^{s(t+\tau)-s(t)} \;=\; \begin{cases} 2^n-1 & \tau=0 \pmod{2^n-1} \\ -1 & \text{else,} \end{cases}$$

(see Fig. 1).

A prime example of such sequences are maximum-length linear-feedback shift-register sequences, or m-sequences for short. Fig 2 shows the shift register that generates the m-sequence of period 2^3-1 satisfying

$$s(t) \;=\; s(t-2)+s(t-3) \Rightarrow \ldots 111\mathbf{0010111}0010111\ldots$$

Let $q=2^n$ and F_q denote the finite field of q elements. An m-sequence of period 2^n-1 is most easily described in terms of the trace function over the finite field F_q given by:

$$T(x) \;=\; \sum_{i=0}^{n-1} x^{2^i}.$$

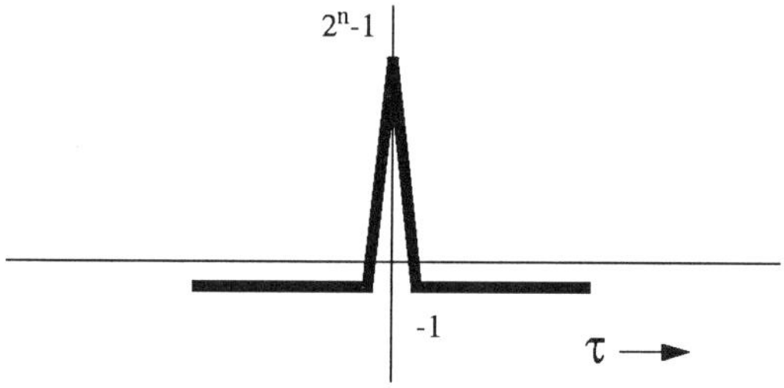

Fig. 1. Ideal Autocorrelation Function

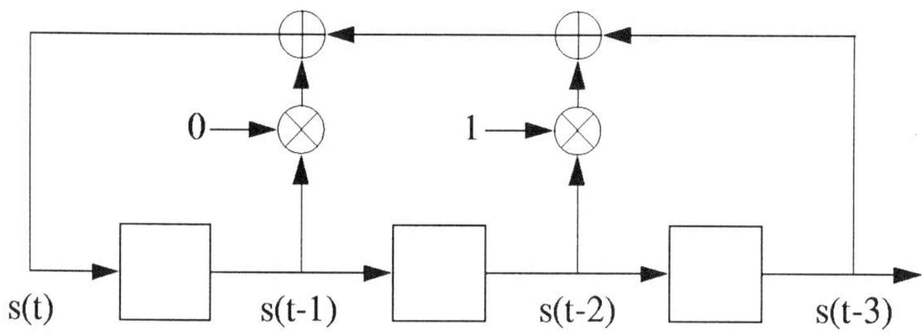

Fig. 2. Shift-register

In terms of the trace function, every m-sequence of length 2^n-1 has an expression of the form

$$s(t) \;=\; T(\alpha^{kt}), \quad t \in Z,$$

where $(k, 2^n-1) = 1$ and α is a primitive element of $GF(2^n)$.

Ideal autocorrelation sequences of period $2^n - 1$ correspond to combinatorial objects known as cyclic Hadamard difference sets. Known constructions for cyclic Hadamard difference sets are tabulated below:

m sequence	1938
Legendre sequence	1933
Hall's sextic residue sequence	1956
Twin prime sequence	1939
GMW sequence	1962

More details can be found for example in [13].

The table also brings out a rather surprising fact, that the most recent construction of cyclic Hadamard difference sets contained in the table dates back to the early sixties. This was the picture until 1997 when Maschietti came up with an observation that led to the discovery of a new family of cyclic Hadamard difference sets. (A parallel exciting development on the topic of ideal autocorrelation sequences that will not be covered here relates to certain recent conjectures that may be found in [24, 23, 6] many of which have subsequently been proven [4, 5].)

First we show an example of how ideal autocorrelation sequences can be constructed using Maschietti's observation. Consider the map $f : F_8 \to F_8$:

$$f(x) \;=\; x + x^6.$$

This map is a $2 \to 1$ map over F_8, i.e, the image of F_8 under f contains 4 elements and each element in the image of f has two pre-images as shown below:

<table><tr><td>x</td><td>0</td><td>1</td><td>α</td><td>α^6</td><td>α^2</td><td>α^5</td><td>α^3</td><td>α^4</td></tr><tr><td>$f(x)$</td><td colspan="2">0</td><td colspan="2">α^5</td><td colspan="2">α^3</td><td colspan="2">α^6</td></tr></table>

The element α in the table is a primitive element of F_8 satisfying $\alpha^3+\alpha+1=0$. Define a binary sequence $\{r(t)\}$ via

$$r(t) \;=\; \begin{cases} 0 & \alpha^t = x + x^6 \\ 1 & \text{else} \end{cases}.$$

Then, $\{r(t)\}$ is as tabulated below

t	0	1	2	3	4	5	6	7	$\cdots$
$r(t)$	1	1	1	0	1	0	0	1	$\cdots$

and can be verified to have ideal autocorrelation.

While Maschietti's construction is novel, in retrospect, it should be noted that even binary m-sequences can be described in terms of such a construction. For instance, for every fixed integer l, $(l,7) = 1$, the binary sequence $\{r(t)\}$ defined via

$$r(t) = \begin{cases} 0 & \alpha^t = x^{2^l} + x, \text{ some } x \in F_8 \\ 1 & \text{else} \end{cases}$$

is the m-sequence of period 7 described previously.

The origin of Maschietti's construction lies in finite geometry, which is the topic of the next section.

2 Hyperovals in a Projective Plane

Let $n \geq 1$, $q = 2^n$ and $PG(2,q)$ be the projective plane corresponding to 2-dimensional projective space over F_q. A $k-$arc of $PG(2,q)$ is a set of k distinct points, no three of which are collinear. The maximum value of k is $q+2$ in which case, the $k-$arc is called a hyperoval. Every hyperoval in $PG(2,q)$ may be written in the form [21]

$$D(f) = \{(1,t,f(t)) \mid t \in F_q\} \bigcup \{(0,1,0),(0,0,1)\},$$

where f is a permutation polynomial of F_q of degree at most $q-2$, satisfying $f(0) = 0, f(1) = 1$ and, for each $s \in F_q$, such that the polynomial

$$f_s(x) = \frac{f(x+s)+f(s)}{x}, \quad f_s(0) = 0$$

is also a permutation polynomial. When $f(x) = x^k$ is a monomial, the hyperoval is called a monomial hyperoval. The following monomial hyperovals are known:

- the hyperovals corresponding to $f(x) = x^{2^k}$, $(k,n) = 1$ known as translation hyperovals
- the Segre hyperoval [26] corresponding to $k = 6$ and n odd, $n \geq 5$
- the Glynn hyperovals [18] :
  - $k = 2^{\frac{n+1}{2}} + \gamma$ for n odd, $n \geq 7$ where $\gamma = 2^l$ if $n = 4l-1$ and $\gamma = 2^{3m+1}$ if $n = 4m+1$
  - $k = 3 \cdot 2^{\frac{n+1}{2}} + 4$ for n odd, $n \geq 7$

A conjecture of Glynn [18] states that these are the only monomial hyperovals.

Monomial hyperovals have the following alternative characterization [16]: $D(x^k)$ is a monomial hyperoval if and only if

- $(k, q-1) = 1$ and
- for every $a \in F_q$, the equation $x^k + x + a = 0$ has either zero or two solutions in F_q

Given a monomial hyperoval $D(x^k)$, let

$$D_k^* = \{x + x^k \mid x \in F_q\} \backslash \{0\}.$$

Maschietti [16] made the following very interesting finding:

Theorem 1. *The $(q+2)$-set $D(x^k)$ is a hyperoval iff D_k^* is a $(q-1, q/2-1, q/4-1)$-difference set in F_q^*, or equivalently, the sequence $\{r(t)\}$ defined by*

$$r(t) = \begin{cases} 0 & \alpha^t \in D_k^* \\ 1 & \text{else} \end{cases}$$

where α is a primitive element of F_q has ideal autocorrelation.

The proof provided by Maschietti is geometric in nature. For the special case when the binary representation of k has Hamming weight 2 (the case of two-bit monomial hyperovals) an alternative algebraic proof is provided in [17] by Xiang. A shorter and more general algebraic proof is contained in the more recent paper by Dillon [4] and is presented in the next subsection. Evans, Hollmann, Krattenthaler and Xiang [7] provide a proof using Jacobi sums (see also [8]).

2.1 Proof of Maschietti's Theorem

The proof by Dillon shows that hyperoval sequences can be constructed as shown in Fig. 3. Ideal autocorrelation then follows from an application of Parseval's theorem. Let α be a primitive element of the finite field $GF(2^n)$. Let $q = 2^n$. Given a function $H(x) : F_q \to \Re$, its Walsh transform $\hat{H}(\lambda)$ is defined by

$$\hat{H}(\lambda) = \frac{1}{\sqrt{q}} \sum_{x \in F_q} H(x)(-1)^{T(x\lambda)}, \quad \lambda \in F_q.$$

The result

$$\begin{aligned} \sum_{\lambda \in F_q} \hat{H}(\lambda)\hat{H}(\lambda y) &= \sum_{\lambda \in F_q} \frac{1}{q} \sum_{x \in F_q} H(x)(-1)^{T(\lambda x)} \sum_{z \in F_q} H(z)(-1)^{T(\lambda y z)} \\ &= \frac{1}{q} \sum_{x \in F_q} H(x) \sum_{z \in F_q} H(z) \sum_{\lambda \in F_q} (-1)^{T(\lambda[x+yz])} \\ &= \sum_{z \in F_q} H(yz)H(z) \end{aligned}$$

is known as Parseval's relation.

Let $g(x)$ satisfy

$$g(x) = \begin{cases} 0 & x = y^k + y, \text{ some } y \in F_q \\ 1 & \text{else} \end{cases}$$

and set $G(x) = (-1)^{g(x)}$. Then

$$\begin{aligned}
\hat{G}(\lambda) &= \frac{1}{\sqrt{q}} \sum_{x \in F_q} G(x)(-1)^{T(x\lambda)} \\
&= 2\frac{1}{\sqrt{q}} \sum_{\substack{x \in F_q \\ x = y + y^k, y \in F_q}} (-1)^{T(\lambda x)} - \frac{1}{\sqrt{q}} \sum_{x \in F_q} (-1)^{T(\lambda x)} \\
&= \frac{1}{\sqrt{q}} \sum_{y \in F_q} (-1)^{T(\lambda[y+y^k])} - 0 \\
&= \frac{1}{\sqrt{q}} \sum_{z \in F_q} (-1)^{T(z^k + z\lambda^{\frac{k-1}{k}})} \\
&= \hat{M}(\lambda^{\frac{k-1}{k}})
\end{aligned}$$

where $\hat{M}(\lambda)$ is the Walsh transform of $M(z) = (-1)^{T(z^k)}$. Let $r(t) = g(\alpha^t)$, then $\{r(t)\}$ is the sequence associated to the hyperoval corresponding to $g(x)$.

Note that if $(k, q-1) = 1$, then $\{m(t)\} = \{T(\alpha^{kt})\}$ is a binary m-sequence. The proof that the sequence $\{r(t)\}$ has ideal autocorrelation then proceeds as follows, (see Fig. 3)

$$\begin{aligned}
\sum_{x \in F_q} (-1)^{T(x^k \theta^k) + T(x^k)} &= \sum_{\lambda \in F_q} M(x) M(x\theta) \\
&= \sum_{\lambda \in F_q} \hat{M}(\lambda) \hat{M}(\lambda\theta) \\
&= \sum_{\lambda \in F_q} \hat{M}(\lambda^{\frac{k-1}{k}}) \hat{M}(\lambda^{\frac{k-1}{k}} \theta) \\
&= \sum_{\lambda \in F_q} \hat{G}(\lambda) \hat{G}(\lambda \theta^{\frac{k-1}{k}}) \\
&= \sum_{x \in F_q} g(x) g(x \theta^{\frac{k-1}{k}}).
\end{aligned}$$

It follows that $\{r(t)\} = \{g(\alpha^t)\}$ has ideal autocorrelation function.

The linear span [25] of a sequence is the smallest degree of a linear recursion satisfied by the sequence. The following conjecture appears in [17]:

Conjecture 1. Let $n \geq 9$ be odd , $k = 6$ and $\{r(t)\}$ be the sequence associated to the Segre hyperoval. Then the linear span L_n of $\{r(t)\}$ satisfies the following recursion:

$$\frac{L_n}{n} = 1 + \frac{L_{n-2}}{n-2} + \frac{L_{n-4}}{n-4} \qquad \text{for all } n \geq 9.$$

Let $(1)^l$ denote a string of l consecutive 1's. Let $\mathcal{A}_1, \mathcal{A}_2$ denote the collection of all binary, $\{0,1\}$ strings of the form $(1)^{4a+1}0$, where a is any integer ≥ 0 and

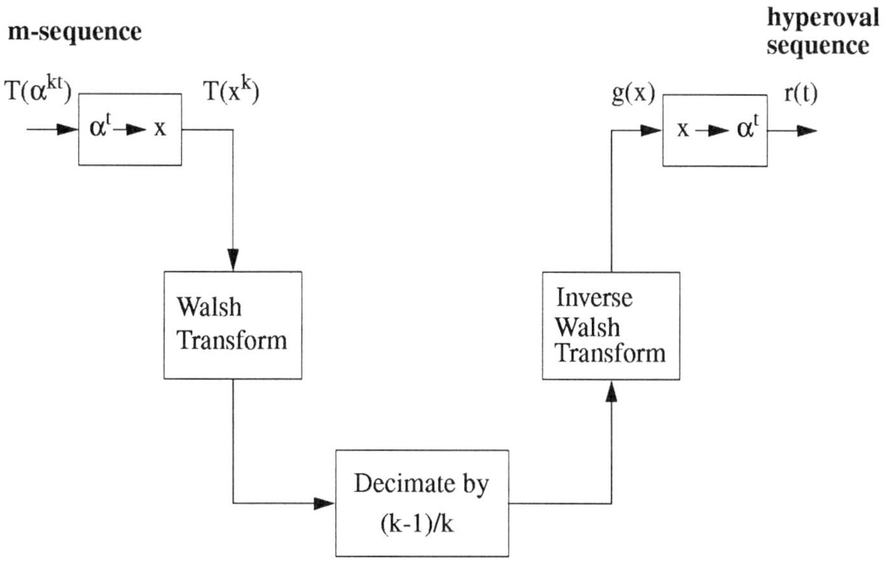

Fig. 3. explaining Ideal Autocorrelation (Dillon)

$(1)^{4b}$, where b is any integer ≥ 0 respectively. Note that the empty string ϕ is contained in $\mathcal{A}_2$. Let

$$\mathcal{A} = \mathcal{A}_1 \bigcup \mathcal{A}_2,$$

and $\mathcal{A}^*$ denote the set of all strings obtained by concatenating one or more strings from $\mathcal{A}$. Let

$$\mathcal{B} = \{(01\ \gamma\ 0(1)^{2s}\ ) \mid s \geq 0,\ \gamma \in \mathcal{A}^*\ \} \bigcup \{(011\ \gamma\ 11) \mid \gamma \in \mathcal{A}^*\ \}.$$

Let $\mathcal{B}_n$ denote the set of all binary $\{0,1\}$ strings in $\mathcal{B}$ of length n, i.e., that consist of n $\{0,1\}$ symbols. Then:

Theorem 2. *Let $k = 6$ and $\{r(t)\}$ be the sequence associated with the Segre hyperoval. Then $r(t)$ has the trace expansion*

$$r(t) = \sum_{(\lambda_0,\lambda_1,\ldots,\lambda_{n-1})\in\mathcal{B}_n} T(\alpha^{-(\sum_{j=0}^{n-1} 2^j \lambda_j)t}).$$

where $T(x) = \sum_{i=0}^{n-1} x^{2^i}$ is the trace function over F_q. For each $\{0,1\}$ string $(\lambda_0, \lambda_1, \cdots, \lambda_{n-1})$ in $\mathcal{B}_n$, the cyclotomic coset containing $\sum_{i=0}^{n-1} 2^i \lambda_i$ is of size n. Thus $\{r(t)\}$ has linear span $L_n = n * |\mathcal{B}_n|$. The sizes $|\mathcal{B}_n|$ of the sets $\mathcal{B}_n$ satisfy the Fibonacci like recursion

$$|\mathcal{B}_n| = |\mathcal{B}_{n-2}| + |\mathcal{B}_{n-4}| + 1\ .$$

Thus, in addition to proving the conjecture, the theorem also provides a trace expansion of the sequence $\{r(t)\}$. Three different proofs of this theorem are now available due to Evans, Hollmann, Krattenthaler and Xiang [7], Chang, Golomb, Gong and Kumar (presented here) and Dobbertin [6]. Evans et. al. [7, 8] also determine the linear span of the ideal autocorrelation sequences associated to the Glynn hyperovals. The remainder of this paper presents the authors' proof of the linear span conjecture. Interestingly, the proof makes use of rings of characteristic 4.

Section II uses transforms over a Galois ring to reduce the linear span to the number of solutions to a certain equation. Section III places the solutions to this equation in one-one correspondence with solutions to a certain combinatorial tiling problem. Section IV presents the solution to the tiling problem and this solution is used in Section V to provide the trace expansion as well as the linear span of the sequence. The section concludes with an example.

3 An algebraic expression for the linear span

3.1 Galois rings

Let Z_4 denote the set of integers modulo 4. Let $n = 2m+1, m \geq 2$ be an integer and set $q = 2^n$. Let F_q be the finite field of 2^n elements. Let α be a primitive element of F_q and let $h(x) = \sum_{i=0}^{n-1} h_i x^i$ be the minimum polynomial of α over F_2. By regarding the coefficients h_i of $h(x)$ as elements in Z_4, $h(x)$ may be regarded as a polynomial over Z_4 having $\{0, 1\}$ coefficients. Let $g(x)$ be the polynomial over Z_4 of degree n specified by

$$g(x^2) = (-1)^n h(-x)h(x)$$

and consider the set $R_n = Z_4[x]/(g(x))$. It turns out that R_n is an example of a Galois ring of size 4^n and characteristic 4. Galois rings of characteristic 4 are defined as Galois extensions of the ring Z_4 of integers modulo 4. They can be shown to be local rings whose unique maximal ideal is the ideal generated by 2. Let R_n^* be the set of units of R_n and let ξ be an element in R_n that is a root of $g(x)$. Then ξ has multiplicative order $2^n - 1$ in R_n. Let $\mathcal{T}_n = \{0, 1, \xi, \ldots, \xi^{2^n-2}\}$. Any element $z \in R_n$ can be expressed uniquely as $z = A + 2B$ for $A, B \in \mathcal{T}_n$. Let μ denote the modulo-2 reduction map. Clearly we may identify $\xi \pmod 2$ with the primitive element α in the finite field F_q (cf. [19], [20]).

3.2 Transform over the Galois ring

Let $s(t) = 1 + r(t)$ be the complement of the sequence $\{r(t)\}$ associated to the Segre hyperoval. Let $\hat{s}(\lambda)$ be the finite field transform of $s(t)$,

$$\hat{s}(\lambda) = \sum_{t=0}^{2^n-2} s(t)\alpha^{\lambda t}.$$

Since there are $q/2 - 1$ 1's in one period of $\{s(t)\}$, it follows that $\hat{s}(0) = 1$. By the Fourier inversion formula, we have

$$s(t) = \sum_{\lambda=0}^{2^n-2} \hat{s}(\lambda)\alpha^{-\lambda t}.$$

It follows that the linear span L_n of $\{s(t)\}$ equals the number of transform coefficients $\hat{s}(\lambda) \neq 0$. Let $u(t)$ be a sequence over Z_4 that only takes on values in the set $\{0,1\}$ and satisfies:

$$u(t) = s(t) \pmod 2.$$

Analogous to the finite-field transform of $s(t)$, we have the Galois ring transform (see [19]) $\hat{u}(\lambda)$ of $\{u(t)\}$ given by

$$\hat{u}(\lambda) = \sum_{t=0}^{2^n-2} u(t)\xi^{\lambda t}.$$

Since $u(t) = s(t) \pmod 2$, it follows that $\hat{u}(\lambda) = \hat{s}(\lambda) \pmod 2$ and therefore,

$$L_n = |\{\lambda \mid 2\hat{u}(\lambda) \neq 0\}|. \quad (1)$$

Let D denote the difference set D_k^* associated to $\{r(t)\}$. Then

$$\begin{aligned} u(t) = 1 &\Leftrightarrow \alpha^t \in D \\ &\Leftrightarrow \alpha^t = x + x^k, \text{for some } x \in F_q \\ &\Leftrightarrow \xi^t = (x + x^k)^q, \text{for some } x \in \mathcal{T}_n. \end{aligned}$$

Since the map $x \to x + x^k$ is 2 to 1, we can write

$$2\hat{u}(\lambda) = \sum_{x \in \mathcal{T}_n \; x \neq 0,1} (x^k + x)^{q\lambda} = \sum_{x \in \mathcal{T}_n} (x^k + x)^{q\lambda}.$$

Let

$$\lambda = \sum_{i=0}^{n-1} \lambda_i 2^i, \;\; \lambda_i \in \{0,1\}$$

and $\Delta := \{\underline{\rho} = (\rho_0, \rho_1, \ldots, \rho_{n-1})\}$ where

$$\rho_i \in \left\{2^i, \; 2^i k \pmod{q-1}, \; (k+1)2^{i-1} \pmod{q-1}\right\}.$$

Since $\hat{s}(0) = 1$, we have $\hat{u}(0) \neq 0$. For $0 < \lambda < q - 1$, we have

$$\begin{aligned} 2\hat{u}(\lambda) &= \sum_{x \in \mathcal{T}_n} (x^k + x + 2x^{(k+1)2^{n-1}})^\lambda \\ &= \sum_{x \in \mathcal{T}_n} \prod_{i=0}^{n-1} (x^{k2^i} + x^{2^i} + 2x^{(k+1)2^{i-1}})^{\lambda_i} \\ &= \sum_{\underline{\rho} \in \Delta} \sum_{x \in \mathcal{T}_n} 2^{\chi(\underline{\rho})} x^{\underline{\lambda} \cdot \underline{\rho}}, \end{aligned} \quad (2)$$

where

$$\underline{\lambda} \cdot \underline{\rho} = \sum_{i=0}^{n-1} \lambda_i \rho_i ,$$

and

$$\chi(\underline{\rho}) = \sum_{i=0}^{n-1} \chi(\rho_i) \quad \text{with} \quad \chi(\rho_i) = \begin{cases} 1 & \text{if } \rho_i = 2^{i-1}(k+1) \\ 0 & \text{else} . \end{cases}$$

For each $0 < \lambda < q-1$, let

$$A_\lambda = \{\underline{\rho} \in \Delta \mid \underline{\rho} \cdot \underline{\lambda} = 0 \pmod{q-1} \text{ and } \chi(\underline{\rho}) \leq 1\} .$$

Since

$$\sum_{x \in \mathcal{T}_n} x^t = \begin{cases} 0 & t \neq 0 \pmod{q-1} \\ q-1 & t \equiv 0 \pmod{q-1}, \end{cases}$$

it follows that

$$2\hat{u}(\lambda) = (q-1) \sum_{\underline{\rho} \in A_\lambda} 2^{\chi(\underline{\rho})} .$$

The last equation tells us that $2\hat{u}(\lambda) \in \{0,2\}$ and consequently, $\hat{s}(\lambda) \in \{0,1\}$. Thus to determine the linear span of the sequence $s(t)$, we need to determine the set of λ for which

$$\sum_{\underline{\rho} \in A_\lambda} 2^{\chi(\underline{\rho})} = 2 \pmod 4 .$$

As we shall see, for any $0 < \lambda < q-1$, $\underline{\lambda} \cdot \underline{\rho} = 0 \pmod{q-1}$ is possible if and only if $\chi(\underline{\rho}) = 1$ and moreover, the set A_λ always has size either 0 or 1. Given the above, we thus have the following alternative expression for the linear span L_n of $\{s(t)\}$

$$L_n = 1 + |\{0 < \lambda < q-1 \mid |A_\lambda| \neq 0\}| .$$

4 An Equivalent Tiling Problem

The problem of determining the set of all pairs

$$\{(\lambda, \underline{\rho}) \mid 0 < \lambda < q-1, \ \underline{\rho} \in A_\lambda\} ,$$

can be rephrased as a tiling problem. From here on, we specialize the discussion to the case of the Segre hyperoval, i.e., we set $k = 6$. Consider the equation $\underline{\lambda} \cdot \underline{\rho} = 0 \pmod{q-1}$. Every nonzero multiple $l(q-1)$, $l = 1, 2, 3, \ldots$ of $q-1$ can be written in the form

$$l(q-1) = aq + b \quad \text{where } a = (l-1) \text{ and } b = (q-1) - (l-1).$$

Thus the binary representation of $l(q-1)$ always has Hamming weight n. Also, (mod $q-1$), we may identify every exponent 2^{n+j}, $j \geq 0$ with 2^j. Under such an identification, every multiple $l(q-1)$ of $q-1$ reduces to the integer $q-1$.

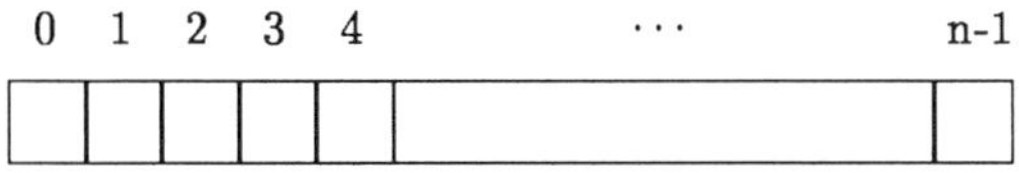

Fig. 4. The strip that to be tiled.

Thus we may regard every pair $(\lambda, \rho), 0 < \lambda < q-1,\ \ \underline{\lambda} \cdot \underline{\rho} = 0 \pmod{q-1}$ as providing a tiling of the strip of length n shown in Fig. 4 by 1's.

Although a linear strip is shown in the figure, the strip should really be regarded as being circular as we work modulo $(q-1)$ and therefore there is wrap-around at the ends. Hereafter, we shall refer to the strip as "the circle with n slots" or simply the "circle". Setting $k = 6$ in the definition of Δ, we see that the building blocks in this tiling are the binary representations of the integers belonging to the sets

$$\left\{\{2^i,\ \ 6 \cdot 2^i,\ \ 7 \cdot 2^{i-1}\},\ i = 0, 1, \ldots n-1\right\},$$

with all integers reduced modulo $q-1$. Since these binary representations consist of either 1, 2 or 3 successive integers, we refer to the three different types of blocks as singles, (S), doubles (D), or triples (T). An example of each is shown in Fig. 5

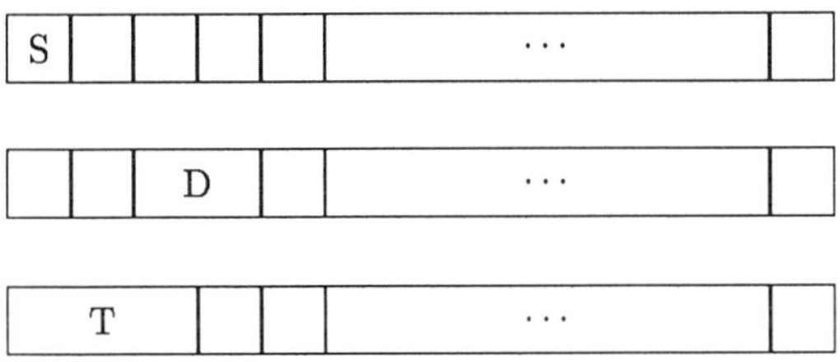

Fig. 5. The three types of blocks.

We will use the phrases 'single at position i", "double at position i" and "triple at position i", to refer to a single occupying position i, a double occupying positions $i, i+1$ and a triple occupying positions $i, i+1, i+2$ respectively, with all addition conducted modulo n. The overlap of two blocks must be resolved using carries as shown in the example overlap in Fig. 6, of a triple and a single. The result of the overlap is the pattern in Fig. 7.

Any tiling of the strip must conform to the 5 rules below.

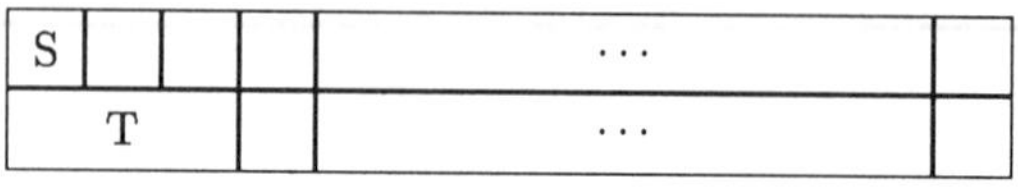

Fig. 6. Overlap of two blocks

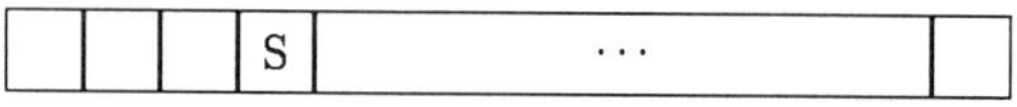

Fig. 7. The result of overlap

Rule 1 Each block can be used at most once. The total number of blocks used cannot exceed $n-1$. (This is because $0 < \lambda < q-1$, so that λ has Hamming weight $\leq n-1$).

Rule 2 At most one triple can be used. (We need that $\chi(\underline{\rho}) \leq 1$.)

The remaining three rules reflect the fact that we can pick at most one term from each of the sets

$$\{\{2^i,\ 6\cdot 2^i,\ 7\cdot 2^{i-1}\}\}.$$

Rule 3 A single S cannot immediately precede a double, i.e., the following arrangement is forbidden:

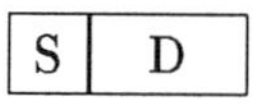

Rule 4 A single S cannot occupy the middle position of a triple, i.e., the following pattern is forbidden:

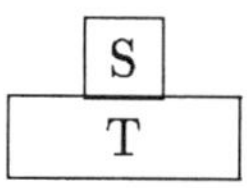

Rule 5 The head of a double can not be overlap the tail of a triple. i.e., the pattern below is forbidden:

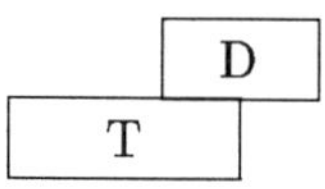

5 Counting the Number of Distinct Tilings

We begin with an observation regarding the number of carries that need to be cleared. Let $\mathcal{P}$ be a tiling of the circle using blocks S, D and T in conjunction with the 5 rules stated earlier. Initially, let us lay down all the blocks that make up the tiling $\mathcal{P}$ on the circle in their respective positions. By Rules 1 and 2, the number b of blocks cannot exceed $n-1$ and therefore the sum h of the Hamming weights of the integers associated to the b blocks cannot exceed $3 + 2(n-2) = 2n-1$. Beginning with the zero position, we clear the carries associated with the 1's in that position. By "clearing a carry" in position i, we mean that we compute the sum, say θ, of the 1's in position i contributed by the various blocks, as well as its decomposition $\theta = \theta_0 + 2\theta_1$, $\theta_0 \in \{0,1\}$, place θ_0 in position i and carry forward θ_1 1's over to position $(i+1)$. If θ itself lies in the set $\in \{0,1\}$, then we say that no clearing of the carry in that position is necessary.

Clearing a carry will result in a loss of Hamming weight of at least 1. Thus after the carry in position 0 has been cleared, there are at most $h-1$ 1's spread around the circle. We next clear the carry in position 1 and continue in this way. Since there are at most $(2n-1) - n = n-1$ carries that can be cleared in all, there must be at least one position i_0, $0 \leq i_0 \leq n-1$ in which it is not necessary to clear a carry. Thus there is no carry that "feeds" into position $i_0 + 1$. We will refer to the position $i_0 + 1$ where i_0 is the first position where it is not necessary to clear carries as a *no-carry* position. By rotating the circle, i.e., by relabelling, we can assume without loss of generality, that the no-carry position is the 0 position. This fact will be used in the proofs to follow.

Lemma 1. *It is impossible to tile the circle with n slots using only singles S and doubles D.*

Proof. Let us assume without loss of generality, that the no-carry position is 0. Since only $n-1$ blocks are allowed by Rule 1, it is not permitted to tile the circle using n singles S. It follows from Rule 3 that if there exists such a tiling, at least two blocks making up the tiling must overlap in some position.

Let i be the smallest index of a position at which there is an overlap of two or more blocks. Clearly, we may assume by our no-carry-at-0 assumption, that there is no carry feeding into position i. Thus, the overlap must necessarily be of the form shown in Fig. 8. Note that by Rule 3, positions 0 thru $i-2$ must necessarily be tiled by doubles. Thus i is odd.

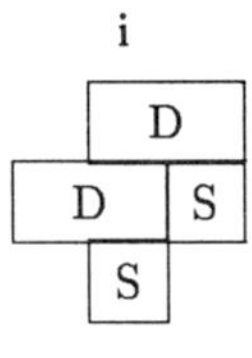

Fig. 8. Formation of a quartet (Q) through overlap.

We name the pattern that results from the overlap a *quartet*, and write Q for short. Note that a quartet in position i results in the placement of 4 1's at positions $i-1, i, i+1$ and $i+2$. Also, if there is a quartet at position i, by Rule 3, there can be no doubles occupying either of the positions $i+1$, $i+2$. A tiling requires the placement of a 1 in every position around the circle. From the figure it is clear that a carry feeds into position $i+2$. Thus there can be no single in position $i+2$ either. As a result, a quartet in position i results in $i+3$ becoming a no-carry position. Since i is odd, $i+3$ is even. Continuing in this way, we will obtain a series $0, i+3, \ldots,$ of even no-carry positions. Consider the no-carry position $j+3$ resulting from the location of the last overlap (i.e., the overlap which occurs at the largest index $j, 0 \leq j \leq n-3$). Then $j+3$ is a no-carry position. Note that $j+3 \neq n$ as j is odd. In the remaining positions $j+3, j+4, \ldots, n-1, 0, 1, \ldots, i-2$ there is by assumption, no overlap. Since there is a double at position $i-1$, this portion of the circle cannot be tiled by singles alone. By Rule 3, it must therefore necessarily be tiled by doubles alone, but this is impossible as there are an odd number $((n-j+3)+(i-1))$ of positions to tile.

Also, since a Q in position i implies there was a double in position $i-1$, we have the following rule:

Rule A1. A Q in position i cannot be preceded by an S in position $i-2$ (mod n). □

It follows from the above lemma that the equation $\underline{\lambda} \cdot \underline{\rho} = 0$ has a solution only if $\chi(\underline{\rho}) = 1$.

Lemma 2. *Let $\mathcal{P}$ be a tiling of the circle using blocks S, D and T in conjunction with Rules 1-5. Let 0 be a no-carry position as before and let there be exactly one triple T placed in position t. Then t is a no-carry position.*

Proof. Let l be the smallest index of a position at which there is an overlap. The lemma clearly holds if $t \leq l$, We now consider three cases:

Case (i) $t \geq l+3$. In this case, the overlap must necessarily result in a quartet. As mentioned in the proof of Lemma 1, $l+3$ is necessarily a no-carry position. If $t = l+3$, we are done. Otherwise, we can relabel the circle by subtracting $l+3$ from the index of each position. We will still have 0 as the no-carry position and since $t > l+3$, we have in effect, moved the triple closer to the 0 position. We let l be the position of first overlap under the new labelling and return to the start of the proof. This process cannot continue forever and sooner or later, the triple position will cause it to fall into one of the other cases.

Case (ii) $t = l+2$. In this case, in addition to the triple at position t, singles S and doubles D in positions $l-1, l, l+1, l+2$ must necessarily appear as shown in Fig. 9 for the pattern to yield a tiling of the circle with n slots. However, by Rules 1-5, neither a single nor a double can be placed at position $t+1$ and thus this case cannot occur.

Case (iii) $t = l+1$. In this case, the blocks in the vicinity of the triple and the first overlap position, can fall into one of two patterns as shown in Fig. 10.

In either of the two cases, it is not possible to complete the pattern to yield a tiling as in both cases, this would be equivalent to tiling a linear strip of odd

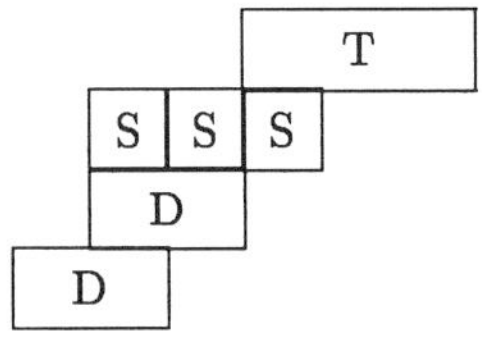

Fig. 9. Overlap in case $t = l + 2$

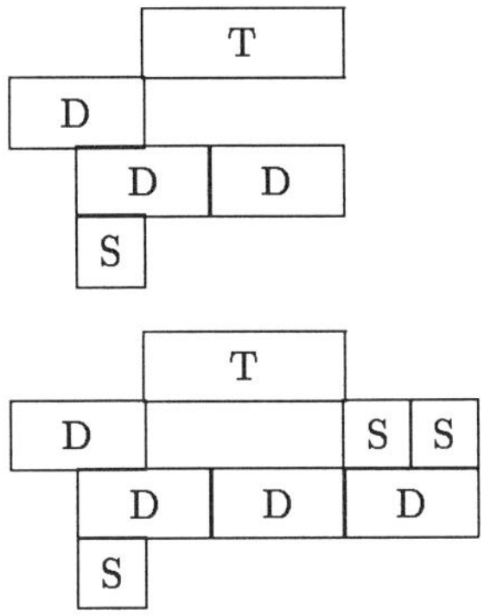

Fig. 10. Overlaps in case $t = l + 1$

lengths $n-6$ (corresponding to positions $t+4, t+5, \ldots, n-1, 0, \ldots, t-3$ in case of the first pattern) and $n-8$ (corresponding to positions $t+6, t+7, \ldots, n-1, 0, \ldots, t-3$ in case of the second pattern) using only doubles and quartets which is impossible. □

Lemma 3. *Let $f(l)$ be the number of different ways to tile a linear strip of length $2l, l > 0$, using only doubles D and quartets Q without overlap of the blocks in any position. Then*

$$f(l) = f(l-1) + f(l-2), \quad l \geq 3, \quad f(1) = 1 \text{ and } f(2) = 2.$$

Proof. The proof follows from considering separately, the cases when the block at the beginning of the linear strip is a double D and a quartet Q respectively. □

We define $f(0) = 1$ and then the recursion $f(l) = f(l-1) + f(l-2)$ is also valid for $l = 2$.

Lemma 4. *Let $\Omega(m)$ be the number of cyclically different ways to tile the circle with $n = 2m + 1$ slots using basic patterns S, D, and T along with Rules 1-5. Then*

$$\Omega(m) = \sum_{k=0}^{m-1} f(k) + f(m-2). \tag{3}$$

where $\{f(k)\}$ *is given by Lemma 3.*

Proof. By Lemma 1, it is not possible to tile a line of size $n = 2m + 1$ by using only doubles and singles. Thus we only need consider the case where exactly one triple is used. By Lemma 2 we can assume as shown in the picture below, that without loss of generality, the triple occupies the 0 position and that this moreover, is a no-carry position.

We consider two cases:

Case (i) There is no overlap of blocks in position 0.

It is easy to see that this implies there is no overlap of blocks in positions $1, 2$ as well. Thus we may pretend for tiling purposes, that we have a linear strip of even length $n - 3$ (obtained by excluding the triple) that is to be tiled using singles and doubles only. By the arguments used in the proof of Lemma 1, overlaps result in quartets that do not overlap with other blocks. Thus we only need count the number of tilings of a linear strip of length $n - 3$ that involve nonoverlapping singles, doubles and quartets and that are in accordance with Rules 1-5 and Rule A1. Let $K_1(m)$ be the number of such tiling patterns:

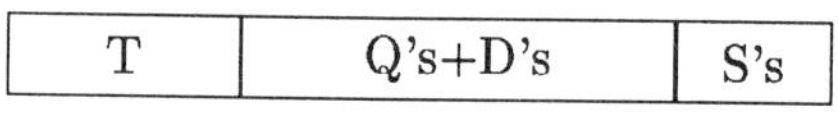

Since $n - 3 = 2(m - 1)$ is even, if singles appear, then they must appear in pairs and only at the end of the linear strip by Rules 3 and A1. Let $G_i, i \geq 0$ be the number of tiling patterns in which $2i$ consecutive singles appear. Then

$$K_1(m) = G_0 + G_1 + \cdots + G_{m-1}.$$

Clearly, $G_{m-1} = 1 = f(0)$ and for $0 \leq i \leq m - 2$, G_i is equal to the number of different ways to tile a linear strip of length $2(m - 1) - 2i$ using only non-overlapping doubles and quartets together with Rules 1-5 and Rule A1. By Lemma 3,

$$G_i = f(m - 1 - i) \qquad i = 0, 1, \cdots, m - 2. \tag{4}$$

As a result,

$$K_1(m) = \sum_{j=0}^{m-1} f(j). \tag{5}$$

Case (ii) Overlap in position 0

By Rules 1-5, the only possible overlapping patterns are those shown in Fig. 11 and Fig. 12.

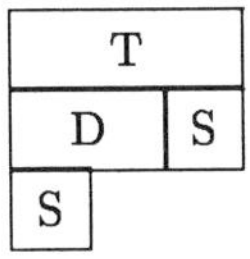

Fig. 11. Overlap in position of triple, first case.

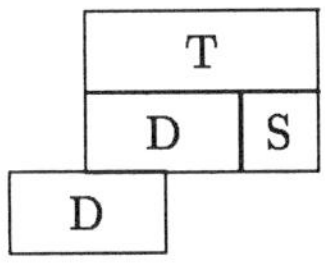

Fig. 12. Overlap in position of triple, second case.

If the pattern shown in Fig. 11 appears, 4 is a no-carry position and the remaining $n-4$ positions $4, 5, \ldots, , n-1$ must be tiled using doubles and singles together with Rules 1-5. As $n-4$ is odd, by Lemma 1, this is impossible. So we only need to consider the case in Fig. 12. In this case, we have to tile $n-5$ positions using nonoverlapping singles, doubles and quartets. By Rules 3 and A1, singles must appear at the end. But the presence of a double at position $n-1$ rules out the presence of any singles. Thus all tiling patterns assume the form:

By Lemma 3, the number $K_2(m)$ of different ways to tile a line of size $n-5$=2(m-2) by using quartets and doubles without overlapping equals $f(m-2)$. Therefore

$$\Omega(m) \;=\; K_1(m) + K_2(m) \;= \sum_{j=0}^{m-1} f(j) + f(m-2).$$

where $\{f(j)\}$ is as defined in Lemma 3. □

6 Recovering λ from a tiling pattern

With respect to the notation $\mathcal{A}^*$ introduced in Section 1, the total number $\mu(l)$ of strings of length $2l$, $l \geq 1$ contained in $\mathcal{A}^*$ is given by the recursion:

$$\mu(l) \;=\; \mu(l-1) + \mu(l-2), \;\; \mu(1) = 1, \;\; \mu(2) = 2.$$

The presence of the empty string ϕ in $\mathcal{A}^*$ causes $\mu(0) = 1$. The recursion follows from noting that every string of length $2l$ in $\mathcal{A}^*$, $l \geq 1$ is either composed of a string of length $2l - 2$ drawn from $\mathcal{A}^*$ followed by the string 10 or else, a string of length $2l - 4$ drawn from $\mathcal{A}^*$ followed by the string 1111. It follows that $\mu(l) = f(l)$ for all $l \geq 0$.

Given a tiling pattern, one can recover from the known positions of the blocks, S, D, T, the corresponding value of λ. The table below provides a dictionary translating tiling patterns to binary representations of the corresponding values of λ.

Table 1. A dictionary.

Lemma 4	Tiling pattern	λ in binary (all strings are of length n)
Case(i)	T | Q's+D's | S's	01(string in $\mathcal{A}^*$)0(1)^{2s}
Case(ii)	1 | 1 | 1 | 1 | Q's+D's | 1	011(string in $\mathcal{A}^*$)11

The integer s appearing in the table is ≥ 0.

Some observations are in order:

1. Given a tiling pattern, i.e., a collection of blocks that tile the circle, along with their respective positions on the circle, by cyclically shifting the blocks in the pattern, i.e., by shifting each block in position i to position $i + \tau$ (mod n), we will arrive at another tiling pattern. Two tiling patterns are cyclically distinct if one cannot be obtained from the other via such a cyclic shift. Since all the tiling patterns listed in the table have at least one triple, and since the triple is in position zero, it follows that all the patterns in the table are cyclically distinct. The patterns listed in the table and those obtained by cyclically shifting the patterns provide an exhaustive listing of all possible tiling patterns.
2. Every tiling pattern corresponds to a certain value of λ, $0 < \lambda < q - 1$. We would like to claim that there is a one-one correspondence between the set of all tiling patterns and the set of λ's shown in the table as well as their conjugates (λ_1 is a conjugate of λ_2 if and only if $\lambda_1 = 2^i\lambda_2 \pmod{q-1}$). Close examination of the number of 1's on either side of a zero in the binary representation of a λ listed in the table (or of its conjugate) will allow the identification of the position of the triple in the associated tiling pattern. Thereafter, the remaining blocks of the tiling are easily identified. Thus, distinct tiling patterns result in distinct λ's.

As a check on the one-one correspondence between tiling patterns and λ's, the number of λ reported in the table can be counted using the result $\mu(l) = f(l)$

given above and will be seen to coincide with $\Omega(m)$. This $1-1$ correspondence is also equivalent to saying that for every λ, $0 < \lambda < q-1$, there is at most one $\underline{\rho} \in \Delta^n$, satisfying $\underline{\lambda} \cdot \underline{\rho} = 0$. Together with the result of Lemma 1, this implies the result mentioned in Section 2, namely that $\underline{\lambda} \cdot \underline{\rho} = 0 \pmod{q-1}$ is possible if and only if $\chi(\underline{\rho}) = 1$ and moreover, the set A_λ always has size either 0 or 1. The λ for which there is such a $\underline{\rho}$ (i.e., a tiling pattern) appear in the table.

It follows that the linear span of the sequence $\{r(t)\}$ is given by

$$L_n = n\Omega(m).$$

By Lemma 4,

$$\Omega(m) = \sum_{j=0}^{m-1} f(j) + f(m-2)$$

$$\Omega(m-1) = \sum_{j=0}^{m-2} f(j) + f(m-3)$$

$$\Omega(m-2) = \sum_{j=0}^{m-3} f(j) + f(m-4).$$

Since $f(0) = f(1) = 1$, and $f(l) = f(l-1) + f(l-2), l \geq 2$, we have that

$$\Omega(m) = \Omega(m-1) + \Omega(m-2) + 1.$$

It also follows that the sequence $\{s(t)\}$ has the following trace representation

$$s(t) \;=\; 1 \;+\; \sum_{\lambda \in \mathcal{S}} T(\alpha^{-\lambda t}), \quad t \in Z$$

where $\mathcal{S}$ is the collection of all λ whose binary representations appear in the table above. This proves Theorem 2 of Section 1.

6.1 Example

In the table below the elements λ in the set $\mathcal{S}$ are shown along with the associated tiling patterns for the case $n = 11$, i.e., $m = 5$.

Example 1. $n = 11$

$$\begin{aligned} s(t) = 1 + T(\alpha^{5t} + \alpha^{25t} + \alpha^{105t} + \alpha^{309t} + \alpha^{469t} + \alpha^{83t} + \alpha^{39t} + \alpha^{29t} \\ + \alpha^{3t} + \alpha^{19t} + \alpha^{73t} + \alpha^{33t} + \alpha^{9t} + \alpha^{17t} + \alpha^{149t}) \qquad t \in Z \end{aligned}$$

ACKNOWLEDGEMENT

The authors would like to acknowledge useful discussions with Peter Gaal and Lloyd Welch.

Table 2. The elements λ and the associated tiling patterns.

Tiling Pattern	λ	coset leader	coset leader for $-\lambda$
T S S S S S S S S	01011111111	767	5
T D S S S S S S	01100111111	447	25
T D D S S S S	01101001111	367	105
T D D D S S	01101010011	347	309
T D D D D	01101010100	171	469
T D D Q	01101011110	491	83
T D Q D	01101111100	251	39
T Q D D	01111110100	191	29
T Q Q	01111111110	511	3
T Q D S S	01111110011	507	19
T D Q S S	01101111011	879	73
T Q S S S S	01111101111	991	33
1 1 1 1 Q D 1	01111111011	895	9
1 1 1 1 D Q 1	01110111111	959	17
1 1 1 1 D D D 1	01110101011	699	149

References

1. S. D. Cohen and R. W. Matthews, "A class of exceptional polynomials", Trans. Amer. Math. Soc. **405**, pp. 897-909, 1994.
2. A. Chang, P. Gaal, S. W. Golomb, G. Gong and P. V. Kumar, "Some results relating to sequences conjectured to have ideal autocorrelation", submitted to *IEEE Trans. Inform. Theory*.
3. A. Chang, T. Helleseth and P. V. Kumar, "Further results on a conjectured 2-level autocorrelation sequence", submitted to *IEEE Trans. Inform. Theory*.
4. J. Dillon, "Multiplicative difference sets via additive characters", Designs, Codes and Cryptography, to appear.
5. J. Dillon, H. Dobbertin, "Cyclic difference sets with Singer parameters," preprint.
6. H. Dobbertin, "Kasami power functions, permutation polynomials and cyclic difference sets", to appear in *Proc. of the NATO ASI Workshop*, Bad Windsheim, Aug. 3-14, 1998.
7. R. Evans, H. D. L. Hollmann, C. Krattenthaler and Q. Xiang, "Gauss sums, Jacobi sums and p-ranks of cyclic difference sets, " to appear in *J. Combin. Theory, Ser. A*.
8. Q. Xiang, "Recent results on difference sets with classical parameters," to appear in *Proc. of the NATO ASI Workshop*, Bad Windsheim, Aug. 3-14, 1998.
9. R. Gold, "Maximal recursive sequence with 3-valued cross-correlation functions", *IEEE Trans. Inform. Theory*, Vol 14, pp. 154-156, 1968.
10. G. Gong, P. Gaal, and S. W. Golomb, "A suspected new infinite class of $(2^n - 1, 2^{n-1} - 1, 2^{n-2} - 1)$ cyclic difference sets", ITW, Longyearbyen, Norway, July 6-12 1997.
11. G. Gong and S. W. Golomb, "Hadamard transforms of three-term sequences", preprint.
12. R. Lidl, G. L. Mullen and G. Turnwald, "Dickson polynomials", Pitman Monographs in pure and applied mathematics, Vol. 65, Addison Wesley, 1993.
13. L. D. Baumert, *Cyclic Difference Sets, Lecture Notes in mathematics*, **182**, Springer-Verlag, 1971.
14. S. W. Golomb, *Shift Register Sequences*, 2nd edition, Aegean Park Press, 1982.
15. T. Helleseth and P. V. Kumar, "Sequences with low correlation", to appear in the *Handbook of coding theory* edited by V. S. Pless and W. C. Huffman, Elsevier Science B.V., 1998.
16. A. Maschietti, "Difference sets and hyperovals", *Designs, Codes and Cryptography*, **14**, pp. 89-98, 1998.
17. Q. Xiang, "On balanced binary sequences with two-level autocorrelation functions", *IEEE Transactions of Information Theory*, **44**, pp. 3153-3156, Nov. 1998.
18. D. G. Glynn, "Two new sequences of ovals in finite Desarguesian planes of even order", *Lecture Notes in Mathematics*, **1036**, Springer-Verlag, pp. 217-229, 1983.
19. A. R. Hammons, P. V. Kumar, A. R. Calderbank, N.J.A. Sloane, and P. Solé, "The Z_4-linearity of Kerdock, Preparata, Goethals and related codes", *IEEE Trans. Inform. Theory*, vol. IT-40, pp. 301-319, March 1994.
20. P. V. Kumar, T. Helleseth, and A. R. Calderbank, "An upper bound for Weil exponential sums over Galois rings and applications", *IEEE Trans. Inform. Theory*, vol. IT-41, pp. 456-468, March 1995.
21. J. W. P. Hirschfeld, "Ovals in a Desarguesian plane of even order", *Annali Mat. Pura Appl.*, vol. 102, pp. 79-89, 1975.
22. F. J. MacWilliams and N. J. A. Sloane, *The Theory of Error-Correcting Codes*. Amsterdam, The Netherlands: North Holland, 1977.

23. J.-S. No, H. Chung and M. -S. Yun, "Binary Pseudorandom sequences of period $2^m - 1$ with ideal autocorrelation generated by the polynomial $Z^d + (Z+1)^d$," *IEEE trans. Inform. Theory*, **44**, pp. 1278-1282, 1998.
24. J-S. No, S. W. Golomb, G. Gong, H-K. Lee and P. Gaal, "New binary pseudorandom sequences of period $2^n - 1$ with ideal autocorrelation", *IEEE Trans. Inform. Theory*, **44**, pp. 814-817, 1998.
25. M. K. Simon, J. K. Omura, R. A. Scholtz and B. K. Levitt, *Spread-Spectrum Communications*, vol. 1, Rockville, Md: Computer Science Press, 1985.
26. B. Segre and U. Bartocci, "Ovali ed altre curve nei piani di Galois di carattereristica due," *Acta Arith.*, vol. 8, pp.423-449, 1971.

Cyclic Hadamard Difference Sets – Constructions and Applications

Solomon W. Golomb

Communication Sciences Institute
University of Southern California
Los Angeles, CA 90089-2565 U.S.A.
Email: c/o milly@mizar.usc.edu

Abstract. A cyclic (v, k, λ) difference set with parameters $v = 4t - 1, k = 2t - 1, \lambda = t - 1$ is called a cyclic Hadamard difference set. In all known cases, $v = 4t-1$ is either a prime, a product of twin primes, or one less than a power of 2. When v is a prime (but not a Mersenne prime), all known examples are obtained by the quadratic residue (Legendre sequence) construction or, in addition, when also $v = 4a^2+27$, from Hall's sextic residue construction. When v is a product of twin primes, all known examples arise from the Jacobi symbol (Stanton-Sprott) construction. However, when $v = 2^n - 1$, in addition to the Singer difference set (PN-sequence) construction and (for composite $n > 4$) the GMW examples, there are several new constructions, all based on the existence of $GF(2^n)$. All inequivalent examples of cyclic Hadamard difference sets with $v = 2^n - 1$ for $n \leq 10$ have now been identified, and all belong to known or suspected infinite classes of examples. (There are ten inequivalent examples at $n = 10$.)
Because of their favorable autocorrelation properties, binary sequences corresponding to cyclic Hadamard difference sets are widely used in communications and radar. Several of these applications are described.

1 Mathematical Overview

A *cyclic difference set* D, with parameters (v, k, λ), is a set $D = \{d_1, d_2, \ldots, d_k\}$ of k integers, distinct modulo v, such that the congruence $d_i - d_j \equiv r \pmod{v}$ has exactly λ solutions $(d_i, d_j) \in D \times D$ for every $r, 1 \leq r \leq v-1$. If $k = \frac{v-1}{2}$ and $\lambda = \frac{k-1}{2}$, then D is called a *cyclic* **Hadamard** *difference set.* Setting $\lambda + 1 = t$, we see that in this case $v = 4t-1, k = 2t-1, \lambda = t-1$, for some positive integer t.

Corresponding to the cyclic (v, k, λ) difference set D, we associate the binary sequence $\{a_i\}$, periodic with period v, by the rule: $\left\{ \begin{array}{l} a_i = 1 \text{ for } i \in D \\ a_i = 0 \text{ for } i \notin D \end{array} \right\}$. Then the periodic autocorrelation function $C_a(\tau)$ of $\{a_i\}$ clearly satisfies:

$$C_a(\tau) = \frac{1}{v}\sum_{i=1}^{v} a_i a_{i+\tau} = \left\{ \begin{array}{l} \frac{k}{v} \text{ if } \tau \equiv 0 \pmod{v} \\ \frac{\lambda}{v} \text{ if } \tau \not\equiv 0 \pmod{v} \end{array} \right\}.$$

We can also define the ± 1-sequence $\{b_i\}$, where $b_i = (-1)^{a_i}$. The periodic autocorrelation function $C_b(\tau)$ of $\{b_i\}$ then satisfies: $C_b(\tau) = \frac{1}{v}\sum_{i=1}^{v} b_i b_{i+\tau}$ $= \begin{Bmatrix} 1 & \text{if } \tau \equiv 0 \pmod{v} \\ 1 - \frac{4n}{v} & \text{if } \tau \not\equiv 0 \pmod{v} \end{Bmatrix}$, where $n = k - \lambda$.

In the cyclic Hadamard difference set case, if $v > 2$ we can take $v = 4t-1$, $k = 2t-1$, $\lambda = t-1$, for some positive integer t. Then we have

$$C_a(\tau) = \begin{Bmatrix} (2t-1)/(4t-1) & \text{if } \tau \equiv 0 \pmod{v} \\ (t-1)/(4t-1) & \text{if } \tau \not\equiv 0 \pmod{v} \end{Bmatrix}$$

and

$$C_b(\tau) = \begin{Bmatrix} 1 & \text{if } \tau \equiv 0 \pmod{v} \\ -\frac{1}{v} & \text{if } \tau \not\equiv 0 \pmod{v} \end{Bmatrix}.$$

When dealing with a pair of two-valued sequences $\{x_i\}$ and $\{y_i\}$ of length n (i.e. sequences of u's and v's, where u and v are any two distinct elements of any kind), an alternative definition of *correlation* between $\{x_i\}$ and $\{y_i\}$ is $C_{xy} = \frac{A-D}{n} = \frac{A-D}{A+D}$, where A is the number of positions i where $\{x_i\}$ and $\{y_i\}$ *agree*, while D is the number of positions where they *disagree*. The same result would be obtained by replacing the value "u" with "+1", and the value "v" with "-1", in the sequences $\{x_i\}$ and $\{y_i\}$.

From this perspective, $C_b(\tau)$ is the preferred way to define the autocorrelation function of the periodic binary sequence corresponding to a cyclic (v, k, λ) difference set. Note that for cyclic **Hadamard** difference sets, $C_b(\tau) = \frac{-1}{v}$ is very close to zero for all $\tau \not\equiv 0 \pmod{v}$, especially for large values of v.

Two cyclic difference sets D_1 and D_2 with the same values of (v, k, λ) are considered *equivalent* if $D_2 \equiv qD_1 + c \pmod{v}$ for any integer c and any integer q with $(q, v) = 1$. Up to equivalence, there is only one cyclic Hadamard difference set with $v = 2^n - 1$ for $n \leq 4$. There are two at $n = 5$ and two at $n = 6$. L.D. Baumert completed a search for all examples at $n = 7$ (i.e. all (127, 63, 31) cyclic difference sets), and found six inequivalent cases, as reported in [2].

Since Baumert's monograph appeared in 1971, complete searches for cyclic Hadamard difference sets with $v = 2^n - 1$ have been completed for $n = 8$ (by U. Cheng, 1983), for $n = 9$ (by R. Dreier, 1991), and for $n = 10$ (by P. Gaal and S.W. Golomb, 1998), finding 4, 5, and 10 inequivalent examples, respectively. Remarkably, all these examples for $n \leq 10$ belong to known or suspected infinite families of cyclic Hadamard difference sets, where some half dozen of these families were only discovered within the past two years.

One of the newly discovered families can be simply described in terms of its corresponding binary (0,1) sequence, as a term-by-term sum of three m-sequences of period $2^n - 1$ for every *odd* $n \geq 5$. Specifically, if $\{a_i\}$ is any m-sequence of period $2^n - 1$ with $n = 2w+1$, let $\{s_i\} = \{a_i\} + \{a_{qi}\} + \{a_{q^2 i}\}$ where $q = 2^w + 1$. Two more new families are term-by-term sums of *five* m-sequences in suitable relation to one another. From each of these two, yet another new family is obtained by a transformation making use of the trace function from

$GF(2^n)$ to $GF(2)$. See [7]. Finally, there are three more new families based on constructions involving "hyperovals" in *finite projective planes.*

2 Applications to Radar, Sonar, and Synchronization

2.1 Overview

In all three of the applications mentioned in the title of this section, one of the objectives (often the major objective) is to determine a point in time with great accuracy. In radar and sonar, we want to determine the round trip time from transmitter to target to receiver very accurately, because the one-way time (half of the round-trip time) is a measure of the distance to the target (called the *range* of the target).

The simplest approach would be to send out a pure "impulse" of energy, and measure the time until it returns. The ideal "impulse" would be virtually instantaneous in duration, but with such high amplitude that the total energy contained in the pulse would be significant, much like a Dirac "delta function". However, the Dirac "delta function" not only fails to exist as a mathematical *function*, but it is also unrealizable as a physical *signal.* Close approximations to it — very brief signals with very large amplitudes — may be valid mathematically, but are impractical to generate physically. Any actual transmitter will have an upper limit on "peak power" output, and hence a short pulse will have a very restricted amount of total energy: at most, the peak power times the pulse duration. More total energy can be transmitted if we extend the duration; but if we transmit at uniform power over an extended duration, we do not get a sharp determination of the round trip time. This dilemma is illustrated in Figure 1.

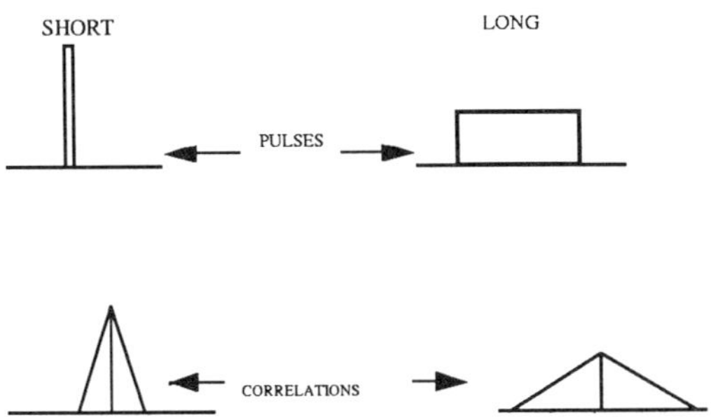

Fig. 1. The shorter the pulse, the sharper the autocorrelation function

In the presence of noise, the time of the peak of the autocorrelation function of the extended pulse becomes quite difficult to determine with high precision.

It is here that clever combinatorial mathematics comes to the rescue. By using a suitably coded pulse pattern, or coded signal pattern, we extend the duration, thereby increasing the total transmitted energy as much as desired, while still maintaining a sharp spike in the autocorrelation function to mark the round trip propagation time with great precision. Such a technique was successfully used in 1961 by the Jet Propulsion Laboratory to bounce a radar signal off the surface of Venus and detect it back on earth. Not only was that the first successful radar ranging of another planet in the solar system, but the distance thus measured made it possible to improve the accuracy of the "Astronomical Unit" (the mean radius of the earth's orbit around the sun, and the basic yardstick for distances within the solar system) by more than three orders of magnitude.

2.2 Types of Signals and Correlations

A *monostatic radar* is one which has transmitter and receiver at the same location, which requires that the transmitter be turned off while the signal is being received. A *bistatic radar* is one which has two separate antennas, often at a considerable distance from one another, for transmitting and receiving. With a bistatic radar, it is possible, in principle, to be transmitting all the time.

A *pulse radar* is one which transmits a succession of pulses, not necessarily uniformly spaced, but which is turned off between pulses. A *CW (continuous wave) radar* is one which stays on the air for an extended period of time, transmitting a sine wave, usually with phase modulation added. A *frequency hop radar* is one which jumps from one to another of a finite set of frequencies, according to some predetermined pattern, thus combining certain features of the pulse radar and the CW radar.

For correlation purposes, most radar patterns are either *finite* (the transmitter is turned off before and after the pattern is sent) or *periodic* (the same basic pattern is repeated periodically a number of times). If a CW radar is modulated in both phase and amplitude, the resulting signal can be modeled as a sequence of complex numbers, which also have both phase and amplitude. If only the amplitude varies, the sequence of complex numbers becomes a sequence of real numbers. If only the phase varies, the sequence of complex numbers reduces to a sequence of values on the unit circle of the complex plane. The real line intersects the unit circle at the points +1 and -1, and binary modulation by +1 and -1 can be regarded either as phase modulation or as amplitude modulation.

The returning signal can be correlated against either an ideal model of itself, or against some other signal (often called a "complementary" signal or sequence), especially designed to highlight some specific feature of the returning signal. In all cases, the ratio of the value of correlation for $\tau = 0$, i.e. when the signal is aligned with itself, versus the maximum value of the correlation for $\tau \neq 0$, is a measure of the clarity with which range can be measured in a noisy environment.

If there is a relative motion (either toward or away) between the transmitter and the target, then the returning radar or sonar signal is shifted not only in time, but also in frequency. The frequency shift ("Doppler shift") is proportional to the time derivative of range — i.e. to the velocity of approach or separation, between communicator and target. The two-dimensional autocorrelation function of the signal, in the time and frequency domains, is called its *ambiguity function*, and the ideal shape of such a function is a spike, or an inverted *thumb-tack* (or *drawing pin*, in the U.K.).

2.3 Barker Sequences

R.H. Barker [1] asked: for what lengths L do binary sequences of +1's and -1's, $\{a_j\}_{j=1}^{L}$, exist, with finite, unnormalized auto-correlation $K(\tau)$, defined by

$$K(\tau) = \sum_{j=1}^{L-\tau} a_j a_{j+\tau}, \tag{1}$$

bounded by 1 in absolute value for $\tau \neq 0$? That is, the requirement is

$$|K(\tau)| \leq 1 \text{ for } 1 \leq |\tau| \leq L-1. \tag{2}$$

Barker gave examples having the following lengths:

L	*Sequence*
2	+1, +1
3	+1, +1, -1
4	+1, +1, -1, +1
5	+1, +1, +1, -1, +1
7	+1, +1, +1, -1, -1, +1, -1
11	+1, +1, +1, -1, -1, -1, +1, -1, -1, +1, -1
13	+1, +1, +1, +1, +1, -1, -1, +1, +1, -1, +1, -1, +1

When $L \equiv -1 \pmod 4$, Barker sequences of length L correspond to cyclic Hadamard designs, and to cyclic Hadamard matrices of order $L+1$. The three values of L in question, $L = 3, 7, 11$, are all primes, and the corresponding Barker sequences can be regarded as Legendre sequences for these primes. Sincer Barker sequences are not periodic, it is critical to start the Legendre sequence at the right place.

The Barker sequence for $L = 4$ corresponds to the only known "totally cyclic" Hadamard matrix. R. Turyn and J. Storer proved [8] that there are no additional examples of Barker sequences of odd length. It is universally believed, but not yet proved, that there is no even length Barker sequence of length $L > 4$. Partial results include that if such an L exists, it must be a perfect square, but not a power of 2, and that its set of cyclic shifts would form another totally cyclic Hadamard matrix.

2.4 Pulse Patterns and "Optimal Rulers"

A pulse radar is able to send one or more pulses of radio-frequency energy toward a target. It is a convenient and rather realistic assumption to require all the pulses to be the same in both duration and amplitude. The signal design problem then reduces to devising *patterns* of these identical pulses, so that the autocorrelation function of the pattern is as impulse-like as possible.

The problem is usually restated as follows: For each positive integer n, what is the shortest length $L = L(n)$ for which there is a sequence $\{a_1, a_2, \ldots, a_n\}$ with $0 = a_1 < a_2 < \ldots < a_n = L$, such that the set of $\binom{n}{2}$ differences $\{a_j - a_i\}$, with $1 \le i < j \le n$, are all distinct?

The model underlying this restatement is the following. At each integer a_i in the sequence, there is a pulse of brief duration and unit amplitude. Thus, the sequence corresponds to a pulse pattern with n pulses, spread out over a total duration of L (or, more precisely L plus one pulse duration) with the unnormalized autocorrelation function $K(\tau)$ satisfying $K(0) = n$, $0 \le K(\tau) \le 1$ for $1 \le \tau \le L$, and $K(\tau) = 0$ for $\tau > L$. (The distinctness of the differences $\{a_j - a_i\}$ guarantees $K(\tau) \le 1$ for all τ, $|\tau| \ge 1$.) Finding the *shortest* length $L = L(n)$ for the sequence achieves the desired signal parameters with the shortest duration for the pulse pattern.

The sequence model can also be described in terms of a certain class of *rulers* (the measuring devices, not the monarchs or autocrats).

A ruler of length L has only n marks on it, at integer positions $a_1, a_2, \ldots, a_n$, where $a_1 = 0$ and $a_n = L$ are the two endpoints of the ruler. If every integer distance $d, 1 \le d \le L$, can be measured in one and only one way as a distance between two of the n marks, then the ruler is called a *perfect ruler*. For a perfect ruler, $L = \binom{n}{2}$, since there are exactly $\binom{n}{2}$ distances between the n marks, and these must be some permutation of $\{1, 2, 3, \ldots, L\}$. In Figure 2, we see perfect rulers for $n = 2, 3$, and 4, with the corresponding radar pulse patterns and their autocorrelation functions $K(\tau)$.

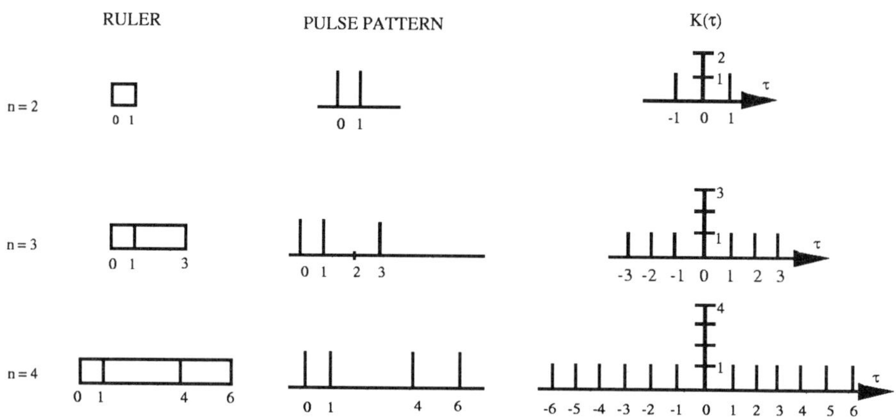

Fig. 2. Perfect rulers for $n = 2, n = 3$, and $n = 4$, with the corresponding pulse radar patterns and their unnormalized autocorrelation functions.

Unfortunately, for $n > 4$, there are no perfect rulers.

Theorem 1. *For $n > 4$, no perfect rulers exist.*

Proof. A ruler with n marks has $n-1$ *intervals* between marks. For a *perfect* ruler, these $n-1$ intervals must all have distinct positive integer lengths, and the sum of these lengths must be $L = \binom{n}{2}$. Hence these intervals must be (in some order) $1, 2, 3, \ldots, n-1$, since any *other* set of $n-1$ distinct positive integers will have a larger sum than $\binom{n}{2}$. Since *all* distances between marks on the ruler must be distinct, the interval of length 1 cannot be next to an interval of length $\leq n-2$ (since the sum of two consecutive intervals is a measured distance of the ruler, and every length from 1 to $n-1$ is *already* measured, as a single interval). This can only be achieved if the interval of length 1 is at one end of the ruler, and is immediately followed by the interval of length $n-1$. Similarly, the interval of length 2 cannot be next to any interval of length $\leq n-3$ (to avoid two consecutive intervals with a total length $\leq n-1$, equaling the length of a single interval), nor can it be next to the interval of length $n-2$ (since $2+(n-2) = n = 1+(n-1)$, and the distance n would be measured in more than one way). This requires that the interval of length 2 must also be at an end of the ruler, and must also be next to the interval of length $n-1$. But then the entire ruler consists of only three intervals: $1, n-1, 2$, which means there are only four marks altogether, and $n = 4$. (Note that with $n = 4$, there really is a perfect ruler with the consecutive intervals of lengths 1,3,2.) □

There are two obvious ways to relax the requirements on a perfect ruler to get objects which exist for all n. A *covering ruler* with n marks and length L measures every distance from 1 to L, as a distance between two marks on the ruler, in *at least one way*; while a *spanning ruler* with n marks and length L measures every distance from 1 to L, as a distance between two marks on the ruler, in *at most one way*. The interesting combinatorial problems are to determine the *longest covering ruler* with n marks, and the *shortest spanning ruler* with n marks, for each positive integer n. Both of these problems have long histories in the combinatorial literature. However, the application to pulse radar involves only finding the *shortest spanning ruler* for each n. (Martin Gardner [4] termed these objects "Golomb rulers", a name which seems subsequently to have been widely adopted.)

The behavior of $L(n)$ as a function of n, for the shortest spanning ruler, is quite erratic in detail, although it is easily shown that asymptotically $\liminf_{n\to\infty} \frac{L(n)}{n^2} \leq 1$. The value of $L(n)$ has been determined by exhaustive computer search for all $n \leq 21$. (For the larger values of n, this search was performed by a distributed world-wide effort coordinated on the Internet [6]. In addition to left-right reversal of the ruler, these rulers are not unique for several of the smaller values of n. One example of a spanning ruler of length $L(n)$, for each $n \leq 19$, is shown in Table 1. (Column "m" in Table 1 shows the number of inequivalent spanning rulers of length $L(n)$ with n marks.)

These minimum spanning rulers have another, very different application in radio astronomy. In radio astronomy, only receiving antennas are used, and the spacing between two antennas generates a difference in the time a signal is received, which is used in making interferometry measurements on the signal. If several antennas are constructed along a straight line, they should be spaced

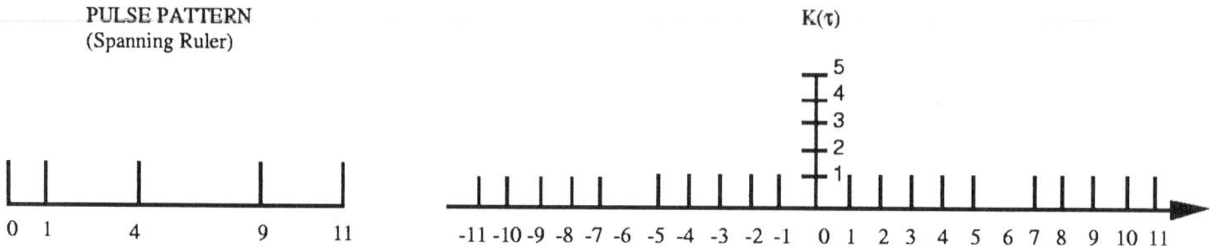

Fig. 3. Radar pulse pattern, and its autocorrelation, for $n = 5$ pulses.

along that line in the pattern of a spanning ruler, to get maximum interferometry information by having all the arrival time differences distinct.

The properties of these rulers also play an important role in x-ray diffraction crystallography. In this technique, one attempts to determine the bonding angles of a crystal by shining x rays at it, and observing the diffraction patterns which emerge. What in fact is measurable by this method is the *differences* of the bonding angles, from which one wishes to reconstruct the bonding angles themselves. This leads to an "inverse problem": given the autocorrelation function $K(\tau)$, what is the set of possible signals which might have produced it? For several decades, crystallographers relied on a "theorem" of S. Piccard which asserted (in our terminology): "If two spanning rulers have the same autocorrelation function $K(\tau)$ the two rulers are either identical or mirror images of each other." A counter-example was found by G.S. Bloom [3] for 6-mark rulers: $\{0, 1, 4, 10, 12, 17\}$ and $\{0, 1, 8, 11, 13, 17\}$, which are in fact *shortest* 6-mark spanning rulers. This generalizes to a two-parameter family of counter-examples, all involving 6-mark spanning rulers. No counter-examples with fewer than 6 marks are possible, and none with more than 6 marks are known. There are partial results which suggest that counter-examples may occur only in the case $n = 6$ [5], and it has been proved that no counter-examples occur with $n \leq 12$ marks except for $n = 6$.

TABLE 1
TABLE OF THE SHORTEST SPANNING RULERS

n	L(n)	m	Sequence of Marks
2	1	1	0,1
3	3	1	0,1,3
4	6	1	0,1,4,6
5	11	2	0,1,4,9,11
6	17	4	0,1,4,10,12,17
7	25	5	0,1,4,10,18,23,25
8	34	1	0,1,4,9,15,22,32,34
9	44	1	0,1,5,12,25,27,35,41,44
10	55	1	0,1,6,10,23,26,34,41,53,55
11	72	2	0,1,4,13,28,33,47,54,64,70,72
12	85	1	0,2,6,24,29,40,43,55,68,75,76,85
13	106	1	0,2,5,25,37,43,59,70,85,89,98,99,106
14	127	1	0,5,28,38,41,49,50,68,75,92,107,121,123,127
15	151	1	0,6,7,15,28,40,51,75,89,92,94,121,131,147,151
16	177	1	0,1,4,11,26,32,56,68,76,115,117,134,150,163,168,177
17	199	1	0,8,31,34,40,61,77,99,118,119,132,143,147,182,192,194,199
18	216	1	0,2,10,22,53,56,82,83,89,98,130,148,153,167,188,192,205,216
19	246	1	0,1,6,25,32,72,100,108,120,130,153,169,187,190,204,231,233,242,246

A systematic construction for *very good* spanning rulers (though not necessarily *best* spanning rulers) with k marks, and length not exceeding $(k-1)^2$, can be obtained from cyclic (v, k, λ)-designs with $v = n^2 + n + 1, k = n + 1, \lambda = 1$ (called *finite projective planes*), which are known to exist whenever n is a prime or a power of a prime. (Whether they ever exist for any other value of n is an unsolved problem.) Think of the cyclic $(v, k, 1)$ design D as a circle of circumference v, with k marks at the k positions d_i, for all $d_i \in D$.

On this circle, every integer arc length from 1 to $v - 1$ can be found in one and only one way as a *distance* between two marks. Remove the longest arc between two consecutive marks, and straighten out what is left. This will be a spanning ruler with k marks, and length at most $v - k = n^2 = (k-1)^2$. For example, $D = \{3, 6, 7, 12, 14\}$ is a $(21, 5, 1)$ cyclic difference set, and the longest arc (modulo 21) is from 14 to 3, which we remove. We then look at $D - 3 = \{0, 3, 4, 9, 11\}$; and a ruler with five marks, at 0, 3, 4, 9, and 11, is a shortest spanning ruler, measuring every length from 1 to 11 inclusive except for length 10. (This is the other shortest spanning ruler with five marks, inequivalent to the one shown if Figure 3.)

References

1. R.H. Barker, Group Synchronization of Binary Digital Systems, *Communication Theory* (Proceedings of the Second London Symposium on Information Theory), London, Butterworth, 1953, 273-287.

2. L.D. Baumert, *Cyclic Difference Sets*, Lecture Notes in Mathematics 182, Springer-Verlag 1971.
3. G.S. Bloom and S.W. Golomb, Applications of Numbered, Undirected Graphs, *Proceedings of the IEEE*, vol. 65, no. 4, April, 1977, 562-571.
4. M. Gardner, *Wheels, Life and Other Mathematical Amusements*, W.H. Freeman and Co., New York, 1983; Chapter 15, 152-165.
5. S.W. Golomb and G. Yovanof, The Polynomial Model in the Study of Counterexamples to S. Piccard's "Theorem", *Ars Combinatoria*, vol. 48, April, 1998, 43-63.
6. B. Hayes, Collective Wisdom, *American Scientist*, vol. 86, no. 2, March-April, 1998, 118-122.
7. J.-S. No, S.W. Golomb, G. Gong, H.-K. Lee and P. Gaal, Binary Pseudorandom Sequences of Period $2^n - 1$ with Ideal Autocorrelation, *IEEE Transactions on Information Theory*, vol. 44, no. 2, March, 1998, 814-817.
8. R. Turyn and J. Storer, On Binary Sequences, *Proceedings of the American Mathematical Society*, vol. 12, no. 3, June, 1961, 394-399.

Correlation of m-Sequences and Related Topics

Tor Helleseth

Department of Informatics
University of Bergen
Høyteknologisenteret
N-5020 Bergen, Norway
Email: torh@ii.uib.no

Abstract. We present a survey of many connections between the auto and crosscorrelation function of m-sequences and related topics. We show connections to perfect (PN) and almost perfect (APN) nonlinear power mappings which are of interest in differential cryptanalysis. Further, we give connections to difference sets, sequences with two-level ideal autocorrelation and families of sequences with optimal correlation properties, as well as to hyperovals and permutation polynomials.

1 Introduction

Let F_q denote the finite field with q elements. Let $f(x) = \sum_{i=0}^{m} f_i x^i \in F_q[x]$ be a primitive polynomial with coefficients from F_q, which is the minimum polynomial of a primitive element $\alpha \in F_{q^m}$, (i.e., α has order $q^m - 1$). A nonzero sequence $\{s(t)\}$ with elements from F_q generated by the recursion having $f(x)$ as its *characteristic polynomial*, i.e.,

$$\sum_{i=0}^{m} f_i s(t+i) = 0, \text{ for all } t \tag{1}$$

is called a maximal-length linear-feedback shift-register sequence (m-sequence).

The sequence is completely determined by the initial values $s(0)$, $s(1)$, $\cdots$, $s(m-1)$ and this recursion. Hence, the recursion (1) generates q^m distinct sequences. Let Tr_m be the *trace function* which is the mapping from the finite field F_{q^m} to the subfield F_q given by

$$Tr_m(x) = \sum_{i=0}^{m-1} x^{q^i}.$$

The sequences determined by

$$s(t) = Tr_m(a\alpha^t),\ a \in F_{q^m} \tag{2}$$

obey the recursion above since

$$\begin{aligned}\sum_{i=0}^{m} f_i s(t+i) &= \sum_{i=0}^{m} f_i Tr_m(a\alpha^{t+i}) \\ &= Tr_m(a\alpha^t \sum_{i=0}^{m} f_i \alpha^i) \\ &= Tr_m(a\alpha^t f(\alpha)) \\ &= 0\end{aligned}$$

and α is a zero of the characteristic polynomial. From properties of the trace function the sequences corresponding to different values of $a \in F_{q^m}$ are distinct. Hence, when a runs through F_{q^m}, (2) generates all the q^m distinct sequences generated by the recursion (1) having $f(x)$ as its *characteristic polynomial.* The period of the m-sequence is $q^m - 1$ and the m-sequences generated by (1) are cyclic shifts of each other since any nonzero $a = \alpha^i$ for some $i, 0 \leq i \leq q^m - 2$.

Example. Let $f(x)$ be the binary primitive polynomial $f(x) = x^3 + x + 1$. Then $s(t+3) + s(t+1) + s(t) = 0$ generates the m-sequence 0010111 ... of period $n = 2^3 - 1 = 7$.

Definition 1. *We define two sequences* $\{s_1(t)\}$ *and* $\{s_2(t)\}$ *to be* cyclically equivalent *if there exists an integer* τ *such that*

$$s_1(t+\tau) = s_2(t)$$

for all t*, otherwise they are said to be* cyclically distinct.

The number of primitive elements in F_{q^m} is $\phi(q^m - 1)$ where ϕ is Euler's function. It is not hard to see that m-sequences of the same length generated by different characteristic polynomials are cyclically distinct. However, since α^{q^i}, $0 \leq i \leq m-1$, all have the same minimum polynomial, the number of cyclically distinct m-sequences of period $q^m - 1$ is precisely $\phi(q^m - 1)/m$.

Some of the important properties of m-sequences are the following.

Lemma 1. *Let* $\{s(t)\}$ *be an m-sequence of period* $q^m - 1$ *over* F_q.
(1) (Balance property) Let N_b *be the number of occurrences of an element* b *in a period of the sequence. Then,*

$$N_b = \begin{cases} q^{m-1} & \text{if } b \in F_q \setminus \{0\} \\ q^{m-1} - 1 & \text{if } b = 0. \end{cases}$$

(2) (Run property) As t *varies over* $0 \leq t \leq q^m - 2$*, the m-tuple*

$$(s(t), s(t+1), \cdots, s(t+m-1))$$

runs through all the elements in F_q^m *exactly once, with the exception of the all zero m-tuple which does not occur.*

(3) (Shift and add property) For any τ, $0 < \tau \leq q^m - 2$, there exists a δ for which

$$s(t) + s(t+\tau) = s(t+\delta), \text{ for all } t.$$

(4) (Constancy on cyclotomic cosets) There exists a cyclic shift τ of $\{s(t)\}$, such that $s(q^i t + \tau) = s(t + \tau)$ for all t.

From now on assume that $q = p$ is a prime. The autocorrelation function of the sequence $\{s(t)\}$, with symbols in the finite field F_p, is defined by

$$\theta_s(\tau) = \sum_{t=0}^{p^m-2} \omega^{s(t+\tau)-s(t)},$$

where ω is a primitive complex pth root of unity. Then an important property of an m-sequence is its two-level and ideal autocorrelation function.

Lemma 2. *Let p be a prime, and let $\{s(t)\}$ be an m-sequence of period $p^m - 1$ over F_p. Then*

$$\theta_s(\tau) = \begin{cases} -1 & \text{if } \tau \not\equiv 0 \ (\text{mod } p^m - 1) \\ p^m - 1 & \text{if } \tau \equiv 0 \ (\text{mod } p^m - 1). \end{cases}$$

Proof. Let the m-sequence be given by $s(t) = Tr_m(a\alpha^t)$, where $a \in F^*_{p^m} = F_{p^m} \setminus \{0\}$. In the case $\tau \equiv 0$ (mod $p^m - 1$), the result is obvious. In the case $\tau \not\equiv 0$ (mod $p^m - 1$), it follows that,

$$\theta_s(\tau) = \sum_{t=0}^{p^m-2} \omega^{s(t+\tau)-s(t)}$$
$$= \sum_{t=0}^{p^m-2} \omega^{Tr_m((a\alpha^\tau - a)\alpha^t)}.$$

Clearly, $a\alpha^\tau - a \neq 0$, since $\tau \not\equiv 0$ (mod $p^m - 1$). By the balance property of the m-sequence, it follows that

$$\theta_s(\tau) = (p^{m-1} - 1) + p^{m-1}(\omega + \omega^2 + \cdots + \omega^{p-1}) = -1.$$

□

The balance, run and optimal autocorrelation properties of m-sequences are properties one would expect of a randomly generated sequence whose symbols are statistically independent and are equally likely to be either 0 or 1. This pseudorandom behavior of m-sequences has led to numerous applications in practice including synchronization, position location, random number generation and multiple-access communication.

There is a close connection between difference sets and binary sequences with two-level ideal autocorrelation exhibited by m-sequences.

Definition 2. *Let G be a finite group with v elements and let D be a subset of G with k elements. Then D is a (v, k, λ) difference set if $d_1 - d_2$, d_1, $d_2 \in D$ represent any nonzero element in G exactly λ times.*

Any m-sequence $\{s(t)\}$ gives a $(2^m - 1, 2^{m-1} - 1, 2^{m-2} - 1)$ difference set by defining G as the integers $\pmod{2^m - 1}$, and let D be given by

$$D = \{t \mid s(t) = 0\}.$$

Example. Let $\{s(t)\} = 0010111$ be the m-sequence of period 7 with zeros in positions $D = \{0, 1, 3\}$. Then D is a $(7, 3, 1)$ difference set since all nonzero differences $\pmod 7$ can be written uniquely as a difference of two distinct elements of D.

Until recently, the known cyclic $(2^m - 1, 2^{m-1} - 1, 2^{m-2} - 1)$ difference sets were the following:

- Singer difference sets for all $m > 0$.
- GMW difference sets for composite m.
- Paley-Hadamard difference sets for $2^m - 1$ a prime.
- Some sporadic difference sets found by computer search.

2 Crosscorrelation of m-sequences

Let $\{u(t)\}$ and $\{v(t)\}$ be two sequences of period n with an alphabet of size p, where p is a prime. Then the crosscorrelation function between the two sequences is defined as

$$\theta_{u,v}(\tau) = \sum_{t=0}^{n-1} \omega^{u(t+\tau)-v(t)}.$$

An important problem in many applications is to find large families of sequences with good correlation properties. Let

$$\mathcal{F} = \{\{s_i(t)\} \mid 1 \le i \le M\}$$

be a family of M sequences over F_p each of period n. Let $\theta_{i,j}(\tau)$ denote the crosscorrelation between the ith and jth sequence at shift τ, i.e.,

$$\theta_{i,j}(\tau) = \sum_{t=0}^{n-1} \omega^{s_i(t+\tau)-s_j(t)}, \quad 0 \le \tau \le n-1.$$

The classical goal in sequence design for CDMA systems has been minimization of the parameter

$$\theta_{max} = \max\{|\theta_{i,j}(\tau)| \mid \text{either } i \neq j \text{ or } \tau \neq 0\}$$

for fixed n and M.

Since m-sequences have ideal autocorrelation properties, it is natural to study the crosscorrelation function between pairs of m-sequences. Many families of low correlation sequences have been constructed from pairs of m-sequences.

Let $\{u(t)\}$ and $\{v(t)\}$ be a pair of m-sequences over F_p of period $p^m - 1$. Without loss of generality we assume the sequences are shifted cyclically such that

$$u(t) = Tr_m(\alpha^t)$$

and

$$v(t) = Tr_m(\beta^t)$$

for some elements α and β of order $p^m - 1$.

Since α and β are primitive elements there exists a d, $1 \leq d \leq p^m - 2$, $gcd(d, p^m - 1) = 1$ such that $\beta = \alpha^d$. The crosscorrelation function between the two m-sequences can be expressed by

$$\begin{aligned} \theta_{u,v}(\tau) &= \sum_{t=0}^{p^m-2} \omega^{u(t+\tau)-v(t)} \\ &= \sum_{t=0}^{p^m-2} \omega^{Tr_m(\alpha^{t+\tau}-\beta^t)} \end{aligned}$$

where ω is a complex, primitive pth root of unity. Since α^t runs through $F_{p^m}^*$ when $t = 0, 1, \cdots, p^m - 2$, it follows that

$$\theta_{u,v}(\tau) = \sum_{x \in F_{p^m}^*} \omega^{Tr_m(cx - x^d)}, \quad c = \alpha^\tau.$$

To evaluate the crosscorrelation function between any two m-sequences we have to evaluate this exponential sum for all $c \in F_{p^m}^*$. Hence, the crosscorrelation function depends only on the decimation d that relates the two m-sequences. We use $C_d(\tau)$ to denote the crosscorrelation function between the m-sequence $\{s(t)\}$ and its decimation $\{s(dt)\}$, i.e.,

$$C_d(\tau) = \sum_{i=0}^{p^m-2} \omega^{s(t+\tau)-s(dt)}.$$

From the properties of the trace function and the results in Trachtenberg [32] and Helleseth [13], we obtain the following:

Lemma 3. *(1) $C_d(\tau)$ is a real number.*

(2) Let $d \cdot d^{-1} = 1 \pmod{p^m - 1}$, then the values and the number of occurrences of each value of $C_d(\tau)$ and $C_{d^{-1}}(\tau)$ are the same.

(3) $C_{dp^j}(\tau) = C_d(\tau)$.

(4) $C_d(p^j\tau) = C_d(\tau)$.

(5) $\sum_{\tau=0}^{p^m-2} C_d(\tau) = 1$.

(6) $\sum_{\tau=0}^{p^m-2} C_d^2(\tau) = p^{2m} - p^m - 1.$

(7) $\sum_{\tau=0}^{p^m-2} C_d^l(\tau) = -(p^m-1)^{l-1} + 2(-1)^{l-1} + a_l p^{2m}$ *where* a_l *is the number of solutions of the equations*

$$x_1 + x_2 + \cdots + x_{l-1} + 1 = 0$$
$$x_1^d + x_2^d + \cdots + x_{l-1}^d + 1 = 0$$

where $x_i \in F_{p^m}^*$ *for* $i = 1, 2, \cdots, l-1$.

Theorem 1. *If* $d \notin \{1, p, \cdots, p^{m-1}\}$, *(i.e., when the two m-sequences are cyclically distinct), then* $C_d(\tau)$ *takes on at least three values as* τ *varies over the set* $\tau = \{0, 1, \cdots, p^m - 2\}$.

Proof. This follows from Lemma 3 (5) and (6) by showing that these two relations can not be satisfied for a two-valued crosscorrelation function, except when $d \in \{1, p, \cdots, p^{m-1}\}$. The full proof is given in Helleseth [13]. □

Hence, the crosscorrelation function between to m-sequences generated by different characteristic polynomials always takes on at least three different values. The following theorem, in part due to Gold [12], Kasami [19] and Welch (see [32]) gives two examples of binary m-sequences that lead to three-valued crosscorrelation functions.

Theorem 2. *Let* $e = gcd(m,k)$ *and* m/e *be odd. Let* $d = 2^k + 1$ *or* $d = 2^{2k} - 2^k + 1$. *Then* $C_d(\tau)$ *takes on the following three values:*

$$\begin{array}{llll} -1 + 2^{\frac{m+e}{2}} & \textit{occurs} & 2^{m-e-1} + 2^{\frac{m-e-2}{2}} & \textit{times} \\ -1 & \textit{occurs} & 2^m - 2^{m-e} - 1 & \textit{times} \\ -1 - 2^{\frac{m+e}{2}} & \textit{occurs} & 2^{m-e-1} - 2^{\frac{m-e-2}{2}} & \textit{times.} \end{array}$$

Let $\{u(t)\}$ be a binary m-sequence of period $2^m - 1$ and $\{v(t)\} = \{u(dt)\}$, where $d = 2^k + 1$ (or $d = 2^{2k} - 2^k + 1$). In the case $gcd(m,k) = 1$, the family of sequences defined by

$$\mathcal{F} = \{\{u(t)\}\} \cup \{\{v(t)\}\} \cup \{\{u(t) + v(t+\tau)\} \mid 0 \le \tau \le 2^m - 2\}$$

is a set of $M = 2^m + 1$ cyclically distinct sequences of period $n = 2^m - 1$ and $\theta_{max} = \sqrt{2n} = \sqrt{2^{m+1}}$. This is known to be optimal for the given M and n. Applications include the Navstar global positioning system.

A pair of binary m-sequences with three-valued crosscorrelation function with values

$$-1, -1 + 2^{\lfloor \frac{m+2}{2} \rfloor}, -1 - 2^{\lfloor \frac{m+2}{2} \rfloor}$$

is called a *preferred pair*.

Theorem 2 can be applied to construct a preferred pair except in the case $m \equiv 0 \pmod 4$. The conjecture in Sarwate and Pursley [31] that for $m \equiv 0$

(mod 4) no pairs of preferred sequences exist was solved by Calderbank and McGuire [24].

In the case when m is even Niho [27] found that the following family of sequences has good crosscorrelation properties. For $m \equiv 0 \pmod 4$ these sequences have a lower value of maximum crosscorrelation magnitude than the ones constructed from Theorem 2.

Theorem 3. *Let $m \equiv 0 \pmod 4$ and $m = 2k$. Let $d = 2^{k+1} - 1$, then $C_d(\tau)$ takes on the following four values:*

$$\begin{array}{lll} -1+2^{k+1} & \textit{occurs } \frac{1}{3}(2^{2k-1}-2^{k-1}) & \textit{times} \\ -1+2^{k} & \textit{occurs } 2^{k} & \textit{times} \\ -1 & \textit{occurs } 2^{2k-1}-2^{k-1}-1 & \textit{times} \\ -1-2^{k} & \textit{occurs } \frac{1}{3}(2^{2k}-2^{k}) & \textit{times.} \end{array}$$

In addition to the previous theorems, the crosscorrelation function has been completely determined for the values of d given below. Here we are dealing with binary sequences and it is implicitly assumed that m has the property that $gcd(d, 2^m - 1) = 1$.

(1) $d = (2^{\frac{m}{2}} + 1)(2^{\frac{m}{4}} - 1) + 2$, $m \equiv 0 \pmod 4$.
(2) $d = 2^{\frac{m}{2}} + 3$, m even.
(3) $d = \frac{1}{3}(2^m - 1) + 2^i$, m even.
(4) $d = 2^{\frac{m}{2}} - 2^{\frac{m}{4}} + 1$, $m \equiv 0 \pmod 8$.
(5) $d = -1$.
(6) $d = 2^{\frac{m}{2}} + 2^{\frac{m+2}{4}} + 1$, $m \equiv 2 \pmod 4$.
(7) $d = 2^{\frac{m+2}{2}} + 3$, $m \equiv 2 \pmod 4$.
(8) $d = \sum_{i=0}^{\frac{m}{2}} 2^{ik}$ $(0 < k < m, gcd(k, m) = 1)$ for $m \equiv 0 \pmod 4$.
(9) $d = 2^{\frac{m}{2}} + 2^{\frac{m}{4}} + 1$ for $m \equiv 0 \pmod 4$.

Case (1) was settled by Niho [27], cases (2)-(4) by Helleseth [13], [14], case (5) by Lachaud and Wolfmann [21], cases (6), (7) by Cusick and Dobbertin [7] and cases (8) and (9) by Dobbertin [10].

The following cases, where m is odd, were long standing conjectures for more than 25 years of decimations having three-valued crosscorrelation function with the values $\{-1, -1 \pm 2^{\frac{m+1}{2}}\}$.

Welch's conjecture:

$$d = 2^{\frac{m-1}{2}} + 3.$$

Niho's conjecture [27]:

$$d = \begin{cases} 2^{\frac{m-1}{2}} + 2^{\frac{m-1}{4}} - 1 & \text{if } m \equiv 1 \pmod 4 \\ 2^{\frac{m-1}{2}} + 2^{\frac{3m-1}{4}} - 1 & \text{if } m \equiv 3 \pmod 4. \end{cases}$$

Recently Welch's conjecture was solved by Canteaut, Charpin and Dobbertin [2]. Further, Hollman and Xiang have announced a proof of both of the conjectures of Welch and Niho.

3 Crosscorrelation of nonbinary m-sequences

In the case of nonbinary m-sequences similar results are given by Helleseth [13] and Trachtenberg [32]. In particular the following generalization of Theorem 2 has been obtained.

Theorem 4. *Let p be an odd prime. Let $e = gcd(m,k)$ and m/e odd. Let $d = \frac{1}{2}(p^{2k}+1)$ or $d = p^{2k} - p^k + 1$, where $d \not\equiv p^j \pmod{p^m-1}$, then $C_d(\tau)$ takes on the following three values:*

$$\begin{array}{lll} -1 + p^{\frac{m+e}{2}} & \textit{occurs } \frac{1}{2}(p^{m-e} + p^{\frac{m-e}{2}}) & \textit{times} \\ -1 & \textit{occurs } p^m - p^{m-e} - 1 & \textit{times} \\ -1 - p^{\frac{m+e}{2}} & \textit{occurs } \frac{1}{2}(p^{m-e} - p^{\frac{m-e}{2}}) & \textit{times.} \end{array}$$

Proof. For a full proof when $e = 1$ see Trachtenberg [32]. The generalization when $e > 1$ is due to Helleseth [13].

The generalization below of Theorem 3 can be found in Helleseth [13].

Theorem 5. *Let p be an odd prime, $m = 2k$, and $p^k \not\equiv 2 \pmod 3$. Let $d = 2p^k - 1$, then $C_d(\tau)$ takes on the following four values*

$$\begin{array}{lll} -1 + 2p^k & \textit{occurs } \frac{1}{6}(p^{2k} - p^k) & \textit{times} \\ -1 + p^k & \textit{occurs } p^k & \textit{times} \\ -1 & \textit{occurs } \frac{1}{2}(p^{2k} - p^k - 2) & \textit{times} \\ -1 - p^k & \textit{occurs } \frac{1}{3}(p^{2k} - p^k) & \textit{times.} \end{array}$$

In addition to these two cases the crosscorrelation function has been determined completely in Helleseth [13] for the following three decimations, where we assume m is selected so that $gcd(d, p^m - 1) = 1$.

(1) $d = 2p^{\frac{m}{2}} - 1, m \equiv 0 \pmod 4$.
(2) $d = \frac{1}{2}(p^m - 1) + p^i$.
(3) $d = \frac{1}{3}(p^m - 1) + p^i$.

Two old conjectures in Helleseth [13] that are still unproved are the following:

Conjecture. (1) If $m = 2^i$, then $C_d(\tau)$ has at least four different values.
(2) If $d \equiv 1 \pmod{p-1}$, then -1 is always one of the values of $C_d(\tau)$.

4 PN power mappings

Let $f(x)$ be a mapping $f : F_{p^m} \to F_{p^m}$. Let $N(a,b)$ denote the number of solutions $x \in F_{p^m}$ of $f(x+a) - f(x) = b$ where $a, b \in F_{p^m}$ and let

$$\Delta_f = \max\{N(a,b) \mid a, b \in F_{p^m},\ a \neq 0\}.$$

Nyberg [29] defined a mapping f to be differentially k-uniform if $\Delta_f = k$. This concept is of interest in cryptography since differential and linear cryptanalysis exploit weaknesses of the uniformity of the functions which are used in DES and in many other block ciphers.

We next will study the differential uniformity of functions of the form $f(x) = x^d$ over F_{p^m}, where p is a prime. For applications in cryptography, one would like functions for which Δ_f is as small as possible. In the binary case the solutions come in pairs so $\Delta_f = 2$ is the smallest possible value. However, in the case when p is odd there exist functions for which $\Delta_f = 1$. These functions are sometimes called *perfect nonlinear* (PN). It is interesting to observe that 1-uniform power mappings have also been studied under the name of planar functions which are functions such that $f(x+a) - f(x)$ is a permutation polynomial for all $a \in F_{p^m}^*$. There are only three known families of 1-uniform power mappings.

Observe that since p is the characteristic of the field F_{p^m}, then Δ_f is constant on the cyclotomic coset dp^i, (i.e. $\Delta_{x^{dp^i}} = \Delta_{x^d}$), for $i = 0, 1, \cdots, m-1$. Further, since $N(a, b) = N(1, b/a^d)$ when $a \neq 0$, it is sufficient to consider the case $a = 1$ when determining Δ_f.

Theorem 6. *Let $f(x) = x^d$ be a mapping over F_{p^m} where $p > 2$. Then $f(x)$ is 1-uniform (or PN) in the following three cases,*

(1) $d = 2$,

(2) $d = p^k + 1$, where $m/\gcd(m, k)$ is odd,

(3) $d = (3^k + 1)/2$, where $p = 3$, k is odd and $\gcd(m, k) = 1$.

Comments on proof: This can be proved in several ways. (1) Follows since $(x+1)^2 - x^2 = b$ is equivalent to $2x + 1 = b$ which has a unique solution for any $b \in F_{p^m}$. (2) Follows since $(x+1)^{p^k+1} - x^{p^k+1} = b$ leads to a linearized equation $x^{p^k} + x + 1 - b = 0$, which since $m/\gcd(m, k)$ is odd can easily be seen to have a unique solution for any $b \in F_{p^m}$.

The proof of (3) given in Coulter and Matthews [5] (and in part with a different proof in Helleseth and Sandberg [18]) also turned out be a counterexample to an old conjecture about planar functions.

In Coulter and Matthews [5] the three values of d above are mentioned as the only known planar functions of the form $f(x) = x^d$.

In Helleseth and Sandberg [18], a close connection is given between families of sequences with optimal correlation properties and planar functions from power mappings. This construction will be given in the next section. In view of this connection, the results in (1) and (2) also follow from known results on sequence designs (for instance (2) can be found in Kumar and Moreno [20]).

5 Sequences with good correlations from PN mappings

Based on any differentially 1-uniform power mapping $f(x) = x^d$ one can construct a family of sequences with good correlation properties.

Let ω be a complex pth root of unity and let $Tr_m(x) = \sum_{i=0}^{m-1} x^{p^i}$ denote the trace mapping from F_{p^m} to F_p. If $a \in F_{p^m}^*$ it holds for any differentially 1-uniform mapping $f(x) = x^d$ that,

$$\sum_{x \in F_{p^m}} \omega^{Tr_m(f(x+a)-f(x))} = \sum_{b \in F_{p^m}} \omega^{Tr_m(b)} = 0$$

since $f(x+a) - f(x)$ is a permutation polynomial of F_{p^m} for all $a \in F_{p^m}^*$. Let $c \in F_{p^m}^*$ and

$$S(c, \lambda) = \sum_{x \in F_{p^m}} \omega^{Tr_m(cf(x)+\lambda x)},$$

then

$$\begin{aligned}
|S(c, \lambda)|^2 &= \sum_{x,y \in F_{p^m}} \omega^{Tr_m(c(f(x)-f(y))+\lambda(x-y))} \\
&= \sum_{y,z \in F_{p^m}} \omega^{Tr_m(c(f(y+z)-f(y))+\lambda z)} \\
&= p^m + \sum_{z \in F_{p^m}^*} \omega^{Tr_m(\lambda z)} \sum_{b \in F_{p^m}} \omega^{Tr_m(cb)} \\
&= p^m.
\end{aligned}$$

It follows that

$$|S(c, \lambda)| = |\sum_{x \in F_{p^m}} \omega^{Tr_m(cf(x)+\lambda x)}| = \sqrt{p^m}$$

for all $\lambda \in F_{p^m}$ whenever $c \in F_{p^m}^*$.

Let α be a primitive $(p^m - 1)$th root of unity in F_{p^m}. Let $\{s_c(t)\}$ be the sequence of period $p^m - 1$ defined by

$$s_c(t) = Tr_m(c\alpha^{dt} + \alpha^t).$$

Then

$$\mathcal{F} = \{\{s_c(t)\} \mid c \in F_{p^m}\}$$

is a family of p^m cyclically distinct sequences with maximum correlation bounded by $1 + \sqrt{p^m}$ in magnitude. This follows since,

$$\begin{aligned}
\theta(\tau) &= \sum_{t=0}^{p^m-2} \omega^{s_{c_1}(t+\tau)-s_{c_2}(t)} \\
&= \sum_{t=0}^{p^m-2} \omega^{Tr_m((c_1\alpha^{d\tau}-c_2)\alpha^{dt}+(\alpha^\tau-1)\alpha^t)} \\
&= -1 + \sum_{x \in F_{p^m}} \omega^{Tr_m(cx^{dt}+\lambda x)},
\end{aligned}$$

where $c = c_1\alpha^{d\tau} - c_2$ and $\lambda = \alpha^\tau - 1$. Hence, $|\theta(\tau)| \leq 1 + \sqrt{p^m}$, except when $c_1 = c_2$ and $\tau = 0$. Thus, family $\mathcal{F}$ has the same parameters as the best presently known families of sequences found in Kumar and Moreno [20]. It is perhaps interesting to note that these sequences correspond to $d = 2$ or $d = p^k + 1$ where $m/gcd(m,k)$ is odd. Observe that $f(x) = x^2$ and $f(x) = x^{p^k+1}$ are also planar permutation polynomials.

6 Binary APN power mappings

In the binary case, the solutions of $f(x+a) - f(x) = b$ come in pairs and therefore $\Delta_f = 2$ is the smallest possible value. For this reason we will mainly consider power mappings $f(x) = x^d$ over F_{2^m} such that $\Delta_f = 2$. Such mappings are called *almost perfect nonlinear* (APN).

In this section we present a short survey of known results on binary APN power mappings. The known examples of APN power mappings over F_{2^m} are given in the following theorem. It is not hard to see that $\Delta_{x^d} = \Delta_{x^{d^{-1}}}$, where d^{-1} is the multiplicative inverse of d modulo $2^m - 1$.

Known binary APN power mappings are given by the following theorem.

Theorem 7. *Let $f(x) = x^d$ be a mapping over F_{2^m}. Then $f(x)$ is APN in the following cases,*

(1) $d = 2^m - 2$, where m is odd,

(2) $d = 2^k + 1$, where $\gcd(m,k) = 1$,

(3) $d = 2^{2k} - 2^k + 1$, where $\gcd(m,k) = 1$,

(4) $d = 2^{\frac{m-1}{2}} + 3$, where m is odd,

(5)

$$d = \begin{cases} 2^{\frac{m-1}{2}} + 2^{\frac{m-1}{4}} - 1 & \text{if } m \equiv 1 \pmod 4 \\ 2^{\frac{m-1}{2}} + 2^{\frac{3m-1}{4}} - 1 & \text{if } m \equiv 3 \pmod 4. \end{cases}$$

Comments on proof: (1) and (2) are due to Nyberg [29] and Beth and Ding [1].

(3) in the case m odd, is essentially due to a result due to Chabaud and Vaudenay [3] or more explicitly formulated by Cusick [6] who showed that if two m-sequences of the same period $2^m - 1$ differ by a decimation of d, and have a three-level crosscorrelation function with values $-1, -1 \pm 2^{\frac{m+1}{2}}$ then the corresponding power mapping is APN. The result then follows from the fact that two m-sequences that differ by the decimations given in ((2) and) (3) are known to have a three-level crosscorrelation function with these values. In the case m even, (3) is proved by Dobbertin [10].

(4) is proved by Dobbertin [10] using a direct approach. The proofs of (3) and (4) give nice and surprisingly connections to permutation polynomials. For example to prove (4) Dobbertin observes that

$$(x+1)^d + x^d = p(x^{2^{\frac{m-1}{2}}} + x) + 1$$

where

$$p(x) = x^{2^{\frac{m+1}{2}}+1} + x^3 + x$$

can be shown to be a permutation polynomial of F_{2^m}. Since $x^{2^{\frac{m-1}{2}}} + x$ is a two-to-one map, it follows that $f(x) = x^d$ is an APN power mapping.

(5) is also proved by Dobbertin using a direct approach. It also follows since this decimation has recently been shown to give a three-level crosscorrelation function with values $-1, -1 \pm 2^{\frac{m+1}{2}}$.

One also observes that if a function is APN it does not necessarily have three-valued crosscorrelation. The prime example for this is (1) (i.e., $d = 2^m - 2$). It is shown by Lachaud and Wolfmann [21] that the crosscorrelation function takes on all possible values $C_d(\tau) \equiv 3 \pmod 4$ where $| C_d(\tau) | \leq 2\sqrt{2^m}$.

A conjecture due to Dobbertin is that the power mapping $f(x) = x^d$ for $d = 2^{4r} + 2^{3r} + 2^{2r} + 2^r - 1$ and $m = 5r$ is an APN mapping.

7 Some APN power mappings over F_{p^m}

In this section, we give examples of APN power mappings over F_{p^m}, where p is an odd prime.

Theorem 8. *Let $f(x) = x^d$ be a mapping over F_{p^m}, where p is odd. Then $f(x)$ is APN in the following cases,*

(1) $d = p^m - 2$, where $p^m \equiv 2 \pmod 3$,

(2) $d = \frac{p^m-1}{2} - 1$, where $p \equiv 3, 7 \pmod{20}$, $p^m > 7$, $p^m \neq 27$, and m is odd,

(3) $d = 3$, where $p \neq 3$.

Proof. (1) This follows since $(x+1)^{p^m-2} - x^{p^m-2} = b$ is equivalent to the equation $(x+1)^{-1} - x^{-1} = b$, i.e., to $bx^2 + bx + 1 = 0$ in the case when $x \notin \{-1, 0\}$. The latter equation has at most two roots for any $b \in F_{p^m}$.

Observe that both $x = 0$ and $x = -1$ are solutions of the original equation when $b = 1$. In this case, the latter equation has roots $x = \frac{-1\pm\sqrt{-3}}{2}$. If -3 is a non-square in F_{p^m} there are only two solutions of $(x+1)^{p^m-2} - x^{p^m-2} = b$ also in the case $b = 1$. Hence, it follows that $\Delta_f = 2$ if -3 is a non-square in F_{p^m}. Since m is odd, -3 is a non-square in F_{p^m} if and only -3 is a non-square in F_p, which by the law of quadratic reciprocity is equivalent to $p \equiv 2 \pmod 3$.

(2) This is proved in Helleseth and Sandberg [18].

(3) This follows directly since $(x+1)^3 - x^3 = b$ is equivalent to $3x^2 + 3x + 1 - b = 0$ which has at most two roots for any $b \in F_{p^m}$ and exactly two roots $x = 0$ and $x = -1$ for $b = 1$. □

The following results are proved in Helleseth, Rong and Sandberg [17].

Theorem 9. *Let $f(x) = x^d$ be a mapping over F_{p^m}, where p is odd. Then $f(x)$ is APN when $p^m > 7$, and*

$$d = \begin{cases} \frac{p^m+1}{4} + \frac{p^m-1}{2} & \text{if } p^m \equiv 3 \pmod 8 \\ \frac{p^m+1}{4} & \text{if } p^m \equiv 7 \pmod 8). \end{cases}$$

Theorem 10. *Let $d = p^k + 2$, where $p > 3$ is a prime. Let $f(x) = x^d$ be a mapping over F_{p^m}, where $m = 2k$. Then*

$$\Delta_f = \begin{cases} 2 & \text{if } p^k \equiv 1 \pmod 3, \text{ (i.e., } (-3)^k \text{ is a square } \pmod p) \\ 4 & \text{if } p^k \equiv 2 \pmod 3, \text{ (i.e., } (-3)^k \text{ is a non-square } \pmod p). \end{cases}$$

Theorem 11. *Let $f(x) = x^d$ be a mapping over F_{p^m}, where $p \geq 3$. Let m be odd and $d = p^{\frac{m+1}{2}} - 1$. Then*

$$\Delta_f = \begin{cases} p-2 & \text{if } p > 3 \\ 2 & \text{if } p = 3. \end{cases}$$

Theorem 12. *Let $d = 3^m - 3$ where $m > 1$ is an odd integer. Then the mapping $f(x) = x^d$ is an APN mapping over F_{3^m}.*

The following result from Helleseth, Rong and Sandberg [17] is a generalization of an observation due to Cusick [6] and in a more general form by Vaudenay and Chaubaud [3]. This is a useful observation to analyze power mappings in cases when the corresponding crosscorrelation function is known.

Theorem 13. *Let $\theta(a,b) = \sum_{x \in F_{p^m}} \omega^{T(ax+bx^d)}$. Then*

$$\sum_{a,b \in F_{p^m}} |\theta(a,b)|^4 = p^{2n} \sum_{u,v \in F_{p^m}} N(u,v)^2.$$

As mentioned in the proof of Theorem 7, Cusick [6] observed in the binary case that $f(x) = x^d$ is APN whenever two m-sequences that differ by a decimation d have a three-level crosscorrelation function with values $-1, -1 \pm 2^{\frac{m+1}{2}}$. For instance, the decimations $d = 2^k + 1$ and $d = 2^{2k} - 2^k + 1$ where m is odd and $\gcd(m,k) = 1$ have this property. From the well known crosscorrelation distribution Cusick derived $\sum_{b \in F_{2^m}} N(1,b)^2 = 2^{m+1}$ which combined with the facts $\sum_{b \in F_{2^m}} N(1,b) = 2^m$, $N(1,b) = 0$ or $N(1,b) \geq 2$ proves the APN property.

In the nonbinary case, Trachtenberg [32] has shown that $d = \frac{p^{2k}+1}{2}$ and $d = p^{2k} - p^k + 1$ in the case m odd and $\gcd(m,k) = 1$ have a three-valued crosscorrelation function with values $-1, -1 \pm p^{\frac{m+1}{2}}$. However, numerical results indicate that $f(x) = x^d$ will usually lead to $\Delta_f \geq 4$ in these cases.

One result for $d = \frac{p^k+1}{2}$ is the following theorem, whose proof is based upon a result in Coulter and Matthews [5].

Theorem 14. *Let $f(x) = x^d$ be a mapping over F_{p^m}, where p is an odd prime and $d = \frac{p^k+1}{2}$. Then $\Delta_f \le \gcd\left(\frac{p^k-1}{2}, p^{2m}-1\right)$.*

In the case $p = 3$ and $\gcd(2m, k) = 1$, then $f(x) = x^d$ is the 1-uniform (PN) or planar function given in Coulter and Matthews [5]. In addition we obtain some new APN mappings for $p = 5$.

Theorem 15. *Let $f(x) = x^d$ be a mapping over F_{5^m}, where $d = \frac{5^k+1}{2}$ and $\gcd(2m, k) = 1$. Then $f(x)$ is an APN mapping.*

8 Sequences with ideal autocorrelation

Some sporadic sequences of period $n = 2^m - 1$ with ideal autocorrelation have been found by computer search. Several conjectures by No et. al. [26] exist. As one example let $\{s(t)\}$ be an m-sequence. Let $r = 2^{\frac{m+1}{2}} + 1$ where m is odd, then numerical results indicate that the sequence $\{a(t)\}$ given by

$$a(t) = s(t) + s(rt) + s(r^2t)$$

has ideal autocorrelation. The linear complexity (i.e., the degree of the polynomial of smallest degree that generates the sequence) of the sequence $\{a(t)\}$ is $3m$. Several other conjectures of sequences of period $2^m - 1$ with two-level ideal autocorrelation and low linear complexity exist. The reader is referred to No et. al. [26].

Let $f(x) = x^d$ be a binary APN mapping. Then

$$(x+1)^d + x^d = b$$

has exactly 2 solutions for 2^{m-1} values of $b \in F_{2^m}$ and no solutions for the remaining 2^{m-1} values of $b \in F_{2^m}$.

Let α denote a primitive element of F_{2^m}. In the case of odd m, No, Chung and Yun [25] defined a binary sequence $\{s(t)\}$ by

$$s(t) = 1 \text{ if and only if } (x+1)^d + x^d = \alpha^t \text{ has 2 solutions.}$$

In the case of even m, they defined a binary sequence $\{s(t)\}$ by

$$s(t) = 0 \text{ if and only if } (x+1)^d + x^d = \alpha^t \text{ has 2 solutions.}$$

In the case $d = 2^k + 1$, $gcd(k, m) = 1$, then $\{s(t)\}$ is the m-sequence given by $s(t) = Tr_m(\alpha^t)$, and thus has two-level ideal autocorrelation.

More surprisingly, in the case $d = 2^{2k} - 2^k + 1$ and $m = 3k \pm 1$, No, Chung and Yun [25] conjectured that $\{s(t)\}$ has ideal autocorrelation. In the case m odd this conjecture was proved by Dillon [8], while the case m even was proved by Dillon and Dobbertin [9].

Dobbertin modified and extended the conjecture, by defining the binary sequence

$$s(t) = 0 \text{ if and only if } (x+1)^d + x^d + 1 = \alpha^t \text{ has 2 solutions.}$$

where $d = 2^{2k} - 2^k + 1$ and $gcd(k, m) = 1$ and conjectured that this always has ideal autocorrelation. This conjecture has recently been proved by Dillon and Dobbertin [9].

In particular the conjecture by No et. al. [26] above turned out to be a special case of Dobbertin's conjecture by letting $k = 1/2 \pmod m$, and thus is proved in Dillon and Dobbertin [9].

9 Hyperovals in $PG(2, 2^m)$

Recently, Maschietti [23] found a new family of $(2^m - 1, 2^{m-1} - 1, 2^{m-2} - 1)$ difference sets, or equivalently new families of sequences of period $n = 2^m - 1$ with two-level ideal autocorrelation. A hyperoval in $PG(2, 2^m)$ is a set of $2^m + 2$ points, no three points on a line. Every hyperoval can be written in the form

$$D(f) = \{(1, t, f(t)) \mid t \in F_{2^m}\} \cup \{(0,1,0), (0,0,1)\}$$

where $f(t)$ is a permutation polynomial over F_{2^m} of degree $\leq 2^m - 2$ and $f(0) = 0, f(1) = 1$ and

$$f_s(x) = \frac{f(x+s) + f(s)}{x}, f_s(0) = 0$$

is also a permutation polynomial for all $s \in F_{2^m}$.

If $f(x) = x^k$ is a monomial then $D(x^k)$ is called a monomial hyperoval. The known monomial hyperovals include:

(1) Translation hyperoval $D(x^{2^i})$ where $gcd(i, m) = 1$.
(2) Segre hyperoval $D(x^6)$ where $m \geq 5$ odd.
(3a) Glynn hyperoval $D(x^k)$ of type I, where

$$k = \begin{cases} 2^{\frac{m+1}{2}} + 2^{\frac{3m+1}{4}} & \text{if } m \equiv 1 \pmod 4,\ m \geq 7, \\ 2^{\frac{m+1}{2}} + 2^{\frac{m+1}{4}} & \text{if } m \equiv 3 \pmod 4,\ m \geq 7. \end{cases}$$

(3b) Glynn hyperoval $D(x^k)$ of type II, where

$$k = 3 \cdot 2^{\frac{m+1}{2}} + 4, m \text{ odd }, m \geq 7.$$

Glynn has conjectured that these are all monomial hypoperovals

Maschietti [23] gave the following alternative characterization of a monomial hyperoval:

Lemma 4. *$D(x^k)$ is a monomial hyperoval if and only if $gcd(k, 2^m - 1) = 1$ and $x^k + x + a = 0$ has 0 or 2 solutions for all $a \in F_{2^m}$.*

Based on hyperovals Maschietti's result can be used to construct monomial hyperovals in the following way.

Theorem 16. *Let $D(x^k)$ be a monomial hyperoval and let α be a primitive element of F_{2^m}. Then the sequence defined by*

$$s(t) = 0 \text{ if and only if } x^k + x = \alpha^t \text{ has (two) solutions}$$

has ideal autocorrelation.

The linear complexity of these two-level ideal autocorrelation sequences has recently been determined. The sequences obtained from Translation hyperovals are m-sequences. The sequences obtained from the Segre hyperovals are shown by Evans et. al. [11] and independently of Chang et. al. [4] to have linear complexity $m(2F_{\frac{m-1}{2}} - 1)$ where F_n is the nth Fibonacci number.

The linear complexity of the sequences obtained from the Glynn hyperovals has recently been determined by Evans et. al. [11].

10 Some ternary sequences with two-level ideal autocorrelation

Recently, Helleseth, Kumar and Martinsen [16] have constructed a new family of ternary sequences with ideal autocorrelation function. Their main result is the following.

Theorem 17. *Let $d = 3^{2k} - 3^k + 1$, $m = 3k$ and let $\{s(t)\}$ be the ternary sequence given by*

$$s(t) = Tr_n(\alpha^t + \alpha^{dt})$$

where α is a primitive element of F_{3^m}. Then the sequence $\{s(t)\}$ has ideal two-level autocorrelation.

It is fascinating that the Kasami exponent $d = 3^{2k} - 3^k + 1$ again plays a significant role. Numerical results indicate that a similar result is not true for the Kasami exponent for sequences over F_p when $p > 3$.

11 Conclusions

We have given a survey of the many connections between the auto and cross-correlation function of m-sequences and related topics about perfect (PN) and almost perfect (APN) nonlinear power mappings, permutation polynomials, difference sets, binary and ternary sequences with two-level ideal autocorrelation and families of sequences with optimal correlation properties.

References

1. T. Beth and C. Ding, On almost perfect nonlinear permutations, *Advances in Cryptology - EUROCRYPT'93*. Lecture Notes in Computer Science, Vol. 765, pp. 65-76, Springer-Verlag, 1994.
2. A. Canteaut, P. Charpin and H. Dobbertin, Binary m-sequences with three-valued crosscorrelation: A proof of Welch's conjecture, preprint, 1998.
3. F. Chabaud and S. Vaudenay, Links between differential and linear cryptanalysis, *Advances in Cryptology - EUROCRYPT'94*. Lecture Notes in Computer Science, Vol. 950, pp. 356-365, Springer-Verlag, 1995.
4. A. Chang, S.W. Golomb, G. Gong and P.V. Kumar, Trace expansion and linear span of ideal autocorrelation sequences associated to the Segre hyperoval, preprint, May 1998.
5. R. S. Coulter and R. Matthews, Planar functions and planes of the Lenz-Barlotti class II, *Designs, Codes and Cryptography*, Vol. 10, pp. 165-185, 1997.
6. T. W. Cusick, Constructing differentially uniform permutations via crosscorrelation functions, unpublished manuscript.
7. T. W. Cusick and H. Dobbertin, Some new three-valued crosscorrelation functions for binary m-sequences, *IEEE Trans. Inform. Theory*, Vol. 42, pp. 1238-1240, 1996.
8. J.F. Dillon, Multiplicative difference sets via additive characters, to appear in *Designs, Codes and Cryptography.*
9. J.F. Dillon and H. Dobbertin, New cyclic difference sets with Singer parameters, preprint, February 1999.
10. H. Dobbertin, Almost perfect nonlinear power functions on $GF(2^n)$, preprint, July 1997.
11. R. Evans, H. Hollmann, C. Krattenthaler and Q. Xiang, Gauss sums, Jacobi sums, and p-ranks of cyclic difference sets, submitted for publication.
12. R. Gold, Maximal recursive sequences with 3-valued recursive cross-correlation functions, *IEEE Trans. Inform. Theory*, Vol. 14, pp. 154-156, 1968.
13. T. Helleseth, Some results about the cross-correlation function between two maximal linear sequences, *Discrete Math.*, Vol. 16, pp. 209-232, 1976.
14. T. Helleseth, A note on the cross-correlation function between two binary maximal length linear sequences, *Discrete Math.*, Vol. 23, pp. 301-307, 1978.
15. T. Helleseth and P. Vijay Kumar, Sequences with low correlation, Chapter in: Handbook of Coding Theory (Eds. V.S. Pless and W.C. Huffman), North-Holland 1998.
16. T. Helleseth, P.Vijay Kumar and H. Martinsen, A new family of ternary sequences with ideal two-level autocorrelation function, submitted for publication.
17. T. Helleseth, C. Rong and D. Sandberg, New families of almost perfect nonlinear power mappings, *IEEE Transactions on Inform. Theory*, Vol. 45, pp. 475-485, 1999.
18. T. Helleseth and D. Sandberg, Some power mappings with low differential uniformity, *Applicable Algebra in Engineering, Communication and Computing*, Vol. 8, pp. 363-370, 1997.
19. T. Kasami, Weight distribution formula for some class of cyclic codes, Technical Report R-285 (AD 632574), Coordinated Science Laboratory, University of Illinois, Urbana, April 1966.
20. P. Kumar and O. Moreno, Prime-phase sequences with periodic correlation properties better than binary sequences, *IEEE Trans. Inform. Theory*, Vol. 37, pp. 603-616, 1991.

21. G. Lachaud and J. Wolfmann, The weights of the orthogonals of the extended quadratic binary Goppa codes, *IEEE Trans. Inform. Theory* Vol. 36, pp. 686-692, 1990.
22. R. Lidl and H. Niederreiter, *Finite Fields.* Vol. 20 Encyclopedia of Mathematics and its Applications, Reading, MA: Addison-Wesley, 1983.
23. A. Maschietti, Difference sets and hyperovals, *Designs, Codes and Cryptography,* Vol. 14, pp. 89-98, 1998.
24. G. McGuire and A.R. Calderbank, Proof of a conjecture of Sarwate and Pursley regarding pairs of binary m-sequences, *IEEE Trans. Inform. Theory,* Vol. 41, pp. 1153–1155, 1995.
25. J.S. No, H. Chung and M.S. Yun, Binary pseudorandom sequences of period $2^m - 1$ with ideal autocorrelation generated by the polynomial $z^d + (z+1)^d$, *IEEE Trans. Inform. Theory,* Vol. 44, pp. 1278-1282, 1998.
26. J.S. No, S.W. Golomb, G. Gong, H.K. Lee and P. Gaal, Binary pseudorandom sequences of period $2^n - 1$ with ideal autocorrelation, *IEEE Trans. Inform. Theory,* Vol. 44, pp. 814-817, 1998.
27. Y. Niho, Multi-valued cross-correlation functions between two maximal linear recursive sequences, Ph.D. Thesis, University of Southern California, 1972.
28. K. Nyberg and L. R. Knudsen, Provable security against a differential attack, *Journal of Cryptology,* Vol. 8, pp. 27-37, 1995.
29. K. Nyberg, Differentially uniform mappings for cryptography, *Advances in Cryptology - EUROCRYPT'93.* Lecture Notes in Computer Science, Vol. 765, pp. 55-64, Springer-Verlag, 1994.
30. D. Sandberg, On the differential uniformity of power mappings, Master Thesis, Department of Informatics, University of Bergen, 1997.
31. D.V. Sarwate and M.B. Pursley, Crosscorrelation properties of pseudorandom and related sequences, *Proc. IEEE* 68 , pp. 593-619, 1980.
32. H. M. Trachtenberg, On the cross-correlation functions of maximal linear sequences, Ph. D. Thesis, University of Southern California, Los Angeles, 1970.

Some Computable Complexity Measures for Binary Sequences

Harald Niederreiter

Institute of Discrete Mathematics, Austrian Academy of Sciences, Sonnenfelsgasse 19, A-1010 Vienna, Austria; e-mail: niederreiter@oeaw.ac.at

Abstract. The well-known Kolmogorov complexity of binary sequences is a beautiful theoretical concept, but it is impractical since it is not computable. This paper will focus on complexity measures that are not only of theoretical interest, but that can also be obtained in an algorithmic manner, such as the linear complexity, the linear complexity profile, and the tree complexity. A survey of such complexity measures will be given, with an emphasis on recent results.

1 Introduction

Complexity measures for binary sequences are of fundamental importance in theoretical computer science. They are also of practical relevance since e.g. in the area of stream ciphers in cryptology they are used as a mathematical tool for assessing the security level offered by a concrete keystream. The aim of this paper is to survey classical as well as recently introduced complexity measures for binary sequences predominantly, but not exclusively, from the viewpoint of stream ciphers. Unless otherwise indicated, the bits 0 and 1 will always be considered as elements of the finite field $\mathbf{F}_2$.

Pretty much on top of the hierarchy of complexity measures we find the *Kolmogorov complexity* which is based on a very general computational model, namely that of universal Turing machines. If S is a finite binary sequence, or in other words a bit string, then its Kolmogorov complexity $K(S)$ is, roughly speaking, the length of the shortest computer program that outputs S. The precise definition that is widely used nowadays works with a special type of universal Turing machine, a so-called self-delimiting Turing machine (see [4, Chap. 3]). The definition of Kolmogorov complexity can be extended to (infinite) binary sequences S by considering $K_n(S)$, the Kolmogorov complexity of the first n terms of S. This leads to various notions of randomness, for instance, S is "random" (that is, of the highest complexity) if there exists a constant c such that

$$K_n(S) \geq n - c \quad \text{for } n = 1, 2, \ldots.$$

For the rich and beautiful theory of these notions of randomness we refer again to [4]. The only hitch in this theory is that it can be proved that $K(S)$ is in general not computable, and so the Kolmogorov complexity is of no practical significance.

We will therefore focus on complexity measures that are of lower rank in the hierarchy since they are based on simpler types of machines, but which have the advantage that they can be obtained in an algorithmic manner. In the world of stream ciphers, a standard complexity measure of this form is the linear complexity where the underlying family of machines is that of linear feedback shift registers. Section 2 is devoted to the linear complexity and related concepts, such as the linear complexity profile, the k-error linear complexity, the maximum-order complexity, and the 2-adic complexity. In Section 3 we discuss complexity measures based on pattern counting, namely the tree complexity and the Lempel-Ziv complexity. Some open problems are stated in Section 4.

2 Linear Complexity and Related Concepts

2.1 Linear Complexity

The linear complexity is a basic complexity measure for keystreams in the system-theoretic approach to stream ciphers (compare with [33]). Most keystream generators used in practice employ linear feedback shift registers (LFSRs) as building blocks, and so it is natural from an engineering point of view to measure the complexity of a keystream by the extent to which it can be simulated by an LFSR.

Definition 1. The *linear complexity* $L(S)$ of an ultimately periodic binary sequence S is the length of the shortest LFSR that can generate S, with the convention that $L(S) = 0$ if S is the zero sequence.

In an equivalent definition, $L(S)$ is the degree of the minimal polynomial $m(x) \in \mathbf{F}_2[x]$ of S. A convenient way of introducing the minimal polynomial is via generating functions. With an arbitrary binary sequence

$$S : s_1, s_2, \ldots \in \mathbf{F}_2$$

we associate its *generating function*

$$G(x) = \sum_{i=1}^{\infty} s_i x^{-i} \in \mathbf{F}_2((x^{-1})),$$

where $\mathbf{F}_2((x^{-1}))$ is the field of formal Laurent series over $\mathbf{F}_2$ in the variable x^{-1}. Then it is well known (see e.g. [21], [24]) that S is an LFSR sequence with minimal polynomial $m(x)$ if and only if its generating function $G(x)$ is a rational function of the form

$$G(x) = \frac{h(x)}{m(x)}$$

with $h(x) \in \mathbf{F}_2[x]$ and $\gcd(h(x), m(x)) = 1$.

It is of great significance for cryptanalysis that if we know that $L(S) \leq k$ for some positive integer k, then $L(S)$ and also the minimal polynomial $m(x)$ can be computed from the first $2k$ terms of S by the Berlekamp-Massey algorithm (see [20, Chap. 8]). A lot of information on the linear complexity of specific keystreams can be found in [33].

2.2 Linear Complexity Profile

The linear complexity is a rather crude complexity measure. A more subtle complexity analysis, which also allows the treatment of nonperiodic sequences, is based on the following notions.

Definition 2. If S is an arbitrary binary sequence and n a positive integer, then $L_n(S)$ is defined to be the length of the shortest LFSR that can generate the first n terms of S, with the convention that $L_n(S) = 0$ if the first n terms of S are all 0. The sequence $L_1(S), L_2(S), \ldots$ is called the ***linear complexity profile*** of S.

The linear complexity profile is a nondecreasing sequence of nonnegative integers, and it is trivial that we always have $L_n(S) \leq n$. We can recursively compute $L_1(S), \ldots, L_n(S)$ by the Berlekamp-Massey algorithm with $O(n^2)$ bit operations. A recent speedup was designed by Blackburn [3] using FFT techniques.

There is a close connection between the linear complexity profile of S and the continued fraction expansion of the generating function of S which has been observed, in different forms, by various authors; see e.g. Dai and Zeng [7], Dornstetter [10], Mills [22], Niederreiter [24], and Welch and Scholtz [35]. This connection was used in Niederreiter [25] to carry out a detailed probabilistic analysis of the linear complexity profile for random binary sequences in a suitable stochastic model. A dynamical system representing the continued fraction transformation was set up and shown to be isomorphic to a Bernoulli shift. This has the important consequence that the partial quotients in a continued fraction expansion in $\mathbf{F}_2((x^{-1}))$ are, in a natural sense, independent and identically distributed. In particular, the step sizes $d_1, d_2, \ldots$ in the linear complexity profile of a binary sequence (which are just the degrees of the partial quotients of the generating function) are i.i.d. random variables with the probability distribution $\mathrm{Prob}(d_j = m) = 2^{-m}$ for all positive integers m. As another consequence we get that for a random binary sequence S we have

$$L_n(S) = \frac{n}{2} + O(\log n) \quad \text{for all } n \geq 2,$$

and furthermore deviations of the order of magnitude $\log n$ must appear for infinitely many n. Completely analogous results are valid when $\mathbf{F}_2$ is replaced by an arbitrary finite field (see [25]). These theorems lead to benchmarks for statistical randomness tests using the linear complexity profile.

An additional requirement for the linear complexity profile of keystreams was introduced by Piper [32], namely that a keystream S should have an acceptable linear complexity profile (i.e., one close to that of a random sequence) for every starting point. In other words, if S has the terms $s_1, s_2, \ldots$ and S_r is the shifted sequence $s_{r+1}, s_{r+2}, \ldots$, then S_r should have an acceptable linear complexity profile for every integer $r \geq 0$. A probabilistic analysis of this requirement was carried out in Niederreiter [26]. A survey of the above results and of related topics can be found in [27].

Piper's requirement leads to the following algorithmic problem: given the first n terms $s_1, s_2, \ldots, s_n$ of the binary sequence S, design an efficient algorithm to compute $L_m(S_r)$ for $1 \leq m \leq n - r$ and $0 \leq r \leq n - 1$, i.e., that part of the linear complexity profile of the shifted sequences S_r that depends just on the knowledge of $s_1, s_2, \ldots, s_n$. With the Berlekamp-Massey algorithm we need $O(n^3)$ bit operations for this task. In order to arrive at a faster algorithm, we first restate the problem in the equivalent language of generating functions and continued fractions. We start from $r = n - 1$, where we have only the bit s_n and the associated generating function

$$G_n(x) = s_n x^{-1}.$$

Next, for $r = n - 2$ we get the bit string s_{n-1}, s_n and the associated generating function

$$G_{n-1}(x) = s_{n-1}x^{-1} + s_n x^{-2} = \frac{G_n(x) + s_{n-1}}{x}.$$

Continuing in this manner, we obtain the generating functions $G_n(x), \ldots, G_1(x)$, and our algorithmic problem reduces to the following.

Problem. Given the continued fraction expansion of a generating function $G(x) \in \mathbf{F}_2((x^{-1}))$, compute the continued fraction expansion of

$$\frac{G(x) + a}{x} \quad \text{with } a \in \mathbf{F}_2$$

in an efficient manner.

This problem was solved by Niederreiter and Vielhaber [30] who developed the following algorithm. The starting point is the continued fraction expansion

$$G(x) = [A_1, A_2, \ldots],$$

where the partial quotients A_k are polynomials over $\mathbf{F}_2$ of positive degree. In the course of the algorithm the constant polynomial 1 is also allowed as a partial quotient.

Algorithm. Compute the continued fraction expansion of $(G(x)+a)/x$ from that of $G(x)$.

Step 1. If $a = 0$, do nothing. If $a = 1$, compute $G(x) + 1$ according to

$$G(x) + 1 = [1, A_1 + 1, A_2, A_3, \ldots].$$

Step 2. Given

$$G(x) + a = [A_1, A_2, \ldots],$$

we make the constant terms of all A_k, k even, equal to 0 by iterating the following procedure for increasing values of k: if the constant term of A_k, k even, is equal to 1, then replace the segment A_k, A_{k+1} in the continued fraction expansion by the equivalent segment $A_k + 1, 1, A_{k+1} + 1$.
Step 3. Given

$$G(x) + a = [A_1, A_2, \ldots]$$

with the constant terms of all A_k, k even, equal to 0, we obtain

$$\frac{G(x) + a}{x} = [xA_1, \frac{A_2}{x}, xA_3, \frac{A_4}{x}, \ldots].$$

Step 4. Each partial quotient 1 in the last continued fraction expansion is eliminated by replacing a segment $A, 1, B$ by $A + 1, B + 1$.
After Step 4 we get the canonical continued fraction expansion of $(G(x) + a)/x$, i.e., the unique continued fraction expansion with all partial quotients of positive degree.

In the special case where $G(x)$ is a rational function, so that it has a finite continued fraction expansion, it is easily seen that the above algorithm requires $O(d)$ bit operations, where d is the degree of the denominator of $G(x)$ in the reduced form. If we further specialize to the rational generating functions $G_n(x), \ldots, G_1(x)$, then we obtain the following theorem from [30].

Theorem 1. *The continued fraction expansions of $G_n(x), \ldots, G_1(x)$ can be simultaneously computed by the above algorithm with $O(n^2)$ bit operations.*

Remark. Consider again the situation where we calculate the continued fraction expansions of $G_n(x), \ldots, G_1(x)$ and assume that $s_n = 1$. Then for $1 \le j \le n$ the reduced form of $G_j(x)$ has the denominator x^{n-j+1}, and so by the theory of continued fractions the sum of the degrees of the partial quotients of $G_j(x)$ is equal to $n - j + 1$. Consequently, we need at least $n - j + 1$ bits to describe the continued fraction expansion of $G_j(x)$. This implies that the description of the continued fraction expansions of $G_n(x), \ldots, G_1(x)$ together requires at least $n(n + 1)/2$ bits. We conclude that our algorithm for computing the continued fraction expansions of $G_n(x), \ldots, G_1(x)$ has, in general, the optimal bit complexity as far as the order of magnitude is concerned.

An efficient implementation of the algorithm on a special kind of automaton called a transducer can also be found in [30]. An extension of the algorithm to generating functions over arbitrary finite fields, which is based on somewhat different principles, is given in Niederreiter and Vielhaber [31].

2.3 The k-Error Linear Complexity

The idea of the k-error linear complexity stems from the stability theory of stream ciphers as described in the book of Ding, Xiao, and Shan [9]. In this theory one studies the behavior of $L_n(S)$ under bit changes in the sequence S. The desirable property of stability holds if a small number of bit changes does not cause a significant drop in the linear complexity. We follow the terminology of Stamp and Martin [34] since the definition below is just a slight variation of theirs.

Definition 3. Let S be an arbitrary binary sequence and let n and k be integers with $n \geq 1$ and $0 \leq k \leq n$. Then the *k-error linear complexity* $L_{n,k}(S)$ of the first n terms of S is defined by

$$L_{n,k}(S) = \min L_n(T),$$

where the minimum is over all sequences T which differ from S in at most k bits among the first n bits.

In the paper of Stamp and Martin [34] an efficient algorithm for the calculation of the k-error linear complexity is developed for the case where S is a periodic sequence whose period length is a power of 2. This algorithm generalizes the well-known algorithm of Games and Chan [11] which corresponds to the case $k = 0$. We note that the Games-Chan algorithm has also been extended in a different direction, namely to the determination of the linear complexity of a periodic sequence over an arbitrary finite field with period length a power of the characteristic; see Blackburn [1], [2] and Ding, Xiao, and Shan [9, App. B]. A generalization of the Stamp-Martin algorithm to arbitrary finite fields was recently developed by Kaida, Uehara, and Imamura [16].

An investigation of the distribution of values of the k-error linear complexity over bit strings of fixed length was initiated by Niederreiter and Paschinger [28]. These results may form the basis of a probabilistic theory of the k-error linear complexity.

In analogy with the linear complexity profile, one can consider the *k-error linear complexity profile* of a binary sequence S, i.e., the sequence of nonnegative integers $L_{n,k}(S), n = 1, 2, \ldots$, for fixed S and k. No work has yet been done on this concept, except of course in the case $k = 0$ which yields the linear complexity profile.

2.4 Higher-Order Complexity Measures

Starting from the linear complexity, one can develop a hierarchy of related complexity measures. For instance, the *quadratic complexity* of a bit string is the length of the shortest feedback shift register with linear or quadratic feedback function that can generate the given bit string. More generally, for a positive integer m the *mth-order complexity* of a bit string is the length of the shortest feedback shift register with a polynomial of degree $\leq m$ as the feedback function that can generate the given bit string. Here we use again the standard convention that these complexities are 0 if all terms of the bit string are 0. The highest in this particular hierarchy is the following complexity measure.

Definition 4. If S is an arbitrary binary sequence and n a positive integer, then the *maximum-order complexity* $M_n(S)$ is the length of the shortest feedback shift register with arbitrary feedback function that can generate the first n terms of S, with the convention that $M_n(S) = 0$ if the first n terms of S are all 0.

The maximum-order complexity was studied by Jansen [12] and Jansen and Boekee [14], [15]. These authors also introduced the *maximum-order complexity profile*, i.e., the sequence $M_1(S), M_2(S), \ldots$ for a given S, and pointed out that a known algorithm for building a directed acyclic word graph can be utilized to compute this profile in an efficient manner. They also showed that $M_n(S)$ is of the order of magnitude $\log n$ for random binary sequences S. This theory was extended to ensembles of sequences in Jansen [13].

2.5 2-Adic Complexity Measures

An interesting analog of linear complexity, but one which is based on 2-adic arithmetic rather than binary arithmetic, was introduced and investigated by Klapper and Goresky [17], [18]. From the system-theoretic viewpoint, this approach replaces LFSRs by "feedback with carry" shift registers (FCSRs). An FCSR can be thought of as an LFSR with ordinary addition instead of addition mod 2 and with auxiliary memory for storing the carry. The parity bit in the memory is fed back into the first cell of the register and the higher-order bits form the new value of the memory. We refer to [18] for a detailed mathematical description of an FCSR. On this basis, Klapper and Goresky [17] defined the *2-adic span* $\lambda_2(S)$ of an ultimately periodic binary sequence S to be the least number of bits in an FCSR that can generate S.

Let S be the sequence $s_0, s_1, \ldots$ of bits, where we now view a bit as an element (of course, 0 or 1) of the ring of integers. We associate with S an object that plays the role of a generating function, but this time it is

$$\sigma = \sum_{i=0}^{\infty} s_i 2^i \in \mathbf{Z}_2,$$

where $\mathbf{Z}_2$ is the ring of 2-adic integers. The sequence S is ultimately periodic if and only if the corresponding σ is a rational number $p/q \in \mathbf{Z}_2$ with $\gcd(p, q) = 1$ and q odd. This gives rise to the following definition in [18].

Definition 5. The *2-adic complexity* of an ultimately periodic binary sequence S is the real number

$$\varphi_2(S) = \log_2 \max(|p|, |q|),$$

where $\log_2$ denotes the logarithm to the base 2.

If it is known that $\varphi_2(S) \leq k$, then p and q, and thus $\varphi_2(S)$, can be computed by an algorithm using the first $2k+2$ terms of S. For any ultimately periodic binary sequence S, the 2-adic span $\lambda_2(S)$ and the 2-adic complexity $\varphi_2(S)$ differ at most by $2 + \log_2 \varphi_2(S)$. A survey of 2-adic complexity measures can also be found in the book of Cusick, Ding, and Renvall [6, Chap. 14].

3 Complexity Measures Based on Pattern Counting

3.1 Tree Complexity

A complexity measure motivated by the theory of automatic sequences was introduced by Niederreiter and Vielhaber [29]. This complexity measure is obtained by counting certain patterns in the given binary sequence S. For typographic reasons it will be convenient to denote the terms of S by $s(1), s(2), \ldots$.

Definition 6. For an arbitrary binary sequence S and a positive integer h, the *tree complexity* $T_h(S)$ is the number of different (2^h-1)-tuples of the form

$$(s(k), s(2k), s(2k+1), s(4k), \ldots, s(4k+3), \ldots, s(2^{h-1}k), \ldots, s(2^{h-1}k+2^{h-1}-1))$$

with $k = 1, 2, \ldots$.

This definition also makes sense for bit strings provided that we make h and k so small that the whole (2^h-1)-tuple is contained in the given bit string. However, the present discussion is restricted to infinite sequences. The following basic properties of the tree complexity were shown in [29]:
(i) $1 \leq T_h(S) \leq 2^{2^h-1}$;
(ii) $T_h(S) \leq T_j(S)$ for $h \leq j$;
(iii) if $T_{h+1}(S) = T_h(S)$ for some $h \geq 1$, then $T_j(S) = T_h(S)$ for all $j \geq h$;
(iv) $T_1(S) = 1$ if and only if $T_h(S) = 1$ for all $h \geq 1$ and S is a constant sequence.

For many special sequences studied in the literature, such as the Rueppel sequence, the Thue-Morse sequence, and the Rudin-Shapiro sequence, we have $T_h(S) = O(1)$, i.e., $T_h(S) = t$ for all $h \geq h_0$ with suitable positive integers t and h_0. The following result from [29] explains this phenomenon. The proof is obtained by using the theory of automatic sequences as developed by Christol *et al.* [5].

Theorem 2. *The sequence S satisfies $T_h(S) = O(1)$ if and only if its generating function is an algebraic function over $\mathbf{F}_2$.*

If $T_h(S)$ is unbounded as a function of h, then by the basic properties listed above, the slowest possible growth is $T_h(S) = h+1$ for all $h \geq 1$. A sequence S with this property was constructed in [29].

An interesting family of special binary sequences is given by *paperfolding sequences* (see [8]). A paperfolding sequence $s(1), s(2), \ldots$ is determined in a specific way by the terms $s(2^k), k = 0, 1, \ldots$. In detail, to determine $s(n), n \geq 1$, we write $n = 2^k b$ with an odd integer b and then put

$$s(n) = \begin{cases} s(2^k) & \text{if } b \equiv 1 \bmod 4, \\ s(2^k) + 1 & \text{if } b \equiv 3 \bmod 4, \end{cases}$$

where the addition is in $\mathbf{F}_2$. It was shown in [29] that for a paperfolding sequence S we have $T_h(S) = O(2^h)$.

In contrast to algebraic sequences, i.e., sequences for which the generating function is an algebraic function, and paperfolding sequences, the tree complexity increases much faster for random sequences. In fact, it is rather easy to prove (see [29]) that for a random binary sequence S we have

$$T_h(S) = 2^{2^h - 1} \quad \text{for all } h \geq 1.$$

This yields, in particular, a doubly exponential gap between the tree complexity of algebraic sequences and of random sequences. The growth rate of $T_h(S)$ as a function of h can be used to classify sequences. Besides the sequences with $T_h(S) = O(1)$, we may e.g. consider the family of sequences with polynomial growth, exponential growth, or doubly exponential growth of the tree complexity $T_h(S)$.

3.2 Lempel-Ziv Complexity

For the sake of completeness we mention the classical complexity measure based on pattern counting that is due to Lempel and Ziv [19]. The *Lempel-Ziv complexity* measures the rate at which new patterns emerge as we move along a given sequence. Let S be the bit string $s_1, \ldots, s_n$. We split up S into adjacent blocks. By definition, the first block consists of s_1. If $s_1, \ldots, s_m$ is a union of blocks, or in other words if s_m is the last bit in a block, then the next block $s_{m+1}, \ldots, s_{m+k}$ is uniquely determined by the following two properties:
(i) the bit string $s_{m+1}, \ldots, s_{m+k-1}$ occurs as a subword in $s_1, \ldots, s_{m+k-2}$;
(ii) the bit string $s_{m+1}, \ldots, s_{m+k}$ does not occur as a subword in $s_1, \ldots, s_{m+k-1}$.
The Lempel-Ziv complexity is then the number of blocks into which S is split up by this procedure.

We note that this procedure plays also an important role in the well-known data compression algorithm of Ziv and Lempel [36]. The Lempel-Ziv complexity was studied from a cryptologic point of view in Mund [23].

4 Open Problems

We state some open problems that arise from the discussions in the previous sections and that are interesting from our perspective.

Open Problem 1. Design a fast algorithm for computing the k-error linear complexity $L_{n,k}(S)$ in the general case.

Open Problem 2. Determine the behavior of $L_{n,k}(S)$ for random sequences S and for meaningful ways in which n and k are related, e.g. for fixed k and $n \to \infty$.

Open Problem 3. Design a fast algorithm for computing the tree complexity of a bit string.

Open Problem 4. Investigate the possible connections between the maximum-order complexity, the tree complexity, and the Lempel-Ziv complexity.

References

[1] S.R. Blackburn, A generalisation of the discrete Fourier transform: determining the minimal polynomial of a periodic sequence, *IEEE Trans. Inform. Theory* **40** (1994), 1702–1704.

[2] S.R. Blackburn, A generalisation of the discrete Fourier transform, in "Applications of Finite Fields" (D. Gollmann, ed.), pp. 111–116, Oxford University Press, Oxford, 1996.

[3] S.R. Blackburn, Fast rational interpolation, Reed-Solomon decoding, and the linear complexity profiles of sequences, *IEEE Trans. Inform. Theory* **43** (1997), 537–548.

[4] C. Calude, "Information and Randomness: An Algorithmic Perspective", Springer-Verlag, Berlin, 1994.

[5] G. Christol, T. Kamae, M. Mendès France, and G. Rauzy, Suites algébriques, automates et substitutions, *Bull. Soc. Math. France* **108** (1980), 401–419.

[6] T.W. Cusick, C. Ding, and A. Renvall, "Stream Ciphers and Number Theory", Elsevier, Amsterdam, 1998.

[7] Z. Dai and K. Zeng, Continued fractions and the Berlekamp-Massey algorithm, in "Advances in Cryptology – AUSCRYPT '90" (J. Seberry and J. Pieprzyk, eds.), Lecture Notes in Computer Science, Vol. 453, pp. 24–31, Springer-Verlag, Berlin, 1990.

[8] M. Dekking, M. Mendès France, and A. van der Poorten, FOLDS!, *The Math. Intelligencer* **4** (1983), 130–138, 173–181, 190–195.

[9] C. Ding, G. Xiao, and W. Shan, "The Stability Theory of Stream Ciphers", Lecture Notes in Computer Science, Vol. 561, Springer-Verlag, Berlin, 1991.

[10] J.-L. Dornstetter, On the equivalence between Berlekamp's and Euclid's algorithms, *IEEE Trans. Inform. Theory* **33** (1987), 428–431.

[11] R.A. Games and A.H. Chan, A fast algorithm for determining the complexity of a binary sequence with period 2^n, *IEEE Trans. Inform. Theory* **29** (1983), 144–146.

[12] C.J.A. Jansen, "Investigations on Nonlinear Streamcipher Systems: Construction and Evaluation Methods", Ph.D. Thesis, Technical University of Delft, 1989.

[13] C.J.A. Jansen, The maximum order complexity of sequence ensembles, in "Advances in Cryptology – EUROCRYPT '91" (D.W. Davies, ed.), Lecture Notes in Computer Science, Vol. 547, pp. 153–159, Springer-Verlag, Berlin, 1991.

[14] C.J.A. Jansen and D.E. Boekee, The shortest feedback shift register that can generate a given sequence, in "Advances in Cryptology – CRYPTO '89" (G. Brassard, ed.), Lecture Notes in Computer Science, Vol. 435, pp. 90–99, Springer-Verlag, Berlin, 1990.

[15] C.J.A. Jansen and D.E. Boekee, On the significance of the directed acyclic word graph in cryptology, in "Advances in Cryptology – AUSCRYPT '90" (J. Seberry and J. Pieprzyk, eds.), Lecture Notes in Computer Science, Vol. 453, pp. 318–326, Springer-Verlag, Berlin, 1990.

[16] T. Kaida, S. Uehara, and K. Imamura, A new algorithm for the k-error linear complexity of sequences over $GF(p^m)$ with period p^n, this volume.

[17] A. Klapper and M. Goresky, 2-adic shift registers, in "Fast Software Encryption" (R. Anderson, ed.), Lecture Notes in Computer Science, Vol. 809, pp. 174–178, Springer-Verlag, Berlin, 1994.

[18] A. Klapper and M. Goresky, Feedback shift registers, 2-adic span, and combiners with memory, *J. Cryptology* **10** (1997), 111–147.

[19] A. Lempel and J. Ziv, On the complexity of finite sequences, *IEEE Trans. Inform. Theory* **22** (1976), 75–81.

[20] R. Lidl and H. Niederreiter, "Finite Fields", Addison-Wesley, Reading, MA, 1983; reprint, Cambridge University Press, Cambridge, 1997.

[21] R. Lidl and H. Niederreiter, Finite fields and their applications, in "Handbook of Algebra" (M. Hazewinkel, ed.), Vol. 1, pp. 321–363, Elsevier, Amsterdam, 1996.

[22] W.H. Mills, Continued fractions and linear recurrences, *Math. Comp.* **29** (1975), 173–180.

[23] S. Mund, Ziv-Lempel complexity for periodic sequences and its cryptographic application, in "Advances in Cryptology – EUROCRYPT '91" (D.W. Davies, ed.), Lecture Notes in Computer Science, Vol. 547, pp. 114–126, Springer-Verlag, Berlin, 1991.

[24] H. Niederreiter, Sequences with almost perfect linear complexity profile, in "Advances in Cryptology – EUROCRYPT '87" (D. Chaum and W.L. Price, eds.), Lecture Notes in Computer Science, Vol. 304, pp. 37–51, Springer-Verlag, Berlin, 1988.

[25] H. Niederreiter, The probabilistic theory of linear complexity, in "Advances in Cryptology – EUROCRYPT '88" (C.G. Günther, ed.), Lecture Notes in Computer Science, Vol. 330, pp. 191–209, Springer-Verlag, Berlin, 1988.

[26] H. Niederreiter, Keystream sequences with a good linear complexity profile for every starting point, in "Advances in Cryptology – EUROCRYPT '89" (J.-J. Quisquater and J. Vandewalle, eds.), Lecture Notes in Computer Science, Vol. 434, pp. 523–532, Springer-Verlag, Berlin, 1990.

[27] H. Niederreiter, Finite fields and cryptology, in "Finite Fields, Coding Theory, and Advances in Communications and Computing" (G.L. Mullen and P.J.-S. Shiue, eds.), pp. 359–373, M. Dekker, New York, 1993.

[28] H. Niederreiter and H. Paschinger, Counting functions and expected values in the stability theory of stream ciphers, this volume.

[29] H. Niederreiter and M. Vielhaber, Tree complexity and a doubly exponential gap between structured and random sequences, *J. Complexity* **12** (1996), 187–198.

[30] H. Niederreiter and M. Vielhaber, Simultaneous shifted continued fraction expansions in quadratic time, *Applicable Algebra Engrg. Comm. Comput.* **9** (1998), 125–138.

[31] H. Niederreiter and M. Vielhaber, An algorithm for shifted continued fraction expansions in parallel linear time, *Theoretical Computer Science*, to appear.

[32] F. Piper, Stream ciphers, *Elektrotechnik und Maschinenbau* **104** (1987), 564–568.

[33] R.A. Rueppel, Stream ciphers, in "Contemporary Cryptology: The Science of Information Integrity" (G.J. Simmons, ed.), pp. 65–134, IEEE Press, New York, 1992.

[34] M. Stamp and C.F. Martin, An algorithm for the k-error linear complexity of binary sequences with period 2^n, *IEEE Trans. Inform. Theory* **39** (1993), 1398–1401.

[35] L.R. Welch and R.A. Scholtz, Continued fractions and Berlekamp's algorithm, *IEEE Trans. Inform. Theory* **25** (1979), 19–27.

[36] J. Ziv and A. Lempel, A universal algorithm for sequential data compression, *IEEE Trans. Inform. Theory* **23** (1977), 337–343.

Meeting the Welch Bound with Equality

Dilip V. Sarwate

Department of Electrical and Computer Engineering
and the Coordinated Science Laboratory
1308 West Main Street
Urbana IL 61801 USA
Email: sarwate@comm.csl.uiuc.edu

Abstract. A signal set whose root-mean-square inner product magnitude equals the Welch lower bound is called a WBE signal set. WBE signal sets are of interest in synchronous CDMA communication systems. This chapter surveys the known results on WBE signal sets and extends them in several ways. In particular, WBE signal sets over signal alphabets whose size is a prime power (the most important case is the quaternary alphabet), and arbitrary-size WBE signal sets over small alphabets are constructed. Constructions are also described for signal sets whose maximum inner product magnitudes equal (or are only very slightly larger than) the Welch bound. Similarly, signal sets whose root-mean-square correlation magnitudes equal the Welch lower bound on correlations are of interest in asynchronous CDMA communication systems. It is shown that the root-mean-square correlation magnitude of all WBE signal sets (and many more) equals the Welch bound. Finally, the signal-to-noise ratio for CDMA communication systems using WBE signal sets is studied.

1 Introduction

A quarter-century ago, Welch [33] published a collection of lower bounds on the maximum magnitude of the inner products of a set of unit-norm complex-valued vectors, and used these results to deduce lower bounds on the maximum magnitudes of the periodic and aperiodic correlation functions for sets of periodic sequences. The method used was a familiar one – Welch found lower bounds on the root-mean-square (RMS) magnitude, and then asserted that the maximum magnitude could not be smaller than these lower bounds. In the years that followed the publication of the Welch bounds, the statement

"The set satisfies the Welch bound with equality"

was generally understood [1], [8], [10], [15], [21] to mean that the *maximum* magnitude of the parameter under consideration was equal to the corresponding lower bound given in [33]. Indeed, this interpretation is supported by both the title and the statement of the main results in [33]. Recently, however, Massey and his colleagues [14], [20], have considered sets of vectors for which the RMS magnitude of the inner product equals the Welch lower bound. The capacity of code-division multiple-access (CDMA) systems in which these so-called WBE signal sets are used as the signature sequences has been studied in [20]. Following this, other authors [31], [32] have considered the signal-to-interference ratio in CDMA systems when WBE signal sets are used as signature sequences.

In this chapter, both the RMS magnitude and the maximum magnitude of the inner product are considered. Signal sets whose maximum inner product magnitude equals the Welch bound (hereafter called MWBE signal sets) form a subclass of WBE signal sets, and are also described. The results in [14] are generalized in several obvious ways from linear binary codes and their cosets to linear nonbinary codes and their cosets, and from WBE signal sets to MWBE signal sets. It is shown that almost all interesting codes are WBE signal sets – indeed, in stark contrast to the folk theorem that "all codes are good except the ones we can think of" one can assert that

"All codes are WBE signal sets except the ones we *don't want* to think of!"

Some of the WBE signal sets described in [14] are actually MWBE signal sets. Large sets of signals whose maximum inner product magnitudes are only very slightly larger than the Welch bound are also described. All these constructions provide signal sets whose sizes and alphabets are restricted in various ways. A new result in this chapter is a Kronecker product construction that provides WBE signal sets of arbitrary size. A construction of WBE signal sets over alphabets whose size is a prime power is also described. The most important practical application of the latter result is to the design of signal sets over quaternary alphabets for use in quadriphase communication systems. Next, it is noted that subsets of WBE signal sets generally are not WBE signal sets. The effect of this on system analyses and information-theoretic analyses of channel capacity is briefly described.

For asynchronous CDMA communication systems, various correlation parameters of the signal sets are more important than the inner products. The Welch lower bounds [33] on the RMS and maximum magnitudes of the periodic, odd, and aperiodic correlation functions have been studied previously [18], [21], [23], [24]. The same bounds also apply to the RMS and maximum magnitudes of the more-recently introduced polyphase correlation functions [4], [28] which include the periodic and odd correlation functions as special cases. Welch also proved more general bounds on the periodic and aperiodic correlation function magnitudes. It is noted that these also apply to the polyphase correlation functions. It is also noted that signal sets (such as WBE signal sets) that meet the Welch bounds on RMS correlation are also complementary signal sets. Finally, the chapter considers the signal-to-noise ratio parameter [17] for asynchronous CDMA communication systems that use WBE signal sets as signature sequences. It is noted that in many cases, better than average performance can be guaranteed if WBE signal sets are used. The chapter concludes with a brief discussion of these results.

2 Bounds on Inner Products

2.1 WBE Signal Sets

Let $\mathcal{X} = \{x^{(1)}, x^{(2)}, \ldots, x^{(K)}\}$ denote a set of K complex-valued vectors of length N where, to avoid special cases, it is assumed that $K \geq N > 1$. The vectors in $\mathcal{X}$ need not be distinct, that is, $\mathcal{X}$ can be a multiset. The signal matrix $S(\mathcal{X})$ of $\mathcal{X}$ is defined to be the $K \times N$ matrix whose k-th row is $x^{(k)} = (x_0^{(k)}, x_1^{(k)}, \ldots, x_{N-1}^{(k)})$, $1 \leq k \leq K$. The set $\mathcal{X}$ will be called a ***signal set*** of size K and length N if all the rows of $S(\mathcal{X})$ in $\mathcal{X}$ have energy N, that is, $\|x^{(k)}\|^2 = \langle x^{(k)}, x^{(k)} \rangle = \sum_{i=0}^{N-1} |x_i^{(k)}|^2 = N$ for all k, $1 \leq k \leq K$. Conversely, any $K \times N$ matrix whose rows all have energy N specifies a signal set $\mathcal{X}$ of size K and length N. The *maximum* inner product magnitude $I_{\max}(\mathcal{X})$ of a signal set $\mathcal{X}$ is defined by

$$I_{\max}(\mathcal{X}) = \max_{k \neq l} |\langle x^{(k)}, x^{(l)} \rangle|.$$

The *root-mean-square* (RMS) inner product magnitude $I_{\mathrm{rms}}(\mathcal{X})$ is the square root of the average value of $|\langle x^{(k)}, x^{(l)} \rangle|^2$ over all $k \neq l$. Thus,

$$I_{\mathrm{rms}}(\mathcal{X}) = \left[\frac{1}{K(K-1)} \sum_{k=1}^{K} \sum_{l \neq k,\ l=1}^{K} |\langle x^{(k)}, x^{(l)} \rangle|^2 \right]^{1/2}.$$

When the signal set is obvious from the context, it is convenient to just write $I_{\max}$ and I_{rms}. Note also that $I_{\max} \geq I_{\mathrm{rms}}$.

Proposition 1. *(Massey and Mittelholzer [14]): For any signal set $\mathcal{X}$,*

$$\sum_{k=1}^{K} \sum_{l=1}^{K} |\langle x^{(k)}, x^{(l)} \rangle|^2 \geq K^2 N \tag{1}$$

with equality if and only if the columns $y^{(0)}, y^{(1)}, \ldots, y^{(N-1)}$ *of $S(\mathcal{X})$ are orthogonal vectors of energy K, that is,*

$$\langle y^{(m)}, y^{(n)} \rangle = \begin{cases} K, & m = n, \\ 0, & m \neq n. \end{cases} \tag{2}$$

Welch [33] used a different argument to prove the more general result

$$\sum_{k=1}^{K} \sum_{l=1}^{K} |\langle x^{(k)}, x^{(l)} \rangle|^{2s} \geq \frac{K^2 N^{2s}}{\binom{N+s-1}{s}}. \tag{3}$$

Note that (1) is the special case $s = 1$ of (3). However, the condition for equality stated in Proposition 1 was not known before.

It follows from (1) and (3) that for a signal set of size K and length N,

$$I_{\text{rms}} \geq \sqrt{N\frac{K-N}{K-1}} \tag{4}$$

which is a form of Welch's bound. Since $I_{\max} \geq I_{\text{rms}}$, the Welch bound [33] is usually stated as

$$I_{\max} = \max_{k \neq l} |\langle x^{(k)}, x^{(l)} \rangle| \geq \sqrt{N\frac{K-N}{K-1}}. \tag{5}$$

If equality holds in (1), then the signal set is said to meet the Welch bound (4) on I_{rms} with equality, and is called a WBE signal set [14]. Similarly, signal sets which meet the Welch bound (5) on $I_{\max}$ with equality can be called MWBE signal sets. The inner product magnitudes of MWBE signal sets satisfy

$$|\langle x^{(k)}, x^{(l)} \rangle| = \sqrt{N\frac{K-N}{K-1}} \quad \text{for all } k \neq l.$$

Examples of MWBE signal sets are the $N \times N$ discrete Fourier transform matrix with (k,l)-th element $\exp(j2\pi kl/N)$ where $j = \sqrt{-1}$, and $\sqrt{N} \cdot I$ where I is the $N \times N$ identity matrix. These sets contain N orthogonal signals of length N and will be called orthogonal MWBE signal sets. These orthogonal MWBE signal sets correspond to frequency-division multiple-access (FDMA) and time-division multiple-access (TDMA) signaling respectively.

Proposition 2. *(Massey and Mittelholzer [14]): If $\mathcal{X}$ and $\hat{\mathcal{X}}$ are WBE signal sets of length N and sizes K and $\hat{K}$, then $\mathcal{X} \cup \hat{\mathcal{X}}$ is a WBE signal set of size $K + \hat{K}$ and length N.*

Note that it is not necessary that $\mathcal{X}$ and $\hat{\mathcal{X}}$ be different signal sets. For example, if the TDMA (or FDMA) signal set is used for both $\mathcal{X}$ and $\hat{\mathcal{X}}$, then the larger (multi)set has $2N$ signals of length N. All inner products are 0 except for $\langle x^{(k)}, \hat{x}^{(k)} \rangle = N$, $1 \leq k \leq K$, and both sides of (1) have value $4N^3$. However, $I_{\max}(\mathcal{X} \cup \hat{\mathcal{X}}) = N$. These signaling schemes in effect divide the $2N$ transmitters into N pairs with each pair using its allotted time slot (or frequency slot) as a two-user adder channel. As a final comment, note that the union of two MWBE signal sets is generally not a MWBE signal set.

Given a WBE signal set $\mathcal{X}$ of size K and length N, Proposition 2 allows the construction of a WBE signal multiset of size LK as the union of L copies of $\mathcal{X}$. It is also possible to transform $\mathcal{X}$ into a different set $\hat{\mathcal{X}}$ so as to ensure that the signals in $\mathcal{X} \cup \hat{\mathcal{X}}$ are distinct. Several "WBE-preserving" transformations such as cyclic shifts, multiplication of rows or columns of $S(\mathcal{X})$ by constants of unit magnitude, deletions of columns of $S(\mathcal{X})$ etc., for binary signal sets are described in [14]. The next two propositions describe general transformations that include all these WBE-preserving transformations as special cases.

Proposition 3. *If $S(\mathcal{X})$ is a WBE (MWBE) signal set of size K and length N, then $DS(\mathcal{X})U$, where D is a $K \times K$ diagonal unitary matrix and U is a unitary matrix, is a WBE (MWBE) signal set $\hat{\mathcal{X}}$ of size K and length N.*

U is called unitary if $U(U^*)^T$ is the identity matrix, and the result is, in essence, a restatement of the fact that unitary transformations preserve inner products.

Proposition 4. *Let $\mathcal{X}$ denote a WBE signal set of size K and length N. If all the entries in a column of $S(\mathcal{X})$ are complex numbers of unit magnitude, then the set $\hat{\mathcal{X}}$ obtained by deleting this column of $S(\mathcal{X})$ is a WBE signal set of size K and length $N-1$.*

The proof is similar to the one given in [14]. Proposition 4 can be applied to the FDMA signal set but not to the TDMA signal set described above. Also, deletions of columns generally do not preserve the MWBE property.

Propositions 2 and 3 together allow the design of very large WBE signal sets with distinct signals (or WBE signal multisets if one so wishes.) For all of these, I_{rms} *equals* the right side of (4) which is slightly smaller than $\sqrt{N}$. However, this does not mean that *all* the inner product magnitudes are smaller than $\sqrt{N}$. In fact, $I_{\max}$ can be considerably larger than $\sqrt{N}$. For example, $I_{\max} = N$ for the union of two TDMA (or FDMA) WBE signal sets. More generally, a result in [33] shows that $I_{\max}$ exceeds $\sqrt{N}$ for large signal sets, even though $I_{\text{rms}} < \sqrt{N}$ for a WBE signal set.

The union construction of Proposition 2 requires the WBE sets to be of the same length. It is also possible to combine WBE sets of different lengths via the Kronecker product construction to produce a larger WBE signal set with longer signals. Recall that if A and B are $K \times N$ and $\hat{K} \times \hat{N}$ matrices respectively, then their Kronecker (or tensor) product is the $K\hat{K} \times N\hat{N}$ matrix $A \otimes B$ given by

$$A \otimes B = \begin{bmatrix} a_{1,1}B & a_{1,2}B & \cdots & a_{1,N}B \\ a_{2,1}B & a_{2,2}B & \cdots & a_{2,N}B \\ \vdots & \vdots & \ddots & \vdots \\ a_{K,1}B & a_{K,2}B & \cdots & a_{K,N}B \end{bmatrix}.$$

WBE signal sets retain the WBE property when combined via the Kronecker product construction because the inner product of two columns of $A \otimes B$ is the product of the inner products of the corresponding columns of A and B.

Proposition 5. *If $S(\mathcal{X})$ and $S(\hat{\mathcal{X}})$ are WBE signal sets of sizes K and $\hat{K}$ respectively and lengths N and $\hat{N}$ respectively, then $S(\mathcal{X}) \otimes S(\hat{\mathcal{X}})$ is a WBE signal set of size $K\hat{K}$ and length $N\hat{N}$.*

Corollary 5.1. *The Kronecker product of two MWBE signal sets is an MWBE signal set if and only if the constituent MWBE signal sets are orthogonal MWBE signal sets.*

2.2 WBE Signal Sets over Small Alphabets

The construction of WBE signal sets over small alphabets is of importance for practical applications. The essential results for binary signals are given in [14]. In this chapter, these results are re-stated and generalized to larger alphabets.

Consider first that the inner product of two signals of length N over the binary alphabet $\{+1, -1\}$ equals $N-2d$ where d is the *Hamming distance* between the signals.

Proposition 6. *There exists a binary WBE signal set of size K if and only if K equals 2 or is a multiple of 4.*

$K \times K$ Hadamard matrices provide binary MWBE signal sets of size K and length K. However, even for those multiples of 4 for which the existence of a $K \times K$ Hadamard matrix has not been demonstrated, it is possible to construct a binary WBE signal set of size K and length $N < K$.

More generally, linear codes are the source of large classes of WBE signal sets over small alphabets, and these are considered next.

2.3 WBE Signal Sets from Linear Codes

Let $\mathcal{C}$ denote a linear (N, k) code over $GF(p)$, p a prime, with the property that each code symbol C_m is a different linear function of the k information symbols. Since there are only p^k different linear functions of k variables, it must be that $p^k \geq N$. Let $\hat{S}(\mathcal{C})$ denote the $p^k \times N$ matrix whose rows are the p^k codewords of $\mathcal{C}$. Since the code symbols are linear functions, then for any given column of $\hat{S}(\mathcal{C})$, either each element of $GF(p)$ occurs p^{k-1} times in the column, or every entry in the column is zero. The latter occurs if and only if the corresponding code symbol is the zero function. Since different code symbols must be different functions, $\hat{S}(\mathcal{C})$ can have *at most one* all-zero column. For $m \neq n$, the code symbols C_m and C_n are different linear functions, and hence $C_m - C_n$ is a *nonzero* linear function of the information symbols. It follows that the difference of two columns of $\hat{S}(\mathcal{C})$ has the property that each field element occurs p^{k-1} times in the difference.

In communications applications, codewords are mapped into complex-valued vectors by mapping $x \in GF(p)$ onto $\exp(j2\pi x/p)$. This mapping is used to convert $\hat{S}(\mathcal{C})$ into a signal matrix $S(\mathcal{C})$ whose elements are the complex p-th roots of unity. The inner product of two columns is

$$\langle y^{(m)}, y^{(n)} \rangle = \sum_{i=1}^{p^k} \exp(j2\pi(C_m^{(i)} - C_n^{(i)})/p) = p^{k-1} \sum_{l=0}^{p-1} \exp(j2\pi l/p) = 0$$

because $C_m^{(i)} - C_n^{(i)}$ takes on each value $0, 1, \ldots, p-1$ an equal number of times as i varies from 1 to p^k, and the sum of the complex p-th roots of unity is 0.

All of the above also applies to linear codes over the integer ring $\mathcal{Z}_q$ with the minor modification that in some cases, $C_m^{(i)} - C_n^{(i)}$ may take on all values $0, 1, \ldots, r-1$ in a subring $\mathcal{Z}_r$ equally often. Hence, the following result holds:

Proposition 7. *(Helleseth and Kumar [8]): If $\mathcal{C}$ is a linear (N, k) code over $\mathcal{Z}_q$ such that the code symbols C_m are different linear functions of the information symbols, then $S(\mathcal{C})$ is a WBE signal set of size $K = q^k$ and length N.*

Corollary 7.1. *Proposition 7 holds for linear codes over $GF(p)$.*

Both these results are straightforward generalizations of the Massey and Mittelholzer [14] result that a linear binary code $\mathcal{C}$ is a WBE signal set if and only if its dual code has no codewords of Hamming weight 2. An equivalent result due to Assmus and Mattson is that linear binary codes satisfying the stated condition have mean-square Hamming weight $N(N+1)/4$ (cf. [34]).

The remainder of this section only considers codes over finite fields, but many of the results also apply to codes over $\mathcal{Z}_q$. For codes over $GF(p)$, it is convenient to think of the k information symbols as an element ζ of the extension field $GF(p^k)$. Since any linear function from $GF(p^k)$ to $GF(p)$ is of the form $z \to \mathrm{Tr}(\alpha \cdot z)$ where $\mathrm{Tr}(z) = z + z^p + z^{p^2} + \ldots + z^{p^{k-1}}$ is the trace function from $GF(p^k)$ to $GF(p)$ and $\alpha \in GF(p^k)$ [13], a codeword in this linear code can be expressed as

$$C^{(\zeta)} = (\mathrm{Tr}(\alpha_1 \cdot \zeta), \mathrm{Tr}(\alpha_2 \cdot \zeta), \mathrm{Tr}(\alpha_3 \cdot \zeta), \ldots, \mathrm{Tr}(\alpha_N \cdot \zeta))$$

where $\alpha_1, \alpha_2, \ldots, \alpha_N$ are N different elements of $GF(p^k)$. Note that at most one of the α_i can be 0.

It is convenient to simply call $\mathcal{C}$ a WBE signal set whenever $S(\mathcal{C})$ is a WBE signal set. A *coset* of $\mathcal{C}$ is $a + \mathcal{C} = \{a + c : c \in \mathcal{C}\}$ where a is a vector of length N over $GF(p)$. Then, $S(a + \mathcal{C}) = S(\mathcal{C})U$ where is a unitary matrix. From Proposition 3, it follows that if $\mathcal{C}$ is a WBE (MWBE) signal set, then so is $a + \mathcal{C}$. This is a minor generalization of a result noted in [14] for binary codes.

Corollary 7.2. *Cosets of codes*

- *If $\mathcal{C}$ is a WBE signal set, then any coset of $\mathcal{C}$ is a WBE signal set.*
- *If $\mathcal{C}$ is a MWBE signal set, then any coset of $\mathcal{C}$ is a MWBE signal set.*

From Proposition 2 it follows that

Corollary 7.3. *If $\mathcal{C}$ is a WBE signal set, then the union of any number of cosets of $\mathcal{C}$ is a WBE signal set.*

In fact, since the entire space $[GF(p)]^N$ is a union of cosets of $\mathcal{C}$, we have

Corollary 7.4. *The (N, N) code over $GF(p)$ is a WBE signal set.*

Of course, this result is also immediate from Corollary 7.1 since each codeword symbol in this code is a different information symbol. Notice that $I_{\mathrm{max}} = N$ for this WBE signal set even though $I_{\mathrm{rms}} < \sqrt{N}$, thus showing once again that the WBE property does not guarantee small values of I_{max}.

The only codes excluded by the hypotheses of Corollary 7.1 are those with repeated symbols (whether zero or nonzero). Codes with repeated symbols are usually of interest only for very noisy channels; generally, better performance is obtained if different parity check equations specify the various code symbols. Thus, colloquially, Corollary 7.1 says that almost all interesting codes are WBE signal sets. Some especially interesting codes are considered next.

2.4 Cyclic and Extended Cyclic Codes

Let $h(z)$ denote an irreducible polynomial of degree k and period N over $GF(p)$. Then, the codewords of the (N, k) cyclic code with parity-check polynomial $h(z)$ can be represented as

$$C^{(\zeta)} = (\mathrm{Tr}(\alpha^0 \cdot \zeta), \mathrm{Tr}(\alpha^1 \cdot \zeta), \mathrm{Tr}(\alpha^2 \cdot \zeta), \ldots, \mathrm{Tr}(\alpha^{N-1} \cdot \zeta)) \tag{6}$$

where α^{-1} is a root of $h(z)$. Since the code symbols are different linear functions, it follows from Corollary 7.1 that the code is a WBE signal set. Now suppose that this code is extended to include the zero function $\mathrm{Tr}(0 \cdot \zeta)$ as a code symbol. This extra symbol is, in fact, an "overall parity check" since

$$\sum_{i=0}^{N-1} \mathrm{Tr}(\alpha^i \cdot \zeta) = \mathrm{Tr}\left(\sum_{i=0}^{N-1} \alpha^i \cdot \zeta\right) = \mathrm{Tr}\left(\left(\sum_{i=0}^{N-1} \alpha^i\right) \cdot \zeta\right) = \mathrm{Tr}(0 \cdot \zeta).$$

Hence, this extended cyclic code is also a WBE signal set which proves

Proposition 8. *If $h(z)$ is an irreducible polynomial of degree k and period N over $GF(p)$, then the (N, k) cyclic code, and the $(N+1, k)$ extended cyclic code, with parity-check polynomial $h(z)$ are WBE signal sets.*

When the parity-check polynomial $h(z)$ is a *primitive* polynomial of degree k, the cyclic $(p^k - 1, k)$ code is called a maximal-length sequence code (henceforth m-sequence code), and the (p^k, k) extended code is called an extended m-sequence code. Binary m-sequence codes and extended binary m-sequence codes are WBE signal sets [14]. In fact, these sets (and more generally, p-ary m-sequence codes and extended m-sequence codes) are MWBE signal sets. Note also that the signal matrices of extended binary m-sequence codes are $2^k \times 2^k$ Hadamard matrices.

Proposition 9. *If $h(z)$ is a primitive polynomial of degree k over $GF(p)$, then*

- m-*sequence code: the $(p^k - 1, k)$ cyclic code with parity-check polynomial $h(z)$ is a nonorthogonal MWBE signal set.*
- *Extended* m-*sequence code: The (p^k, k) extended cyclic code with parity-check polynomial $h(z)$ is an orthogonal MWBE signal set.*

The signal matrix for the extended m-sequence code is a $p^k \times p^k$ orthogonal matrix whose entries are complex p-th roots of unity. In contrast, the discrete Fourier transform (FDMA) signal matrix for this length is a $p^k \times p^k$ matrix whose entries are complex p^k-th roots of unity.

The m-sequence codes and the extended m-sequence codes are well-known in the literature as simplex signal sets and orthogonal signal sets. They also form the basis for the following observations about WBE signal sets derived from linear codes.

Proposition 10. *An (N, k) code is a WBE signal set if and only if it is equivalent under column permutation to a punctured extended* m-*sequence code with k information symbols.*

Here, puncturing means deletion of symbols, that is, columns of the signal matrix, of the extended m-sequence code with k information symbols. Next, consider cyclic codes of length N and extended cyclic codes of length $N+1$. If the parity-check polynomial $h(z)$ is of period N, then the residues of x^i mod $h(z)$, $0 \le i \le N-1$ are distinct, and hence each code symbol is a different linear function of the information symbols. Thus we have the following result.

Proposition 11. *Suppose that N is not a multiple of p. Then,*

- *A cyclic code of length N over $GF(p)$ is a WBE signal set if and only if its parity-check polynomial $h(z)$ has period N.*
- *An extended cyclic code of length $N+1$ over $GF(p)$ is a WBE signal set if and only if its parity-check polynomial $h(z)$ has period N.*

One way of ensuring that $h(z)$ has period N is to require that $h(z)$ be divisible by an irreducible polynomial of period N.

Corollary 11.1. *If N is not a multiple of p, then cyclic codes of length N and extended cyclic codes of length $N+1$ over $GF(p)$ are WBE signal sets if the parity-check polynomial $h(z)$ is divisible by an irreducible polynomial of period N over $GF(p)$.*

As an example, the binary Gold *codes* are $(2^n-1, 2n)$ cyclic codes with parity-check polynomials $M_\alpha(x)M_\beta(x)$ where $M_\alpha(x)$ and $M_\beta(x)$ denote the minimal polynomials of primitive elements α, $\beta \in GF(2^n)$ and the m-sequences u and v generated by $M_\alpha(x)$ and $M_\beta(x)$ are a *preferred pair* of m-sequences [24]. The representatives of the cyclic equivalence classes of the codewords in this code can be expressed as $\{v, u, u+v, u+Tv, u+T^2v, \ldots, u+T^{N-1}v\}$ where T is the left cyclic shift operator and $N = 2^n-1$, and constitute the well-known class of Gold *sequences* [24]. According to Corollary 11.1, Gold codes are WBE signal sets, as indeed are all supercodes of the m-sequence codes and the extended m-sequence codes. In fact, the extended $(2^n, 2n)$ Gold codes and Gold-like codes [24] are WBE signal sets of length $N = 2^n$ and size $N^2 = 2^{2n}$ for which $I_{\max} = 2^{\lfloor (n+2)/2 \rfloor}$ which equals $2\sqrt{N}$ or $\sqrt{2N}$ according as n is even or odd.

2.5 Some Interesting Cosets of Cyclic Codes

Massey and Mittelholzer [14] noted that the Gold *sequence* set described above is very nearly a coset of an m-sequence code.

Proposition 12. *(Massey and Mittelholzer [14]): The expurgated set of Gold sequences $\{u, u+v, u+Tv, u+T^2v, \ldots, u+T^{N-1}v\}$ is a WBE signal set.*

In fact, from Proposition 9 and Corollary 7.2, it follows that

Corollary 12.1. *The expurgated set of Gold sequences is an MWBE signal set.*

Of course, as noted in [14], *any* coset of an m-sequence code is an WBE signal set. More strongly, it follows from Corollary 7.2 that

Corollary 12.2. Any *coset of an* m*-sequence code or extended* m*-sequence code is an MWBE signal set.*

Also, as noted above, the *entire* Gold code is a WBE signal set. Constructions similar to the one in Proposition 12 can also be applied to various other classes of codes described in [24] to obtain WBE signal sets.

2.6 Binary Very Nearly MWBE Signal Sets

Cosets of the m-sequence codes and extended m-sequence codes are MWBE signal sets, but *supercodes* of these codes generally are not MWBE signal sets. In two interesting cases, however, certain supercodes of binary MWBE signal sets are *very nearly* MWBE signal sets in the sense that $I_{\max} = \sqrt{N}$ and equality very nearly holds in (5).

The extended small Kasami codes are $(2^{2n}, 3n)$ extended binary cyclic codes whose parity-check polynomials are of the form $M_\alpha(x)M_\gamma(x)$ where $\alpha \in GF(2^{2n})$ and $\gamma \in GF(2^n)$ are primitive elements and $\gamma = \alpha^{2^n+1}$ (cf. Chapter 8.7 of [13]). The inner products take on values $\pm 2^n = \pm\sqrt{N}$ and 0 only. Thus, equality *very nearly* holds in (5) for these WBE signal sets. One could even say that these *small Kasami signal sets* are asymptotically MWBE signal sets in the sense that the ratio of the maximum inner product magnitude to the right side of (5) approaches 1 as $N \to \infty$.

The $(2^{2n}, 4n)$ binary *nonlinear* Kerdock code (cf. Chapter 15 of [13]) is a union of cosets of the extended m-sequence code and is thus a WBE signal set. Unfortunately, the complement of each codeword is also a codeword, and the inner product of the signals corresponding to these two codewords is $-N$. The inner product of the signals corresponding to any two noncomplementary codewords takes on values $\pm 2^n = \pm\sqrt{N}$ and 0. It follows that the $(2^{2n}, 4n-1)$ subcode obtained by choosing either a codeword or its complement from the Kerdock code gives a signal set of size $N^2/2$ and length N with $I_{\max} = 2^n = \sqrt{N}$ and these *Kerdock signal sets* are asymptotically MWBE signal sets. The Kerdock signal sets are also optimum in the sense of being the largest possible binary signal sets for which $I_{\max} = \sqrt{N}$. This was noted by Levenshtein [11] who showed that if $I_{\max} \le \sqrt{N}$, then $K \le N^2/2$ for binary signal sets. In contrast to the Kerdock signal sets, the extended Gold codes of the same length have N^2 signals and thus are twice the size. However, these codes have $I_{\max} = 2\sqrt{N}$ which is also twice as large. In fact, the $(2^{2n}, 5n)$ extended large Kasami codes give an even larger signal set with $N^{5/2}$ signals for which $I_{\max} = 2\sqrt{N}$ also [24].

The signal sets described above have lengths 2^{2n}. The Kasami construction cannot be used for lengths 2^{2n+1}, while the Kerdock codes of length $N = 2^{2n+1}$ are linear and have the same parameters as the extended Gold codes, viz. $K = N^2$ and $I_{\max} = \sqrt{2N}$. Of course, since $\sqrt{N} = 2^{(2n+1)/2}$ is not an integer, there are no nearly MWBE signal sets for these lengths. However, since Sidelnikov [26] has shown that for *binary* signal sets of size N^2, $I_{\max} > \sqrt{2N-2}$, the Kerdock and Gold code sets are optimal with respect to this bound [24].

2.7 WBE Signal Sets over Prime Power Alphabets

Many of the results stated above are also applicable to codes over prime-power fields $GF(p^s)$, but often do not produce any new classes of signal sets. This is because the results hold only when the elements of $GF(p^s)$ are mapped to the complex p-th roots of unity (and not to the p^s-th roots of unity) via $x \to \exp(j2\pi \mathrm{Tr}(x)/p)$ where $\mathrm{Tr}(\cdot)$ denotes the trace function from $GF(p^s)$ to $GF(p)$.

In practical applications, however, there is often the need to design signals over prime-power alphabets such as the quaternary alphabet $\{+1, j, -1, -j\}$. Unfortunately, codes over $GF(2^2)$, when used as described above, yield binary signals and not quaternary signals. One solution to this difficulty is to use codes over $\mathcal{Z}_4$ as described in Proposition 7. Codes over $\mathcal{Z}_4$ have attracted a great deal of research interest in recent years [6],[8], and many new results are available. A somewhat different solution is described below wherein signals over a q-ary alphabet are constructed from short codes over $GF(q)$. Interestingly, this construction allows the use of an ***arbitrary*** one-one mapping from $GF(q)$ to the complex q-th roots of unity.

Proposition 13. *Let C denote a linear (N, k) code over the field $GF(q)$ with the property that no code symbol C_m is a nonzero multiple of another code symbol C_n. Let $S(C)$ be a signal matrix obtained by mapping the codeword matrix $\hat{S}(C)$ to the complex q-th roots of unity via an* arbitrary *one-one mapping. Then $S(C)$ is a WBE signal set of size $K = q^k$ and length N.*

Thus, one can use codes over $GF(2^2)$ to construct WBE signal sets over the quaternary alphabet just as one can use codes over $\mathcal{Z}_4$. However, although both types of signal sets have the WBE property, codes over $\mathcal{Z}_4$ often can be designed to have smaller $I_{\max}$ than the codes over $GF(2^2)$. Finally, note that N cannot exceed $(q^k - 1)/(q - 1)$ if this construction is used.

2.8 WBE Signal Sets of Arbitrary Size

WBE signal sets of obtained from codes are of size size q^k (or $L \cdot q^k$) where $q^k \geq N$. It is thus of interest to consider whether a WBE signal set of size K and length N exists for *any* given $K \geq N > 1$. The answer is trivially Yes if one uses a $(N, 1)$ code over $\mathcal{Z}_K$. A more interesting construction is as follows.

Proposition 14. *Given any $N > 1$ and any $K = p_1^{k_1} p_2^{k_2} \ldots p_n^{k_n} \geq N$, there exists a WBE signal set of size K and length N over the complex r-th roots of unity where r is a divisor of $p_1 p_2 \ldots p_n$.*

The idea here is to apply Proposition 5 to the $(p_i^{k_i}, k_i)$ extended cyclic codes to construct a WBE signal set of size K and length K and then shorten it in accordance with Proposition 4. Thus, given any $N > 1$, one can construct a WBE signal set of size K and length N for each $K \geq N$. Note, however, that since K and $K + 1$ are relatively prime, the WBE signal sets of these sizes provided by Proposition 14 are over completely different alphabets. Of course, a little thought about the condition for equality in Proposition 1 shows that it is not possible to augment *any* WBE signal set by just one signal and still retain the WBE property.

Just as it is not possible to add one signal to a WBE signal set and still retain the WBE property, it is not possible to delete one signal from a WBE signal set and still retain the WBE property. More generally, while there might exist subsets of WBE signal sets that are themselves WBE signal sets (for example, when a WBE signal set is the union of two WBE signal sets), *not all* subsets of a WBE signal set are WBE signal sets.

Proposition 15. *Let $\mathcal{X}$ denote a WBE signal set of size K and length N. Then, for each $L < K$, at least one subset $\mathcal{Y} \subset \mathcal{X}$ of size L is not a WBE signal set. However, if $\mathcal{Y}$ is a WBE signal set, then so is its complement $\mathcal{X} - \mathcal{Y}$.*

In fact, for at least one subset $\mathcal{Y}$,

$$I_{\text{rms}}(\mathcal{Y}) \geq I_{\text{rms}}(\mathcal{X}) = \sqrt{N\frac{K-N}{K-1}} > \sqrt{N\frac{L-N}{L-1}}$$

and thus this $\mathcal{Y}$ is not a WBE signal set. Indeed, for most values of L, *no* subset of size L is a WBE signal set.

2.9 Random Signal Sets

The standard model for a randomly chosen signal set $\mathcal{X}$ over a p-ary alphabet is that the entries in the $K \times N$ signal matrix $M(\mathcal{X})$ are KN independent identically distributed complex-valued discrete random variables uniformly distributed on the complex p-th roots of unity. It is easily verified that the signals have energy N and that for $k \neq l$, $E[|\langle x^{(k)}, x^{(l)}\rangle|^2] = N$. Thus, for a randomly chosen signal set $\mathcal{X}$,

$$E\left[\sum_{k=1}^{K}\sum_{l=1}^{K} |\langle x^{(k)}, x^{(l)}\rangle|^2\right] = K(K-1)N + KN^2 > K^2N$$

so that the average randomly chosen set of signals is not a WBE signal set. Note that $E[(I_{\text{rms}}(\mathcal{X}))^2] = N$ which is slightly larger than the minimum value $N(K-N)/(K-1)$ given by the Welch bound (4).

The standard model for randomly chosen signal sets is convenient because the independence of the symbols greatly simplifies system analyses. An equivalent model is that the K signals are a *random sample with replacement* from the (N,N) code over $GF(p)$ described in Corollary 7.4. Sampling with replacement allows for the possibility that the same signal might be chosen more than once and thus $I_{\text{rms}}(\mathcal{X}) = \sqrt{N}$. On the other hand, a more appropriate model might be to construct $\mathcal{X}$ by random sampling *without replacement* so that all the signals in $\mathcal{X}$ are guaranteed to be distinct. Since the (N,N) code is a WBE signal set, Proposition 15 shows that

$$E[(I_{\text{rms}}(\mathcal{X}))^2] = N\frac{p^N - N}{p^N - 1} = (I_{\text{rms}}(\mathcal{C}))^2 > N\frac{K-N}{K-1}$$

where C denotes the (N, N) code. Although the average random subset $\mathcal{X}$ is not a WBE signal set under this interpretation either, it is worth noting that the RMS inner product magnitude for randomly chosen signals *is* smaller than $\sqrt{N}$ under this interpretation.

2.10 Information Theory and WBE Signal Sets

The most important point to keep in mind in the analysis of communication systems that use WBE signal sets is that the various properties of WBE signal sets do not always apply to subsets. In particular, subsets of WBE signal sets are not necessarily WBE signal sets. Consider then, the result in [14] that if a WBE signal set is used, then the mean-square interference in a correlation receiver for the l-th transmitted signal is the same for all values of l, that is,

$$\sum_{k \neq l} |\langle x^{(k)}, x^{(l)} \rangle|^2 = KN - N^2, \quad 1 \leq l \leq K. \tag{7}$$

Now suppose that only $L < K$ transmitters are active. Since some signals are no longer present at the receiver input, the mean-square interference in all the L active receivers is reduced, and is smaller than $KN - N^2$. Thus, absence of some signals improves the performance of the rest. However, unless the active subset happens to be a WBE signal set also, the mean-square interference in all the receivers will not be *uniformly* reduced to $LN - N^2$. Indeed, it is easy to show that for all choices of l and L, the *average* mean-square interference in the l-th receiver when $L-1$ *other* transmitters are active is $LN - N^2 + N(N-1)(K-L)/(K-1) > LN - N^2$, and thus for at least one choice of $L-1$ other transmitters, the mean-square interference in the l-th receiver is worse than that promised by the WBE signal theory. In short, one cannot simply replace K with L in (7) when only L transmitters are active. The WBE property is quite fragile and does not hold for subsets or for variations from the stated conditions.

A fixed assignment of a WBE signal set to transmitters does not retain the WBE property when some transmitters are silent. Thus, suppose that K transmitters use a WBE signal set and can simultaneously transmit at rate R so that the *symmetric* rate vector $(R, R, \ldots, R)$ is achieved. If the first transmitter is silent, the rest can still transmit via their assigned signals from the WBE signal set at rate R so that rate vector $(0, R, R, \ldots, R)$ can be achieved. In fact, since the interference is less, higher rates should be achievable, but without a detailed study of the WBE signal set, it is not clear what "symmetric" rate vector $(0, R', R', \ldots, , R')$ (where $R' > R$) can be achieved. Indeed, the entire rate region achievable by a fixed assignment of a WBE signal set (that is, the set of rate vectors $(R_1, R_2, \ldots, R_K)$ at which it is possible to transmit via use of the WBE signal set) seems difficult to compute without detailed knowledge of the structure of the WBE signal set. A variable assignment of signals to transmitters (depending on which transmitters have data to send at any given time) does not work either because there may not be a WBE signal set of the right size. For example, the size of a binary WBE signal set is either 2 or a multiple of 4 which is not helpful if exactly 13 transmitters wish to be active.

In studying the capacity region of multiple-access channels, one might wish to use different WBE signal sets to achieve different rate vectors. For any given $K \geq N$, Proposition 14 can be used to construct WBE signal sets of sizes $K, K-1, K-2, \ldots$. Thus, no matter which $L < K$ transmitters are active, one could assign a WBE signal set of size L to the active transmitters, and each active receiver would uniformly suffer a mean-square multiple-access interference of $LN - N^2$. Unfortunately, these various signal sets are over different alphabets, so that the achievable rates are not directly comparable since they apply to different multiple-access channels. On the other hand, if the different alphabets are combined into a single alphabet, then this large alphabet has size $p_1 p_2 \ldots p_s$ where $p_1, p_2, \ldots, p_s$ are all the primes that do not exceed K. In the absence of multi-user interference and channel noise, any transmitter can send $\log_2 p_1 p_2 \ldots p_s$ bits per symbol over the channel, so the capacity is quite large. On the other hand, only a small subset of the alphabet is used by the WBE signal sets described above, so that the rate vectors that can be achieved by the use of WBE signal sets may well turn out to be a very small subset of the actual capacity region.

In summary, an insistence on always using a WBE signal set may be a hindrance rather than an advantage, and the simplifications in the analysis due to the uniform interference level might be insufficient to overcome the disadvantages. Subsets of WBE signal sets do not enjoy the WBE properties and results on WBE signal sets are not always applicable to the subsets.

3 Bounds on Correlation Functions

The bounds on inner products discussed in the previous section are of interest in studies of synchronous CDMA communication systems. Analyses of asynchronous CDMA communication systems lead to the consideration of various correlation functions of the signals (see, for example, [24]), and bounds on the magnitudes of these are studied in this section.

3.1 The Correlation Functions

As before, let $\mathcal{X}$ denote a set of K signals of length N and energy N, and let $S(\mathcal{X})$ denote the signal matrix. Now, however, it is not necessary to insist that $K \geq N$. For signals $x^{(k)}, x^{(l)} \in \mathcal{X}$, the *aperiodic* crosscorrelation function $C_{x^{(k)},x^{(l)}}(m)$ for $x^{(k)}$ and $x^{(l)}$ (more simply, $C_{k,l}(m)$) is defined as

$$C_{k,l}(m) = C_{x^{(k)},x^{(l)}}(m) = \begin{cases} \sum_{n=0}^{N-1-m} x_n^{(k)} (x_{n+m}^{(l)})^*, & 0 \leq m \leq N-1, \\ \sum_{n=0}^{N-1+m} x_{n-m}^{(k)} (x_n^{(l)})^*, & 1-N \leq m < 0, \\ 0, & |m| \geq N. \end{cases} \tag{8}$$

For $0 \leq m < N$, the *periodic* crosscorrelation function $\theta_{x^{(k)},x^{(l)}}(m) = \theta_{k,l}(m)$ and the *odd* crosscorrelation function $\hat{\theta}_{x^{(k)},x^{(l)}}(m) = \hat{\theta}_{k,l}(m)$ are given by

$$\theta_{k,l}(m) = C_{k,l}(m) + C_{k,l}(m-N) \text{ and } \hat{\theta}_{k,l}(m) = C_{k,l}(m) - C_{k,l}(m-N), \quad (9)$$

and have been studied extensively in the context of asynchronous CDMA systems using biphase modulation. Consideration of Q-ary modulation in CDMA systems leads to the study of the *polyphase* crosscorrelation function [4], [28] which is defined as

$$\theta_{k,l}^{(\gamma)}(m) = C_{k,l}(m) + \gamma C_{k,l}(m-N). \quad (10)$$

where γ is a unit-magnitude complex number. Note also that $\theta_{k,l}^{(\gamma)}(0) = \langle x^{(k)}, x^{(l)} \rangle$ for all γ. If $k = l$, we get the autocorrelation functions which are usually denoted by $C_k(m)$, $\theta_k(m)$, $\hat{\theta}_k(m)$, and $\theta_k^{(\gamma)}(m)$ respectively. All the autocorrelation functions have the same value for zero offset: $\theta_k(0) = \hat{\theta}_k(0) = C_k(0) = \theta_k^{(\gamma)}(0) = \langle x^{(k)}, x^{(k)} \rangle = N$.

Many results in this section apply to all of the correlation functions described above. Let $g(\cdot)$ denote any of the correlation functions and let the index set $\mathcal{M}$ be defined as

$$\mathcal{M} = \begin{cases} \{m : 1-N \leq m \leq N-1\}, & \text{if } g(\cdot) = C(\cdot), \\ \{m : 0 \leq m \leq N-1\}, & \text{if } g(\cdot) = \theta^{(\gamma)}(\cdot),\ \theta(\cdot),\ \hat{\theta}(\cdot). \end{cases}$$

Let $\mathcal{M}^* = \mathcal{M} - \{0\}$, and $M = |\mathcal{M}|$ denote the cardinality of $\mathcal{M}$. The maximum crosscorrelation value $g_{c,\max}(\mathcal{X})$ and maximum out-of-phase autocorrelation value $g_{a,\max}(\mathcal{X})$ are defined to be

$$g_{c,\max}(\mathcal{X}) = \max_{k \neq l} \max_{m \in \mathcal{M}} |g_{k,l}(m)| \quad \text{and} \quad g_{a,\max}(\mathcal{X}) = \max_{k} \max_{m \in \mathcal{M}^*} |g_k(m)|$$

The maximum correlation value is $g_{\max}(\mathcal{X}) = \max\{g_{c,\max}(\mathcal{X}), g_{a,\max}(\mathcal{X})\}$. The RMS correlation value is obtained by averaging all $K(K-1)M + K(M-1) = K(KM-1)$ correlation magnitudes and is thus

$$\begin{aligned}(g_{\text{rms}}(\mathcal{X}))^2 &= \frac{1}{K(KM-1)} \left[\sum_{k=1}^{K} \sum_{l=1}^{K} \sum_{m \in \mathcal{M}} |g_{k,l}(m)|^2 - KN^2 \right] \\ &= \frac{1}{K(KM-1)} \left[K(K-1)M(g_{c,\text{rms}}(\mathcal{X}))^2 + K(M-1)(g_{a,\text{rms}}(\mathcal{X}))^2 \right]\end{aligned}$$

where $g_{c,\text{rms}}(\mathcal{X})$ is the RMS crosscorrelation value obtained by averaging over the $K(K-1)M$ crosscorrelation magnitudes, and $g_{a,\text{rms}}(\mathcal{X})$ is the RMS autocorrelation value obtained by averaging over the $K(M-1)$ out-of-phase autocorrelation magnitudes. Note also that $g_{c,\max} \geq g_{c,\text{rms}}$, $g_{a,\max} \geq g_{a,\text{rms}}$, and $g_{\max} \geq g_{\text{rms}}$.

3.2 The Welch Bounds

It was shown in [18] that

$$\sum_{m\in\mathcal{M}} |g_{k,l}(m)|^2 = \sum_{m\in\mathcal{M}} g_k(m)(g_l(m))^*. \tag{11}$$

holds for the aperiodic, periodic, and odd correlation functions. More recently, it has been noted [28] that the same proof shows that polyphase correlation functions also satisfy this identity. Thus, as described in [21], it follows from (11) that

$$\sum_{k=1}^{K}\sum_{l=1}^{K}\sum_{m\in\mathcal{M}} |g_{k,l}(m)|^2 = K^2N^2 + \sum_{m\in\mathcal{M}^*}\left|\sum_{k=1}^{K} g_k(m)\right|^2. \tag{12}$$

Thus the following generalization of Proposition 1 holds [16], [21], [28].

Proposition 16. *The aperiodic and polyphase correlation functions (which include the periodic and odd correlation functions as special cases) for a signal set satisfy the following lower bound:*

$$\sum_{k=1}^{K}\sum_{l=1}^{K}\sum_{m\in\mathcal{M}} |g_{k,l}(m)|^2 \geq K^2N^2 \tag{13}$$

where equality holds in (13) if and only if

$$\sum_{k=1}^{K} g_k(m) = 0 \text{ for all } m \in \mathcal{M}^*. \tag{14}$$

Since the left side of (13) includes K in-phase autocorrelations with total value KN^2, the Welch bounds on the RMS correlation are as follows:

$$g_{\text{rms}} \geq \sqrt{N\frac{NK-N}{MK-1}} \tag{15}$$

while the more familiar form is

$$g_{\max} \geq \sqrt{N\frac{NK-N}{MK-1}}. \tag{16}$$

The bounds for the polyphase correlations (including periodic and odd correlations) are slightly larger than the right side of (4) but are nonetheless still slightly smaller than $\sqrt{N}$. The bounds for aperiodic correlation are slightly smaller than $\sqrt{N/2}$ since $M = 2N-1$ in this case.

Welch obtained the bounds (15) and (16) for the periodic and aperiodic correlation functions as a direct consequence of the inner product bound (3). Thus, in [33], it was shown that for periodic and aperiodic correlations functions,

$$\sum_{k=1}^{K}\sum_{l=1}^{K}\sum_{m\in\mathcal{M}} |g_{k,l}(m)|^{2s} \geq \frac{K^2N^{2s}M}{\binom{M+s-1}{s}}. \tag{17}$$

Mow [15] has shown that a straightforward modification of Welch's argument proves (17) for odd correlation functions as well, and a further simple modification shows that (17) holds for polyphase correlations as well.

3.3 Complementary Signal Sets

Sets of signals satisfying the relation

$$\sum_{k=1}^{K} C_k(m) = 0 \text{ for all } m \neq 0 \tag{18}$$

are called complementary signal sets [5], [22], [30].

Proposition 17. *The aperiodic and polyphase correlation functions (which include the periodic and odd correlation functions as special cases) for a complementary signal set satisfy*

$$\sum_{k=1}^{K}\sum_{l=1}^{K}\sum_{m} |g_{k,l}(m)|^2 = K^2N^2$$

and thus meet the Welch bound (15) on RMS correlation with equality.

Various designs for complementary signal sets are given in [5], [22], and [30]. However, in hindsight, it seems that it is really very easy to construct a complementary signal set. The author did not find an explicit reference to cite for the following construction, but its elementary nature suggests that it has undoubtedly been noted by several researchers already.

Proposition 18. *Every WBE signal set is a complementary signal set.*

Proposition 18 and the results of Section II imply that a surprisingly large number of signal sets are actually complementary signal sets. For example, almost all "interesting" codes and their cosets are complementary signal sets. As a special case, it was noted in [22] that irreducible cyclic codes and their cosets are complementary signal sets, and in particular, the expurgated set of Gold sequences of Proposition 12 is a complementary signal set.

4 WBE Signal Sets in Asynchronous CDMA Systems

4.1 The Signal-to-Noise Ratio Parameters

The signal-to-noise ratio (SNR) for asynchronous binary CDMA communication systems was defined by Pursley [17] and has been the subject of considerable study since. Let

$$\mu_{k,i}(n) = \sum_{l=1-N}^{N-1} C_{k,i}(l)[C_{k,i}(l+n)]^* = \sum_{l=1-N}^{N-1} C_k(l)[C_i(l+n)]^* \tag{19}$$

where the second equality follows from a generalization of (11) (see, e.g. Eq. (5.24) in [24].) In [17] it is shown that the variance of the multiple-access interference in a receiver that is synchronized to the i-th signal is proportional to the *SNR parameter*

$$r_i = \sum_{k\neq i} 2\mu_{k,i}(0) + \mathrm{Re}[\mu_{k,i}(1)] \approx R_i = \sum_{k\neq i} 2\mu_{k,i}(0).$$

Now suppose that the set of signature sequences is a WBE signal set or, more generally, a complementary signal set. Then, from (18), (19), and the property $C_i(-l) = [C_i(l)]^*$ of the aperiodic autocorrelation function, it follows that

$$r_i = 2(K-1)N^2 - 4\sum_{l=1}^{N-1} |C_i(l)|^2$$
$$+2(K-2)N\mathrm{Re}[C_i(1)] - 2\mathrm{Re}\left(\sum_{l=1}^{N-2} C_i(l)[C_i(l+1)]^*\right) \tag{20}$$

and thus

$$R_i = 2(K-1)N^2 - 4\sum_{l=1}^{N-1} |C_i(l)|^2. \tag{21}$$

The sum $\sum_l |C_i(l)|^2$ in (21) is called the sidelobe energy of the signal $x^{(i)}$ and is nonzero (except for TDMA signal sets). Hence, $R_i < 2(K-1)N^2$ (and generally, $r_i < 2(K-1)N^2$ also) for all interesting WBE signal sets. Consider instead a CDMA communication system in which the K signature sequences are chosen via random sampling (with replacement) from the (N, N) code as in Section II.J. Then, both R_i and r_i have expected value $2(K-1)N^2$. Thus, since each R_i is smaller than $2(K-1)N^2$ for WBE signal sets, these signal sets appear to be somewhat better than random sequences with respect to this measure of performance. Another curious aspect of (21) is that the *larger* the sidelobe energy, the smaller the variance of the multiple-access interference. Signals with *small* sidelobe energy are usually preferred as signature sequences in CDMA communication systems on the grounds that such signals perform better over multipath channels and are easy to acquire. Hence, considerable effort has been expended in the past on searches for signals with small sidelobe energy. Paradoxically, Eq. (21) implies that from the point of view of multiple-access interference, it would have been better if the searches had tried to find signals with the largest sidelobe energy instead of the smallest!

In most instances, R_i or r_i must be determined via numerical evaluation of the autocorrelation function values needed in (20) or (21). It has already been shown that $R_i < 2(K-1)N^2$ for complementary signal sets, but it is not obvious without computation whether $r_i < 2(K-1)N^2$ also. In some cases, however, it is possible to state more specific results.

- The complementary pairs of binary sequences described by Taki *et al.* [29] have the property that $C_{k,i}(l) = 0$ for all even values of l. It follows from (19) that $r_i = R_i < 2(K-1)N^2$ for any collection of such complementary pairs of sequences.
- If the sequence $x^{(i)}$ alternately takes on real and imaginary values, then $C_i(l)$ is real for even values of l and imaginary for odd values of l. It follows that $\mathrm{Re}[C_i(l)C_i^*(l+1)] = 0$ and hence $r_i = R_i < 2(K-1)N^2$. Of course, the signal set must be a WBE signal set (or complementary signal set) for this to hold, but this requirement is easily satisfied. If alternate columns of the signal matrix of a real-valued WBE signal set $\mathcal{X}$ are multiplied by $\pm j$, then, according to Proposition 3, the resulting signal set $\hat{\mathcal{X}}$ is also a WBE signal set. In particular, if $\mathcal{X}$ is a *binary* WBE signal set, then $\hat{\mathcal{X}}$ is a WBE signal set over the quaternary alphabet $\{+1, -1, +j, -j\}$. Such quaternary signals are a special case of the EOE sequences described by Fukumasa, Kohno, and Imai [4] and can also be thought of as being derived from binary sequences via the elementary Method B of [9].

4.2 SNR Parameters for Some Cyclic WBE Signal Sets

Many of the constructions of WBE signal sets of length N in Section II are based on cyclic codes of length N. The signature sequences used in CDMA applications are often required to have full period N, and WBE signal sets of this type are considered in the next result.

Proposition 19. *If a WBE signal set $\mathcal{X}$ is either a coset or a union of cosets of a linear cyclic code C all of whose nonzero codewords have period N, then,*

$$\sum_{i=1}^{K} |C_i(l)|^2 = K(N-l), \quad 1 \le l \le N-1, \tag{22}$$

$$\bar{R}(\mathcal{X}) = \frac{1}{K}\sum_{i=1}^{K} R_i = 2(K-1)N^2 - 2N^2 + 2N, \tag{23}$$

$$\textit{and} \quad \bar{r}(\mathcal{X}) = \frac{1}{K}\sum_{i=1}^{K} r_i \le 2(K-1)N^2 - N^2 - N + 2. \tag{24}$$

This result was proved earlier for cosets of irreducible binary cyclic codes. Recently, it has been proved for a class of quaternary sequences.

Corollary 19.1. *(Sarwate [23]): Proposition 19 holds for any coset of an irreducible binary cyclic code. In particular, it holds for the coset which forms the expurgated set of Gold sequences described in Proposition 12.*

Corollary 19.2. *(Sun and Leib [28]): Proposition 19 holds for $\mathcal{A}^-$, an expurgated quadriphase signal set obtained from the Family $\mathcal{A}$ described by Boztaş, Hammons and Kumar [3].*

Family $\mathcal{A}$ is a *quaternary* supercode of a *binary* m-sequence code, and the signal set $\mathcal{A}^-$, which consists of one nonbinary signal from each nonbinary cyclic equivalence class of $\mathcal{A}$, is a quaternary coset of the binary m-sequence code. As an example of this, note that Table II of [3] lists the 9 nonzero cycle representatives in a Family $\mathcal{A}$ of period 7, of which the first is the binary m-sequence $v = 2002022$ and the second a quaternary sequence $u = 3221211$. Adding the cyclic shifts of the first sequence to the second produces (cyclic shifts of) the rest of the sequences in the Table. In other words, if different cycle representatives are chosen, the signals in Table II of [3] can be expressed as the set $\{v, u, u+v, u+Tv, \ldots, u+T^6v\}$ while the signal set $\mathcal{A}^-$ is $\{u, u+v, u+Tv, \ldots, u+T^6v\}$. Thus, $\mathcal{A}^-$ is a quaternary coset of the binary m-sequence code $\{0, v, Tv, \ldots, T^6v\}$, and the expurgation is remarkably similar to that used to obtain the expurgated set of Gold sequences.

Sun and Leib [28] have also asserted with respect to $\mathcal{A}^-$ that

> "... it is remarkable that the average user interference of this family at optimal phases outperforms the ideal random sequences."

Actually, according to Proposition 19, a very large class of signal sets enjoys this property, and the fact that (24) holds for $\mathcal{A}^-$ is not remarkable at all. Indeed, the property is a result of the underlying cyclic structure and does not depend at all on the coset representative – an *arbitrary* choice of quaternary sequence u in the above would have worked just as well! Nor is it necessary to find optimal phases for the family – any choice of phase of the quaternary sequence u gives the same result. What is actually remarkable is that Proposition 19 can be used to construct a signal set for which *all* the SNR parameters r_i are smaller than $2(K-1)N^2$. The construction is as follows. Choose alternate symbols of the coset representative from the subsets $\{0, 2\}$ and $\{1, 3\}$ of Z_4 respectively. Then the signals are alternately real-valued and imaginary-valued as described in the previous subsection. Hence, $\bar{r}(\mathcal{X}) = \bar{R}(\mathcal{X})$ equals the right side of (23). Furthermore, each $r_i = R_i < 2(K-1)N^2$, which is a better result than can be guaranteed via Proposition 19 for $\mathcal{A}^-$. However, such a signal set is not necessarily a good choice for CDMA applications. The two corollaries above are of importance because the signal sets are interesting because of their *maximum* correlation properties – the good SNR parameter properties are just an added attraction. The SNR parameter properties by themselves do not provide guidance in signal selection. Note, for example, that Proposition 19 also holds for a multiset formed by the union of a coset with itself, but the resulting signal set is of no interest for CDMA applications.

4.3 Optimal Shifts of WBE Signal Sets

Since $\bar{R}(\mathcal{X}) = 2(K-1)N^2 - 2N^2 + 2N$ for a cyclic WBE signal satisfying the hypotheses of Proposition 19, at least one signal has R_i no larger than this. However, other R_k may be somewhat larger though (21) guarantees that each is smaller than $2(K-1)N^2$. Now, $T^m\mathcal{X}$, the m-th left cyclic shift of $\mathcal{X}$, is also a WBE signal set satisfying the hypotheses of Proposition 19. But $R_i^{(m)}$, the SNR parameter for $T^m x^{(i)}$, need not be the same as $R_i^{(0)} = R_i$ [18], [19], [24]. Thus, for any given i, one can search for the shift m that minimizes $R_i^{(m)}$. Unfortunately, since $\bar{R}(T^m\mathcal{X}) = \bar{R}(\mathcal{X})$ is fixed, a decrease in the SNR parameter for one signal is always accompanied by an increase in the SNR parameter for other signals. Thus, a better solution might be to search for that m which minimizes $\max_{1\le i\le K} R_i^{(m)}$.

Since $\bar{r}(\mathcal{X}) \le 2(K-1)N^2 - N^2 - N + 2$ for the class of WBE signal sets satisfying the hypotheses of Proposition 19, at least one r_i is as small as this. In fact, since this bound is quite loose, one can hope to find that at least one r_i is as small as $\bar{R}(\mathcal{X}) = 2(K-1)N^2 - 2N^2 + 2N$. It sometimes happens that all the r_i are smaller than $2(K-1)N^2$, but, in contrast to the situation for R_i, there is no guarantee that this will always happen. As in the previous paragraph, for any given i, one can choose an m that minimizes $r_i^{(m)}$. Unlike the situation with $\bar{R}$, however, $\bar{r}(T^m\mathcal{X}) \ne \bar{r}(\mathcal{X})$ in general, so that it is possible to choose m to minimize $\bar{r}(T^m\mathcal{X})$ and obtain the best average performance, or to minimize $\max_{1\le i\le K} r_i^{(m)}$ so as to obtain the best worst-case performance.

Thus, given a WBE signal set $\mathcal{X}$ satisfying the hypotheses of Proposition 19, one can choose a cyclically shifted set $T^m\mathcal{X}$ to minimize various SNR parameters. The performance improvement is not very large, but is often large enough to make the computational effort worthwhile. It is also worth keeping in mind that the kinds of searches discussed here shift all the signals in $\mathcal{X}$ by the same m. If the shifts applied to the signals are not the same, the resulting signal set is no longer a WBE signal set, and the above results may no longer apply. Nonetheless, it is possible that SNR parameters might be reducible to the same level as those of a WBE signal set or even further. Indeed, even for nonWBE signal sets, it is usually possible to reduce the SNR parameters below $2(K-1)N^2$ but the computational effort in searching over K^N shifts is formidable, and only restricted searches are generally feasible. The advantage of beginning with a WBE signal set is that some measure of goodness of performance is guaranteed by the results described above *without* any search, and the results can be improved somewhat by a much narrower search.

4.4 Subsets of WBE Signal Sets

WBE signal sets have too many signals to be of use in practical CDMA systems since the number of simultaneous transmissions that can be supported is generally much smaller than the signature sequence length N. On the other hand, the results proved above require the cancellations described in (18) and these hold only when the entire set is used. Since subsets of WBE signal sets are generally not WBE signal sets, they do not enjoy the properties described in Proposition 19 except in an average sense – a situation that is similar to that discussed in Proposition 15. For any given i, there are $\binom{K-1}{L-1}$ subsets of size L that include $x^{(i)}$. Any other $x^{(k)}$ belongs to $\binom{K-2}{L-2}$ of these subsets. Hence, (23) and (24) imply that the following result holds:

Proposition 20. *For any signal $x^{(i)}$ in a WBE signal set $\mathcal{X}$ satisfying the hypotheses of Proposition 19, and any L, $2 \leq L \leq K$, there exists a $\mathcal{Y} \subset \mathcal{X}$ of size L such that $x^{(i)} \in \mathcal{Y}$ and*

$$R_i \leq 2(L-1)N^2 - \frac{L-1}{K-1}\left(2N^2 - 2N\right)$$
$$r_i \leq 2(L-1)N^2 - \frac{L-1}{K-1}\left(N^2 + N - 2\right)$$

The right sides of the above inequalities are just $(L-1)/(K-1)$ times the right sides of (23) and (24). In fact, these quantities are respectively the average values of R_i and the bound on r_i over all subsets of $\mathcal{X}$ that contain $x^{(i)}$ and $L-1$ other signals. Note that the values obtained are smaller than $2(L-1)N^2$ but the improvement over random sequences is not as much as with the full WBE signal set. Unfortunately, although the i-th signal is guaranteed to achieve a good SNR, the other signals in the set do not necessarily also achieve a good SNR. In contrast, for a WBE signal set, *all* the K signals achieve values of R_i smaller than $2(K-1)N^2$. Note that Sun and Leib [28] have proved a slightly different version of Proposition 20. Their result is that, averaged over all subsets $\mathcal{Y} \subset \mathcal{A}^-$ of size L, the value of $\bar{r}(\mathcal{Y})$ is bounded by $(L-1)/(K-1)$ times the right side of (24), just as in Proposition 20 above. It is easily shown that the Sun-Leib result also holds for the class of WBE signals described in Proposition 19.

5 Concluding Remarks

Signal sets that meet the Welch bounds on the maximum magnitude of the inner product or the correlation functions are not easy to find. In contrast, finding signal sets that meet the Welch bounds on the RMS magnitudes is almost trivially easy – *almost all* linear codes (and their cosets) meet the latter bounds, with rates and minimum distances hardly being of any concern. Thus, meeting the Welch bound on the RMS magnitudes is far too broad a criterion for signal design, and the traditional measures for signal performance should not be discarded. The recent interest in proving various information-theoretic results via the use of WBE signal sets is also somewhat puzzling. It would seem that almost any result which holds only when the signals form a WBE signal set could have been proved long ago since (at least in terms of mean-square Hamming weights) it has been known for nearly forty years that most binary linear codes are WBE signal sets! Finally, it is worth re-emphasizing that subsets of WBE signal sets usually are not WBE signal sets, and thus, careful analysis is needed to determine the performance of systems that use WBE signal sets.

6 Acknowledgments

This research was supported in part by the Joint Services Electronics Program under Grant N00014-90-J-1270, and in part by the U. S. Army Research Office under Focused Research Initiative Grant DAAH01-95-1-0216.

References

1. W. O. Alltop, "Complex sequences with low periodic correlations," *IEEE Trans. Inform. Theory*, vol. IT-26, pp. 350–354, May 1980.
2. I. F. Blake and R. C. Mullin, *The Mathematical Theory of Coding*, New York: Academic Press, 1975.
3. S. Boztaş, R. Hammons, and P. V. Kumar, "4-phase sequences with near-optimum correlation properties," *IEEE Trans. Inform. Theory*, vol. 38, pp. 1101-1113, May 1992.
4. H. Fukumasa, R. Kohno, and H. Imai, "Design of pseudonoise sequences with good odd and even correlation properties for DS/CDMA," *IEEE J. Selected Areas Commun.*, vol. 12, pp. 828-836, June 1994.
5. M. J. E. Golay, "Complementary series," *IRE Trans. Inform. Theory*, vol. IT-7, pp. 82-87, April 1961.
6. A. R. Hammons, Jr., P. V. Kumar, A. R. Calderbamk. N. J. A. Sloane, and P. Solé, "The Z_4 linearity of Kerdock, Preparata, Goethals, and related codes," *IEEE Trans. Inform. Theory*, vol. 40, pp. 301-319, March 1994.
7. T. Helleseth and P. V. Kumar, "Pseudonoise sequences," in *The Communications Handbook*, J. D. Gibson (ed.), Piscataway, NJ: IEEE Press, 1997.
8. T. Helleseth and P. V. Kumar, "Sequences with low correlation," in *Handbook of Coding Theory*, V. Pless (ed.), Elsevier Press (1998).
9. S. M. Krone and D. V. Sarwate, "Quadriphase sequences for spread-spectrum multiple-access communication," *IEEE Trans. Inform. Theory*, vol. IT-30, pp. 520-529, May 1984.
10. P. V. Kumar and C.-M. Liu, "On lower bounds to the maximum correlation of complex, roots-of-unity sequences," *IEEE Trans. Inform. Theory*, vol. 36, pp. 633-640, May 1990.
11. V. I. Levenshtein, "Bounds on the maximum cardinality of a code with bounded modulus of the inner product," *Soviet Math. Dokl.*, vol. 25, pp. 526-531, 1982.
12. F. J. MacWilliams and N. J. A. Sloane, "Pseudorandom sequences and arrays," *Proc. IEEE*, vol. 64, pp. 1715-1729, December 1976.
13. F. J. MacWilliams and N. J. A. Sloane, *Theory of Error Correcting Codes*. Amsterdam: North Holland, 1977.
14. J. L. Massey and T. Mittelholzer, "Welch's bound and sequence sets for code-division multiple-access systems," in *Sequences II: Methods in Communication, Security, and Computer Science*, R. Capocelli, A. De Santis, and U. Vaccaro, (eds.), pp. 63-78, New York: Springer-Verlag, 1993.
15. W. H. Mow, "On the bounds on odd correlation of periodic sequences," *IEEE Trans. Inform. Theory*, vol. 40, pp. 954-955, May 1994.
16. W. H. Mow, "Optimal sequence sets meeting Welch's lower bound," *Proc. 1995 IEEE Int'l. Symp. Inform. Theory*, (Whistler, Canada: September 17–22, 1995) p. 89.

17. M. B. Pursley, "Performance evaluation for phase-coded spread-spectrum multiple-access communication – Part I: System analysis," *IEEE Trans. Commun.*, vol. COM-25, pp. 795-799, August 1977.
18. M. B. Pursley and D. V. Sarwate, "Evaluation of correlation parameters for periodic sequences," *IEEE Trans. Inform. Theory*, vol. IT-23, pp. 508-513, July 1977.
19. M. B. Pursley and D. V. Sarwate, "Performance evaluation for phase-coded spread-spectrum multiple-access communication – Part II: Code sequence analysis," *IEEE Trans. Commun.*, vol. COM-25, pp. 800-803, August 1977.
20. M. Rupf and J. L. Massey, "Optimum sequence multisets for synchronous code-division multiple-access channels," *IEEE Trans. Inform. Theory*, vol. 40, pp. 1261–1266, July 1994.
21. D. V. Sarwate, "Bounds on crosscorrelation and autocorrelation of sequences," *IEEE Trans. Inform. Theory*, vol. IT-25, pp. 720–724, November 1979.
22. D. V. Sarwate, "Sets of complementary sequences," *Electronics Letters*, vol. 19, pp. 711-712, September 1983.
23. D. V. Sarwate, "Mean-square correlation of shift register sequences," *IEE Proceedings*, vol. 131, Part F, pp. 101–106, April 1984.
24. D. V. Sarwate and M. B. Pursley, "Crosscorrelation properties of pseudorandom and related sequences," *Proc. IEEE*, vol. 68, pp. 593–619, May 1980.
25. H. D. Schotten, "Mean-square correlation parameters of code sequences in DS-CDMA systems," *Proc. 1997 IEEE Int'l. Symp. Inform. Theory*, (Ulm, Germany: June 29–July 4, 1997), p. 486.
26. V. M. Sidelnikov, "On mutual correlation of sequences," *Soviet Math. Dokl.*, vol. 12, pp. 197-201, 1971.
27. S. G. S. Shiva, "Some results on binary codes with equidistant words," *IEEE Trans. Inform. Theory*, vol. IT-15, pp. 328-329, March 1969.
28. F.-W. Sun and H. Leib, "Optimal phases for a family of quadriphase CDMA sequences," *IEEE Trans. Inform. Theory*, vol. 43, pp. 1205-1217, July 1997.
29. Y. Taki, H. Miyakawa, M. Hatori, and S. Namba, "Even-shift orthogonal sequences," *IEEE Trans. Inform. Theory*, vol. IT-15, pp. 295-300, March 1969.
30. C.-C. Tseng and C. L. Liu, "Complementary sets of sequences," *IEEE Trans. Inform. Theory*, vol. IT-18, pp. 644-652, September 1972.
31. P. Viswanath, V. Anantharam, and D. Tse, "Optimal sequences, power control and capacity of synchronous CDMA systems with linear multiuser receivers," *Proc. 1998 IEEE Inform. Theory Workshop*, (Killarney, Ireland: June 22-26, 1998), pp. 134-135.
32. P. Viswanath and V. Anantharam, "Optimal sequences and sum capacity of synchronous CDMA systems," *Proceedings of the 36th Allerton Conference on Communication, Control, and Computing*, (Monticello, Illinois: September 23-25, 1998), pp. 272-281.
33. L. R. Welch, "Lower bounds on the maximum cross correlation of signals," *IEEE Trans. Inform. Theory*, vol. IT-20, pp. 397-399, May 1974.
34. N. Zierler, "A note on the mean square weight for group codes," *Information and Control*, vol. 5, pp. 87-89, January 1962.

My Favorite Integer Sequences

N. J. A. Sloane

Information Sciences Research
AT&T Shannon Lab
Florham Park, NJ 07932-0971 USA
Email: njas@research.att.com

Abstract. This paper gives a brief description of the author's database of integer sequences, now over 35 years old, together with a selection of a few of the most interesting sequences in the table. Many unsolved problems are mentioned.

1 How it all began

I started collecting integer sequences in December 1963, when I was a graduate student at Cornell University, working on perceptrons (or what are now called neural networks). Many graph-theoretic questions had arisen, one of the simplest of which was the following.

Choose one of the n^{n-1} rooted labeled trees with n nodes at random, and pick a random node: what is its expected height above the root? To get an integer sequence, let a_n be the sum of the heights of all nodes in all trees, and let $W_n = a_n/n$. The first few values $W_1, W_2, \ldots$ are

$$0, 1, 8, 78, 944, 13800, 237432, \ldots ,$$

a sequence engraved on my memory. I was able to calculate about ten terms, but I needed to know how W_n grew in comparison with n^n, and it was impossible to guess this from so few terms. So instead I tried to guess a formula for the nth term. Again I was unsuccessful. Nor could I find this sequence in Riordan's book [86], although there were many sequences that somewhat resembled it.

So I started collecting all the sequences I could find, entering them on punched cards, thinking that if any of these sequences came up in another problem, at least I would know what *they* were.

I never did find that sequence in the literature, but I learned Pólya's theory of counting and (with John Riordan's help) obtained the answer, which appears in [87]. There *is* a simple formula, although maybe not simple enough to be guessed:

$$W_n = (n-1)! \sum_{k=0}^{n-2} \frac{n^k}{k!} .$$

An old formula of Ramanujan [99] then implies that

$$\frac{W_n}{n^n} \sim \sqrt{\frac{2\pi}{n}}, \quad \text{as} \quad n \to \infty ,$$

which was what I needed. That sequence became number A435 in the collection.

The idea of a "dictionary" of integer sequences was received with enthusiasm by many people, and in 1973 I published [91], containing about 2400 sequences, arranged lexicographically. One correspondent commented on the book by saying "There's the Old Testament, the New Testament and the Handbook of Integer Sequences".

Over the next twenty years an enormous amount of new material arrived, preprints, reprints, postcards, typewritten letters, handwritten letters, etc., and it was not until 1995 that — with Simon Plouffe's help — a sequel [93] appeared. This contained 5500 sequences.

Around the same time I set up two services that can be used to consult the database via electronic mail. The first has the address

`sequences@research.att.com` ,

and simply looks up a sequence in the table. The second

`superseeker@research.att.com` ,

tries hard to find an explanation even for a sequence not in the table.

A large number of new sequences started arriving as soon as [93] appeared, and when in 1996 the total number reached 16,000, three times the number in the book, I decided to set up a web site for the database [92]. New sequences still continue to pour in, at about 10000 per year. At present, in April 1999, the database contains about 50000 sequences, and the web site receives over 2500 hits per day.

The main reason for this rapid expansion is that in the two books I only included sequences that had been or were about to be published. For the on-line version, where storage space is no longer a limitation, any well-defined and sufficiently interesting sequence is eligible for inclusion.

There is now also an electronic *Journal of Integer Sequences* [40] and a mailing list for sequence fans [89].

The following sections describe how the database is used (Section 2) and the kinds of sequences it contains (Section 3). Section 4 discusses a few "hard" sequences and Section 5 some recursive examples. Then Sections 6–8 describe sequences associated with meandering rivers and stamp-folding, extremal codes and lattices, and Levine's sequence.

Besides integer sequences, the table also contains many examples of arrays of numbers, Pascal's triangle, for example — see Section 9. The final three sections discuss sequences associated with the Wythoff array, the boustrophedon transformation of sequences, and Tchoukaillon solitaire.

2 How the database is used

The main applications of the database are in identifying sequences or in finding out the current status of a known sequence.

The database has been called a mathematical analogue of a "fingerprint file" [10]. You encounter a number sequence, and wish to know if anyone has ever come across it before. If your sequence is in the database, the reply will provide a description, the first 50 or so terms (usually enough to fill three lines on the screen), and, when available, formulae, recurrences, generating functions, references, computer code for producing the sequence, links to relevant web sites, etc.

Let me illustrate how the database is used with a typical story. Last summer the following question arose at AT&T Labs in connection with a quantization problem [90]. Given an n-dimensional lattice Λ, for which integers N does Λ have a sublattice of index N that is geometrically similar to N?

For the two-dimensional root lattice A_2, for example, it is easy to see that a necessary and sufficient condition is that N be of the form $a^2 + ab + b^2$.

However, for the four-dimensional lattice A_4 the situation is more complicated. The first thing we did was to run a computer search, which showed that A_4 has a similar sublattice of index $N = c^2$ if and only if c is one of the numbers

$$1, 4, 5, 9, 11, 16, 19, 20, 25, 29, 31, 36, \ldots$$

This turned out to be sequence A31363 in the database, with a reference to Baake [3], where it appears in an apparently different context as the indices of coincidence site sublattices in a certain three-dimensional quasicrystal. [3] identifies these numbers as those positive integers in which all primes congruent to 2 or 3 (modulo 5) appear to an even power. This was a very useful hint in getting started on our problem. We were able to show that this is the correct condition for the A_4 lattice, and to find analogous results for a number of other lattices [15]. (However, we have not yet found a direct connection between A_4 and the quasicrystal problem. Nevertheless, the occurrence of the same numbers in the two problems cannot be entirely coincidental.)

My files contain many similar stories. Some other examples can be found in Chapter 3 of [93].

Most of the applications however are less dramatic. One encounters a sequence in the middle of a calculation, perhaps

$$1, 1, 2, 3, 6, 10, 20, 35, 70, 126, 252, \ldots ,$$

and one wants to know quickly what it is — preferably a formula (in this case it is $\binom{n}{[n/2]}$) or generating function.

It is worth emphasizing a special case of this: the simplification of binomial coefficient sums. Powerful methods are available for simplifying such sums by computer [75], [77]. But if one is in a hurry, one can first try evaluating the initial terms and looking up the sequence in the table. E.g. the sum

$$a(n) = \sum_{k=0}^{n} \binom{2n-2k}{n-k}^2 \binom{2k}{k}$$

produces

$$1, 8, 88, 1088, 14296, 195008, \ldots$$

(A36917), and the entry in the table supplies a recurrence

$$n^3 a(n) = 16\left(n - \frac{1}{2}\right)(2n^2 - 2n + 1)a(n-1) - 256(n-1)^3 a(n-2)$$

and a reference ([77], p. ix, although the sum is misstated in the first printing). I have begun entering into the database all the sequences corresponding to the singly-indexed identities in Gould's table [32] of binomial coefficient identities.

A related application is in identifying arithmetic inequalities. Suppose you suspect that

$$\sigma(n) \geq d(n) + \phi(n), \quad \text{for} \quad n \geq 2\,, \tag{1}$$

where σ, d and ϕ are respectively the sum of divisors, number of divisors, and Euler totient functions. You could evaluate the sequence formed by the difference of the two sides, which for $n \geq 1$ is

$$-1, 0, 0, 2, 0, 6, 0, 7, 4, 10, 0, 18, 0, 14, 12, \ldots$$

(A46520). The table then points you to a reference ([73], §I.3.1) where this is stated as a theorem. (Again I would like to get more examples of such sequences.)

Another important use for the database is in finding out the present state of knowledge about some problem. A second story will illustrate this. The number of Latin squares of order n (Q_n) is given by sequence A315 (see Fig. 1). Computing Q_n is one of the famous hard problems in combinatorics.

n	Q_n
1	1
2	1
3	1
4	4
5	56
6	9408
7	16942080
8	535281401856
9	377597570964258816
10	7580721483160132811489280

Fig. 1. Sequence A315, the number of Latin squares of order n (McKay and Rogoyski [71]).

In 1991 Brendan McKay, at the Australian National University in Camberra, computed Q_{10} (which as can be seen is a rather large number). When he checked the database he found that the same value had recently been obtained by Eric Rogoyski of Cadence Design Systems in San Jose, California. As it turned out the two methods were similar but not identical, and he and Rogoyski ended up writing a joint paper [71], acknowledging the database for bringing them together.

3 Types of Sequences

The database contains sequences from all branches of science, including

- enumeration problems (combinatorics, graph theory, lattices, etc.)
- number theory (number of solutions to $x^2 + y^2 + z^2 = n$, etc.)
- game theory (winning positions, etc.)
- physics (paths on lattices, etc.)
- chemistry (sizes of clusters of atoms, etc.)
- computer science (number of steps to sort n things, etc.)
- communications (m-sequences, weight distributions of codes, etc.)
- puzzles
- etc.

To be accepted into the database, a sequence must be well-defined and interesting. However, when in doubt, my tendency is to accept rather than to reject. The amazing coincidences of the Monstrous Moonshine investigations [14] make it difficult to say that a particular sequence, no matter how obscure, will never be of interest.

Readers are urged to send me any sequences they come across that are not at present in the database. There is a convenient electronic form for this purpose in [92]. Some of the reasons for sending in your sequence are as follows:

- this stakes your claim to it
- your name is immortalized
- the next person who comes across it will be grateful and, not least,
- you may benefit from this yourself, when you come across the same sequence some weeks from now.

Often one finds that a particular project may involve dozens of sequences, all variants of a few basic ones. Ideally you should send them all to the database!

4 Hard sequences

One of the keywords used in the database is "hard", which indicates that the term following those given is not known. Besides the Latin squares problem mentioned above, some other classic hard sequences are the following:

Projective planes. The number of projective planes of orders $n = 2, 3, \ldots, 10$ (A1231):

$$1, 1, 1, 1, 0, 1, 1, 4, 0,$$

where the last term refers to the result of Clement Lam et al. [58] (completing work begun in [66]) that there is no projective plane of order 10.

The Poincaré conjecture. The number of differential structures on the n-sphere, for $n = 1, 2, \ldots, 16$ (A1676):

$$1, 1, 1?, 1, 1, 1, 28, 2, 8, 6, 992, 1, 3, 2, 16256, 2,$$

as given by Kervaire and Milnor [41]. The *Poincaré conjecture* is that the third term is 1.

Dedekind's problem. The number of monotone Boolean functions of n variables, for $n = 1, 2, 3, \ldots, 8$ (A7153):

$$1, 4, 18, 166, 7579, 7828352, 2414682040996,$$
$$56130437228687557907786$$

where the last term was computed by Wiedemann [100]. (As is the case for many of these examples, there are several other versions of this sequence in the database.)

The Hadamard maximal determinant problem. What is the maximal determinant of an $n \times n$ $\{0, 1\}$-matrix? The values for $n = 1, 2, \ldots, 13$ are (A3432):

$$1, 1, 2, 3, 5, 9, 32, 56, 144, 320, 1458, 3645, 9477,$$

where the last two terms are due to Ehlich, and Ehlich and Zeller [25], [26]. For $n \equiv -1 \pmod 4$, Hadamard showed that the nth term is equal to

$$(n+1)^{(n+1)/2}/2^n\,,$$

provided that what is now called a Hadamard matrix of order n exists. In some cases conference matrices give the answer when $n \equiv +1 \pmod 4$, but the problem of finding the other terms in the sequence seems to have been untouched for over 35 years. It would be nice to have confirmation of the above values as well as some more terms!

Enumerating Hadamard matrices. The number of Hadamard matrices of orders $n = 4, 8, 12, \ldots, 28$ (A7299) is

$$1, 1, 1, 5, 3, 60, 487,$$

where the last entry is the work of Kimura [50], [51], [52], [53]. The *Hadamard conjecture* is that such a matrix always exists if n is a multiple of 4. Judging by how rapidly these numbers are growing, this should not be hard to prove, yet it has remained an open question for a century. Of course, as the example in Section 7 shows, such numerical evidence can be misleading.

The kissing number problem. How many spheres can touch another sphere of the same size? For arrangements that occur as part of a lattice packing, the answers are known for $n = 1, 2, \ldots, 9$ (A1116):

$$2, 6, 12, 24, 40, 72, 126, 240, 272,$$

the last term being due to Watson (see [16]). For nonlattice packings, all we know is

$$2, 6, 12, ?, ?, ?, ?, 240, \geq 306 \ .$$

The best bounds known in dimensions 4, 5, 6 and 7 are respectively

$$24\text{–}25,\ 40\text{–}46,\ 72\text{–}82 \quad \text{and} \quad 126\text{–}140$$

— see [16] for further information.

Sorting by prefix reversal. If you can only reverse segments that include the initial term of the current permutation, how many reversals are needed to transform an arbitrary permutation of n letters to the identity permutation? To state this another way [24]:

> The chef in our place is sloppy, and when he prepares a stack of pancakes they come out all different sizes. Therefore, when I deliver them to a customer, on the way to the table I rearrange them (so that the smallest winds up on top, and so on, down to the largest at the bottom) by grabbing several from the top and flipping them over, repeating this (varying the number I flip) as many times as necessary. If there are n pancakes, what is the maximum number of flips (as a function $f(n)$ of n) that I will ever have to use to rearrange them?

The only exact values known are $f(1), \ldots, f(9)$:

$$0, 1, 3, 4, 5, 7, 8, 9, 10,$$

due to Garey, Johnson and Lin, and Robbins (see [24], [29]). It is also known that $f(n) \geq n+1$ for $n \geq 6$, $f(n) \geq 17n/16$ if n is a multiple of 16 (so $f(32) \geq 34$), and $f(n) \leq (5n + 5)/3$, the last two bounds being due to Gates and Papadimitriou [29]. At present this sequence is not even in the database, since insufficient terms are known to distinguish it from other sequences with the same beginning. Again it would be nice to have more terms.

5 Recursive sequences

Whereas the sequences in the previous section enumerated some class of objects, the following are self-generated.

Differences = complement (A5228):

$$1, 3, 7, 12, 18, 26, 35, 45, 56, 69, 83, 98, 114, \ldots$$

The differences (A30124)

$$2, 4, 5, 6, 8, 9, 10, 11, 13, 14, 15, 16, 17, 19, \ldots$$

are the terms not in the sequence! This is one of many fine self-generating sequences from Hofstadter [38].

Golomb's sequence. The nth term is the number of times n appears (A1462):

$$1, 2, 2, 3, 3, 4, 4, 4, 5, 5, 5, 6, 6, 6, 6, 7, 7, 7, 7, 8, \ldots$$

The nth term is the nearest integer to (and converges to)

$$\tau^{2-\tau} n^{\tau-1} ,$$

where $\tau = (1+\sqrt{5})/2$ [31], [33, Section E25].

Wilson's primeth recurrence. a_{n+1} is the a_n-th prime (A7097), shown in Fig. 2. The sequence was sent in by R. G. Wilson V [102], and the last few terms were computed by P. Zimmermann [103] and M. Deléglise [18]. Their algorithm is a slightly speeded up version of an algorithm for computing $\pi(x)$, the number of primes not exceeding x, due to J. C. Lagarias, V. S. Miller and A. M. Odlyzko (see [57]). It is quite remarkable that it is possible to compute so many terms of this sequence.

Recamán's sequences. (i) $a_n = a_{n-1} - n$ if $a_{n-1} - n > 0$ and $a_{n-1} - n$ has not already occurred in the sequence, otherwise $a_n = a_{n-1} + n$ (A5132):

$$1, 3, 6, 2, 7, 13, 20, 12, 21, 11, 22, 10, 23, 9, \ldots$$

(ii) $a_{n+1} = a_n/n$ if n divides a_n, otherwise $a_{n+1} = na_n$ (A8336):

$$1, 1, 2, 6, 24, 120, 20, 140, 1120, 10080, \ldots$$

These were send in by B. Recamán [84]. How fast do they grow?

The $10,000 sequence. In a colloquium talk at AT&T Bell Labs [12], John Conway discussed the sequence (A4001)

$$1, 1, 2, 2, 3, 4, 4, 4, 5, 6, 7, 7, 8, 8, 8, 8, 9, \ldots$$

defined by (for $n \geq 3$)

$$a(n+1) = a(a(n)) + a(n+1-a(n)) .$$

(In words, $a(n+1)$ is the $a(n)$th term in from the left plus the $a(n)$th term in from the right.) This sequence seems to have been introduced by either David

1
2
3
5
11
31
127
709
5381
52711
648391
9737333
174440041
3657500101
88362852307
2428095424619
75063692618249
2586559730396077

Fig. 2. a_{n+1} is the a_n-th prime.

Newman or Douglas Hofstadter around 1986. In his talk, Conway said that he could prove that $\frac{a(n)}{n} \to \frac{1}{2}$, and offered $10,000 for finding the exact n at which $\left|\frac{a(n)}{n} - \frac{1}{2}\right|$ last exceeds $\frac{1}{20}$. My colleague Colin Mallows did not take long to analyze the sequence, and came up with an answer of 6083008742 [67].

Colin tells me that in fact the problem is actually much easier than either he or John Conway had believed, and the true answer is 1489. A recent paper by Kubo and Vakil [56] also studies this sequence and its generalizations. See also [33, Section E31].

Many variants of this sequence have not yet been analyzed. Even the rate of growth of this one is not known: $a(1) = a(2) = 1$, $a(n) = a(a(n-2)) + a(n - a(n-2))$, A5229 [67], which begins

$$1, 1, 2, 3, 3, 4, 5, 6, 6, 7, 7, 8, 9, 10, 10, 10, 11, 12, 12, \ldots$$

The Prague clock sequence. This is not really recursive in the same sense as the preceding sequences, but it seems to fit in here. It is, in any case, delightful. This is A28354: 1, 2, 3, 4, 32, 123, 43, 2123, 432, 1234, 32123, 43212, 34321, 23432, 123432, 1234321, 2343212, 3432123, 4321234, 32123432, 123432123, 43212343, 2123432123, 432123432, 1, 2, 3, 4, 32, ... According to [39], the sequence indicates how the astronomical clock in Prague strikes the hours. There is a single bell which (i) makes from 1 to 4 strokes at a time, (ii) the number of strokes follows the sequence

$$\ldots 3212343212343 \ldots,$$

(iii) at the nth hour, for $n = 1, 2, \ldots, 24$, the strokes add to n, and (iv) at the 25th hour there is a single stroke (so the sequence has period 24). As the reader will see by studying the sequence, its existence depends on two coincidences!

Cayley's mistake. Since the sequences in the database are numbered A1, A2, A3, ..., several people humorously proposed the "diagonal" sequence (now A31135) in which the nth term is equal to the nth term of An:

$$1, 2, 1, 0, 2, 3, 0, 6, 8, 4, 63, 1, 316, 42, 16, \ldots,$$

and the even less well-defined sequence (now A37181) with nth term equal to A31135(n) + 1. I resisted adding these sequences for a long time, partly out of a desire to maintain the dignity of the database, and partly because A22 was only known to 11 terms!

Sequence A22 gives the number of "centered hydrocarbons with n atoms", and is based on an 1875 paper of Cayley [9]. The paper is extremely hard to follow, and gives incorrect values for $n = 12$ and 13. The errors it contains were reproduced in [7], [91] and [93], even though Herrmann [37] had pointed out these errors in 1880. As far as we can tell, a correct verion of this sequence was never published until Eric Rains and I did so in 1999 [83]. Although Henze and Blair, Pólya and many others have written articles enumerating related families of chemical compounds (see [83] for a brief survey), this sequence seems to have been forgotten for over 100 years.

Once we determined what it was that Cayley was trying to enumerate, Pólya's counting theory quickly gave the answer, and the correct version of A22 is now in the database (as are the two diagonal sequences mentioned above – mostly because users of the database kept proposing them).

6 Meanders and stamp-folding

The meandric and stamp-folding numbers are similar to the better-known Catalan numbers (sequence A108) in that they are fundamental, easily described combinatorial quantities that arise in many different parts of mathematics, but differ from them in that there is no known formula, and in fact seem to be quite hard to compute.

The stamp-folding problem has a history going back to at least Lucas [63] in the nineteenth century, while the meandric problem seems to have been first mentioned by Poincaré [80]. The meandric problem asks: in how many different ways can a river (starting in the South-West and flowing East) cross a road n times? For $n = 5$ crossings there are $M_5 = 8$ possibilities, shown in Fig. 3. The sequence M_1, M_2, M_3, ... (A5316) begins

$$1, 1, 2, 3, 8, 14, 42, 81, 262, 538, 1828, \ldots$$

These are called *meandric* numbers, since the river *meanders* across the road. The even-numbered terms M_2, M_4, $M_6, \ldots$ give the number of different ways an

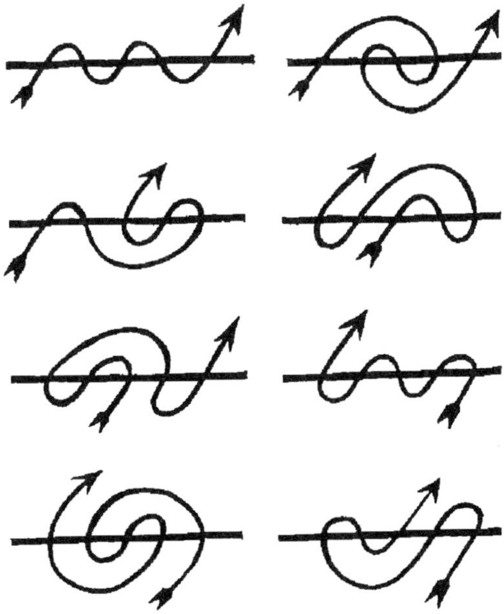

Fig. 3. The eight ways a river (going from South-West to North-East) can cross a road five times.

oriented line can cross a Jordan curve (A5315). There are several other interpretations, one of which is the number of "simple alternating transit mazes" [78], [79].

The stamp-folding problem asks the same question, but now the line is only semi-infinite. Equivalently, how many ways are there to fold a strip of n stamps? Pictures illustrating the first few terms of the stamp-folding sequence (A1011) can be found on the front cover of [91] and in Figure M4587 of [93]. The sequence begins

$$1, 1, 2, 5, 14, 38, 120, 353, 1148, 3527, \ldots$$

No polynomial-time algorithm is known for computing either sequence (on the other hand it is not known that such algorithms do not exist). The best algorithms known require on the order of nC_n steps, where $C_n = \frac{1}{n+1}\binom{2n}{n}$ is the nth Catalan number (A108): these algorithms are due to Koehler [55] for the stamp-folding problem and to Knuth and Pratt [54] and Reeds [85] for the meandric problem. They are too complicated to describe here.

Using these algorithms, Stephane Legendre [61] has extended the stamp-folding sequence to 26 terms and the meandric sequence to 25 terms. Knuth and Pratt [54] have computed 17 terms of the M_2, M_4, $M_6, \ldots$ subsequence. Lando and Zvonkin [59], [60] and Di Francesco, Golinelli and Guitter [21], [20], [22] have also studied these sequences.

The exact rate of growth of these sequences is not known. The best bounds for M_{2n} presently known appear to be due to Reeds, Shepp and McIlroy [85]. It is easy to see that M_{2n} is submultiplicative, and that $C_n \leq M_{2n} \leq C_n^2$, which implies that

$$\mu = \lim_{n\to\infty} M_{2n}^{1/n}$$

exists and satisfies $4 \leq \mu \leq 16$. In [85] it is shown that

$$8.8 \leq \mu \leq 13.01 .$$

Besides the papers already mentioned, the meandric and stamp-folding problems have recently been discussed by Arnold [1], Di Francesco [19], Harris [36], Lando and Zvonkin [59], [60], Lunnon [64], Sade [88], Smith [95], and Touchard [98].

7 Extremal codes and lattices

Let C be a binary linear self-dual code of length n in which the weight of every codeword is a multiple of 4, and let

$$W_C(x,y) = \sum_{c\in C} x^{n-wt(c)} y^{wt(c)}$$

be its weight enumerator. Examples are the Hamming code of length 8, with weight enumerator

$$f = x^8 + 14x^4y^4 + y^8 ,$$

and the Golay code of length 24, with weight enumerator

$$g = x^{24} + 759x^{16}y^8 + 2576x^{12}y^{12} + 759x^8y^{16} + y^{24} .$$

A remarkable theorem of Gleason says that the weight enumerator of any such code C is a polynomial in f and g. (References for this section are [16], [65], [82].) E.g. if C has length $n = 72$, its weight enumerator can be written as

$$W_C = a_0 f^9 + a_1 f^6 g + a_2 f^3 g^2 + a_3 g^3 ,$$

for rational numbers $a_0, \ldots, a_3$. If we choose these numbers so that the minimal distance of this (hypothetical) code C is as large as possible, we find that

$$W_C = 1 + 0x^4 + 0x^8 + 0x^{12} + 249849x^{16} + 18106704x^{20} + 462962955x^{24} + \cdots ,$$

so that C would have minimal distance 16 (the coefficients form sequence A18236). The coefficients in this "extremal" weight enumerator are all nonnegative integers, but whether a code exists with this weight enumerator is an important unsolved question.

One can perform this calculation for any length that is a multiple of 8, and in [70] it was shown that when as many initial terms as possible are set to zero, the next term in the extremal weight enumerator is always positive (so the minimal distance of the hypothetical extremal code is precisely $4[n/24]+4$).

Lengths that are multiples of 24 are especially interesting. Figure 4 shows the leading term in the extremal weight enumerator for a binary self-dual code of length $n = 24m$ (this is sequence A34414). It is "obvious" that these numbers

n	Coefficient
0	1
24	759
48	17296
72	249849
96	3217056
120	39703755
144	481008528
168	5776211364

Fig. 4. Number of codewords of minimal weight $n/6+4$ in extremal weight enumerator of length n.

are growing rapidly, and in fact it is shown in [70] that if $n = 24m$ then the leading coefficient is

$$\binom{24m}{5}\binom{5m-2}{m-1}\Big/\binom{4m+4}{5}.$$

The next term in the extremal weight enumerator (the number of codewords of weight $4[n/24]+8$) also grows rapidly — see Fig 5 (sequence A34415).

n	Coefficient
0	1
24	2576
48	535095
72	18106704
96	369844880
120	6101289120
144	90184804281
168	1251098739072

Fig. 5. Number of codewords of next-to-minimal weight $n/6+8$ in extremal weight enumerator of length n.

Again it is "obvious" that these numbers also grow rapidly, especially if one examines a more extensive table, where the number of digits continues to increase with each term. Yet — this shows the danger of drawing conclusions just from numerical data — it is proved in [69] that this sequence goes negative at $n = 3696$ and stays negative forever. (See Theorem 29 of [82] for more precise information.)

So codes corresponding to these extremal weight enumerators certainly do not exist for $n \geq 3696$ (since the weight enumerator of a genuine code must have nonnegative coefficients). They are known to exist for $n = 24$ (the Golay code) and 48 (a quadratic residue code), but every case from 72 to 3672 is open. For further details, and in particular for a description of the analogous situation when n is a multiple of 8 but not of 24, see [82].

The situation for lattices is similar. Consider an even unimodular lattice Λ of dimension n, with theta series

$$\Theta_\Lambda(q) = \sum_{u \in \Lambda} q^{u \cdot u} .$$

Examples are the eight-dimensional root lattice E_8, with theta series

$$\begin{aligned} f &= 1 + 240q^2 + 2160q^4 + 6720q^6 + \cdots \\ &= 1 + 240 \sum_{m=1}^{\infty} \sigma_3(m) q^{2m} , \end{aligned}$$

where $\sigma_3(m)$ is the sum of the cubes of the divisors of m, and the 24-dimensional Leech lattice, with theta series

$$\begin{aligned} g &= 1 + 196560q^4 + 16773120q^6 + 398034000q^8 + \cdots \\ &= f^3 - 720q^2 \prod_{m=1}^{\infty} (1 - q^{2m})^{24} . \end{aligned}$$

The Taylor series expansion of the last product defines the famous Ramanujan numbers. (The coefficients of f and g give sequences A4009 and A8408, respectively; the Ramanujan numbers form sequence A594.)

Then a theorem due essentially to Hecke says that the theta series of any such lattice Λ is a polynomial in f and g. We may then define extremal theta series just as we defined extremal weight enumerators. It is known [69] that the leading term in the extremal theta series is positive, but that again the next coefficient eventually becomes negative (even though the sequence begins with about 1700 exponentially growing terms). The dimension where the first negative coefficient appears is around 41000. The leading terms are shown in Fig. 6 (sequence A34597).

Lattices corresponding to the extremal theta series exist for $n = 24$ (the Leech lattice) and 48 (at least three inequivalent lattices exist), but every case from 72 to about 41000 is open. The 24- and 48-dimensional extremal lattices provide record kissing numbers in those dimensions; it would be nice to have a 72-dimensional example.

n	Coefficient
0	1
24	196560
48	52416000
72	6218175600
96	565866362880
120	45792819072000

Fig. 6. Number of shortest vectors u (with $u \cdot u = n/12 + 2$) in extremal theta series at dimension n.

8 Levine's sequence

In the summer of 1997 Lionel Levine [62] submitted a new sequence to the table, a sequence of such beauty that it immediately caught the attention of several of my colleagues. It is constructed via the array in Fig. 7. If a row of the array

1 1
1 2
1 1 2
1 1 2 3
1 1 1 2 2 3 4
1 1 1 1 2 2 2 3 3 4 4 5 6 7
1 1 1 1 1 1 1 2 2 2 2 2 2 3 3 3 3 3 4 4 4 4 5 5 5 5 6 6 6 7 7 7 8 8 9 9 10 11 12 13 14

Fig. 7. The array that produces Levine's sequence $1, 2, 2, 3, 4, 7, 14, \ldots$.

contains the numbers

$$a_1 \;\; a_2 \;\; a_3 \;\; \ldots \;\; a_{k-1} \;\; a_k$$

then the next row contains

$$a_k \text{ 1's}, \quad a_{k-1} \text{ 2's}, \quad a_{k-2} \text{ 3's}, \; \ldots$$

Levine's sequence (A11784) is obtained by taking the last term in each row:

$$\begin{aligned} &1, 2, 2, 3, 4, 7, 14, 42, 213, 2837, 175450, 139759600, 6837625106787, \\ &266437144916648607844, 508009471379488821444261986503540, \ldots \end{aligned} \tag{2}$$

The terms grow unexpectedly rapidly! The nth term L_n is

(i) the sum of the elements in row $n - 2$
(ii) the number of elements in row $n - 1$
(iii) the last element in row n
(iv) the number of 1's in row $n + 1$
...

Furthermore, if $s(n,i)$ denotes the sum of the first i elements in row n, then we have

(v) $L_{n+2} = s(n, L_{n+1})$

(vi) $L_{n+3} = \sum_{i=1}^{L_{n+1}} s(n,i)$

(vii) $L_{n+4} = \sum_{i=1}^{L_{n+1}} \binom{s(n,i)+1}{2}$.

The latter identity was found by Allan Wilks [101], who also found a more complicated formula for L_{n+5}, and used it to compute the last two terms shown in (2). No other terms are known!

As to the rate of growth, we have only a crude estimate. Bjorn Poonen and Eric Rains [81] showed that

$$\log L_n \sim c\tau^n \ , \qquad (3)$$

where $\tau = (1+\sqrt{5})/2$. *Sketch of Proof.* (a) $L_{n+2} \le L_{n+1}L_n$, and so $\log L_n$ is bounded above by a Fibonacci-like sequence. (b) The sum of the $(n+1)$st row is at most

$$\binom{\left[\frac{L_{n+2}}{L_n}\right]+1}{2} L_n \ ,$$

which implies

$$\frac{L_{n+3}}{2L_{n+2}} \ge \frac{L_{n+2}}{2L_{n+1}} \frac{L_{n+1}}{2L_n}$$

and so $\log(L_{n+1}/2L_n)$ is bounded below by a Fibonacci-like sequence.

Colin Mallows [68] has determined numerically that a reasonably good approximation to L_n is given by

$$\frac{1}{c_1} e^{c_2 \tau^n}$$

where $c_1 \approx 0.277$, $c_2 \approx 0.05427$. It would be nice to have better estimates for L_n, and one or more additional terms.

9 Arrays of Numbers

Besides number *sequences*, the database also contains *arrays* of numbers that have been converted to sequences. Triangular arrays are read by rows, in the obvious way. E.g. Pascal's triangle of binomial coefficients

1
1 1
1 2 1
1 3 3 1
1 4 6 4 1

becomes sequence A7318:

$$1, 1, 1, 1, 2, 1, 1, 3, 3, 1, 1, 4, 6, 4, 1, \ldots$$

Square arrays are read by antidiagonals, usually in this order:

$$\begin{array}{l} a_0\ a_2\ a_5\ a_9\ \ldots \\ a_1\ a_4\ a_8\ \ldots \\ a_3\ a_7\ \ldots \\ a_6\ \ldots \\ \ldots \end{array}$$

E.g. the Nim-addition table [11]

$$\begin{array}{l} 0\,1\,2\,3\,4 \ldots \\ 1\,0\,3\,2\,5 \ldots \\ 2\,3\,0\,1\,6 \ldots \\ 3\,2\,1\,0\,7 \ldots \\ \ldots \end{array}$$

becomes sequence A3987:

$$0, 1, 1, 2, 0, 2, 3, 3, 3, 3, 4, 2, 0, 2, 4, \ldots$$

Other classical arrays are the Stirling numbers of both kinds, Eulerian numbers, etc.

A less well-known array arises from **Gilbreath's conjecture**. This conjecture states that if one writes down the primes in a row, and underneath the absolute values of the differences, as in Fig. 8, then the leading terms (shown underlined) of all rows except the first are equal to 1 ([33], §A10). The corresponding sequence (A36262) is

$$2, 1, 3, 1, 2, 5, 1, 0, 2, 7, 1, 2, 2, 4, 11, 1, \ldots$$

Odlyzko [76] has verified the conjecture out to 3×10^{11}.

$$\begin{array}{l} 2\quad 3\quad 5\quad 7\quad 11\quad 13\quad 17\quad 19\quad 23\quad \ldots \\ \quad \underline{1}\quad 2\quad 2\quad 4\quad 2\quad 4\quad 2\quad 4\quad \ldots \\ \quad\quad \underline{1}\quad 0\quad 2\quad 2\quad 2\quad 2\quad 2\quad \ldots \\ \quad\quad\quad \underline{1}\quad 2\quad 0\quad 0\quad 0\quad 0\quad \ldots \\ \quad\quad\quad\quad \underline{1}\quad 2\quad 0\quad 0\quad 0\quad \ldots \\ \quad\quad\quad\quad\quad \underline{1}\quad 2\quad 0\quad 0\quad \ldots \\ \quad\quad\quad\quad\quad\quad \ldots \end{array}$$

Fig. 8. Gilbraith's conjecture is that the leading terms of all rows in this array except the first are always 1 (the top row contains the primes, subsequent rows are the absolute values of the differences of the previous row).

10 The Wythoff array

This array shown in Fig. 9 has many wonderful properties, some of which are mentioned here. I learned about most of these properties from John Conway [13], but this array has a long history — see Fraenkel and Kimberling [28], Kimberling [43], [44], [45], [46], [47], [48], [49], Morrison [74], and Stolarsky [96], [97]. It is related to a large number of sequences in the database (the main entry is A35513).

$$\begin{array}{rr|rrrrrrrrr}
0 & 1 & 1 & 2 & 3 & 5 & 8 & 13 & 21 & 34 & 55 \\
1 & 3 & 4 & 7 & 11 & 18 & 29 & 47 & 76 & .. & \\
2 & 4 & 6 & 10 & 16 & 26 & 42 & 68 & .. & & \\
3 & 6 & 9 & 15 & 24 & 39 & 63 & .. & & & \\
4 & 8 & 12 & 20 & 32 & 52 & .. & & & & \\
5 & 9 & 14 & 23 & 37 & 60 & .. & & & & \\
6 & 11 & 17 & 28 & 45 & 73 & .. & & & & \\
7 & 12 & 19 & 31 & 50 & 81 & .. & & & & \\
. & .. & .. & .. & .. & .. & .. & & & &
\end{array}$$

Fig. 9. The Wythoff array.

Construction (1). The two columns to the left of the vertical line consist respectively of the nonnegative integers n, and the *lower Wythoff sequence* (A201), whose nth term is $[(n+1)\tau]$. The rows are then filled in by the Fibonacci rule that each term is the sum of the two previous terms. The entry n in the first column is the *index* of that row.

Definition. The *Zeckendorf expansion* of a number n is obtained by repeatedly subtracting the largest possible Fibonacci number until nothing remains. E.g. $100 = 89+8+3 = F_{11}+F_6+F_4$. The *Fibonacci successor* to n, Sn, say, is found by replacing each F_i in the Zeckendorf expansion by F_{i+1}. E.g. the Fibonacci successor to 100 is $S100 = F_{12}+F_7+F_5 = 144+13+5 = 162$.

Construction (2). The two columns to the left of the vertical line in Fig. 9 read n, $1+Sn$; then after the vertical line the row continues

$$m \quad Sm \quad SSm \quad SSSm \quad SSSSm \quad \ldots,$$

where $m = n+1+Sn$.

Construction (3). Let $\{S1, S2, S3, \ldots\} = \{2, 3, 5, 7, 8, 10, 11, \ldots\}$ be the sequence of Fibonacci successors (A22342). The first column to the right of the line consists of the numbers not in that sequence: 1, 4, 6, 9, 12, ... (A7067). The rest of each row is filled in by repeatedly applying S.

Construction (4). The entry in row n and column k is

$$[(n+1)\tau F_{k+2}] + F_{k+1}n$$

(where $k = 0$ indicates the first column to the right of the vertical line).

Some properties of the array to the right of the line are the following:

(i) The first row consists of the Fibonacci sequence $1, 2, 3, 5, 8, \ldots$

(ii) Every row satisfies the Fibonacci recurrence.

(iii) The leading term in each row is the smallest number not found in any earlier row.

(iv) Every positive integer appears exactly once.

(v) The terms in any row or column are monotonically increasing.

(vi) Every positive Fibonacci-type sequence (i.e. satisfying $a(n) = a(n-1) + a(n-2)$ and eventually positive) appears as some row of the array.

(vii) The terms in any two adjacent rows alternate.

There are infinitely many arrays with Properties 1–7, see [47].

The nth term of the *vertical para-Fibonacci sequence*

$$0, 0, 0, 1, 0, 2, 1, 0, 3, 2, 1, 4, 0, 5, 3, 2, 6, 1, 7, 4, 0, 8, 5, \ldots$$

(A019586) gives the index (or parameter) of the row of the Wythoff array that contains n. This sequence also has some nice fractal-like properties:

(a) If you delete the first occurrence of each number, the sequence is unchanged. Thus if we delete the underlined numbers from

$$\underline{0}, 0, 0, \underline{1}, 0, \underline{2}, 1, 0, \underline{3}, 2, 1, \underline{4}, 0, \underline{5}, 3, 2, \underline{6}, 1, \underline{7}, 4, 0, \underline{8}, 5, \ldots$$

we get

$$0, 0, 0, 1, 0, 2, 1, 0, 3, 2, 1, 4, 0, 5, 3, 2, 6, 1, 7, 4, 0, 8, 5, \ldots$$

again!

(b) Between any two consecutive 0's we see a permutation of the first few positive integers, and these nest, so the sequence can be rewritten as (read across the rows):

0												
0												
0								1				
0					2			1				
0			3		2			1			4	
0		5	3		2		6	1		7	4	
0	8	5	3	9	2	10	6	1	11	7	4	12

The nth term of the *horizontal para-Fibonacci sequence*

$$1, 2, 3, 1, 4, 1, 2, 5, 1, 2, 3, 1, 6, 1, 2, 3, 1, 4, 1, 2, 7, 1, 2, \ldots$$

(A035612) gives the index (or parameter) of the column of the Wythoff array that contains n. This sequence also has some nice properties.

I hope I have said enough to convince you that the Wythoff array is well worth studying and full of surprises.

11 The Boustrophedon transform

The Taylor series for $\sin x$ and $\cos x$ are easily remembered, but most people have trouble with

$$\tan x = 1x + 2\frac{x^3}{3!} + 16\frac{x^5}{5!} + 272\frac{x^7}{7!} + \cdots ,$$

$$\sec x = 1 + 1\frac{x^2}{2!} + 5\frac{x^4}{4!} + 61\frac{x^6}{6!} + \cdots .$$

However, their coefficients can be calculated from the array in Fig. 10. The

$$\begin{array}{ccccccccccccccc}
 & & & & & & & 1 & & & & & & & \\
 & & & & & & 0 & & 1 & & & & & & \\
 & & & & & 1 & & 1 & & 0 & & & & & \\
 & & & & 0 & & 1 & & 2 & & 2 & & & & \\
 & & & 5 & & 5 & & 4 & & 2 & & 0 & & & \\
 & & 0 & & 5 & & 10 & & 14 & & 16 & & 16 & & \\
 & 61 & & 61 & & 56 & & 46 & & 32 & & 16 & & 0 & \\
0 & & 61 & & 122 & & 178 & & 224 & & 256 & & 272 & & 272 \\
 & & & & & & & \cdots & & & & & & &
\end{array}$$

Fig. 10. The secant-tangent triangle.

nonzero entries on the left are the *secant numbers* (A364):

$$1, 1, 5, 61, 1385, 50521, 2702765, \ldots,$$

and those on the right are the *tangent numbers* (A182)

$$1, 2, 16, 272, 7936, 353792, 22368256, \ldots$$

while the combination of the two sequences (A111):

$$1, 1, 1, 2, 5, 16, 61, 272, 1385, 7936, \ldots \tag{4}$$

are usually called the *Entringer numbers.* The latter count permutations of $\{1, 2, \ldots, n\}$ that alternately fall and rise.

This array is filled by a rule somewhat similar to that for Pascal's triangle: the rows are scanned alternately from right to left and left to right, the leading entry in each row is 0, and every subsequent entry is the sum of the previous entry in the same row and the entry above it in the previous row. (This is the *boustrophedon* or "ox-plowing" rule.) The earliest reference I have seen to this triangle is Arnold [2], who calls it the Euler-Bernoulli triangle. However, it may well be much older origin. [72] gives many other references.

Richard Guy [34] observed that if the entries at the beginnings of the rows are changed from $1, 0, 0, \ldots$ to say $1, 1, 1, 1, 1, \ldots$ or to $1, 2, 4, 8, 16, \ldots$ then the numbers that appear at the ends of the rows form interesting-looking sequences that were not to be found in [93], and asked if they had a combinatorial interpretation. Using $1, 1, 1, \ldots$ for example the triangle becomes

$$\begin{array}{ccccccccccc} & & & & & 1 & & & & & \\ & & & & 1 & & 2 & & & & \\ & & & 4 & & 3 & & 1 & & & \\ & & 1 & & 5 & & 8 & & 9 & & \\ & 24 & & 23 & & 18 & & 10 & & 1 & \\ 1 & & 25 & & 48 & & 66 & & 76 & & 77 \\ & & & & & \ldots & & & & & \end{array}$$

yielding the sequence (A667)

$$1, 2, 4, 9, 24, 77, 294, 1309, \ldots \tag{5}$$

We may regard this process as carrying out a transformation (the *boustrophedon transform*) of sequences: if the numbers at the beginnings of the rows are a_0, a_1, $a_2, \ldots$ (the input sequence) then the numbers at the ends of the rows, b_0, b_1, $b_2, \ldots$ (say) are the output sequence. In [72] we showed that there is a simple relationship between the input and output sequences: their exponential generating functions

$$\mathcal{A}(x) = \sum_{n=0}^{\infty} a_n \frac{x^n}{n!}, \quad \mathcal{B}(x) = \sum_{n=0}^{\infty} b_n \frac{x^n}{n!}$$

are related by

$$\mathcal{B}(x) = (\sec x + \tan x)\mathcal{A}(x) .$$

We also give a combinatorial interpretation of the $\{b_n\}$. E.g. in (5), b_n is the number of up-down subsequences of $\{1, \ldots, n\}$, so that $b_3 = 9$ corresponds to $\emptyset, 1, 2, 3, 12, 13, 23, 132, 231$.

The Entringer sequence (4) then has the property that it shifts one place left under the boustrophedon transform. The lexicographically earliest sequence that shifts *two* places left under this transform (A661) is

$$1, 0, 1, 1, 2, 6, 17, 62, 259, 1230, 6592, \ldots$$

We do not know what this enumerates!

Many examples of similar "eigen-sequences" for other transformations of sequences can be found in Donaghey [23], Cameron [8], and especially [4]. E.g. the sequence giving the number of planted achiral trees [30], [35] (A3238):

$$1, 1, 2, 3, 5, 6, 10, 11, 16, 19, 26, \ldots$$

has the property that it shifts left one place under the "inverse Möbius transformation" given by

$$b_n = \sum_{d|n} a_d .$$

12 Tchoukaillon solitaire (or Mancala, or Kalahari)

These are ancient board games, with hundreds of variants and many different names. The version to be described here is called Tchoukaillon solitaire. It has been studied by several authors (see for example Betten [5] and Broline and Loeb [6]). It is played on a board with a row of holes numbered $0, 1, 2, \ldots$ (see Fig. 11).

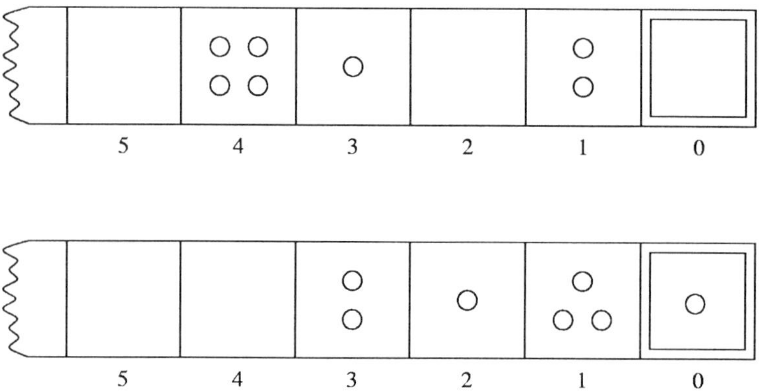

Fig. 11. A move in Tchoukaillon solitaire.

The game begins with n stones placed anywhere except in hole 0. A move consists in picking up the stones in some hole and placing one in each lower-numbered hole. If the last stone falls in hole 0 then play continues, otherwise the game is lost. The objective is to get all the stones into hole 0.

The game is interesting because there is a unique winning position for any number of stones. These winning positions are shown in Fig. 12 (sequence A28932), and can be found by playing the game backwards.

The array can be more explicitly constructed by the rule that if the first 0 in a row (counting from the right) is in position i, then the next row is obtained by writing i in position i and subtracting 1 from all earlier positions. The sequence of successive values of i (A28920) is

$$1, 2, 1, 3, 1, 4, 1, 2, 1, 5, 1, 6, 1, 2, \ldots$$

Let $t(k)$ denote the position where k occurs for the first time in this sequence. The values of $t(1)$, $t(2)$, $t(3), \ldots$ are (sequence A2491):

$$1, 2, 4, 6, 10, 12, 18, 22, 30, 34, 42, \ldots$$

This sequence has some very nice properties. It has been investigated by (in addition to the references mentioned above) David [17], Erdős and Jabotinsky [27] and Smarandache [94].

n	Position
0	0
1	1
2	20
3	21
4	310
5	311
6	4200
7	4201
8	4220
9	4221
10	53110
11	53111
12	642000
13	642001
..	...

Fig. 12. The unique winning position for n stones in Tchoukaillon solitaire.

(i) $t(n)$ can be obtained by starting with n and successively rounding up to the next multiple of $n-1, n-2, \ldots, 2, 1$. E.g. if $n = 10$, we obtain

$$10 \to 18 \to 24 \to 28 \to 30 \to 30 \to 32 \to 33 \to 34 \to 34 \ ,$$

so $t(10) = 34$.

(ii) The sequence can be obtained by a sieving process: write $1, 2, \ldots$ in a column. To get the second column, cross off 1 and every second number. To get the third column, cross off the first and every third number. Then cross off the first and every fourth number, and so on (see Fig. 13). The top number in column n is $t(n)$. Comparison of Figures 12 and 13 shows that connection with the solitaire game.

(iii) Finally, Broline and Loeb [6] (extending the work of the other authors mentioned) show that, for large n,

$$t(n) = \frac{n^2}{\pi} + O(n) \ .$$

It is a pleasant surprise to see π emerge from such a simple game.

References

1. V. I. Arnold, A branched covering of $\mathbb{CP}^2 \to S^4$, hyperbolicity and projective topology [Russian], *Sibir. Mat. Zhurn.* **29** (No. 2, 1988) 36–47 = *Siberian Math. J.* **29** (1988), 717–725.
2. V. I. Arnold, Bernoulli-Euler updown numbers associated with function singularities, their combinatorics and arithmetics, *Duke Math. J.* **63** (1991), 537–555.

1	~~1~~							
2	2	~~2~~						
3	~~3~~							
4	4	4	~~4~~					
5	~~5~~							
6	6	6	6	~~6~~				
7	~~7~~							
8	8	~~8~~						
9	~~9~~							
10	10	10	10	10	~~10~~			
11	~~11~~							
12	12	12	12	12	12	~~12~~		
13	~~13~~							
14	14	~~14~~						
15	~~15~~							
16	16	16	~~16~~					
17	~~17~~							
18	18	18	18	18	18	18	~~18~~	
19	~~19~~							
20	20	~~20~~						

Fig. 13. A sieve to generate the sequence $t(1), t(2), \ldots = 1, 2, 4, 6, 10, 12, 18, \ldots$. At stage n, the first number and every nth are crossed off.

3. M. Baake, Solution of the coincidence problem in dimensions $d \le 4$, pp. 199–237 of R. V. Moody, ed., *The Mathematics of Long-Range Periodic Order*, Kluwer, Dordrecht, 1997.
4. M. Bernstein and N. J. A. Sloane, Some canonical sequences of integers, *Linear Alg. App.* **226–228** (1995), 57–72.
5. D. Betten, Kalahari and the sequence "Sloane No. 377", *Ann. Discrete Math.* **37** (1988), 51–58.
6. D. M. Roline and D. E. Loeb, The combinatorics of Mancala-type games: Ayo, Tchoukaillon, and $1/\pi$, *J. Undergrad. Math. Applic.* **16** (1995), 21–36.
7. R. G. Busacker and T. L. Saaty, *Finite Graphs and Networks*, McGraw-Hill, NY, 1965, see p. 201.
8. P. J. Cameron, Some sequences of integers, *Discrete Math.* **75** (1989), 89–102.
9. A. Cayley, Ueber die analytischen Figuren, welche in der Mathematik Bäume genannt werden und ihre Anwendung auf die Theorie chemischer Verbindungen, *Ber. deutsch. chem. Ges.* **8** (1875), 1056–1059.
10. B. Cipra, Mathematicians get an on-line fingerprint file, *Science* **265** (22 July 1994), 473.
11. J. H. Conway, *On Numbers and Games*, Academic Press, NY, 1976.
12. J. H. Conway, Some crazy sequences, videotaped talk at AT&T Bell Labs, July 15, 1988.
13. J. H. Conway, personal communication, 1999.
14. J. H. Conway and S. P. Norton, Monstous moonshine, *Bull. London Math. Soc.* **11** (1979), 308–339.
15. J. H. Conway, E. M. Rains, N. J. A. Sloane, On the existence of similar sublattices, *Canad. J. Math.*, to appear.
16. J. H. Conway and N. J. A. Sloane, *Sphere Packing, Lattices and Groups*, Springer-Verlag, NY, 3rd. ed., 1998.

17. Y. David, On a sequence generated by a sieving process, *Riveon Lematematika* **11** (1957), 26–31.
18. M. Deléglise, Computation of large values of $\pi(x)$, preprint, 1996.
19. P. Di Francesco, Meander determinants, *Commun. Math. Phys.* **191** (1998), 543–583; LANL e-print hep-th/9612026.
20. P. Di Francesco, O. Golinelli and E. Guitter, Meanders: a direct enumeration approach, *Nucl. Phys.* **B 482** (1996), 497–535; LANL e-print hep-th/9607039.
21. P. Di Francesco, O. Golinelli and E. Guitter, Meander, folding and arch statistics, *Math. Comp. Modeling* **26** (1997), 97–147; LANL e-print hep-th/950607039.
22. P. Di Francesco, O. Golinelli and E. Guitter, Meanders and the Temperley-Lieb algebra, *Commun. Math. Phys.* **186** (1997), 1–59, LANL e-print hep-th/9602025.
23. R. Donaghey, Binomial self-inverse sequences and tangent coefficients, *J. Combin. Theory* **A 21** (1976), 155–163.
24. H. Dweighter, Problem E2569, *Amer. Math. Monthly* **82** (1975), 1010. Comments by M. R. Garey, D. S. Johnson and S. Lin, *loc. cit.* **84** (1977), 296.
25. H. Ehlich, Determinantenabschaetzungen für binäre Matrizen, *Math. Z.* **83** (1964), 123–132.
26. H. Ehlich and K. Zeller, Binäre Matrizen, *Zeit. Angew. Math. Mech.* **42** (1962), 20–21.
27. P. Erdős and E. Jabotinsky, On a sequence of integers generated by a sieving process, *Indag. Math.* **20** (1958), 115–128.
28. A. Fraenkel and C. Kimberling, Generalized Wythoff arrays, shuffles and interspersions, *Discrete Math.* **126** (1994), 137–149.
29. W. H. Gates and C. H. Padadimitriou, Bounds for sorting by prefix reversal, *Discrete Math.* **27** (1979), 47–57.
30. G. Gati, F. Harary and R. W. Robinson, Line colored trees with extendable automorphisms, *Acta Math. Scientia* **2** (1982), 105–110.
31. S. W. Golomb, Problem 5407, *Amer. Math. Monthly* **73** (1966), 674; **74** (1967), 740–743.
32. H. W. Gould, *Combinatorial Identities*, Morgantown, WV, 1972.
33. R. K. Guy, *Unsolved Problems in Number Theory*, Springer-Verlag, NY, 2nd ed., 1994.
34. R. K. Guy, personal communication, 1995.
35. F. Harary and R. W. Robinson, The number of achiral trees, *J. Reine Angewandte Math.* **278** (1975), 322–335.
36. M. G. Harris, A diagrammatic approach to the meander problem, LANL e-print hep-th/9807193.
37. F. Hermann, Ueber das Problem, die Anzahl der isomeren Paraffine der Formel C_nH_{2n+2} zu bestimmen, *Ber. deutsch. chem. Ges.* **13** (1880), 792.
38. D. R. Hofstadter, *Gödel, Fscher, Bach: An Eternal Golden Braid*, Vintage Books, NY, 1980.
39. D. Horský, *Pražský Orloj*, Panorama, Prague, 1988, pp. 76–78.
40. *Journal of Integer Sequences*, www.research.att.com/~njas/sequences/JIS/.
41. M. A. Kervaire and J. W. Milnor, Groups of homotopy sphere I, *Ann. Math.* **77** (1963), 504–537.
42. C. Kimberling, Problem 1615, *Crux Mathematicorum* **17** (No. 2, 1991), 44; **18** (1993), 82–83.
43. C. Kimberling, Orderings of the set of all positive Fibonacci sequences, in G. E. Bergum et al., editors, *Applications of Fibonacci Numbers* **5** (1993), 405–416.
44. C. Kimberling, Interspersions and dispersions, *Proc. Amer. Math. Soc.* **117** (1993), 313–321.

45. C. Kimberling, The first column of an interspersion, *Fib. Quart.* **32** (1994), 301–314.
46. C. Kimberling, Numeration systems and fractal sequences, *Acta Arithmetica* **73** (1995), 103–117.
47. C. Kimberling, Stolarsky interspersions, *Ars Combinatoria* **39** (1995), 129–138.
48. C. Kimberling, The Zeckendorf array equals the Wythoff array, *Fib. Quart.* **33** (1995), 3–8.
49. C. Kimberling, Fractal sequences and interspersions, *Ars Combinatoria* **45** (1997), 157.
50. H. Kimura, Hadamard matrices of order 28 with automorphism groups of order two, *J. Combin. Theory* **A 43** (1986), 98–102
51. H. Kimura, Classification of Hadamard matrices of order 28 with Hall sets, *Discrete Math.* **128** (1994), 257–268.
52. H. Kimura, Classification of Hadamard matrices of order 28, *Discrete Math.* **133** (1994), 171–180.
53. H. Kimura and H. Ohmori, Construction of Hadamard matrices of order 28, *Graphs Combin.* **2** (1986), 247–257.
54. D. E. Knuth and V. R. Pratt, personal communication, 1989.
55. J. E. Koehler, Folding a strip of stamps, *J. Combin. Theory* **5** (1968), 135–152.
56. T. Kubo and R. Vakil, On Conway's recursive sequence, *Discrete Math.* **152** (1996), 225–252.
57. J. C. Lagarias and A. M. Odlyzko, New algorithms for computing $\pi(x)$, pp. 176–193 of D. V. Chudnovsky et al., eds., *Number Theory (New York, 1982)*, Lecture Notes Math. **1052**, Springer-Verlag, NY.
58. C. W. H. Lam, L. Thiel, and S. Swiercz, The non-existence of finite projective planes of order 10, *Canad. J. Math.* **41** (1989), 1117–1123.
59. S. K. Lando and A. K. Zvonkin, Meanders, *Selecta Mathematica Sovietica* **11** (No. 2, 1992), 117–144.
60. S. K. Lando and A. K. Zvonkin, Plane and projective meanders, *Theoretical Computer Science* **117** (1993), 227–241.
61. S. Legendre, personal communication.
62. L. Levine, personal communication, 1997.
63. E. Lucas, *Théorie des Nombres*, Gauthier-Villars, Paris, 1891.
64. W. F. Lunnon, A map-folding problem, *Math. Comp.* **22** (1968), 193–199.
65. F. J. MacWilliams and N. J. A. Sloane, *The Theory of Error-Correcting Codes*, North-Holland, Amsterdam, 1977.
66. F. J. MacWilliams, N. J. A. Sloane, and J. G. Thompson, On the existence of a projective plane of order 10, *J. Combin. Theory* **A 14** (1973), 66–78.
67. C. L. Mallows, Conway's challenge sequence, *Amer. Math. Monthly* **98** (1991), 5–20; **99** (1992), 563–564.
68. C. L. Mallows, personal communication, 1997.
69. C. L. Mallows, A. M. Odlyzko and N. J. A. Sloane, Upper bounds for modular forms, lattices and codes, *J. Algebra* **36** (1975), 68–76.
70. C. L. Mallows and N. J. A. Sloane, An upper bound for self-dual codes, *Information and Control* **22** (1973), 188–200.
71. B. D. McKay and E. Rogoyski, Latin squares of order 10, *Electron. J. Combin.* **2** (1995), #N3.
72. J. Millar, N. J. A. Sloane and N. E. Young, A new operation on sequences: the boustrophedon transform, *J. Combin. Theory* **A 76** (1996), 44–54.
73. D. S. Mitrinović, J. Sándor and B. Crstici, *Handbook of Number Theory*, Kluwer, Dordrecht, 1996.

74. D. R. Morrison, A Stolarsky array of Wythoff pairs, in *A Collection of Manuscripts Related to the Fibonacci Sequence*, Fibonacci Assoc., Santa Clara, CA, 1980, pp. 134–136.
75. I. Nemes, M. Petkovšek, H. S. Wilf and D. Zeilberger, How to do Monthly problems with your computer, *Amer. Math. Monthly* **104** (1997), 505–519.
76. A. M. Odlyzko, Iterated absolute values of differences of successive primes, *Math. Comp.* **61** (1993), 373–380.
77. M. Petkovšek, H. S. Wilf and D. Zeilberger, $A = B$, Peters, Wellesley, MA, 1996.
78. A. Phillips, Simple alternating transit mazes, preprint, June 1988. A shorter version appears on the author's home page at www.math.sunysb.edu/~tony/mazes.
79. A. Phillips, La topologia dei labirinti, pp. 57–67 of M. Emmer, ed., *L'occhio di Horus: itinerari nell'imaginario matematico*, Istit. Enciclopedia Ital., Rome, 1989.
80. H. Poincaré, Sur un théorème de géométrie, *Rend. Circ. Mat. Palermo* **33** (1912), 375–407 = *Oeuvres*, VI, pp. 499–538.
81. B. Poonen and E. M. Rains, personal communication, 1997.
82. E. M. Rains and N. J. A. Sloane, Self-dual codes, pp. 177–294 of *Handbook of Coding Theory*, ed. V. S. Pless and W. C. Huffman, Elsevier, 1998.
83. E. M. Rains and N. J. A. Sloane, On Cayley's enumeration of alkanes (or 4-valent trees), *J. Integer Sequences* **2** (1999), Article 99.1.1.
84. B. Recamán, personal communication.
85. J. A. Reeds, L. A. Shepp and M. D. McIlroy, Numerical bounds for the Arnold "meander" constant, manuscript, 1991.
86. J. Riordan, *An Introduction to Combinatorial Analysis*, Wiley, NY, 1958.
87. J. Riordan and N. J. A. Sloane, The enumeration of rooted trees by total height, *J. Austral. Math. Soc.* **10** (1969), 278–282.
88. A. Sade, *Sur les Chevauchements des Permutations*, published by the author, Marseille, 1949.
89. *SeqFan Mailing List*, home page www.ccr.jussieu.fr/gmpib/seqfan/seqfan.html.
90. S. D. Servetto, V. A. Vaishampayan and N. J. A. Sloane, Multiple description lattice vector quantization, pp. 13–22 of *Proceedings DCC '99: Data Compression Conference (Snowbird, 1999)*, ed. J. A. Storer and M. Cohn, IEEE Computer Society Press, Los Alamitos, CA, 1999.
91. N. J. A. Sloane, *A Handbook of Integer Sequences*, Academic Press, NY, 1973.
92. N. J. A. Sloane, *The On-Line Encyclopedia of Integer Sequences*, published electronically at www.research.att.com/~njas/sequences.
93. N. J. A. Sloane and S. Plouffe, *The Encyclopedia of Integer Sequences*, Academic Press, 1995.
94. F. Smarandache, *Only Problems, Not Solutions!*, Xiquan. Publ., Phoenix and Chicago, 1993.
95. W. D. Smith, *Studies in Computational Geometry Motivated by Mesh Generation*, Ph.D. Dissertation, Dept. Appl. Math., Princeton Univ., Sept. 1988.
96. K. B. Stolarsky, Beatty sequences, continued fractions, and certain shift operators, *Canad. Math. Bull.* **19** (1976), 472–482.
97. K. B. Stolarsky, A set of generalized Fibonacci sequences such that each natural number belongs to exactly one, *Fib. Quart.* **15** (1977), 224.
98. J. Touchard, Contributions à l'étude du problème des timbres postes, *Canad. J. Math.* **2** (1950), 385–398.
99. G. N. Watson, Theorems stated by Ramanujan (V): approximations connected with e^x, *Proc. London Math. Soc.* **29** (1929), 293–308.
100. D. H. Wiedemann, personal communication.

101. A. R. Wilks, personal communication, 1997.
102. R. G. Wilson, V, personal communication.
103. P. Zimmermann, personal communication, 1988.

Complementary Interpolants and a Welch-Berlekamp-style Algorithm

Marc André Armand

Algebraic Coding Research Group,
Centre for Communications Research, University of Bristol, U.K.
Email: m.a.armand@bristol.ac.uk

Abstract. A new justification of a Welch-Berlekamp-style algorithm is given using the notion of complementary interpolants of [1]. We also show its relation to another algorithm which uses a more succinct formula for computing complementary interpolants. Both these algorithms can be used to solve the Welch-Berlekamp key equation and with simple modifications using the approach of [2, 3], may also be used to compute the linear complexity profile of a sequence.

Keywords: Weak rational interpolation, rank, complementary interpolants.

1 Introduction

Over the real or complex field, the rational interpolation problem may be viewed as that of finding a rational function with specified low order coefficients of its Taylor series expansion about certain (not necessarily distinct) points. In [2], the problem was generalized to arbitrary fields, including fields of finite characteristic, where the concept of Taylor series does not necessarily make sense, and was hence called the generalized rational interpolation problem. This problem concerns the "smallest" rational function P/Q with specified low order coefficients of its formal power series expansion about various points, (where "smallest" is taken to mean that the quantity $\max\{\deg P + 1, \deg Q\}$ is of the smallest possible value), and arises in Coding Theory as well as Cryptography. As noted in [2, 3], the generalized rational interpolation problem includes as a special case, the solution of the classical key equation in the decoding of Reed-Solomon ($\mathcal{RS}$) and $\mathcal{BCH}$ codes. Moreover, the interpretation of the solution of the key equation in the language of linear-feedback-shift-registers [7] also allows the problem of finding the linear complexity profile of a sequence in stream cipher applications to be viewed as a special case of the generalized rational interpolation problem. In [2], the weak generalized rational interpolation problem, a version of the generalized rational interpolation problem, was also formulated so as to avoid problems with singularities when studying the latter. This problem includes as a special case the solution to the Welch-Berlekamp (WB) key equation, ie. the polynomial pair (P, Q) of lowest complexity satisfying

$$s_i Q(x_i) = P(x_i)$$

for $i = 0, \ldots, L-1$ where the L points $x_0, \ldots, x_{L-1}$ are distinct. This is the key equation that appears in the remainder-decoding technique of $\mathcal{RS}$ codes of [9] and it is this special case that we will be focussing on in this paper.

Since the introduction of the WB algorithm of [9] to solve the WB key equation, several other authors have derived similar algorithms. See eg. [2, 4–6]. We give new insight into this much-studied problem by expanding the concept of complementary interpolants of [1], leading to a theoretically clear account of two different yet closely related iterative formulas (Algorithms 1 and 2) which can be used to solve the WB key equation. In particular, our minimality theorem, namely Theorem 1, indicates that the *rank* of an interpolant (P, Q), (given by the quantity $\max\{2\deg P + 1, 2\deg Q\}$), is a more accurate measure of its complexity than the above notion of complexity. Also, we define the main control parameter J of the algorithm of [4] as well as Algorithm 1 – see Definition 3. We remark that the analogue of this parameter also appears in the algorithms of [2, 5], but were not defined. Through our definition of J, we are readily able to prove its use as a control parameter in both Algorithm 1 and the algorithm of [4]. (See Theorem 2.)

2 Preliminaries

Let $\mathcal{F}$ be a field, $\mathcal{F}^* = \mathcal{F} \setminus \{0\}$ and assume all polynomials are in $\mathcal{F}[X]$ and all symbols in $\mathcal{F}$. Also, let λf and δf denote the leading term and degree of a non-zero polynomial f respectively. By convention, we take $\delta 0 = -\infty$.

For $L \geq 1$, let $(s, x|L)$ denote the sequence of pairs of points $(s_0, x_0), \ldots, (s_{L-1}, x_{L-1})$ where the x_i are assumed to be distinct. Note that with this assumption, L is upper-bounded by the order $|\mathcal{F}|$ of $\mathcal{F}$. We will assume this upper-bound on L throughout this paper.

Let the polynomial pair, (P, Q) (not both zero) denote a weak rational interpolant of $(s, x|L)$. For polynomials g and f, we write $(gP + fV, gQ + fW)$ as $g(P, Q) + f(V, W)$, for short.

Definition 1. *For $L \geq 1$, let* $\mathrm{WI}(s, x|L)$ *(*WI *– Weak Interpolant) be the set of all non-zero weak rational interpolants of $(s, x|L)$, ie.*

$$\mathrm{WI}(s, x|L) = \{(P, Q) : \ s_i Q(x_i) = P(x_i) \ \textit{for} \ 0 \leq i \leq L-1, (P, Q) \neq (0, 0)\}.$$

Moreover, we say that $(P, Q) \in \mathrm{WI}(s, x|L)$ is monic *if $\lambda Q = 1$.*

Clearly, for any polynomial k and $(P, Q) \in \mathrm{WI}(s, x|L)$, $k(P, Q)$ is in $\mathrm{WI}(s, x|L)$. In particular, for $L = 0$, there is no i in the range $0 \leq i \leq L-1$ so that the equation $s_i Q(x_i) = P(x_i)$ is satisfied vacuously. Hence, we define the set $\mathrm{WI}(s, x|0)$ as

$$\mathrm{WI}(s, x|0) = \mathcal{F}[X] \times \mathcal{F}[X] \setminus \{(0, 0)\}.$$

The following result is well-known and hence a proof is omitted.

Proposition 1. *For $L \geq 1$, let (P, Q), $(V, W) \in \mathrm{WI}(s, x|L)$. Then $x_0, \ldots, x_{L-1}$ are roots of $VQ - PW$.*

Suppose that $(P,Q) \in \mathrm{WI}(s,x|L)$ can be written as $k(F,G)$ where k is a polynomial, $\delta k \geq 1$. If $(F,G) \in \mathrm{WI}(s,x|L)$, then k is said to be *removable*, otherwise (P,Q) is said to be *irreducible*, [1].

The *complexity* $\kappa(P,Q)$ of $(P,Q) \in \mathrm{WI}(s,x|L)$ is $\max\{\delta P+1, \delta Q\}$. If the complexity of $(P,Q) \in \mathrm{WI}(s,x|L)$ is minimal amongst all polynomial pairs in $\mathrm{WI}(s,x|L)$, then we say that (P,Q) is a minimal weak rational interpolant of $(s,x|L)$. Let $\mathrm{MWI}(s,x|L)$ (MWI - *Minimal Weak Interpolant)* be the set of minimal weak rational interpolants of $(s,x|L)$. In particular, we can define the set $\mathrm{MWI}(s,x|0)$ as

$$\mathrm{MWI}(s,x|0) = \{(P,Q): \text{ all } (P,Q) \text{ such that } \kappa(P,Q) = 0\}.$$

Remark 1. Clearly, (P,Q) is irreducible if $(P,Q) \in \mathrm{MWI}(s,x|L)$. The converse however need not be true as indicated by Proposition 2(iii) and Theorem 1 below.

Definition 2. *For* $L \geq 0$, *the* rank $\mathrm{Rank}(P,Q)$ *of* $(P,Q) \in \mathrm{WI}(s,x|L)$ *is* $\max\{2\delta P+1, 2\delta Q\}$, *and* $(P,Q), (V,W) \in \mathrm{WI}(s,x|L)$ *are said to be* complementary *if* $\mathrm{Rank}(P,Q) + \mathrm{Rank}(V,W) = 2L+1$, *[1].*

For complementary interpolants $(P,Q), (V,W) \in \mathrm{WI}(s,x|L)$ with $\mathrm{Rank}(P,Q) < \mathrm{Rank}(V,W)$, it is trivial that $\mathrm{Rank}(P,Q) < 2L+1-\mathrm{Rank}(P,Q)$ so that $2\mathrm{Rank}(P,Q) \leq 2L$ and hence

$$\mathrm{Rank}(P,Q) < L+1 \leq \mathrm{Rank}(V,W).$$

For $L=0$, it is not difficult to verify the set of complementary interpolants to be

$$\{(0,f), (g,h): \text{ all } f,g \in \mathcal{F}^* \text{ and } h \in \mathcal{F}\}.$$

Proposition 2 states some basic properties of complementary interpolants.

Proposition 2. *For* $L \geq 0$, *suppose that* (P,Q), $(V,W) \in \mathrm{WI}(s,x|L)$ *are complementary. Then*
(i) supposing $\mathrm{Rank}(P,Q) < \mathrm{Rank}(V,W)$, *if* $\mathrm{Rank}(P,Q)$ *is even, then* $\delta P < \delta V$, *otherwise* $\delta Q < \delta W$,
(ii) $VQ - PW = u\prod_{i=0}^{L-1}(X - x_i)$ *for some* $u \in \mathcal{F}^*$.
(iii) (P,Q) *and* (V,W) *are both irreducible,*
(iv) if $\mathrm{Rank}(P,Q) < \mathrm{Rank}(V,W)$, *then* $Q \neq 0$.

Proof. (i) If $\mathrm{Rank}(P,Q)$ is even so that $\mathrm{Rank}(V,W)$ is odd, then $2\delta P+1 < \mathrm{Rank}(P,Q) < \mathrm{Rank}(V,W) = 2\delta V+1$, ie. $\delta P < \delta V$. If $\mathrm{Rank}(P,Q)$ is odd so that $\mathrm{Rank}(V,W)$ is even, then $2\delta Q < \mathrm{Rank}(P,Q) < \mathrm{Rank}(V,W) = 2\delta W$, ie. $\delta Q < \delta W$.

(ii), (iii) and (iv) are trivial for $L=0$ and so we only prove these parts for $L \geq 1$.

(ii) The proof is based on that of Lemma 1 of [1]. Let $\mathrm{Rank}(P,Q) = R_1$ and $\mathrm{Rank}(V,W) = R_2$. Then $1+2\delta P \leq R_1$, $2\delta Q \leq R_1$, $1+2\delta V \leq R_2$ and $2\delta W \leq R_2$. We assert that $\delta(VQ-PW) = \max\{\delta(VQ), \delta(PW)\} = L$. The proof is as follows: Firstly, $\delta(VQ) = \delta V + \delta Q \leq \frac{R_2-1}{2} + \frac{R_1}{2} = \frac{R_1+R_2-1}{2} = L$ with equality if Rank(P,Q) is even and Rank(V,W) is odd. Similarly, $\delta(PW) = \delta P + \delta W \leq \frac{R_1-1}{2} + \frac{R_2}{2} = \frac{R_1+R_2-1}{2} = L$ with equality if Rank(P,Q) is odd and Rank(V,W) is even. Then since the sum Rank(P,Q)+Rank(V,W) is odd, we have that either Rank(P,Q) even and Rank(V,W) odd or vice-versa, so that $\delta(VQ) \neq \delta(PW)$. It follows immediately that $\delta(VQ-PW) = \max\{\delta(VQ), \delta(PW)\} = L$. Next, from Proposition 1, the polynomial $VQ - PW$ has roots $x_0, \ldots, x_{L-1}$ so that it is a scalar multiple of $\prod_{i=0}^{L-1}(X - x_i)$ since $\delta(VQ - PW) = L$. Thus, $VQ - PW = u\prod_{i=0}^{L-1}(X - x_i)$ for some $u \in \mathcal{F}^*$.

(iii) Suppose that we can write (P,Q) as $k(F,G)$ where $k \in \mathcal{F}[X]$ with $\delta k \geq 1$ is a removable factor of (P,Q). Since $\delta(VQ - PW) = L$, we have that $VQ - PW = k(VG - FW)$ where $\delta(VG - FW) < L$. From Proposition 1, $VG - FW$ has roots $x_0, \ldots, x_{L-1}$ so that it is a multiple of $\prod_{i=0}^{L-1}(X - x_i)$. But $\delta(VG - FW) < L$ implies that $VG - FW = 0$. Yet, from (ii), we have that $u\prod_{i=0}^{L-1}(X - x_i) = VQ - PW = k(VG - FW)$ where $u \neq 0$. Thus, we have a contradiction. Likewise, (V,W) is irreducible.

(iv) Assuming Rank$(P,Q) <$ Rank(V,W), it suffices to consider the case Rank(P,Q) is odd, since if Rank(P,Q) is even, $\delta P < \delta Q$ which implies $Q \neq 0$.

Since $(P,Q) \in \mathrm{WI}(s,x|L)$, $Q = 0$ implies that $s_jQ(x_j) - P(x_j) = -P(x_j) = 0$ for $j = 0, \ldots, L-1$, ie. P has roots $x_0, \ldots, x_{L-1}$. Therefore, $\prod_{j=0}^{L-1}(X - x_j)|P$ so that $\delta P \geq L$. Yet, $2\delta P + 1 =$ Rank$(P,Q) \leq L$, ie. $\delta P \leq \frac{L-1}{2}$ - a contradiction. Thus, $Q \neq 0$. ‡

We remark that our definition of complementary interpolants differs from that of [1] in that Proposition 2(ii) forms part of their definition which strictly speaking, is superfluous. In the next section, we develop the theory pivotal to the formulation of Algorithms 1 and 2.

3 The general theory

For polynomials P_i and Q_i, we simply write $(P,Q)_i$ for (P_i, Q_i). We fix this notation for the remainder of this paper. Note that if $(P,Q)_i \in \mathrm{WI}(s,x|i)$, then for $0 \leq j < i$, $(P,Q)_i \in \mathrm{WI}(s,x|j)$.

Let $i \geq 1$. For $-1 \leq j \leq i-2$, let

$$M_{j,i-2} = \prod_{k=\max\{0,j\}}^{i-2} (X - x_k).$$

In particular, $M_{-1,-1} = 1$. In this way, $M_{-1,i-2} = M_{0,i-2}$ for all i. For $(P,Q)_j \in \mathrm{WI}(s,x|\max\{0,j\})$, $(P,Q)_j M_{j,i-2} \in \mathrm{WI}(s,x|i-1)$ since $M_{j,i-2}(x_k) = 0$ for $\max\{0,j\} \leq k \leq i-2$ which implies $(s_kQ_j(x_k) - P_j(x_k))M_{j,i-2}(x_k) = 0$ for $0 \leq k \leq i-2$. Also, let

$$c_{j,i} = (s_{i-1}Q_j(x_{i-1}) - P_j(x_{i-1}))M_{j,i-2}(x_{i-1})$$

and

$$d_i = s_{i-1}Q_{i-1}(x_{i-1}) - P_{i-1}(x_{i-1})$$

where $(P,Q)_{i-1} \in \mathrm{WI}(s,x|i-1)$. The quantity d_i is the obstruction to $(P,Q)_{i-1}$ also being an element of $\mathrm{WI}(s,x|i)$ and is thus called the *discrepancy* of $(P,Q)_{i-1}$ at the ith-position. Similarly, $c_{j,i}$ is called the *discrepancy* of $M_{j,i-2}(P,Q)_j \in \mathrm{WI}(s,x|i-1)$ at the ith-position. We note that it is possible for $c_{j,i}$ to be equal to zero.

Proposition 3. *Let $i \geq 1$. For all j, $-1 \leq j \leq i-1$, suppose that $(P,Q)_j \in \mathrm{WI}(s,x|\max\{0,j\})$. If $d_i \neq 0$, then for j, $-1 \leq j \leq i-2$, such that $\delta P_{i-1} \neq \delta(P_j M_{j,i-2})$ or $\delta Q_{i-1} \neq \delta(Q_j M_{j,i-2})$,*

$$(P,Q)_i = c_{j,i}(P,Q)_{i-1} - d_i M_{j,i-2}(P,Q)_j \in \mathrm{WI}(s,x|i).$$

Moreover, for $j \leq i-1$, $M_{j,i-1}(P,Q)_j \in \mathrm{WI}(s,x|i)$.

Proof. By imposing the restriction $\delta P_{i-1} \neq \delta(P_j M_{j,i-2})$ or $\delta Q_{i-1} \neq \delta(Q_j M_{j,i-2})$ for some j, $-1 \leq j \leq i-2$, it is clear that $(P,Q)_i \neq (0,0)$. Next, we show that $s_k Q_i(x_k) - P_i(x_k) = 0$ for $i \geq 2$ and $0 \leq k \leq i-2$. (Recall that the last relation is satisfied vacuously for $i = 1$.) For $0 \leq k \leq i-2$,

$$\begin{aligned} s_k Q_i(x_k) - P_i(x_k) = c_{j,i}(s_k Q_{i-1}(x_k) - P_{i-1}(x_k)) - \\ d_i(s_k Q_j(x_k) - P_j(x_k))M_{j,i-2}(x_k). \end{aligned}$$

Since $(P,Q)_{i-1} \in \mathrm{WI}(s,x|i-1)$, $s_k Q_{i-1}(x_k) - P_{i-1}(x_k) = 0$ for $0 \leq k \leq i-2$. Also, $(s_k Q_j(x_k) - P_j(x_k))M_{j,i-2}(x_k) = 0$ for $0 \leq k \leq i-2$ since $M_{j,i-2}(P,Q)_j \in \mathrm{WI}(s,x|i-1)$. Thus, $s_k Q_i(x_k) - P_i(x_k) = 0$ for $0 \leq k \leq i-2$. Finally, we show that $s_{i-1}Q_i(x_{i-1}) - P_i(x_{i-1}) = 0$:

$$\begin{aligned} s_{i-1}Q_i(x_{i-1}) - P_i(x_{i-1}) &= c_{j,i}(s_{i-1}Q_{i-1}(x_{i-1}) - P_{i-1}(x_{i-1})) \\ &\quad -d_i(s_{i-1}Q_j(x_{i-1}) - P_j(x_{i-1}))M_{j,i-2}(x_{i-1}) \\ &= c_{j,i}d_i - d_i c_{j,i} \\ &= 0. \end{aligned}$$

This completes the proof that $(P,Q)_i \in \mathrm{WI}(s,x|i)$. That $M_{j,i-1}(P,Q)_j$ is in $\mathrm{WI}(s,x|i)$ for $j \leq i-1$ is trivial. ‡

Lemma 1 was proven in [1] for monic interpolants, (P,Q), $(V,W) \in \mathrm{WI}(s,x|i)$. Obviously, any non-monic interpolant whose second vector component is non-zero, can be made monic by multiplying it by a suitable scalar quantity. Thus, by Proposition 2(iii) and (iv), the lemma also holds for non-monic interpolants.

Lemma 1. *Let $i \geq 1$. If (P,Q), $(V,W) \in \mathrm{WI}(s,x|i)$ are such that (P,Q) is irreducible and $\mathrm{Rank}(P,Q) + \mathrm{Rank}(V,W) \leq 2i$, then (V,W) can be reduced to (P,Q) by cancelling removable factors.*

With Lemma 1, we can prove the following theorem.

Theorem 1. *(Minimality) For $i \geq 0$, suppose that (P,Q), $(V,W) \in \mathrm{WI}(s,x|i)$ are complementary such that* $\mathrm{Rank}(P,Q) < \mathrm{Rank}(V,W)$. *Then*
(i) $(P,Q) \in \mathrm{MWI}(s,x|i)$ *and (P,Q) is unique (up to a non-zero scalar multiple),*
(ii) if $\mathrm{Rank}(P,Q) = i$ *and i is odd, then* (P,Q), $(V,W) \in \mathrm{MWI}(s,x|i)$.

Proof. For $i = 0$, (i) follows directly from the fact that $\kappa(M,N) = 0$ for $(M,N) \in \mathrm{MWI}(s,x|0)$. We next consider $i \geq 1$. Recall, a minimal solution is irreducible. Therefore, that (P,Q) is unique (up to a non-zero scalar multiple) if $(P,Q) \in \mathrm{MWI}(s,x|i)$ is a simple consequence of Lemma 1. We show that $(P,Q) \in \mathrm{MWI}(s,x|i)$:

Suppose $(M,N) \in \mathrm{MWI}(s,x|i)$ so that $\kappa(M,N) \leq \kappa(P,Q)$ and consider three cases: (a) $\mathrm{Rank}(M,N)$ odd, $\mathrm{Rank}(P,Q)$ even, (b) $\mathrm{Rank}(M,N)$ even, $\mathrm{Rank}(P,Q)$ odd, and (c) $\mathrm{Rank}(M,N)$ and $\mathrm{Rank}(P,Q)$ either both even or both odd. Then, using the fact that for any $(A,B) \in \mathrm{WI}(s,x|i)$, $\mathrm{Rank}(A,B) = 2\delta B = 2\kappa(A,B)$ if $\mathrm{Rank}(A,B)$ is even, or $\mathrm{Rank}(A,B) = 2\delta A + 1 = 2\kappa(A,B) - 1$ otherwise, we obtain the following:
(a) Here, $\kappa(M,N) \leq \kappa(P,Q)$ implies $\mathrm{Rank}(M,N) = 2\kappa(M,N) - 1 \leq 2\kappa(P,Q) - 1 = \mathrm{Rank}(P,Q) - 1$, ie. $\mathrm{Rank}(M,N) < \mathrm{Rank}(P,Q)$.
(b) Here, $\kappa(M,N) \leq \kappa(P,Q)$ implies $\mathrm{Rank}(M,N) = 2\kappa(M,N) \leq 2\kappa(P,Q) = \mathrm{Rank}(P,Q) + 1$ with equality if $\kappa(M,N) = \kappa(P,Q)$, ie. $(P,Q) \in \mathrm{MWI}(s,x|i)$.
(c) Here, $\kappa(M,N) \leq \kappa(P,Q)$ implies $\mathrm{Rank}(M,N) \leq \mathrm{Rank}(P,Q)$.

Since the theorem presupposes that $\mathrm{Rank}(P,Q) < \mathrm{Rank}(V,W)$, $\mathrm{Rank}(P,Q) \leq i$ so that with the exclusion of the case $\kappa(M,N) = \kappa(P,Q)$ in (b), we have that $\mathrm{Rank}(M,N) + \mathrm{Rank}(P,Q)$ is either less than or at most $2i$. By Lemma 1, (P,Q) can be reduced to (M,N), ie. $(P,Q) = k(M,N)$ for some non-zero polynomial k (since $(0,0) \notin \mathrm{WI}(s,x|i)$). But (P,Q) is part of a complementary pair and is irreducible, by Proposition 2(iii). This implies that k is a constant so that $(P,Q) \in \mathrm{MWI}(s,x|i)$ and this completes the proof for (i).

(ii) If $\mathrm{Rank}(P,Q) = i$ and i is odd, then $\delta P = \frac{i-1}{2} \geq \delta Q$ so that $\kappa(P,Q) = \frac{i+1}{2}$. Moreover, $\mathrm{Rank}(V,W) = i + 1$ and is even so that $\delta V < \delta W = \frac{i+1}{2} = \kappa(V,W) = \kappa(P,Q)$. From (i), it follows that $(P,Q), (V,W) \in \mathrm{MWI}(s,x|i)$ in this case. ‡

By definition, the sum of the ranks of a complementary pair is unique. The following result indicates that the absolute value of their difference in ranks is also unique.

Corollary 1. *For $i \geq 0$, suppose that (P,Q), $(V,W) \in \mathrm{WI}(s,x|i)$ are complementary such that* $\mathrm{Rank}(P,Q) < \mathrm{Rank}(V,W)$. *Then* $\mathrm{Rank}(V,W) - \mathrm{Rank}(P,Q)$ *is unique.*

Proof. Let (P,Q) and (V,W) be as defined in the corollary. From Theorem 1, we have that (P,Q) is unique (up to a scalar multiple). Therefore, for all other complementary interpolants (P',Q'), $(V',W') \in \mathrm{WI}(s,x|i)$ such that $\mathrm{Rank}(P',Q') < \mathrm{Rank}(V',W')$, we have that $\mathrm{Rank}(P',Q') = \mathrm{Rank}(P,Q)$ so that $\mathrm{Rank}(V',W') - \mathrm{Rank}(P',Q') = 2i + 1 - 2\mathrm{Rank}(P,Q)$ is unique, (since $\mathrm{Rank}(P,Q)$ is unique). ‡

Theorem 1 suggests that a feasible way to compute $(P, Q)_L \in \mathrm{MWI}(s, x|L)$ iteratively would be to begin with an initial complementary pair, and then update it in a way such that the updates are also complementary. Lemma 2 is pivotal to computing complementary interpolants.

Lemma 2. *For $i \geq 0$, let (P, Q), $(V, W) \in \mathrm{WI}(s, x|i)$ be complementary. Then for q, $r \in \mathcal{F}^*$,*

$$\mathrm{Rank}(q(P,Q) + r(V,W)) = \max\{\mathrm{Rank}(P,Q),\ \mathrm{Rank}(V,W)\}.$$

Proof. Without loss of generality, we assume that $\mathrm{Rank}(P,Q) < \mathrm{Rank}(V,W)$. Thus, we need to show that $\mathrm{Rank}(qP + rV,\ qQ + rW) = \mathrm{Rank}(V,W)$ for $q, r \in \mathcal{F}^*$. We have two cases to consider: (i) $\mathrm{Rank}(P,Q)$ even, and (ii) $\mathrm{Rank}(P,Q)$ odd.

(i) Since $\mathrm{Rank}(V,W)$ is odd, we have that $2\delta P + 1 < \mathrm{Rank}(P,Q) = 2\delta Q < \mathrm{Rank}(V,W) = 2\delta V + 1$ and hence $\delta(qP) = \delta P < \delta V = \delta(rV)$. Moreover, that $\mathrm{Rank}(V,W)$ is odd implies $\mathrm{Rank}(V,W) > 2\delta W$ and hence $\max\{2\delta(qQ) = 2\delta Q,\ 2\delta(rW) = 2\delta W\} < \mathrm{Rank}(V,W)$. Thus, $\mathrm{Rank}(qP + rV,\ qQ + rW) = \max\{2\delta(qP + rV) + 1 = 2\delta V + 1,\ 2\delta(qQ + rW) \leq \max\{2\delta(qQ),\ 2\delta(rW)\}\} = \mathrm{Rank}(V,W)$, as required. (ii) Since $\mathrm{Rank}(V,W)$ is even, we have that $2\delta Q < \mathrm{Rank}(P,Q) = 2\delta P + 1 < \mathrm{Rank}(V,W) = 2\delta W$ and hence $\delta(qQ) = \delta Q < \delta W = \delta(rW)$. Moreover, that $\mathrm{Rank}(V,W)$ is even implies $\mathrm{Rank}(V,W) > 2\delta V$ and hence $\max\{2\delta(qP) = 2\delta P,\ 2\delta(rV) = 2\delta V\} < \mathrm{Rank}(V,W)$. Thus, $\mathrm{Rank}(qP + rV,\ qQ + rW) = \max\{2\delta(qP + rV) + 1 \leq \max\{2\delta(qP),\ 2\delta(rV)\} + 1,\ 2\delta(qQ + rW) = 2\delta W\} = \mathrm{Rank}(V,W)$, as required. ‡

4 Deriving a WB-style algorithm

Here, we combine the results of Section 3 to yield a WB-style algorithm, namely Algorithm 1. The following definition introduces a variable which will serve as the main control parameter of this algorithm.

Definition 3. *For $i \geq 0$, suppose that $(E,F)_i, (G,H)_i \in \mathrm{WI}(s, x|i)$ are complementary such that $\mathrm{Rank}(E,F)_i < \mathrm{Rank}(G,H)_i$. We then define J_i as follows:*

$$J_i = (\mathrm{Rank}(G,H)_i - \mathrm{Rank}(E,F)_i - 1)/2.$$

Note that since $(E,F)_i, (V,W)_i \in \mathrm{WI}(s, x|i)$ are complementary such that $\mathrm{Rank}(E,F)_i < \mathrm{Rank}(G,H)_i$, we can easily deduce that $J_i = \mathrm{Rank}(G,H)_i - i - 1$ and since $\mathrm{Rank}(G,H)_i \geq i+1$, J_i is a non-negative integer. Further, by Corollary 1, J_i is independent of any complementary pair, and is rather, a function of i. In particular, by setting $(E,F)_0 = (0, f)$ and $(G,H)_0 = (g,h)$ for any $f, g \in \mathcal{F}^*$ and $h \in \mathcal{F}$, we have $J_0 = 0$.

The initial complementary pair: For convenience, we take $(E,F)_0 = (0,1)$ and $(G,H)_0 = (1,1)$ as the initial complementary pair.

Theorem 2 below gives iterative formulas for computing complementary interpolants $(E,F)_i$, $(G,H)_i \in \mathrm{WI}(s, x|i)$ for $i \geq 1$.

Theorem 2. *For $i \geq 1$, let $(E,F)_{i-1}$, $(G,H)_{i-1} \in \mathrm{WI}(s,x|i-1)$ be complementary such that* $\mathrm{Rank}(E,F)_{i-1} < \mathrm{Rank}(G,H)_{i-1}$, *and let $e_i = s_{i-1}F_{i-1}(x_{i-1}) - E_{i-1}(x_{i-1})$ and $g_i = s_{i-1}H_{i-1}(x_{i-1}) - G_{i-1}(x_{i-1})$.*

If $e_i = 0$ and

$$(E,F)_i = (E,F)_{i-1}$$
$$(G,H)_i = (G,H)_{i-1}(X - x_{i-1}),$$

then $J_i = J_{i-1} + 1$.

If $e_i \neq 0$, $J_{i-1} = 0$, and

$$(E,F)_i = g_i(E,F)_{i-1} - e_i(G,H)_{i-1}$$
$$(G,H)_i = (E,F)_{i-1}(X - x_{i-1}),$$

then $J_i = J_{i-1}$ and $\mathrm{Rank}(E,F)_i = \mathrm{Rank}(G,H)_{i-1}$.

If $i > 1$, $e_i \neq 0$, $J_{i-1} > 0$ (so that $i > 1$), and

$$(E,F)_i = (E,F)_{i-1}(X - x_{i-1})$$
$$(G,H)_i = g_i(E,F)_{i-1} - e_i(G,H)_{i-1},$$

then $J_i = J_{i-1} - 1$ and $\mathrm{Rank}(G,H)_i = \mathrm{Rank}(G,H)_{i-1}$.

Moreover, in all three cases $(E,F)_i$, $(G,H)_i \in \mathrm{WI}(s,x|i)$ are complementary such that $\mathrm{Rank}(E,F)_i < \mathrm{Rank}(G,H)_i$ *and so $(E,F)_i \in \mathrm{MWI}(s,x|i)$.*

Proof. Consider the case $e_i = 0$ and suppose $(E,F)_i$ and $(G,H)_i$ are as defined in the theorem for this case. Since $(E,F)_{i-1} \in \mathrm{WI}(s,x|i-1)$ and $(E,F)_i = (E,F)_{i-1}$, we have that $(E,F)_i \in \mathrm{WI}(s,x|i)$ and by Proposition 3, $(G,H)_i \in \mathrm{WI}(s,x|i)$. Then $J_i = (\mathrm{Rank}(G,H)_i - \mathrm{Rank}(E,F)_i - 1)/2 = (\mathrm{Rank}(G,H)_{i-1} + 2 - \mathrm{Rank}(E,F)_{i-1} - 1)/2 = J_{i-1} + 1$ with $\mathrm{Rank}(E,F)_i < \mathrm{Rank}(G,H)_i$.

Now, suppose that $e_i \neq 0$ and $J_{i-1} = 0$ and let $(E,F)_i$ and $(G,H)_i$ be as defined in the theorem for this case. Using Proposition 3 (by assigning $(P,Q)_{i-1} = (E,F)_{i-1}$, $(P,Q)_j M_{j,i-2} = (G,H)_{i-1}$, $d_i = e_i$ and $c_{j,i} = g_i$), we have that $(E,F)_i$ is in $\mathrm{WI}(s,x|i)$. Also, $(G,H)_i \in \mathrm{WI}(s,x|i)$ by the same proposition. Next, since the theorem presupposes that $\mathrm{Rank}(E,F)_{i-1} < \mathrm{Rank}(G,H)_{i-1}$, we have that $\mathrm{Rank}(E,F)_i = \mathrm{Rank}(G,H)_{i-1}$, by Lemma 2. Also, $\mathrm{Rank}(G,H)_i = \mathrm{Rank}(E,F)_{i-1} + 2$ and so $J_i = (\mathrm{Rank}(G,H)_i - \mathrm{Rank}(E,F)_i - 1)/2 = (\mathrm{Rank}(E,F)_{i-1} + 2 - \mathrm{Rank}(G,H)_{i-1} - 1)/2 = 0 = J_{i-1}$ since $J_{i-1} = 0$ implies that $\mathrm{Rank}(G,H)_{i-1} - \mathrm{Rank}(E,F)_{i-1} = 1$. Moreover, $J_i = 0$ implies that $\mathrm{Rank}(E,F)_i < \mathrm{Rank}(G,H)_i$.

Finally, consider the case $e_i \neq 0$ and $J_{i-1} > 0$ for $i > 1$ and let $(E,F)_i$ and $(G,H)_i$ be as defined in the theorem for this case. Using Proposition 3, we have that $(E,F)_i$ and $(G,H)_i$ are in $\mathrm{WI}(s,x|i)$. Next, since the theorem presupposes that $\mathrm{Rank}(E,F)_{i-1} < \mathrm{Rank}(G,H)_{i-1}$, we have that $\mathrm{Rank}(G,H)_i = \mathrm{Rank}(G,H)_{i-1}$ by Lemma 2. Also, $\mathrm{Rank}(E,F)_i = \mathrm{Rank}(E,F)_{i-1} + 2$. Then $J_i = (\mathrm{Rank}(G,H)_i - \mathrm{Rank}(E,F)_i - 1)/2 = (\mathrm{Rank}(G,H)_{i-1} - \mathrm{Rank}(E,F)_{i-1} - 2 - 1)/2 = J_{i-1} - 1$ with $\mathrm{Rank}(E,F)_i < \mathrm{Rank}(G,H)_i$, since $J_i \geq 0$.

In all three cases, it is easy to see that $\text{Rank}(E,F)_i + \text{Rank}(G,H)_i = 2i + 1$ so that $(E,F)_i$, $(G,H)_i$ are complementary and by Theorem 1, $(E,F)_i \in \text{MWI}(s,x|i)$. This completes the proof for the theorem. ‡

Remark 2. Since $\text{Rank}(G,H)_i = J_i+i+1$, it is easy to verify that $\text{Rank}(E,F)_i = i - J_i$. We note that these expressions are analogous to the assertions "$(G,H)_i$ lies in L_{i+J_i+1}" and "$(E,F)_i$ lies in L_{i-J_i}" in [4] respectively in the justification of their algorithm. (In [4], L_{i+J_i+1} and L_{i-J_i} denote line sets in a plane with Cartesian coordinates.)

Proposition 4 is not required in the derivation of Algorithm 1 below. Nevertheless, it comes in useful in proving Theorem 4.

Proposition 4. *For $i \geq 1$, let e_i and g_i be as defined in Theorem 2. Then e_i and g_i are not both zero.*

Proof. For $i \geq 1$, assume that $(E,F)_{i-1}$ and $(G,H)_{i-1}$ have been determined according to Theorem 2. To prove the proposition, it suffices to consider the case $e_i = 0$ in which $(G,H)_i = (G,H)_{i-1}(X - x_{i-1}) \in \text{WI}(s,x|i)$ and forms part of a complementary pair, by Theorem 2. Then if $g_i = 0$, this would imply that $(G,H)_{i-1} \in \text{WI}(s,x|i)$, meaning that $(X - x_{i-1})$ is a removable factor of $(G,H)_i$. Yet, by Proposition 2(iii), since $(G,H)_i$ forms part of a complementary pair, it is irreducible and so we have a contradiction. It follows that e_i and g_i are never both zero. ‡

We have thus verified Algorithm 1 below which is essentially the algorithm of [4]. Note however that we have suppressed the indices of $(E,F)_i$, $(G,H)_i$, J_i, e_i and g_i, since from Theorem 2, we only need their current values.

Algorithm 1. *(Weak Rational Interpolation of $(s,x|L)$)*
Input: $1 \leq L \leq |\mathcal{F}|$, *$\mathcal{F}$ a field, $(s,x|L)$ where $s_i, x_i \in \mathcal{F}$, x_i distinct.*
Output: $(E,F) \in \text{MWI}(s,x|L)$.

```
(E,F) := (0,1); (G,H) := (1,1); J := 0;

for i := 1 to L do
    { e := s_{i-1}F(x_{i-1}) - E(x_{i-1});
      if (e = 0) { (G,H) := (G,H)(X - x_{i-1});
                   J := J + 1; }
      else { g := s_{i-1}H(x_{i-1}) - G(x_{i-1});
             (G,H) := g(E,F) - e(G,H);
             (E,F) := (E,F)(X - x_{i-1});
             if (J = 0) { swap (E,F) and (G,H); }
             else { J := J - 1; } }
    }
return (E,F).
```

Example 1. Let $\mathcal{F} = \mathbb{Z}_7$. Applying Algorithm 1 to the input sequence $(s,x|6) = (3,1),(3,3),(6,2),(3,6),(3,4),(5,5)$, we obtain

i	e_i	g_i	$(E,F)_i$	$(G,H)_i$	J_i
0	–	–	$(0,1)$	$(1,1)$	0
1	3	2	$(4,6)$	$(0,X+6)$	0
2	0	–	$(4,6)$	$(0,X^2+3X+3)$	1
3	4	1	$(4X+6,6X+2)$	$(4,3X^2+2X+1)$	0
4	0	–	$(4X+6,6X+2)$	$(4X+4,3X^3+5X^2+3X+1)$	1
5	0	–	$(4X+6,6X+2)$	$(4X^2+2X+5,3X^4+4X^2+3X+3)$	2
6	1	1	$(4X^2+5,6X^2+4)$	$(3X^2+2X+1,4X^4+3X^2+3X+6)$	1

We will see from Theorem 4 below that Algorithm 1 can in fact be regarded as an expanded version of another algorithm, namely Algorithm 2 which we derive in Section 5.

5 Deriving a closely related algorithm

In this section, we give a different formula to Theorem 2 for iteratively computing complementary interpolants, leading to Algorithm 2. Its relation to Algorithm 1 will be given in Section 6. Throughout this section, let d_i, $c_{j,i}$ and $M_{j,i-2}$ be as defined in Section 3.

Theorem 3. *For $i \geq 1$, let $(P,Q)_j \in \mathrm{WI}(s,x|\max\{0,j\})$ for $-1 \leq j \leq i-1$ and suppose that $(P,Q)_{-1}$, $(P,Q)_0 \in \mathrm{WI}(s,x|0)$ are complementary. Put*

$$\alpha = \alpha(i-1) = \max_{-1 \leq j \leq i-2} \{j : \mathrm{Rank}(P,Q)_j \neq \mathrm{Rank}(P,Q)_{i-1}\}$$

and define

$$(P,Q)_i = \begin{cases} (P,Q)_{i-1} & \text{if } d_i = 0 \\ c_{\alpha,i}(P,Q)_{i-1} - d_i M_{\alpha,i-2}(P,Q)_\alpha & \text{otherwise.} \end{cases}$$

Then $(P,Q)_i$, $M_{\alpha(i),i-1}(P,Q)_{\alpha(i)} \in \mathrm{WI}(s,x|i)$ are complementary.

Proof. The proof is by induction on i. We note firstly that since $(P,Q)_{-1}$ and $(P,Q)_0$ are complementary, $\mathrm{Rank}(P,Q)_{-1} \neq \mathrm{Rank}(P,Q)_0$ so that $\alpha(0) = -1$ and $M_{\alpha(0),-1}(P,Q)_{\alpha(0)} = (P,Q)_{-1}$ and $(P,Q)_0$ are complementary. Next, we take $i \geq 1$. Note that since $\mathrm{Rank}(P,Q)_{-1} \neq \mathrm{Rank}(P,Q)_0$, $\alpha = \alpha(i-1)$ is well-defined and that $M_{\alpha,i-2}(P,Q)_\alpha$ is in $\mathrm{WI}(s,x|i-1)$, by Proposition 3. Thus we assume that $(P,Q)_{i-1} \in \mathrm{WI}(s,x|i-1)$ has been defined such that $(P,Q)_{i-1}$ and $M_{\alpha,i-2}(P,Q)_\alpha$ are complementary, and define $(P,Q)_i$ as in the theorem.

We assert that $(P,Q)_i \in \mathrm{WI}(s,x|i)$. From Proposition 2(i), we know that $\delta P_{i-1} \neq \delta(P_\alpha M_{\alpha,i-2})$ or $\delta Q_{i-1} \neq \delta(Q_\alpha M_{\alpha,i-2})$. If $d_i \neq 0$, it follows from Proposition 3 that $(P,Q)_i \in \mathrm{WI}(s,x|i)$. If $d_i = 0$, since $(P,Q)_{i-1} \in \mathrm{WI}(s,x|i-1)$ so that $(P,Q)_{i-1} \neq (0,0)$, it follows that $(P,Q)_i = (P,Q)_{i-1} \in \mathrm{WI}(s,x|i)$.

By Proposition 3, $M_{\alpha(i),i-1}(P,Q)_{\alpha(i)}$ is also in $\mathrm{WI}(s,x|i)$. It remains to show that $(P,Q)_i$, $M_{\alpha(i),i-1}(P,Q)_{\alpha(i)}$ are complementary, ie. that the sum of their ranks is equal to $2i+1$. Since $(P,Q)_{i-1}$ and $M_{\alpha,i-2}(P,Q)_\alpha$ are complementary, $\mathrm{Rank}(P,Q)_{i-1} \neq \mathrm{Rank}(P,Q)_\alpha M_{\alpha,i-2}$. Also, using Lemma 2, $\mathrm{Rank}(P,Q)_i$

is either equal to $\text{Rank}(P,Q)_{i-1}$ (if $d_i = 0$, or $d_i \neq 0$ and $c_{\alpha,i} \neq 0$ and $\text{Rank}(P,Q)_{i-1} > \text{Rank}(P,Q)_\alpha M_{\alpha,i-2}$) or $\text{Rank}(P,Q)_\alpha M_{\alpha,i-2}$ (if $d_i \neq 0$ and $c_{\alpha,i} = 0$, or $d_i \neq 0$ and $\text{Rank}(P,Q)_\alpha M_{\alpha,i-2} > \text{Rank}(P,Q)_{i-1}$). We divide the remainder of the proof into these two cases.

If $\text{Rank}(P,Q)_i = \text{Rank}(P,Q)_{i-1}$, then by definition, $\alpha(i) = \alpha(i-1) = \alpha$ and hence $M_{\alpha(i),i-1}(P,Q)_{\alpha(i)} = M_{\alpha,i-2}(P,Q)_\alpha(X - x_{i-1})$ and $\text{Rank}(P,Q)_i + \text{Rank}(P,Q)_{\alpha(i)}M_{\alpha(i),i-1} = \text{Rank}(P,Q)_{i-1} + \text{Rank}(P,Q)_\alpha M_{\alpha,i-2} + 2 = 2(i-1) + 1 + 2 = 2i+1$. If $\text{Rank}(P,Q)_i = \text{Rank}(P,Q)_\alpha M_{\alpha,i-2} (\neq \text{Rank}(P,Q)_{i-1})$, then $\alpha(i) = i-1$ and hence $M_{\alpha(i),i-1}(P,Q)_{\alpha(i)} = M_{i-1,i-1}(P,Q)_{i-1} = (P,Q)_{i-1}(X - x_{i-1})$ and $\text{Rank}(P,Q)_i + \text{Rank}(P,Q)_{\alpha(i)}M_{\alpha(i),i-1} = \text{Rank}(P,Q)_\alpha M_{\alpha,i-2} + \text{Rank}(P,Q)_{i-1}(X - x_{i-1}) = 2(i-1)+1+2 = 2i+1$.

Thus in both cases we see that $(P,Q)_i$ and $M_{\alpha(i),i-1}(P,Q)_{\alpha(i)}$ are complementary. This completes the proof for the theorem. ‡

Since Theorem 3 requires that $(P,Q)_0$, $(P,Q)_{-1} \in \text{WI}(s,x|0)$ are complementary, we arbitrarily fix $\text{Rank}(P,Q)_0 < \text{Rank}(P,Q)_{-1}$ and take $(P,Q)_0 = (0,1)$ and $(P,Q)_{-1} = (1,1)$, the same initial complementary pair used in Algorithm 1. (The reason for using identical initial complementary interpolants will be apparent in Theorem 4.)

We have thus justified Algorithm 2 below. As with Algorithm 1, note that we have suppressed the indices of $\alpha(i)$, d_i and $c_{\alpha,i}$ since we only need their current values.

Algorithm 2. *(Weak Rational Interpolation of $(s,x|L)$ – naive version)*
Input: $1 \leq L \leq |\mathcal{F}|$, $\mathcal{F}$ *a field,* $(s,x|L)$ *where* $s_i, x_i \in \mathcal{F}$, x_i *distinct.*
Output: $(P,Q)_L \in \text{MWI}(s,x|L)$.

```
α := −1; (P,Q)_0 := (0,1); (P,Q)_α := (1,1); M_{−1,−1} := 1;

for i := 1 to L do
    { d := s_{i−1}Q_{i−1}(x_{i−1}) − P_{i−1}(x_{i−1});
      if (d ≠ 0)
        { c := (s_{i−1}Q_α(x_{i−1}) − P_α(x_{i−1}))M_{α,i−2}(x_{i−1}); |* α = α(i − 1) *|
          (P,Q)_i := c(P,Q)_{i−1} − dM_{α,i−2}(P,Q)_α;
          if (Rank(P,Q)_{i−1} ≠ Rank(P,Q)_i) { α := i − 1; } } |* α = α(i) *|
      else { (P,Q)_i := (P,Q)_{i−1}; }
    }
if (Rank(P,Q)_L ≥ L + 1) { (P,Q)_L := M_{α,L−1}(P,Q)_α; } |* α = α(i) *|
return (P,Q)_L.
```

Remark 3. We remark that $\alpha(i)$ can be regarded as an antecedent of $\text{Rank}(P,Q)_i$ and is analogous to the definition of the antecedent of the complexity of an $\mathcal{F}$-sequence of length i in [8, Definition 3.12]. In fact, Algorithm 2 resembles Algorithm MP of [8] with its use of an analogous antecedent. As such, Algorithm 1, which does not require the use of α, may be viewed as an analogous expansion of Algorithm MP into Algorithm MR of [8], which dispenses with the use of the antecedent used in Algorithm MP.

Example 2. Fix $\mathcal{F} = \mathbb{Z}_7$. Applying Algorithm 2 to the input sequence of Example 1, we obtain

i	d_i	$c_{\alpha,i}$	$(P,Q)_i$	$\alpha(i)$
0	–	–	$(0,1)$	-1
1	3	2	$(4,6)$	0
2	0	–	$(4,6)$	0
3	4	1	$(4,3X^2+2X+1)$	2
4	2	0	$(6X+2,2X+3)$	3
5	0	–	$(6X+2,2X+3)$	3
6	5	1	$(X^2+3X+5,6X^4+X^2+X+2)$	5

i	$M_{\alpha(i),i-1}(P,Q)_{\alpha(i)}$
0	$(1,1)$
1	$(0,X+6)$
2	$(0,X^2+3X+3)$
3	$(4X+6,6X+2)$
4	$(4X+4,3X^3+5X^2+3X+1)$
5	$(4X^2+2X+5,3X^4+4X^2+3X+3)$
6	$(6X^2+4,2X^2+6)$

Comparing the complementary interpolants generated by both our algorithms for $1 \leq i \leq L$, we obtain

i	Observation
0	$(P,Q)_i = (E,F)_i$ and $M_{\alpha(i),i-1}(P,Q)_{\alpha(i)} = (G,H)_i$
1	$(P,Q)_i = (E,F)_i$ and $M_{\alpha(i),i-1}(P,Q)_{\alpha(i)} = (G,H)_i$
2	$(P,Q)_i = (E,F)_i$ and $M_{\alpha(i),i-1}(P,Q)_{\alpha(i)} = (G,H)_i$
3	$(P,Q)_i = (G,H)_i$ and $M_{\alpha(i),i-1}(P,Q)_{\alpha(i)} = (E,F)_i$
4	$(P,Q)_i = 5(E,F)_i$ and $M_{\alpha(i),i-1}(P,Q)_{\alpha(i)} = (G,H)_i$
5	$(P,Q)_i = 5(E,F)_i$ and $M_{\alpha(i),i-1}(P,Q)_{\alpha(i)} = (G,H)_i$
6	$(P,Q)_i = 5(G,H)_i$ and $M_{\alpha(i),i-1}(P,Q)_{\alpha(i)} = 5(E,F)_i$

The last table suggests a strong connection between Algorithms 1 and 2, motivating Section 6.

6 Relating the two algorithms

In this section, we establish the connection between Algorithms 1 and 2 suggested in Example 2. A useful tool to us here is Lemma 3 which was proven in [1].

Lemma 3. *For $i \geq 1$, suppose that $(P,Q),(V,W) \in \mathrm{WI}(s,x|i)$, both monic, are complementary. Then Q and W are relatively prime.*

With Lemma 3, we can prove Theorem 4 which states the relation between Algorithms 1 and 2.

Theorem 4. *For $i \geq 0$, let $(P,Q)_i$, $M_{\alpha(i),i-1}(P,Q)_{\alpha(i)} \in \mathrm{WI}(s,x|i)$ be as defined in Theorem 3 and let e_i and $(E,F)_i$, $(G,H)_i \in \mathrm{WI}(s,x|i)$ be as defined in Theorem 2. If $(P,Q)_0 = (E,F)_0$ and $(P,Q)_{-1} = (G,H)_0$, then for $i \geq 1$ and some constants $u, v \in \mathcal{F}^*$,*

$$(P,Q)_i = \begin{cases} u(E,F)_i & \text{if } J_{i-1} = 0 \text{ or } e_i = 0 \\ u(G,H)_i & \text{if } J_{i-1} > 0 \text{ and } e_i \neq 0 \end{cases}$$

and

$$M_{\alpha(i),i-1}(P,Q)_{\alpha(i)} = \begin{cases} v(G,H)_i & \text{if } J_{i-1} = 0 \text{ or } e_i = 0 \\ v(E,F)_i & \text{if } J_{i-1} > 0 \text{ and } e_i \neq 0. \end{cases}$$

Proof. The proof is by induction on i. Firstly, note that the theorem presupposes that $(P,Q)_0 = (E,F)_0$ and $(P,Q)_{-1} = (G,H)_0$. Recall that Theorem 2 presupposes that $(E,F)_0$ and $(G,H)_0$ form an initial complementary pair with $\mathrm{Rank}(E,F)_0 < \mathrm{Rank}(G,H)_0$. In this way, $(P,Q)_0$ and $(P,Q)_{-1}$ form an identical initial complementary pair to $(E,F)_0$ and $(G,H)_0$ with $\mathrm{Rank}(P,Q)_0 < \mathrm{Rank}(P,Q)_{-1}$. Next, we prove the theorem for $i = 1$. We consider two cases: $e_1 \neq 0$ and $e_1 = 0$. Since $(P,Q)_0 = (E,F)_0$ and $(P,Q)_{-1} = (G,H)_0$, we have $e_1 = d_1$ and $g_1 = c_{-1,1}$. Then if $e_1 \neq 0$, since $J_0 = 0$, using Theorem 2 we have $(E,F)_1 = g_1(E,F)_0 - e_1(G,H)_1 = c_{-1,1}(P,Q)_0 - d_1(P,Q)_{-1}$ and $(G,H)_1 = (E,F)_0(X - x_0) = (P,Q)_0(X - x_0)$. By Theorem 3, $(P,Q)_1 = c_{-1,1}(P,Q)_0 - d_1(P,Q)_{-1} = (E,F)_1$ so that by Lemma 2, $\mathrm{Rank}(P,Q)_1 = \mathrm{Rank}(P,Q)_{-1} \neq \mathrm{Rank}(P,Q)_0$ and hence $\alpha(1) = 0$, meaning $M_{\alpha(1),0}(P,Q)_{\alpha(1)} = (P,Q)_0(X - x_0) = (G,H)_1$. Now, if $e_1 = d_1 = 0$, then using Theorems 2 and 3, we have $(E,F)_1 = (E,F)_0 = (P,Q)_0 = (P,Q)_1$ and $(G,H)_1 = (G,H)_0(X - x_0) = (P,Q)_{-1}(X - x_0)$. Furthermore, since $\mathrm{Rank}(P,Q)_1 = \mathrm{Rank}(P,Q)_0$, $\alpha(1) = \alpha(0) = -1$ so that $M_{\alpha(1),0}(P,Q)_{\alpha(1)} = (P,Q)_{-1}(X - x_0) = (G,H)_1$. This completes the proof of the theorem for $i = 1$. Next, we take $i \geq 2$ and consider two cases: $J_{i-1} = 0$ and $J_{i-1} > 0$.

If $J_{i-1} = 0$, then either (i) $e_{i-1} \neq 0$ and $J_{i-2} = 0$ or (ii) $e_{i-1} \neq 0$ and $J_{i-2} = 1$, and if $J_{i-1} > 0$, then either (iii) $e_{i-1} = 0$ and $J_{i-2} \geq 0$ or (iv) $e_{i-1} \neq 0$ and $J_{i-2} > 1$.

For case (i), by Theorem 2, we have $(E,F)_{i-1} = g_{i-1}(E,F)_{i-2} - e_{i-1}(G,H)_{i-2}$ and $(G,H)_{i-1} = (E,F)_{i-2}(X - x_{i-2})$ and $J_{i-1} = 0$. For case (iii), using Theorem 2, we have $(E,F)_{i-1} = (E,F)_{i-2}$ and $(G,H)_{i-1} = (G,H)_{i-2}(X - x_{i-2})$ and $J_{i-1} = J_{i-2} + 1 > 0$. By Lemma 3, $\gcd(Q_{i-1}, Q_{\alpha(i-1)}M_{\alpha(i-1),i-2}) = 1$ so that $(X - x_{i-2}) \not| Q_{i-1}$. Thus, for cases (i) and (iii), we have that $(P,Q)_{i-1} = u(E,F)_{i-1}$ and $M_{\alpha,i-2}(P,Q)_\alpha = v(G,H)_{i-1}$ (where $\alpha = \alpha(i-1)$) for some $u, v \in \mathcal{F}^*$. Without loss of generality, we assume that $u = v = 1$. Then since Theorem 2 gives $\mathrm{Rank}(E,F)_{i-1} < \mathrm{Rank}(G,H)_{i-1}$, we have $\mathrm{Rank}(P,Q)_{i-1} < \mathrm{Rank}(P,Q)_\alpha M_{\alpha,i-2}$. Also, since $(P,Q)_{i-1} = (E,F)_{i-1}$ and $M_{\alpha,i-2}(P,Q)_\alpha = (G,H)_{i-1}$, we have $e_i = d_i$ and $g_i = c_{\alpha,i}$. We next consider two subcases: $e_i \neq 0$ and $e_i = 0$.

By Theorem 3, for $d_i = e_i \neq 0$, we have

$$(P,Q)_i = c_{\alpha,i}(P,Q)_{i-1} - d_i M_{\alpha,i-2}(P,Q)_\alpha = g_i(E,F)_{i-1} - e_i(G,H)_{i-1}.$$

Then by Theorem 2, for $e_i \neq 0$, we have for case (i),

$$(E,F)_i = g_i(E,F)_{i-1} - e_i(G,H)_{i-1} = (P,Q)_i$$

since $J_{i-1} = 0$, and for case (iii),

$$(G,H)_i = g_i(E,F)_{i-1} - e_i(G,H)_{i-1} = (P,Q)_i$$

since $J_{i-1} > 0$. Then by Lemma 2, since $\text{Rank}(P,Q)_{i-1} < \text{Rank}(P,Q)_\alpha M_{\alpha,i-2}$, we have that $\text{Rank}(P,Q)_i = \text{Rank}(P,Q)_\alpha M_{\alpha,i-2}$ and hence $\alpha(i) = i-1$ so that

$$M_{\alpha(i),i-1}(P,Q)_{\alpha(i)} = M_{i-1,i-1}(P,Q)_{i-1} = (E,F)_{i-1}(X - x_{i-1}).$$

Then by Theorem 2, for case (i), since $J_{i-1} = 0$,

$$(G,H)_i = (E,F)_{i-1}(X - x_{i-1}) = M_{\alpha(i),i-1}(P,Q)_{\alpha(i)}$$

and for case (iii), since $J_{i-1} > 0$,

$$(E,F)_i = (E,F)_{i-1}(X - x_{i-1}) = M_{\alpha(i),i-1}(P,Q)_{\alpha(i)}.$$

Now suppose that $d_i = e_i = 0$. By Theorems 2 and 3, for both cases (i) and (iii), we have

$$(P,Q)_i = (P,Q)_{i-1} = (E,F)_{i-1} = (E,F)_i$$

so that $\alpha(i) = \alpha$ and

$$M_{\alpha(i),i-1}(P,Q)_{\alpha(i)} = M_{\alpha,i-1}(P,Q)_\alpha = (G,H)_{i-1}(X - x_{i-1}) = (G,H)_i.$$

This completes the proof for cases (i) and (iii). In a similar way, we can prove the theorem for (ii) and (iv) and so we omit the proof for these remaining cases. ‡

Theorem 4 shows that provided Algorithms 1 and 2 use identical initial complementary interpolants, the subsequent complementary interpolants evaluated by either algorithm are simply scalar multiples of the other. For this reason, we may view Algorithm 2 as a naive version of Algorithm 1.

Acknowledgments: This research forms part of the author's Ph.D. dissertation at the University of Bristol supervised by Dr. Graham Norton. The author would like to thank the Centre for Communications Research for financial assistance, and Dr. Graham Norton, Dr. Tim Blackmore, Prof. Harald Niederreiter and the anonymous referee for their invaluable comments, suggestions and corrections. In particular, the author acknowledges a shorter proof of Theorem 1(ii) given by the referee.

References

1. Berlekamp, E. (1996). Bounded distance +1 soft-decision Reed-Solomon decoding. *IEEE Trans. Inform. Theory*, **42**-3, 704–720.
2. Blackburn, S.R. (1997). A generalized rational interpolation problem and the solution of the Welch-Berlekamp key equation. *Designs, Codes and Cryptography*, **11**-3, 223–234.
3. Blackburn, S.R. (1997). Fast rational interpolation, Reed-Solomon decoding, and the linear complexity profiles of sequences. *IEEE Trans. Inform. Theory*, **43**-2, 537–548.
4. Chambers, W.G., Peile, R.E., Tsie, K.Y., Zein, N. (1993). Algorithm for solving the Welch-Berlekamp key-equation, with a simplified proof. *Electronic Letters*, **29**-18, 1620–1621.
5. Jennings, S.M. (1995). Gröbner basis view of Welch-Berlekamp algorithm for Reed-Solomon codes. *IEEE Proc. Comm.*, **142**-6, 349–351.
6. Liu, T.H. (1984). A new decoding algorithm for Reed-Solomon codes. *PhD. Thesis*, University of Southern California, Los Angeles, CA.
7. Massey, J.L. (1969). Shift-register synthesis and BCH decoding. *IEEE Trans. Inform. Theory*, **15**, 122–127.
8. Norton, G.H. (1995). On the minimal realizations of a finite sequence. *J. Symbolic Computation* **20**, 93–115.
9. Welch, L., Berlekamp, E.R. (1983). Error correction for algebraic block codes. *US Patent*, 4 633 470.

Multiscale Coarse-graining Invariant Sequences

André Barbé

Katholieke Universiteit Leuven
Department of Electrical Engineering , Division SISTA/COSIC
Kardinaal Mercierlaan, 94, B-3001 Leuven / Belgium
Email: Andre.Barbe@esat.kuleuven.ac.be

Abstract. We introduce the concept of coarse-graining invariant sequences (in a commutative ring), and consider their relationship with Mahler equations and automatic sequences.

1 Introduction

Coarse-graining is a particular way of rescaling a pattern. It is a standard technique in renormalization, a concept used in the analysis of critical phenomena in physics [13], where, at the critical point, the coarse-grained versions of certain patterns are somehow similar to the original pattern. The patterns considered here are one-dimensional and two-dimensional sequences with values in a finite commutative ring $\mathcal{R}$ with unity. The two-dimensional sequences are orbits of certain one-dimensional linear cellular automata (LCA) that satisfy the so-called p-Fermat property ([11, 12, 3], see also further). p^s-coarse-graining invariant orbits of such p-Fermat LCA for which $\mathcal{R}$ is the ring of integers modulo p^d (p prime) were introduced earlier in [5–7]. The coarse-graining invariance (CGI) condition on the orbit-sequence could hereby be reduced to certain CGI-conditions concerning the one-dimensional initial state configuration sequence $\underline{a} = (a(k))_{k\in\mathbb{Z}}$ of the automaton. This condition is of the form

$$\forall k \in \mathbb{Z} \text{ and } \forall \sigma \in \{0,1,\ldots,p-1\}: \qquad \sum_{i=0}^{N} c_i a(kp+\sigma+i) = a(k+\kappa_\sigma), \quad (1)$$

where $c_i \in \mathcal{R}$ and $\kappa_\sigma \in \mathbb{Z}$.
We propose to call any sequence that satisfies such an equation "coarse-graining invariant". Denoting by $(a(k+\lambda))_{k\in\mathbb{N}}$ the right-side tail of $\underline{a}$ starting at position λ, it was shown for some particular case in [5], and more generally in [7], that the corresponding generating function $F_\lambda^+(X) \stackrel{\text{def}}{=} \sum_{k\in\mathbb{N}} a(k+\lambda)X^k$ satisfies –for some $\lambda \geq \lambda_0$– a so-called Mahler equation of the particular type

$$Q(X)F_\lambda^+(X) = P_\lambda(X)F_\lambda^+(X^{p^s}) + R_\lambda(X), \quad (2)$$

where $Q(X),\ P_\lambda(X),\ R_\lambda(X)$ are polynomials with coefficients in $\mathcal{R}$. When $\mathcal{R}$ is the ring of integers modulo p^d (p prime), this Mahler equation was used to prove

p-automaticity of $\underline{a}$ and of the corresponding CGI-orbit [7]. p-Automaticity of a sequence means that any element of the sequence can be obtained from a finite automaton whose input is the digit-sequence of the base-p representation of its position (or, equivalently, that the set of all possible p^k-decimations of the given sequence is finite; or still, that the sequence can be generated by a substitution of fixed length p; see for example [1–3], and also the contribution of Allouche and Shallit [4] in this book). Automaticity of sequences which satisfy a Mahler-type equation in $\mathbb{F}_q$, the finite field with q elements, was discussed in [8]. See also the contribution of von Haeseler and Jürgensen [10] in this book for other links between automaticity and Mahler equations.

Actually, the above-mentioned automaticity proof when $\mathcal{R}$ is the ring of integers modulo p^d, was based on a Mahler equation of the more general type

$$Q(X)F(X) = \sum_{i=1}^{s} P_i(X)F(X^{p^i}) + R(X), \tag{3}$$

with $Q(X)$, $P_i(X)$ and $R(X)$ polynomials.

A natural question that presents itself now, is how equation (1) should be generalized in order that its solutions also satisfy this more general Mahler equation (3). The answer will be given in this paper. But first, we recapitulate some basic definitions and facts.

2 Preliminaries

Let $\mathcal{R}$ be a commutative ring with unity.
A one-dimensional linear cellular automaton (LCA) over $\mathcal{R}$ consists of a bidirectionally infinite one-dimensional lattice of "*cells*", with natural addresses in $\mathbb{Z}$, where each of the cells is attributed a state $a \in \mathcal{R}$, and where the cell states are synchronously updated at discrete "time moments" $r \in \mathbb{N}$, according to some uniform local linear state evolution rule (or state transition rule). Formally, it can be considered as a triplet $(\Sigma_1, \underline{a}_0, A)$, where
• $\Sigma_1 = \mathcal{R}^{\mathbb{Z}}$ is the set of possible states of the CA,
• let $a(r,c) \in \mathcal{R}$, with $(r,c) \in \mathbb{N} \times \mathbb{Z}$, denote the state of the cell at position c at time r, then the state of the CA at time r is denoted by $\underline{a}_r = (a(r,c))_{c\in\mathbb{Z}} \in \Sigma_1$.
$\underline{a}_0 \stackrel{\text{def}}{=} (a(0,c))_{c\in\mathbb{Z}}$ is the *initial (state) configuration* of the CA.
$a(0,c)$ will further be written as $a(c)$.
• $A: \Sigma_1 \to \Sigma_1$ is the *linear* state transition rule (or CA-rule) which maps the CA-state at time r into the state at time $r+1$, as follows:

$$a(r+1,c) = \sum_{i=d_1}^{d_2} \rho_i a(r, c-i), \text{ where } \rho_i \in \mathcal{R}, \tag{4}$$

with $d_1, d_2 \in \mathbb{Z}$; $d_1 \leq d_2$.

The *orbit* of the CA is the map $\mathbb{N} \times \mathbb{Z} \to \mathcal{R} : (r,c) \mapsto a(r,c)$, and thus belongs to the set of two-dimensional sequences $\Sigma_2 = \mathcal{R}^{\mathbb{N}\times\mathbb{Z}}$. Such an orbit can be represented graphically as a pattern of colored cells on the regular square lattice associated with the set $\mathbb{N} \times \mathbb{Z}$ (as an example, see Fig. 1 for a partial view on the orbit of the CA with two states in $\mathbb{F}_2$ and with rule $a(r+1,c) = (a(r,c-1)+a(r,c)) \bmod 2$, without considering the tiling which will be introduced later). Iterating (4) gives

$$a(r+k,c) = \sum_{i=kd_1}^{kd_2} \rho(k,i)a(r,c-i), \tag{5}$$

where, for all $i \in \mathbb{Z}$:

$$\rho(k+1,i) = \sum_{j\in\mathbb{Z}} \rho(1,j)\rho(k,i-j), \tag{6}$$

with $\rho(1,i) = \rho_i$ when $d_1 \leq i \leq d_2$, else $\rho(1,i) = 0$.

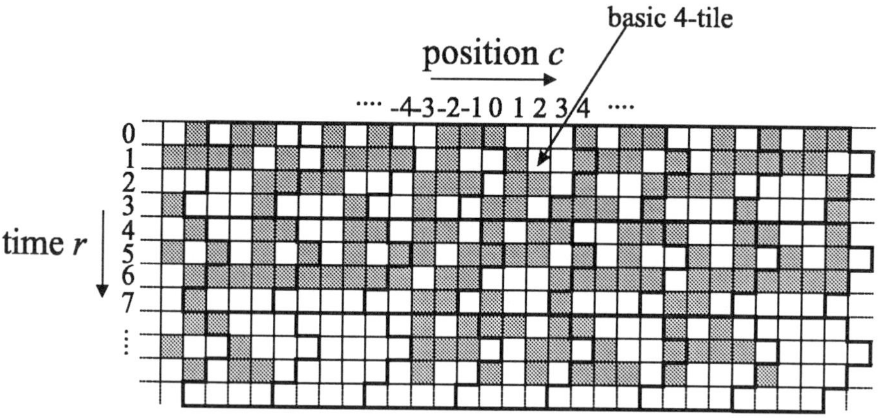

Fig. 1. Part of the two-dimensional orbit of a one-dimensional CA and the basic tiling from Fig. 2 ($\beta_0 = 0, \beta_1 = 1, \beta_2 = 0, \beta_3 = -1,\ \gamma_{i,j} = 0$ for $i,j = 0,1,2,3$). There are 3 other tilings: these are obtained by a horizontal shift of the tiling shown above over 1,2,3 cells respectively.

Another formalism uses the generating Laurent series $F_r(X) = \sum_{i\in\mathbb{Z}} a(r,i)X^i$ of the CA-state at time r, and the generating Laurent polynomial of the CA-rule: $A(X) = \sum_{i=-d_2}^{-d_1} \rho_{-i}X^i$. Then $A(X)^k = \sum_{i=-kd_2}^{-kd_1} \rho(k,-i)X^i$, and the equations (4) and (5) respectively translate into $F_{r+1} = A(X)F_r(X)$ and $F_{r+k} = A(X)^k F_r(X)$.

An LCA with generating Laurent polynomial $A(X)$ is called *m-Fermat* when $A(X)$ is m-Fermat, i.e., when

$$A(X)^m = A(X^m), \tag{7}$$

(see [11, 12, 3]). The following proposition allows the construction of p-Fermat polynomials over Z_{p^d}, the ring of integers modulo p^d (p prime).

Proposition 1. *(see[9, 11]) Let $L(X) \in Z_{p^d}[X]$, then $A(X) = L(X)^{p^d-1}$ mod p^d is a p-Fermat polynomial.*

z-tiling the lattice $\mathbb{N} \times \mathbb{Z}$

Let $[z] \stackrel{\text{def}}{=} \{0, 1, \ldots, z-1\}$ for $z \in \mathbb{N} \setminus \{0\}$.

A z-tile of the regular square lattice $\mathbb{N} \times \mathbb{Z}$ is any arrangement of cells in z consecutive rows and with z cells in each row, such that the lattice can be completely covered by shifting this tile in horizontal and vertical direction over cz and rz cells respectively ($c \in \mathbb{Z}, r \in \mathbb{N}$), without overlap. Any such tiling is defined by a *basic z-tile*, denoted by $T^{(0)}_{z,\Xi}(0,0) \stackrel{\text{def}}{=} \{i, j + \beta_i + \gamma_{i,j}z \mid i, j \in [z]\}$, where β_i and $\gamma_{i,j}$ are integers, and where Ξ is the $z \times z$-matrix with elements $\xi_{i,j} = \beta_i + \gamma_{i,j}z$. The matrix Ξ determines the shape of the tile. Fig. 2 displays a few examples of basic tiles of size $z = 4$ for different parameters. Observe that $\gamma_{i,j} \neq 0$ may lead to disconnected tiles.

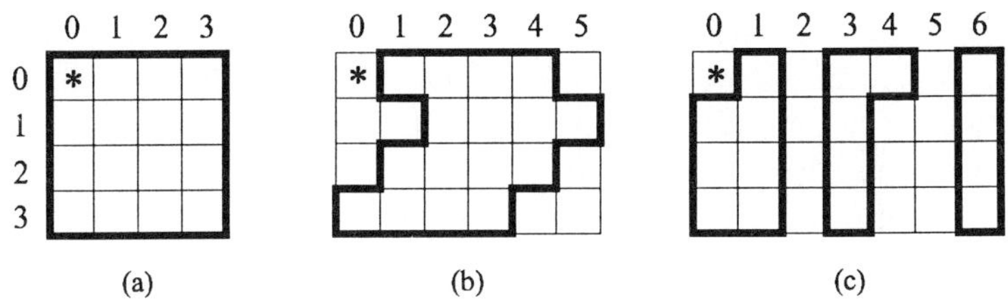

Fig. 2. Some basic tiles for $z = 4$. Tiles have thick boundary lines and contain the cells with addresses $(i, j + \beta_i + 4\gamma_{i,j})$, $i, j \in 0, 1, 2, 3$. The starred cell is the reference cell at position $(0, 0)$. The β_i and$\gamma_{i,j}$-parameters are, for (a): $\beta_i = 0$, $\gamma_{i,j} = 0$; for (b): $\beta_0 = 0$, $\beta_1 = 1$, $\beta_2 = 0$, $\beta_3 = -1$, $\gamma_{i,j} = 0$; for (c): $\beta_0 = \beta_1 = \beta_2 = \beta_3 = 0$, $\gamma_{0,0} = \gamma_{1,2} = \gamma_{2,2} = \gamma_{3,2} = 1$, all other $\gamma_{i,j} = 0$.

Shifting the basic tile over rz rows and cz columns produces a new tile denoted by $T^{(0)}_{z,\Xi}(r, c) \stackrel{\text{def}}{=} \{rz + i, cz + j + \xi_{i,j} \mid i, j \in [z]\}$, $(r, c) \in \mathbb{N} \times \mathbb{Z}$. The set of all these tiles is called the *basic tiling* $\mathcal{T}^{(0)}_{z,\Xi} = \{T^{(0)}_{z,\Xi}(r, c) \mid (r, c) \in \mathbb{N} \times \mathbb{Z}\}$ (a basic tiling of a CA orbit with the tile of Fig. 2(b) is shown in Fig. 1). For a given basic tile, there are z different tilings, each of which can be obtained by shifting the basic tiling horizontally over a distance σ, $\sigma \in [z]$. The σ-shifted tiling will be denoted by $\mathcal{T}^{(\sigma)}_{z,\Xi} = \{T^{(\sigma)}_{z,\Xi}(r, c) \mid (r, c) \in \mathbb{N} \times \mathbb{Z}\}$, where $T^{(\sigma)}_{z,\Xi}(r, c) = \{rz + i, cz + j + \xi_{i,j} + \sigma \mid i, j \in \mathbb{Z}\}$.

z-coarse graining of a CA orbit

Let $O_A(\underline{a}_0)$ be a CA-orbit. Then the $\mathcal{T}^{(\sigma)}_{z,\Xi}$-coarse-graining of this orbit [called the (z,σ)-coarse-graining for short] is the map

$$C^{(\sigma)}_{z,\Xi,W} : \Sigma_2 \to \Sigma_2 : O_A(\underline{a}_0) \mapsto C^{(\sigma)}_{z,\Xi,W} O_A(\underline{a}_0) :$$

$$(C^{(\sigma)}_{z,\Xi,W} O_A(\underline{a}_0))(r,c) = \sum_{i,j\in[z]} w_{i,j} a(rz+i, cz+j+\xi_{i,j}+\sigma),$$

where W is a $z\times z$-matrix with elements $w_{i,j}\in\mathcal{R}$, and where the arithmetic is the one proper to $\mathcal{R}$. This coarse-graining replaces each tile in $\mathcal{T}^{(\sigma)}_{z,\Xi}$ by a single cell, the state of which is a weighted sum of the states of the orbit cells covered by that tile.

We now formulate a crucial property.

Proposition 2. *(see [6]): Let $O_A(\underline{a}_0)$ be an orbit of an m-Fermat LCA with (linear) state transition rule A. Then its coarse-graining , $C^{(\sigma)}_{m^s,\Xi,W} O_A(\underline{a}_0)$ is also the orbit of an LCA with the same transition rule A, the initial state being given by the $C^{(\sigma)}_{m^s,\Xi,W}$- coarse-graining of the rows 0 to m^s-1 of the orbit $O_A(\underline{a}_0)$.*

m^s-coarse-graining invariance of a CA-orbit Let ${}^{\kappa}\underline{a}_0$, κ integer, denote the κ-shift of $\underline{a}_0$, i.e., ${}^{\kappa}\underline{a}_0 = (a(k+\kappa))_{k\in\mathbb{Z}}$. Then ${}^{\kappa}O_A(\underline{a}_0) = O_A({}^{\kappa}\underline{a}_0)$ is the horizontal κ-shift of the orbit $O_A(\underline{a}_0)$. All ${}^{\kappa}O_A(\underline{a}_0)$ with different κ will be considered equivalent (or equal modulo a shift). Then Proposition 2 above allows the formulation of the so-called *coarse-graining invariance problem* for m-Fermat CA: do orbits $O_A(\underline{a}_0)$ exist which –for a given m^s-tiling– are invariant under *all* (m^s,σ)-coarse-graining operations, i.e., for which

$$\boxed{C^{(\sigma)}_{m^s,\Xi,W} O_A(\underline{a}_0) = {}^{\kappa_\sigma}O_A(\underline{a}_0) \text{ for all } \sigma\in[m^s],} \tag{8}$$

with $m^s, \Xi, W, A, \kappa_\sigma$ fixed problem parameters, and κ_σ possibly depending on σ.

This question was positively answered in [6] for $\mathcal{R} = Z_p$, and in [7] for $\mathcal{R} = Z_{p^d}$, with $m=p$ (prime). It suffices to make the coarse-grainings of the first m^s rows of $O_A(\underline{a}_0)$ to be equal to ${}^{\kappa_\sigma}\underline{a}_0$ (this follows from Proposition 2 and from the fact that the orbit is completely determined by both the initial configuration $\underline{a}_0$ and the CA-rule). Making explicit all elements in these m^s first rows in function of the elements of $\underline{a}_0$ [invoking the CA rule (5)] allows the CGI-condition (8) to be solved for the initial configuration $\underline{a}_0$. In general, $\underline{a}_0$ satisfies the following set of m^s-*coarse-graining invariance* equations:

$$\forall k\in\mathbb{Z} \text{ and } \forall\sigma\in[m^s] : \sum_{i=0}^{N} c_i a(m^s k+\sigma+\rho+i) = a(k+\kappa_\sigma), \tag{9}$$

with coefficients $c_i\in\mathcal{R}$ and $\rho\in\mathbb{Z}$, and a parameter $N\in\mathrm{N}$, which depend on the problem parameters Ξ, W, A.

3 Multiscale coarse-graining invariant sequences

Definition 1. Let z be an integer ≥ 2. A *z^s-coarse-graining invariant sequence* $\underline{a} = (a(k))_{k\in\mathbb{Z}}$ (over $\mathcal{R}$) is a sequence that satisfies the following set of *coarse-graining invariance equations*:

$$\boxed{\forall\ k \in \mathbb{Z} \text{ and } \forall\sigma \in [z^s]: \ \sum_{i=0}^{N} c_i a(z^s k + \sigma + i) = b_\sigma a(k + \kappa_\sigma),} \tag{10}$$

with b_σ an element of $\mathcal{R}$, possibly dependent on σ, such that $b_\sigma \underline{a}$ is a one-to-one symbolic transcript of $\underline{a}$ (if this is not the case, we call the CGI-equations and corresponding sequences degenerate).

Remark 1. By a proper transformation, equation (9) can be brought in the form (10). According to the definition above, sequences will be called coarse-graining invariant, independently of whether they are associated with a CA or not. Indeed, by this definition, a CGI-sequence is such that for all $\sigma \in [z^s]$, its (z^s, σ)-coarse-graining over the *local neighbourhood* $\{\sigma+i | i = 0, \ldots, N\}$ [the left-side expression in (10)] equals the original sequence [the right side of (10)] when $b_\sigma = 1$, [or, when $b_\sigma \neq 1$, it equals a symbolic transcript of the original sequence when the equation is not degenerate]. The meaning of the CGI-equation (10) is illustrated in Fig. 3.

Recursions like (10) can be solved starting from a *seed*. This seed can be found using a procedure that was developed in [6] when both c_0 and c_N are units of $\mathcal{R}$ (this will be illustrated in Example 1 at the end of this Section). It was shown in [7] that under these conditions, the generating function associated with $\underline{a}$ satisfies a Mahler equation of the form (2). This in itself implies p-automaticity of $\underline{a}$ under the additional condition that $\mathcal{R}$ is the set of integers modulo p^d (p prime).

The CGI-equations (10) can also be interpreted as enforcing a linear dependence (over $\mathcal{R}$) between a sequence and its z^s-coarse-grained versions. This observation will be the basis for the following generalization which enforces linear dependence between coarse-grained versions of a sequence for different z^j-scales. The generalization is also such that it will lead to the more general Mahler equations of the form (3) [see further Propositions 4 and 5].

Definition 2. Let z be an integer ≥ 2. A *multiscale z^s-coarse-graining invariant sequence* $\underline{a} = (a(k))_{k\in\mathbb{Z}}$ (over $\mathcal{R}$) is a sequence that satisfies the following recursive CGI-equations

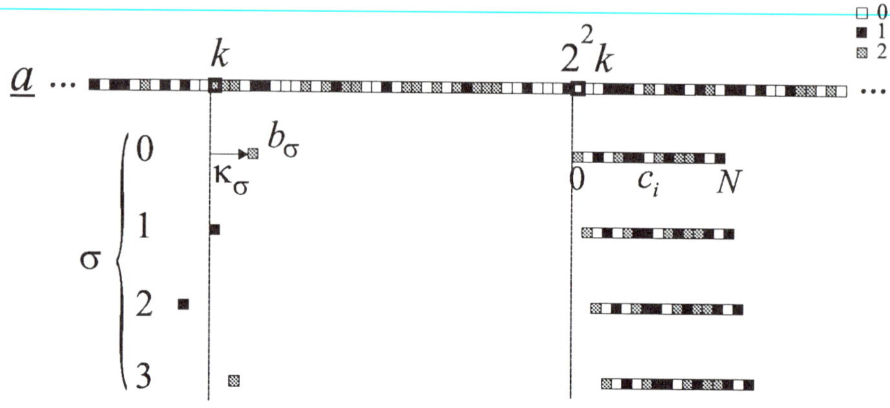

Fig. 3. Illustrating the meaning of the CGI-equation (10) for a 3-valued sequence $\underline{a}$ and $z^s = 2^2$. The sequence is represented as a string of cells whose states can take one of three possible values (represented by different cell contents). 2^2-Coarse-graining relates elements of the sequence in neighbourhoods of the positions k and 2^2k. The set of weights $\{c_i \mid i = 0, \ldots, N\}$ is here represented as a mask positioned at $2^2k + \sigma$ for $\sigma = 0, 1, 2, 3$. This mask also indicates the size of the neighbourhood considered. κ_σ represents the shift with respect to position k, where a single-cell mask containing the value b_σ has to be placed. Coarse-graining invariance means that for all σ it should hold that for all k, the coarse-graining produced by the σ-related mask at position 2^2k should equal the value of the sequence at the b_σ-mask position (near position k) multiplied by b_σ.

$\forall k \in \mathbb{Z}$ and $\forall \sigma_0, \sigma_1, \ldots, \sigma_{s-1} \in [z]$:

$$\boxed{\begin{aligned} &\textstyle\sum_{i=0}^{N} c_i\, a(kz^s + \sigma_{s-1}z^{s-1} + \sigma_{s-2}z^{s-2} \cdots + \sigma_1 z + \sigma_0 + \kappa + i) \\ &+ \textstyle\sum_{i=0}^{N_{\sigma_0}} c_{\sigma_0,i}\, a(kz^{s-1} + \sigma_{s-1}z^{s-2} + \sigma_{s-2}z^{s-3} \cdots + \sigma_2 z + \sigma_1 + \kappa_{\sigma_0} + i) \\ &+ \textstyle\sum_{i=0}^{N_{\sigma_1\sigma_0}} c_{\sigma_1\sigma_0,i}\, a(kz^{s-2} + \sigma_{s-1}z^{s-3} + \sigma_{s-2}z^{s-4} \cdots + \sigma_3 z + \sigma_2 + \kappa_{\sigma_1\sigma_0} + i) \\ &\cdots \qquad\qquad\qquad\qquad \cdots \\ &+ \textstyle\sum_{i=0}^{N_{\sigma_{s-1}\sigma_{s-2}\cdots\sigma_0}} c_{\sigma_{s-1}\sigma_{s-2}\cdots\sigma_0,i}\, a(k + \kappa_{\sigma_{s-1}\sigma_{s-2}\cdots\sigma_0} + i) = 0, \end{aligned}} \tag{11}$$

with $N, N_{\sigma_j\cdots\sigma_0}$ nonnegative integers, c_i, $c_{\sigma_j\cdots\sigma_0,i}$ $(j = 0, 1, .., s)$ coefficients from $\mathcal{R}$ and κ, $\kappa_{\sigma_j\cdots\sigma_0}$ $(j = 0, 1, ..., s)$ integer "shift"-parameters that possibly depend on the values of the σ-parameters displayed in the indices.

The meaning of this rather complex looking equation is elucidated in Fig. 4.

Remark 2. If we introduce, for convenience, the integer number

$$\sigma_{s|l} \stackrel{\text{def}}{=} \sigma_{s-1}z^{l-1} + \sigma_{s-2}z^{l-2} \ldots + \sigma_{s-l} \in [z^l] \text{ when } 0 < l \le s;\ \sigma_{s|0} = 0,$$

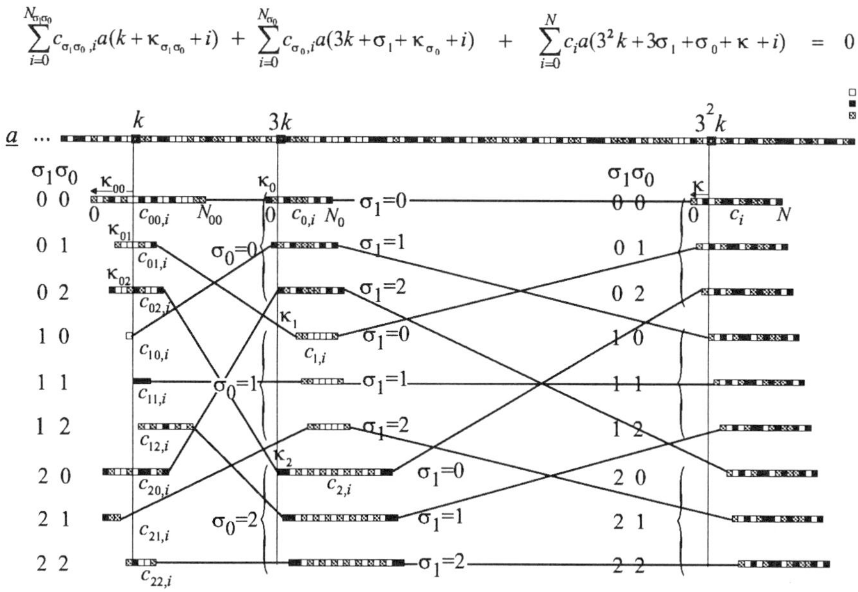

Fig. 4. Illustrating the meaning of the multiscale coarse-graining invariance equation (11) for the particular example given on top (3-valued, $z^s = 3^2$). It relates the coarse-grainings at scales $k, 3k, 3^2k$ as represented by the different masks appearing in the neighbourhoods of these positions. Observe that there is only one mask at position 3^2k (but shifted over 9 different positions as given by $3\sigma_1+\sigma_0$; $\sigma_1, \sigma_0 = 0, 1, 2$). At position $3k$, there are 3 possible masks (one for each σ_0), each shifted over 3 places ($\sigma_1 = 0, 1, 2$). At position k, there are 9 possibly different masks. Coarse-graining invariance requires that the coarse-grainings corresponding to the connected masks (i.e. for each given $\sigma_1\sigma_0$) at the scales $k, 3k, 3^2k$ all add up to 0 (and this for all $k \in \mathbb{Z}$).

and the string

$$\overline{\sigma}_l \overset{\text{def}}{=} \sigma_l\sigma_{l-1}\cdots\sigma_0 \ (\overline{\sigma}_{-1} = \text{ empty string}),$$

and define $\alpha_{k,\sigma_{s|l}} \overset{\text{def}}{=} kz^l + \sigma_{s|l}$ as the *local neighbourhood address* at scale z^l, then (11) can be written in the more compact form

$$\sum_{l=0}^{s} \sum_{i=0}^{N_{\overline{\sigma}_{s-l-1}}} c_{\overline{\sigma}_{s-l-1},i}\, a(\alpha_{k,\sigma_{s|l}} + \kappa_{\overline{\sigma}_{s-l-1}} + i) = 0. \tag{12}$$

Note that the mask-size parameter $N_{\overline{\sigma}_{s-l-1}}$, the $c_{\overline{\sigma}_{s-l-1},i}$-coefficients and the $\kappa_{\overline{\sigma}_{s-l-1}}$-shift-parameters for the z^l-coarse-graining term *may* depend on the σ_j-values that do *not* appear in the associated local neighbourhood address $\alpha_{k,\sigma_{s|l}}$.

The next two properties are easy to prove.

Lemma 1. *If* $\underline{a}$ *is a (multiscale) CGI-sequence, then so is its* β*-shifted version* ${}^{\beta}\underline{a}$*, which also satisfies equation (11), with new shift-parameters* $\kappa'_{\overline{\sigma}_j} = \kappa_{\overline{\sigma}_j} - \beta$.

Lemma 2. *Let* $\underline{a}$ *and* $\underline{b}$ *be two (multiscale)* z^s*-CGI sequences satisfying (12), but with different coefficients* $c_{\overline{\sigma}_j,i}$ *and different parameters* $N_{\overline{\sigma}_j}$ *and* $\kappa_{\overline{\sigma}_j}$ *. Then* $q_1\underline{a} + q_2\underline{b}$*, with* $q_1, q_2 \in \mathcal{R}$*, is also a (multiscale)* z^s*- CGI sequence.*

The following is a direct consequence of the previous Lemmas 1-2 and of the state evolution rule for LCAs [equation (5)].

Proposition 3. *If the initial state configuration* $\underline{a}_0$ *of an LCA is a CGI-sequence, then the state configuration* $\underline{a}_r$ *at any time* r *(i.e., any row-sequence in the orbit), is also a CGI-sequence.*

The link between CGI-sequences and Mahler equations is given by the next two propositions.

Proposition 4. *Let the formal Laurent series* $F_\lambda(X) = \sum_{k\in\mathbb{Z}} a(k+\lambda)X^k = \sum_{k\in\mathbb{Z}} a(k)X^{k-\lambda}$ *be the generating function of* ${}^\lambda\underline{a}$*, where* $\underline{a}$ *is a CGI-sequence satisfying (11) with invertible* c_N*. Then there is an integer* λ_0 *such that for* $\lambda \geq \lambda_0$*,* $F_\lambda(X)$ *satisfies a Mahler equation*

$$Q(X)F_\lambda(X) = \sum_{i=1}^{s} P_{\lambda,i}(X)F_\lambda(X^{z^i}), \tag{13}$$

with $Q(X)$*,* $P_{\lambda,i}(X)$ *polynomials over* $\mathcal{R}$*,* $Q(0) = 1$ *.*

Proof: Equation (11) can be written in the so-called right propagating form [6] [using the more concise notation (12)]:

$$\begin{aligned} a(\alpha_{k,\sigma_{s|s}} + N + \kappa) = &\textstyle\sum_{i=1}^{N} c'_{N-i}\, a(\alpha_{k,\sigma_{s|s}} + N + \kappa - i) \\ &+ \textstyle\sum_{l=0}^{s-1} \sum_{i=0}^{N_{\overline{\sigma}_{s-l-1}}} c'_{\overline{\sigma}_{s-l-1},i}\, a(\alpha_{k,\sigma_{s|l}} + \kappa_{\overline{\sigma}_{s-l-1}} + i), \end{aligned} \tag{14}$$

with $c'_{*,i} = -c_N^{-1} c_{*,i}$. With m an arbitrary integer, observe that $F_\lambda(X)$ can also be written as

$$F_\lambda(X) = \sum_{\sigma_0,\ldots,\sigma_{s-1}\in[z]} \sum_{k\in\mathbb{Z}} a(\alpha_{k,\sigma_{s|s}} + N + m)X^{\alpha_{k,\sigma_{s|s}}+N+m-\lambda}. \tag{15}$$

Insert (14) into (15), the last taken with $m = \kappa$, to obtain

$$F_\lambda(X) = \sum_{i=1}^{N} c'_{N-i}X^i \sum_{\sigma_0,\ldots,\sigma_{s-1}\in[z]} \sum_{k\in\mathbb{Z}} a(\alpha_{k,\sigma_{s|s}} + N + \kappa - i)X^{\alpha_{k,\sigma_{s|s}}+N+\kappa-i-\lambda}$$

$$+X^{N+\kappa} \sum_{l=0}^{s-1} X^{(z^{s-l}-1)\lambda} \sum_{\sigma_0,\ldots,\sigma_{s-l-1}} X^{-z^{s-l}\kappa_{\overline{\sigma}_{s-l-1}}+\sigma_{s-l|s-l}} \sum_{i=0}^{N_{\overline{\sigma}_{s-l-1}}} c'_{\overline{\sigma}_{s-l-1},i} X^{-z^{s-l}i}$$

$$\times \sum_{\sigma_{s-l},\ldots,\sigma_{s-1}} \sum_{k\in\mathbb{Z}} a(\alpha_{k,\sigma_{s|l}} + \kappa_{\overline{\sigma}_{s-l-1}} + i)X^{z^{s-l}(\alpha_{k,\sigma_{s|l}}+\kappa_{\overline{\sigma}_{s-l-1}}+i-\lambda)}. \tag{16}$$

Invoking (15), and observing that the expression on the last line of (16) equals $F_\lambda(X^{z^{s-l}})$, we find that (16) transforms into (13), with

$$\begin{aligned} Q(X) &= 1 - \sum_{i=1}^{N} c'_{N-i} X^i, \\ P_{\lambda,j}(X) &= X^{N+\kappa+(z^j-1)\lambda} \sum_{\sigma_0,\ldots,\sigma_{j-1}} X^{-\kappa_{\overline{\sigma}_{j-1}} z^j + \sigma_{j|j}} \sum_{i=0}^{N_{\overline{\sigma}_{j-1}}} c'_{\overline{\sigma}_{j-1},i} X^{-z^j i}. \end{aligned}$$

$Q(X)$ and $P_{\lambda,j}$ are polynomials when the least degree appearing in them is ≥ 0. The minimal value of λ for which this is the case is given by

$$\lambda_0 = \max_{\sigma_0,\ldots,\sigma_{j-1};j=1,\ldots,s} \{ \frac{z^j(N_{\sigma_{j-1}\ldots\sigma_0} + \kappa_{\sigma_{j-1}\ldots\sigma_0}) - N - \kappa - (\sigma_{j-1} z^{j-1} \ldots + \sigma_0)}{z^j - 1} \}.$$

□

Remark 3. The proof shows that, in order to obtain a Mahler equation from (11), it is necessary that the N_*,c_*- and κ_*-parameters have the dependencies as shown, i.e., N_*,c_*- and κ_* may only depend on the σ_i appearing in $*$, and not on the others.

Proposition 5. *Let $F_\lambda^+(X) = \sum_{k\in\mathbb{N}} a(k+\lambda) X^k$ be the generating function of the right side tail $(a(k+\lambda))_{k\in\mathbb{N}}$ of a multiscale z^s-CGI-invariant sequence satisfying (11) with invertible c_N. Then there is an integer λ_0^+ such that for $\lambda \geq \lambda_0^+$, $F_\lambda^+(X)$ satisfies the Mahler equation*

$$Q(X) F_\lambda^+(X) = \sum_{i=1}^{s} P_{\lambda,i} F_\lambda^+(X^{z^i}) + R_\lambda(X), \tag{17}$$

with $Q(X)$, $P_{\lambda,i}$, R_λ polynomials over $\mathcal{R}$. Moreover, $Q(X)$ and $P_{\lambda,i}(X)$ are identical to the $Q(X)$ and $P_{\lambda,i}(X)$-polynomials in (13).

The proof of this proposition is essentially similar to the one for Proposition 6, although more involved because of the deletion of the left-side tail of $\underline{a}$ (what actually also causes $R_\lambda(X)$ to appear).

The previous proposition allows to apply a theorem (Proposition 6) from [7], which leads to

Proposition 6. *Let $\mathcal{R} = Z_{p^d}$, p prime and $d, s \in \mathbb{N}$, $d, s \geq 1$, then a multiscale p^s-CGI sequence (over Z_{p^d}) is p-automatic.*

The next proposition relates a multiscale CGI-condition on the orbit of an m-Fermat LCA to multiscale CGI of the initial-state configuration.

Proposition 7. *Let $O_A(\underline{a}_0)$ be the orbit of an m-Fermat LCA with rule A, defined over $\mathcal{R}$. Then the following linear dependency condition (in $\mathcal{R}$) between different m^l-coarse-grainings of the orbit leads to an initial-state configuration that is multiscale m^s-CGI (i.e., satisfies an equation like (11)):*
$\forall \sigma_0, \ldots, \sigma_{s-1} \in [m]$:

$$\boxed{\sum_{l=0}^{s} \sum_{i=0}^{M_{\overline{\sigma}_{s-l-1}}} C^{(\sigma_{s|l})}_{m^l, \Xi_{\overline{\sigma}_{s-l-1},i}, W_{\overline{\sigma}_{s-l-1},i}} [^{(i+\kappa_{\overline{\sigma}_{s-l-1}})} O_A(\underline{a}_0)] = 0.} \tag{18}$$

Proof: use the CA-rule(5) to write each $a(r,c)$-element of the orbit that appears in (18) as a function of the initial-state configuration. The l-related terms above will then lead to z^l-coarse-graining terms of the form given in (11). □

Example 1.

Consider the 2-Fermat LCA over $\mathbb{F}_2$ with rule $A(X) = X^{-1} + X$, and the following multiscale CGI-condition for its orbit:
$\forall \sigma_0, \sigma_1 \in \{0,1\}$:

$$C^{(2\sigma_1+\sigma_0)}_{2^2,\Xi,W}O_A(\underline{a}_0) + C^{(\sigma_1)}_{2,\Xi_{\sigma_0,1},W_{\sigma_0,1}}{}^{\kappa_{\sigma_0}}O_A(\underline{a}_0) + C^{(\sigma_1)}_{2,\Xi_{\sigma_0,2},W_{\sigma_0,2}}{}^{(\kappa_{\sigma_0}+4)}O_A(\underline{a}_0) \\ +{}^{\kappa_{\sigma_1,\sigma_0}}O_A(\underline{a}_0) = 0,$$

with all Ξ_* having all elements equal to 0 (i.e., all tiles considered are square), and

$$W = \begin{bmatrix} 1\,0\,1\,0 \\ 0\,0\,0\,0 \\ 1\,0\,1\,0 \\ 0\,0\,0\,0 \end{bmatrix}, W_{0,1} = W_{1,1} = \begin{bmatrix} 1\,1 \\ 1\,1 \end{bmatrix}, W_{0,2} = W_{1,2} = \begin{bmatrix} 1\,0 \\ 0\,0 \end{bmatrix},$$

and $\kappa_0 = \kappa_1 = \kappa_{0,0} = \kappa_{0,1} = \kappa_{1,0} = \kappa_{1,1} = 0$. The corresponding CGI-equation for the initial state configuration $\underline{a}_0 = (a(k))_{k\in\mathbb{Z}}$ is:
$\forall k \in \mathbb{Z}$, $\forall \sigma_1, \sigma_0 \in \{0,1\}$:

$$a(4k+2\sigma_1+\sigma_0-2)+a(4k+2\sigma_1+\sigma_0+4)+a(2k+\sigma_1-1)+a(2k+\sigma_1+2) \\ +a(2k+\sigma_1+4)+a(k) = 0. \quad (19)$$

The associated Mahler equation for $F_\lambda(X) = \sum_{k\in\mathbb{Z}} a(k+\lambda)X^k$, $\lambda \geq \lambda_0 = 4$, is

$$(1+X^6)F_\lambda(X) = \\ X^{\lambda-4}(1+X)(1+X^4+X^{10})F_\lambda(X^2) + X^{4+3\lambda}(1+X)(1+X^2)F_\lambda(X^4),$$

and for $F^+_\lambda(X) = \sum_{k\in\mathbb{N}} a(k+\lambda)X^k$, $\lambda \geq \lambda^+_0 = 6$:

$$(1+X^6)F^+_\lambda(X) = X^{\lambda-4}(1+X)(1+X^4+X^{10})F^+_\lambda(X^2) \\ +X^{4+3\lambda}(1+X)(1+X^2)F^+_\lambda(X^4) + R_\lambda(X),$$

with

$$R_\lambda(X) = \sum_{j=0}^{5} a(\lambda-6+j)X^j + U_\lambda(X) + V_\lambda(X),$$

$$U_\lambda(X) = \textstyle\sum_{j=0}^{\lambda+5} a(\frac{\lambda-8}{2}+1+j)X^j + \sum_{j=0}^{\lambda-1} a(\frac{\lambda-2}{2}+1+j)X^j \\ +\textstyle\sum_{j=0}^{\lambda-5} a(\frac{\lambda+2}{2}+1+j)X^j$$

when $\lambda \bmod 4 = 0, 2$

$$= \textstyle\sum_{j=0}^{\lambda+5} a(\frac{\lambda-7}{2}+1+j)X^j + \sum_{j=0}^{\lambda-1} a(\frac{\lambda-1}{2}+1+j)X^j \\ +\textstyle\sum_{j=0}^{\lambda-5} a(\frac{\lambda+3}{2}+1+j)X^j$$

when $\lambda \bmod 4 = 1, 3$;

$$
\begin{aligned}
V_\lambda(X) &= \textstyle\sum_{j=0}^{3\lambda+3} a(\frac{\lambda-8}{4}+1+j)X^j \qquad\qquad \text{when } \lambda \bmod 4 = 0 \\
&= \textstyle\sum_{j=0}^{3\lambda+3} a(\frac{\lambda-4-\lambda \bmod 4}{4}+1+j)X^j \text{ when } \lambda \bmod 4 \neq 0.
\end{aligned}
$$

Equation (19) can be solved recursively from knowledge of the *seed* $[a(-1), a(0), a(1), a(2), a(3), a(4)]$, using a procedure presented in [6] which is briefly recapitulated here for this example. First observe that the CGI-equation (19) can be rewritten as (arithmetic in $\mathbb{F}_2$!) :
$\forall \sigma_1, \sigma_0 \in \{0,1\}$ and $\forall k \in \mathbb{Z}$:

$$
\begin{aligned}
a(4k+2\sigma_1+\sigma_0+4) = a(4k+2\sigma_1+\sigma_0-2) + a(2k+\sigma_1-1) \\
+a(2k+\sigma_1+2) + a(2k+\sigma_1+4) + a(k). \quad (20)
\end{aligned}
$$

This is the so-called *right-propagating* form of the CGI-equation: the argument of the left side term in this equation becomes eventually greater than all arguments in the right side when k increases. This actually happens when $4k+2\sigma_1+\sigma_0+4 > \max(2k+\sigma_1+4, k)$ (the so-called *strictly right-propagating condition*), what ultimately leads to the condition $4k+2\sigma_1+\sigma_0+4 > \overline{\kappa} = 4$. This is the condition under which (20) can be used to extend the sequence $(a(\kappa))_{\kappa\in\mathbb{Z}}$ for $\kappa > \overline{\kappa}$ when it is known for $\kappa \leq \overline{\kappa}$ (extension "to the right").

In the same way, (19) can be rewritten in its *left-propagating* form:
$\forall \sigma_1, \sigma_0 \in \{0,1\}$ and $\forall k \in \mathbb{Z}$:

$$
\begin{aligned}
a(4k+2\sigma_1+\sigma_0-2) = a(4k+2\sigma_1+\sigma_0+4) + a(2k+\sigma_1-1) \\
+a(2k+\sigma_1+2) + a(2k+\sigma_1+4) + a(k). \quad (21)
\end{aligned}
$$

Now the argument in the left side of this equation is the one that becomes eventually less than the ones in the right side. This happens when $4k+2\sigma_1+\sigma_0-2 > \min(2k+\sigma_1-1, k)$ –the so-called *strictly left-propagating condition*– which ultimately results in the condition $4k+2\sigma_1+\sigma_0-2 < \underline{\kappa} = -1$. This allows (21) to propagate the $a(\kappa)$-elements for $\kappa < \underline{\kappa}$ when the $a(\kappa)$ are known for $\kappa \geq \underline{\kappa}$. Notice that knowledge of $[a(\underline{\kappa}), a(\underline{\kappa}+1), \ldots, a(\overline{\kappa}-1), a(\overline{\kappa})]$ might be sufficient to initiate the left- and right-propagation of $\underline{a}$ using (20-21). This is the case in the present example, and hence $[a(-1), \ldots, a(4)]$ is the seed. If it were not the case, then the *preliminary seed* $[a(\underline{\kappa}), a(\underline{\kappa}+1), \ldots, a(\overline{\kappa}-1), a(\overline{\kappa})]$ can always be extended so that it becomes a seed (i.e., the seed should also satisfy the *condition of sufficient width* in addition to the above-mentioned strictly left- and right-propagating conditions). Note also that a seed –which is meant to be of minimal size– is not necessarily unique. Moreover, the seed-supporting integer interval $[\underline{\kappa}, \overline{\kappa}]$ can generally be so large that the seed contains all sequence-elements that appear in the CGI-equation for one or several values of the k-variable. This enforces dependencies of the elements in the seed. In the present example, this is not the case: all elements of the seed $[a(-1), \ldots, a(4)]$ can be freely chosen, and so there are 64 possible solutions.

Figure 5 displays a part of four possible solutions, two of which with a partial view on a *shift-add correlation pattern* (SAP, [5,6]) which may uncover hidden structures. In view of the properties presented above, all the horizontal sequences in both the CGI-orbits and their SAPs are 2-automatic coarse-graining invariant sequences. According to a theorem in [7], the two-dimensional patterns themselves (orbits and SAPs) are also two-dimensionally 2-automatic (the SAPs are also orbits of an LCA –with rule $A(X)$– but they are generally not coarse-graining invariant as orbits).

4 Conclusion

One-dimensional multiscale coarse-graining invariant (MSCGI) sequences have been defined as sequences satisfying a particular set of difference equations that have terms with multiple scalings in their arguments (see (11)). They were shown to be the proper generalization of the simpler single-scale CGI-sequences that were introduced before, in such a way that these MSCGI-sequences satisfy a Mahler equation –a functional equation in the sequence's "generating function" domain, see (3)– of a more general type than was the case for the single-scale CGI-sequences. The relationship with Mahler equations implies that these MSCGI-sequences are automatic sequences. Whereas one-dimensional single-scale CGI-sequences were originally obtained from imposing invariance of the two-dimensional orbit of certain cellular automata under some coarse-graining operation, it has now also become clear that multiscale coarse-graining invariant two-dimensional CA-orbits have a one-dimensional MSCGI-sequence as initial state configuration. The reverse questions are still open. Can any one-dimensional MSCGI-sequence be considered as the initial state configuration of some CA with an orbit that is MSCGI? Can any Mahler equation of type (3) be linked to a MSCGI-equation of type (11)?

Acknowledgment This work was partially supported by The Concerted Action Project GOA-MIPS of the Flemish Community, and by the Belgian Program on Interuniversity Attraction Poles of the Belgian Prime Minister's Office for Science, Technology and Culture (IUAP P4-02).

References

1. J.-P. Allouche, Automates finis en théorie des nombres, *Exposition.Math.*, **5** (1987), 239-266.
2. J.-P. Allouche and M. Mendès-France, Automata and automatic sequences, in : F.Axel et al., eds.," Beyond Quasicrystals," Les Editions de Physique and Springer-Verlag, Berlin, 1994, 293-367.
3. J.-P. Allouche, E. Cateland, W. J. Gilbert, H.-O. Peitgen, J. O. Shallit and G. Skordev, Automatic maps in exotic numeration systems, *Theory of Computing Systems*, **30** (1997), 285-331.
4. J.-P. Allouche and J.Shallit, The ubiquitous Prouhet-Thue-Morse sequence, *this book.*

5. A. Barbé, F. von Haeseler, H.-O. Peitgen and G. Skordev, Coarse-graining invariant orbits of one-dimensional two-state linear cellular automata, *Internat. J. Bifurcation and Chaos*, **5** (1995), 1611-1631.
6. A. Barbé, Coarse-graining invariant orbits of one-dimensional $\mathbb{Z}_p$-linear cellular automata, *Internat. J. Bifurcation and Chaos*, **6** (1996), 2237-2297.
7. A. Barbé, H.-O. Peitgen and G. Skordev , Automaticity of coarse-graining invariant orbits of one-dimensional linear cellular automata, to appear in *Internat. J. of Bifurcation and Chaos*, **9** (1999).
8. G. Christol, T. Kamae, M. Mendès France and G. Rauzy, Suites algébriques, automates et substitutions, *Bull. Soc. math. France*, **108** (1980), 401-419.
9. A. Robinson, Fast Computation of additive cellular automata, *Complex Systems*, **1** (1987), 211-216.
10. F. von Haeseler and W. Jürgensen, Automaticity of solutions of Mahler equations, *this book.*
11. F. von Haeseler, H.-O. Peitgen and G. Skordev, On the fractal structure of rescaled evolution set , Report 278, Institut für Dynamische Systeme, Univ. Bremen, 1992.
12. F. von Haeseler, H.-O. Peitgen and G. Skordev, Cellular automata, matrix substitutions, and fractals, *Ann. Math. and Art. Intel.*, **8** (1993), 345-362.
13. K. G. Wilson, Problems in physics with many scales of length, *Sci. Amer.*, **241**, 2 (1979), 236-267.

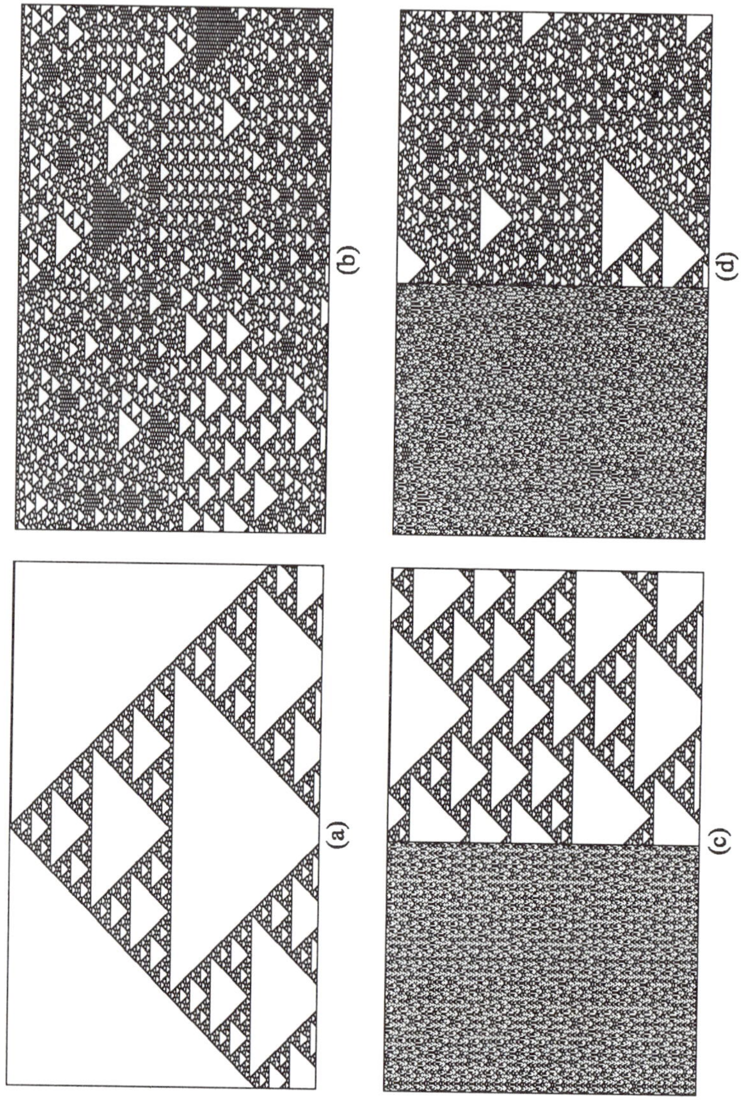

Fig. 5. Four solutions of the CGI-problem of Example 1 (display size: $0 \leq r \leq 244$, $-200 \leq c \leq 200$; black=1, white=0; rotate the page over 90° to view the figures properly).
(a): seed $[a(-1)a(0)a(1)a(2)a(3)a(4)] = [011111]$; (b): seed [001100]; (c): seed [000101]; (d): seed [000010]. For (c) and (d), the left parts show the solution, while the right parts shows the addition mod 2 of the original solution and a horizontally shifted version of it (the so-called shift-add correlation pattern (SAP), here with a horizontal shift over 192 cells). Solution (c) is quasiperiodic.

Regular Cosets and Upper Bounds on the Linear Complexity of Certain Sequences

P. Caballero-Gil*

Department of Statistics, Operations Research and Computation
University of La Laguna. Tenerife. Spain. Email: pcaballe@ull.es

Abstract. New upper bounds on the linear complexity of binary sequences produced by certain families of nonlinear filter functions are derived. They improve upon Key's upper bound and allow to deduce a simple recommendation for the design of secure filter generators. As a tool, a new class of cyclotomic cosets whose degeneration is relatively easy to prove in specific conditions is introduced and analysed.

1 Introduction

Binary sequences generated by nonlinear filter functions applied to the stages of a linear feedback shift register (LFSR) of length L have important applications as running keys in stream ciphers. To provide secure encryption, the key stream must be unpredictable. The linear complexity of a sequence is defined as the length of the shortest linear feedback shift register that can be used to generate it. Since some algorithms make possible to determine the linear recursion of a sequence having linear complexity l just from observing $2l$ consecutive bits of the sequence [8], [12], the linear complexity of a sequence is a widely accepted measure of its unpredictability.

Linear complexity has been studied by various authors [3], [4], [6], [13]. Concretely, upper bounds on the linear complexity of de Bruijn sequences [1] and bent-function sequences [10] have been derived, and a technique for finding upper bounds on the linear complexity of sequences produced by general feedback registers has been developed in [2]. Although it does not appear explicitly in [9], the following result is generally known as Key's upper bound on the linear complexity:

"The linear complexity of the sequences obtained by any kth-order filter function is upper bounded by $\sum_{l=1}^{k} \binom{L}{l}$."

On the other hand, Rueppel's root presence test for the product of distinct phases [14] allows to analyse the contribution of every cyclotomic coset to the linear complexity of the sequences obtained by any kth-order filter function f with

* This research has been partially supported by Spanish TEL98-1020 Project.

a single maximum order term $s_{n+t_0} s_{n+t_1} \cdots s_{n+t_{k-1}}$. This test can be stated as follows:

" Let $\alpha \in GF(2^L)$ be a root of the minimal polynomial of the sequence produced by the LFSR of length L. Then $\alpha^e = \alpha^{2^{e_0}+2^{e_1}+\cdots+2^{e_{k-1}}}, 0 \leq e_0 < e_1 < \cdots < e_{k-1} < L$ is a root of the minimal polynomial of the sequence generated by f if and only if the determinant

$$A_e = \begin{vmatrix} \alpha^{t_0 2^{e_0}} & \alpha^{t_1 2^{e_0}} & \ldots & \alpha^{t_{k-1} 2^{e_0}} \\ \alpha^{t_0 2^{e_1}} & \alpha^{t_1 2^{e_1}} & \ldots & \alpha^{t_{k-1} 2^{e_1}} \\ \cdot & \cdot & \ldots & \cdot \\ \alpha^{t_0 2^{e_{k-1}}} & \alpha^{t_1 2^{e_{k-1}}} & \ldots & \alpha^{t_{k-1} 2^{e_{k-1}}} \end{vmatrix} \neq 0."$$

If the determinant A_e equals zero, then the corresponding cyclotomic coset is said to be degenerate for the function f. The determinant A_e depends on three factors: the nonlinear filter function f, the minimal polynomial of the maximal-length LFSR and the cyclotomic coset whose contribution is analysed.

This work provides new upper bounds on the linear complexity of binary sequences produced by certain families of nonlinear filter functions with a single maximum order term. The bases of the presented results are the root presence test and a new broad collection of cyclotomic cosets that are introduced and studied in Section 2. The degeneration of these new cosets is easily proved for certain families of nonlinear filter functions, fact that is used to obtain, in Section 3, new upper bounds for the linear complexity of the resulting sequences that improve upon Key's upper bound. The most remarkable contribution of the present work is that the proved results provide a practical and simple recommendation for the design of secure filter generators.

2 Regular Cosets

In this section we introduce and analyse a new class of cyclotomic cosets, the so-called regular cosets, which constitute the starting point of the work. Some fundamental notation and a few definitions are introduced before the results can be stated and proved.

Since our concern is with binary sequences, most of the expressions discussed in this work will be over *GF(2)*, the finite field with two elements.

Let $(L, k)_c$ denote the cth common divisor of L and k.

It is well known [7] that the integers in $\{1, 2, ..., 2^L - 2\}$ that are relatively prime to $2^L - 1$ form a group under multiplication modulo $2^L - 1$, and the subset $\{1, 2, 2^2, ..., 2^{L-1}\}$ forms a subgroup. This subgroup, when multiplied by any other element of the group, yields a so-called *proper coset*. In addition to these proper cosets, there are always one or more *improper cosets* that result from multiplying all the elements of the subgroup by an integer which is not relatively prime to $2^L - 1$. The set of all cosets (proper and improper) of the multiplier subgroup constitutes the so-called *cyclotomic cosets* modulo $2^L - 1$.

Let *coset* e denote the cyclotomic coset $\{e,\ e2,\ e2^2,\ ...,\ e2^{L-1}\}$(modulo $2^L - 1$) that contains the integer e. Note that, in the corresponding binary representation, coset e consists of all the successive circular shifts of any of its

L-bit strings. Throughout this work, the decimal form of the cosets' elements $2^{e_0}+2^{e_1}+\cdots+2^{e_{k-1}}$, $0 \leq e_0 < e_1 < \cdots < e_{k-1} < L$, or their binary representation as L-bit strings of Hamming weight k will be used indistinctly.

Definition: A cyclotomic coset modulo $2^L - 1$ whose cardinality is less than L is called a ***regular*** coset. In other words, a regular coset e is a set of the form $\{e,\ e2,\ ...,\ e2^{m-1}\}\ (mod\ \ 2^L - 1)$, where the smallest positive integer m such that $e2^m \equiv e\ (mod\ 2^L - 1)$ is smaller than L.

Next result establishes a simple relation between regular and improper cosets.

Lemma 1: Every regular coset is an improper coset.

Proof: Let us proceed by contradiction by assuming that a regular coset e were a proper coset. Thus, since e would be relatively prime to $2^L - 1$, then for all $m < L$, we have that $e(2^m - 1) \not\equiv 0\ (mod\ 2^L - 1)$, and the coset e would not be a regular coset. □

The converse of Lemma 1 may not be true (see Example 1).

One of the most important points of any regular coset is that its elements can be represented as binary strings of length L containing k ones whose period is smaller than L. Regular cosets' elements are called regular strings. This name is due to the regular distribution of the k ones in the corresponding binary representations.

Next remark is useful for characterizing regular strings.

Remark 1: Note that the period of the ones in any regular string strictly divides $\gcd(L,k)$. Concretely, every regular L-bit string of Hamming weight k is composed of $(L,k)_c$ repetitions of the same non-regular $\frac{L}{(L,k)_c}$-bit string of Hamming weight $\frac{k}{(L,k)_c}$, and belongs to a regular coset whose cardinality is $\frac{L}{(L,k)_c}$. Thus, if L and k are relatively prime, there does not exist any regular coset. On the other hand, if L and k have some factor in common, then each common divisor of L and k, $(L,k)_c$, determines a different collection E^c of regular strings e of the form:

$$e = \ 2^{e_0} + 2^{e_1} + \cdots + 2^{e_{k-1}} \text{ such that } \ 0 \leq e_0 < e_1 < \cdots < e_{k-1} < L,$$

$$\text{for all } i = 0,1,..., \tfrac{k}{(L,k)_c} - 1 \text{ and for all } j = 1,2,...,(L,k)_c - 1,$$

$$e_{i+j\frac{k}{(L,k)_c}} = e_i + \ j\frac{L}{(L,k)_c}\ .$$

In particular, one of those strings, e^c, is considered here the representative element of the collection E^c, and verifies the following expression:

$$e^c = \ 2^{e_0} + 2^{e_1} + \cdots + 2^{e_{k-1}} \text{ such that } \ 0 \leq e_0 < e_1 < \cdots < e_{k-1} < L,$$

for all $i = 0, 1, ..., \frac{k}{(L,k)_c} - 1$ and for all $j = 1, 2, ..., (L, k)_c - 1$,

$$e_i = i \text{ and } e_{i+j\frac{k}{(L,k)_c}} = i + j\frac{L}{(L,k)_c}.$$

□

From the precedent remark, a simple characterization of e^c can be derived.

Lemma 2: Let L and k be integers. Then, the regular string e^c can be expressed as the integer

$$e^c = \left(2^{\frac{k}{(L,k)_c}} - 1\right) \frac{2^L - 1}{2^{\frac{L}{(L,k)_c}} - 1}.$$

Proof: From the previous remark, it is easily deduced that the binary representation of e^c is composed of $(L, k)_c$ repetitions of the same $\frac{L}{(L,k)_c}$-bit string with $\frac{k}{(L,k)_c}$ consecutive ones. The trick of the proof is that the contribution of the j-th string ($j = 1, 2, ..., (L, k)_c$) to the integer e^c is given by $\left(2^{\frac{k}{(L,k)_c}} - 1\right)$ $2^{(j-1)\frac{L}{(L,k)_c}}$ for all $j = 1, 2, ..., (L, k)_c$. Since $1 + \frac{2^{\frac{L}{(L,k)_c}}\left(2^{L - \frac{L}{(L,k)_c}} - 1\right)}{2^{\frac{L}{(L,k)_c}} - 1}$ is the sum of the $(L, k)_c$ first terms of a geometric progression with ratio $2^{\frac{L}{(L,k)_c}}$, then $e^c = \left(2^{\frac{k}{(L,k)_c}} - 1\right) \frac{2^L - 1}{2^{\frac{L}{(L,k)_c}} - 1}$. □

From the particular structure of regular strings, it may be easily deduced that there exists a 1-1 correspondence between each regular L-bit string of Hamming weight k determined by $(L,k)_c$ and each non regular $\frac{L}{(L,k)_c}$-bit string of Hamming weight $\frac{k}{(L,k)_c}$. This idea leads us to next proposition that provides the total number of distinct regular strings related to a common divisor of L and k.

Proposition 1: The number of distinct regular L-bit strings of Hamming weight k belonging to the collection E^c is given by the recursive expression

$$|E^c| = \binom{L/(L, k)_c}{k/(L, k)_c} - \sum_{d \in D} |E^d|,$$

where $D = \{ d : (L, k)_d > (L, k)_c \text{ such that } (L, k)_c \mid (L, k)_d \}$, and if $(L, k)_c = \gcd(L, k)$, then $|E^c| = \binom{L/\gcd(L, k)}{k/\gcd(L, k)}$.

In the expression of this last result, for every common divisor $(L, k)_c$, the total number of regular strings related to each common divisor that is a multiple of $(L, k)_c$ is sustracted to obtain the number of regular strings belonging to the collection E^c. The recursive formula of Proposition 1 may be easily expressed in a non-recursive way by means of the prime factorization of the greatest common divisor of L and k.

Proposition 2: Let $\gcd(L,k) = \prod_{r=1}^{m} p_r^{e_r}$ be the prime factorization of the greatest common divisor of L and k. Then, the total number of distinct regular L-bit strings of Hamming weight k is given by

$$\sum_{J\subseteq\{1,2,\ldots,m\}} (-1)^{|J|+1} \binom{L/\prod_{j\in J} p_j}{k/\prod_{j\in J} p_j}.$$

Proof: The total number of distinct regular L-bit strings of Hamming weight k is given by $\sum_c |E^c|$. Since $\gcd(L,k) = \prod_{r=1}^{m} p_r^{e_r}$, then every common divisor of L and k is a product of powers of prime numbers p_r. Then, by Proposition 1 we have that

$$\sum_c |E^c| =$$

$$\sum_{r=1}^{m} \binom{L/p_r}{k/p_r} + \left[1-\binom{2}{1}\right] \sum_{\substack{\{d\,:\,(L,k)_d = \prod_{j\in J} p_j^{f_j},\\ |J|=2,\, f_j\le e_j\}}} |E^d| + \cdots +$$

$$\left[1-\binom{m}{1}\right] \sum_{\substack{\{d\,:\,(L,k)_d = \prod_{r=1}^{m} p_r^{f_r},\\ f_r\le e_r\}}} |E^d| =$$

$$\sum_{r=1}^{m} \binom{L/p_r}{k/p_r} + \left[1-\binom{2}{1}\right] \sum_{\substack{J\subseteq\{1,2,\ldots,m\}\\ |J|=2}} \binom{L/\prod_{j\in J} p_j}{k/\prod_{j\in J} p_j} + \cdots +$$

$$\left[1-\binom{m}{1}+\cdots+(-1)^{m-1}\binom{m}{m-1}\right] \binom{L/\prod_{r=1}^{m} p_r}{k/\prod_{r=1}^{m} p_r}.$$

Thus, since $1-\binom{r}{1}+\binom{r}{2}-\cdots+(-1)^{r-1}\binom{r}{r-1} = (-1)^{r-1}$, the number of regular strings is given by

$$\sum_{r=1}^{m} \binom{L/p_r}{k/p_r} - \sum_{\substack{J\subseteq\{1,2,\ldots,m\}\\ |J|=2}} \binom{L/\prod_{j\in J} p_j}{k/\prod_{j\in J} p_j} + \cdots + (-1)^{m-1} \binom{L/\prod_{r=1}^{m} p_r}{k/\prod_{r=1}^{m} p_r}.$$

□

Finally, certain values of L and k allow to obtain an extremely simple expression of the total number of regular strings, that will be useful in the next section to establish a tight and suitable upper bound on the linear complexity of certain sequences.

Corollary 1: If the greatest common divisor of L and k is the power of a prime number p, then the total number of distinct regular L-bit strings of Hamming weight k is given by the binomial coefficient

$$\binom{L/p}{k/p}.$$

The previous results are used in the next section to provide upper bounds on the linear complexity of sequences produced by certain nonlinear filter generators. This section ends with an example.

Example 1: Consider the case $L = 12, k = 6$. First, note that not all the improper cosets are regular cosets. A counterexample of this could be the cyclotomic coset {63, 126, 252, 504, 1008, 2016, 4032, 3969, 3843, 3591, 3087, 2079}, that is improper because 63 is not relatively prime to $2^{12} - 1$, but that is not regular because the coset's cardinality is 12.

The numbers 12 and 6 have three common divisors, that are $(12,6)_1{=}2$, $(12,6)_2{=}3$ and $(12,6)_3{=}6$, so there are three different collections of regular strings, represented respectively by the elements $e^1 = 2^0+2^1+2^2+2^6+2^7+2^8 = 455$, $e^2 = 2^0+2^1+2^4+2^5+2^8+2^9 = 819$ and $e^3 = 2^0+2^2+2^4+2^6+2^8+2^{10} = 1365$. From Proposition 1, the number of distinct regular strings in each collection is given respectively by $|E^1| = \binom{6}{3} - 2 = 18$, $|E^2| = \binom{4}{2} - 2 = 4$ and $|E^3| = \binom{2}{1} = 2$.

These collections are composed of the following regular cosets. There are three regular cosets in E^1, that are $\{455, 910, 1820, 3640, 3185, 2275\} \simeq$ {000111000111, 001110001110, ..., 100011100011}, {715, 1430, 2860, 1625, 3250, 2405} $\simeq$ {001011001011, 010110010110, ..., 100101100101} and {1235, 2470, 845, 1690, 3380, 2665} $\simeq$ {010011010011, 100110100110, ..., 101001101001}.

Belonging respectively to E^2 and E^3 are the regular cosets {819, 1638, 3276,2457}$\simeq${001100110011, 011001100110, 110011001100, 100110011001} and {1365, 2730}$\simeq$ {010101010101, 101010101010}.

Thus, by Proposition 2 it is known that the total number of regular 12-bit strings of Hamming weight 6 is $\binom{6}{3} + \binom{4}{2} - \binom{2}{1} = 24.$ □

3 Upper Bounds

In this section, the root presence test is applied to regular cosets introduced in the last section, establishing some conditions under which the corresponding determinants are equal to zero. In this way, the last results on the numbers of regular cosets allow to derive new upper bounds on the linear complexity of different families of nonlinear filter functions with a single maximum order term.

First, next result provides a slight improvement of Key's upper bound for a large range of nonlinear functions.

Theorem 1: Let f be a nonlinear filter function whose single maximum order term is the product $s_{n+t_0}s_{n+t_1}\cdots s_{n+t_{k-1}}$ and $\alpha \in GF(2^L)$ be a root of the minimal polynomial of the sequence produced by the LFSR of length L. If $k|L$ and there exist two different integers $i, j \in \{0, 1, ..., k-1\}$ such that $\alpha^{t_i}, \alpha^{t_j} \in GF(2^{L/k})$, then the linear complexity of the sequences produced by f satisfies the upper bound

$$\left[\sum_{l=1}^{k}\binom{L}{l}\right] - \frac{L}{k}.$$

Proof: According to the hypothesis, $\alpha^{t_i}, \alpha^{t_j} \in GF(2^{L/k})$, so we have that $\alpha^{t_i} \equiv \alpha^{t_i 2^{L/k}} \equiv \alpha^{t_i 2^{2L/k}} \equiv \cdots \equiv \alpha^{t_i 2^{(k-1)L/k}}$ in $GF(2^L)$ and $\alpha^{t_j} \equiv \alpha^{t_j 2^{L/k}} \equiv \alpha^{t_j 2^{2L/k}} \equiv \cdots \equiv \alpha^{t_j 2^{(k-1)L/k}}$ in $GF(2^L)$. Consequently, if the root presence test is applied to the regular coset e, where e is of the form $e = 2^0 + 2^{L/k} + 2^{2L/k} + \cdots + 2^{(k-1)L/k}$, then the particular determinant A_e equals zero since it has two columns linearly dependent over $GF(2^L)$. This fact implies that the regular coset e does not contribute to the linear complexity of f and the Key's upper bound may be diminished in its cardinality, that is $\frac{L}{k}$. □

Example 2: Now consider $L = 8$, $k = 4$, the product $s_n\, s_{n+17}\, s_{n+18}\, s_{n+85}$, and $\alpha \in GF(2^8)$ a root of the minimal polynomial of a sequence produced by the LFSR of length 8. Since $4|8$ and α^0, $\alpha^{85} \in GF(2^2)$, then the regular coset corresponding to the common divisor 4, $\{85, 170\}$ is degenerate for f. □

Now, by using Proposition 2, the upper bound of Theorem 1 can be improved for a family of filter functions that satisfy stronger restrictions.

Theorem 2 (Main Theorem): Let f be a nonlinear filter function whose single maximum order term is the product $s_{n+t_0}s_{n+t_1}\cdots s_{n+t_{k-1}}$, $\alpha \in GF(2^L)$ be a root of the minimal polynomial of the sequence produced by the LFSR of length L, and $\gcd(L,k) = \prod_{r=1}^{m} p_r^{e_r}$ be the prime factorization of the greatest common divisor of L and k. If for every $r \in \{1, 2, ..., m\}$ there exist $\frac{k}{p_r} + 1$ different values of $i \in \{0, 1, ..., k-1\}$ such that $\alpha^{t_i} \in GF(2^{L/p_r})$, then the linear complexity of the sequences produced by f satisfies the upper bound

$$\sum_{l=1}^{k}\binom{L}{l} - \sum_{J\subseteq\{1,2,...,m\}}(-1)^{|J|+1}\binom{L/\prod_{j\in J}p_j}{k/\prod_{j\in J}p_j}.$$

Proof: Consider all the corresponding regular cosets related to each common divisor of L and k. By Remark 1, we know that for every regular coset e corresponding to a prime number p_r or one of its multiples, for all $i = 0,1,...,\frac{k}{p_r}-1$ and for all $j = 1,2,...,p_r-1$, $e_{i+j\frac{k}{p_r}} = e_i + j\frac{L}{p_r}$. On the other hand, by hypothesis we have that for every $r \in \{1,2,...,m\}$ there exist $\frac{k}{p_r}+1$ different values of $i \in \{0,1,...,k-1\}$ such that $\alpha^{t_i} \in GF(2^{L/p_r})$, so the determinant A_e can be computed by expanding it along $k-\frac{k}{p_r}-1$ columns. In this way, A_e may be written as a linear combination of adjuncts of order $\frac{k}{p_r}+1$, every one with two identical rows. Consequently, every A_e equals zero and no regular coset contributes to the linear complexity of the sequences produced by f. □

When the relation between L and k is of a particular form, the hypothesis of the previous theorem can be easily relaxed to involve a broader family of nonlinear functions. This is stated in the following result that makes use of Corollary 1.

Corollary 2: Let f be a nonlinear filter function whose single maximum order term is the product $s_{n+t_0}s_{n+t_1}\cdots s_{n+t_{k-1}}$, $\alpha \in GF(2^L)$ be a root of the minimal polynomial of the sequence produced by the LFSR of length L, and L and k such that their greatest common divisor is the power of a prime number p. If for every $i = 0,1,...,\frac{k}{p}$, the power $\alpha^{t_i} \in GF(2^{L/p})$, then the linear complexity of the sequences produced by f satisfies the upper bound

$$\sum_{l=1}^{k}\binom{L}{l} - \binom{L/p}{k/p}.$$

By using simple tools of combinatorial calculus, it is easy to prove that the number of functions under the conditions of Theorem 2 is very large.

Proposition 3: The number of products under the conditions of Theorem 2 is

$$\binom{2^L-2-\sum_{r=1}^{m}\frac{k}{p_r}}{k-1-\sum_{r=1}^{m}\frac{k}{p_r}}\prod_{r=1}^{m}\binom{2^{\frac{L}{p_r}}-2}{\frac{k}{p_r}}.$$

Example 3: Let $L = 8,\ k = 4, \alpha$ be a primitive element of $GF(2^8)$, $GF(2^4)$= $\{0, 1, \alpha^{17}, \alpha^{34}, \alpha^{68}, \alpha^{136}, \alpha^{85}, \alpha^{170}, \alpha^{102}, \alpha^{204}, \alpha^{153}, \alpha^{51}, \alpha^{187}, \alpha^{119}, \alpha^{238}, \alpha^{221}\}$ and f be any nonlinear filter function whose single maximum order term is one of the $\binom{15}{3}\binom{252}{1} = 114.660$ products of the set: $\{\{s_n s_{n+17}\ s_{n+34}\ s_{n+t_3}/\ t_3 \notin \{0,17,34\}, ..., \{s_{n+119} s_{n+238}\ s_{n+221}\ s_{n+t_3}/\ t_3 \notin \{119,238,221\}\}$. Then, by Corollary 2, the linear complexity of the sequences produced by f is guaranteed to be at most $\binom{8}{1} + \binom{8}{2} + \binom{8}{3} + \binom{8}{4} - \binom{4}{2} = 156.$ □

Remark 2: From the above results we may conclude that using nonlinear filter generators whose values of L (length of the LFSR) and k (order of the filter function) are not relatively prime could be indeed dangerous, because in that case, regular cosets described in this work could produce a great decrease in the linear complexity of the resulting sequences. This decrease implies not only the non-optimality of generators under the hypothesis of the theorems, but also of other generators with the same LFSRs and similar nonlinear filter functions because the difference between sequences produced by both types of generators may be in only one position of a period, whereas the difference in their linear complexities may be high. Thus, many filter generators with values of L and k not relatively prime could have bad linear-complexity stability [5]. Table 1 shows some values of L and k, and the approximate decrease of Key's upper bound derived from the degeneration of all the regular cosets. Note that, for practical values such as $L = 128$ and $k = 64$, this approximate decrease is large, 10^{18}. Furthermore, decreases described here could be added to any other decrease produced by the degeneration of other kinds of cyclotomic cosets.

Table 1. Decrease of Key's upper bound

L	24	60	84	126	128	140	210	420
k	12	30	42	42	64	70	70	210
Decrease	988	10^8	10^{11}	10^{16}	10^{18}	10^{20}	10^{27}	10^{61}

□

4 Conclusions

This paper gives some specific conditions under which Key's upper bound on the linear complexity of nonlinear filter generators cannot be reached. The main result is based on the proved degeneration of a new broad collection of cyclotomic cosets introduced in this work. A slight improvement of Key's upper bound is presented and its effectiveness is suggested by a simple recommendation for the choice of the length of the LFSR and the order of the filter function, that is proposed as a design principle for filter generators. The problem of finding other kinds of cyclotomic cosets whose degeneration could be proved for the same families of nonlinear filter functions described here is a part of a work in progress.

Acknowledgment

The author thanks Amparo Fúster-Sabater for her valuable comments and suggestions about this work.

References

1. A.H. Chan, R.A. Games and E.L. Key, On the Complexity of de Bruijn Sequences, *Journal of Combinatorial Theory, Series A* **33** (1982), 233–246.
2. A.H. Chan, M. Goresky and A. Klapper, On the Linear Complexity of Feedback Registers, *IEEE Transactions on Information Theory* Vol. IT-**36**, No. **3**, May (1990).
3. Z.D. Dai, T. Beth and D. Gollmann, Lower Bounds for the Linear Complexity of Sequences over Residue Rings, *Advances in Cryptology-Eurocrypt'90, Lecture Notes in Computer Science* Vol. **473**, Springer-Verlag, 1991.
4. C. Ding, Lower Bounds on the Weight Complexities of Cascaded Binary Sequences, *Advances in Cryptology-Asiacrypt'90, Lecture Notes in Computer Science* Vol. **453**, Springer-Verlag, 1990.
5. C. Ding, G. Xiao and W. Shan, "The Stability Theory of Stream Ciphers, Lecture Notes in Computer Science Vol. **561**, Springer-Verlag, 1991.
6. A. Fúster-Sabater and P. Caballero-Gil, On the Linear Complexity of Nonlinearly Filtered PN-Sequences, *Advances in Cryptology-Asiacrypt'94, Lecture Notes in Computer Science* Vol. **917**, Springer-Verlag, 1995.
7. S.W. Golomb, "Shift Register Sequences, Holden-Day, San Francisco, California, 1967. Revised edition, Aegean Park Press, Laguna Hills, California, 1982.
8. K. Imamura and W. Yoshida, A Simple Derivation of the Berlekamp-Massey Algorithm and Some Applications, *IEEE Transactions on Information Theory* Vol. IT-**33** No. **1**, Jan (1987).
9. E.L. Key, An Analysis of the Structure and Complexity of Nonlinear Binary Sequence Generators, *IEEE Transactions on Information Theory* Vol. IT-**22** (1976), 732–736.
10. P.V. Kumar and R.A. Scholtz, Bounds on the Linear Span of Bent Sequences, *IEEE Transactions on Information Theory*, Vol. IT-**29**, No. **6**, Nov (1983).
11. R. Lidl and H. Niederreiter, "Introduction to Finite Fields and Their Applications, Cambridge University Press, 1986.
12. J.L. Massey, Shift Register Synthesis and BCH Decoding, *IEEE Transactions on Information Theory*, Vol. IT-**15** (1969), 122–127.
13. H. Niederreiter, The Linear Complexity Profile and the Jump Complexity of Keystream Sequences,em Advances in Cryptology-Asiacrypt'90, Lecture Notes in Computer Science Vol. **473**, Springer-Verlag, 1991.
14. R.A. Rueppel, "Analysis and Design of Stream Ciphers, Springer-Verlag, 1986.

Hadamard Matrices, Self-Dual Codes Over the Integers Modulo 4 and their Gray Images

C. Charnes

SITACS University of Wollongong
Northfields Ave. Wollongong NSW 2522, Australia
e-mail: charnes@cs.uow.edu.au

Abstract. We describe a new construction of codes over Z_4 from the *H-equivalence classes* of Hadamard matrices. We obtain by this method *self-orthogonal* and *self-dual* quaternary and binary codes which are the images of the quaternary codes under the Gray map.

1 Introduction

In this paper we study codes over Z_4, the ring of integers modulo 4 and the associated binary codes which arise from the *H-equivalence classes* of Hadamard matrices by a new construction. We derive the main structural properties of these codes, and survey the quaternary and binary codes arising from the H-equivalence classes of Hadamard matrices of orders up to 20. Providing a detailed account of the codes which are generated by the Hadamard matrices of order 16.

There has been a lot of research recently devoted to quaternary codes; for example: [1], [3], [5], [8], [7], [12]. Klemm [7] established the analogues of the MacWilliams relations for quaternary codes. The *Gray map* was used by Hammons, Kumar, Calderbank, Sloane, and Solé [5] to explain (among other things) a long standing puzzle of coding theory, why certain pairs of non-linear codes are *formally dual.* Quaternary codes have connections with such diverse areas as lattice sphere packings [6] and the theory of *bent* functions [2].

We construct quaternary codes $C(H_n)$ of length n from the representatives of H-equivalence classes of Hadamard matrices of order n. These codes are generated by the rows of H_n, which we Internet as quaternary vectors. We also investigate the properties of the associated linear binary codes $\phi(C(H_n))$, which are the images of the $C(H_n)$ under the Gray map ϕ. These ***binary images*** are Z_4***-linear*** codes, in the terminology of [5]. Our definition of the quaternary and binary codes does not depend on the particular representatives of the H-equivalence class used to define the codes. (Cf. Section 5; see also [11].)

The quaternary codes $C(H_n)$ and their binary counterparts $\phi(C(H_n))$ are either *self-orthogonal* or *self-dual*, depending whether $n \equiv 0 \pmod 8$ or $n \equiv 4 \pmod 8$ respectively. In the self-dual case the quaternary codes are of Type IV. But these turn out to be equivalent to codes which were defined by Klemm [7].

While the binary codes $\phi(C(H_n))$ are ***doubly even***, or Type II codes. (See [12] for a survey self-dual codes over rings and fields.)

Our construction does however produce new self-orthogonal quaternary and binary codes, some of which have interesting parameters. For example, the codes obtained from the five equivalence classes of Hadamard matrices of order 16. These equivalence classes yield five self-orthogonal quaternary codes and binary codes. The ***Euclidean weights*** of the codewords of the quaternary codes are divisible by 8. This is also true for the ***Hamming weights*** of codewords of the corresponding binary images. The binary images have parameters: $[32, k, 8]$, where $k = 6, \ldots, 9$. While their dual codes have parameters: $[32, 32 - k, 4]$, for $k = 6, \ldots, 9$. The (binary) dual codes have the highest possible minimal distance. With the exception of the Reed-Muller code $R(1, 5)$ and the associated quaternary code [5], these codes do not appear to have been previously known.

The connection between quaternary codes and ***quaternary sequences*** is this. A complex roots-of-unity sequence $s = i^a = (i^{a_1}, \ldots, i^{a_n})$, $i = \sqrt{-1}$, is associated with a Z_4-valued vector $a = (a_1, \ldots, a_n)$. Thus to a quaternary code C there corresponds a set of complex sequences $\Omega(C) = \{i^a, a \in C\}$. The effectiveness of C regarded as a set of CDMA signature sequences depends on the ***complex correlations*** of the sequences in $\Omega(C)$ – low correlation sequences correspond to codes with large minimum ***Euclidean distance.***

The rest of the material is organized as follows. Section 2 contains preliminary information. The quaternary codes are defined in Section 3. The structural properties of the quaternary codes are derived in Section 4. In Section 5 we prove that the definition of the quaternary and binary codes respects H-equivalence. Weight enumerators are discussed in Sections 6 and 7. Type IV codes are considered in Section 8. The binary image codes are discussed Section 9. Examples of codes generated by the equivalence classes of orders up to 20 are given Section 10. Section 11 is the conclusion.

Patrick Solé suggested [14] that the Z_4-spans of Hadamard matrices should be investigated. I am grateful to him for suggesting this problem and for his continuing interest in this work.

2 Preliminaries

A Hadamard matrix H_n of order n is an $n \times n$ matrix with rational entries ± 1, such that $H_n H_n^T = nI_n$. The necessary condition for the existence of a Hadamard matrix H_n is that $n = 1, 2$ or $n \equiv 0 \pmod 4$; see e.g. [9], [16]. Two Hadamard matrices of the same order are H-equivalent, if one can be obtained from the other by a sequence of operations consisting of either: (i) exchanging rows (or columns), or (ii) multiplying some of the rows (or columns) by -1. We denote H-equivalence classes by $\{H_n\}$.

A quaternary code C of length n over the ring Z_4 is an additive subgroup of Z_4^n. A Z_4-valued inner product is defined for quaternary codes by $a \cdot b = a_1 b_1 + \cdots + a_n b_n \pmod 4$ for $a, b \in C$. Self-orthogonality and self-duality of quaternary codes is defined with respect to this inner product.

The Gray map $\phi : Z_4^n \to Z_2^{2n}$ maps a quaternary code C of length n onto a binary (but not necessarily linear) code $\phi(C)$ length $2n$. This map is defined as follows. We represent the elements of the ring Z_4 by the integers 0,1,2, and 3, then ϕ maps these integers onto binary vectors as:

$$0 \to (0,0),\ 1 \to (0,1),\ 2 \to (1,1),\ 3 \to (1,0).$$

This map extends to quaternary vectors: $\phi(c_1, \ldots, c_n) = (\phi(c_1), \ldots, \phi(c_n))$. We also require the map $\alpha : Z_4 \to Z_2$, which is defined by

$$\alpha(c) = \begin{cases} 0 & \text{if } c = 0,\ 2; \\ 1 & \text{if } c = 1,\ 3. \end{cases}$$

This map is also defined for quaternary vectors.

Theorem 1 (Hammons et al. [5]). *The binary image $\phi(C)$ of a quaternary linear code C is linear if and only if $a, b \in C \Rightarrow 2\alpha(a) * \alpha(b) \in C$.*

The binary operation '$*$' in Theorem 1 denotes the componentwise product of two (binary) vectors.

3 Definition of the Codes

Since $1 \equiv 1 \pmod 4$ and $-1 \equiv 3 \pmod 4$, we can use this to identify the rational entries ± 1 of H_n with the corresponding elements of Z_4.

Definition 1 (Solé [14]). *Identify the entries of H_n with the elements of Z_4 as described above. Let $C(H_n)$ be the Z_4-module – a quaternary code of length n, whose elements are the Z_4-linear combinations of the rows of H_n.*

Excluding the orders $n = 1, 2$, the necessary condition for the existence of Hadamard matrices immediately implies that the quaternary codes obtained by Definition 1 are self-orthogonal, i.e. $C(H_n) \subseteq C(H_n)^\perp$. Patrick Solé [14], asked when are these codes self-dual, i.e. $C(H_n) = C(H_n)^\perp$. We will answer his question in Section 4.

As well as the quaternary codes $C(H_n)$, we study the binary image codes which are obtained from the $C(H_n)$ by Gray map. Thus to a quaternary code of length n, there corresponds a binary code $\phi(C(H_n))$ of length $2n$.

In what follows we investigate the properties of the quaternary and their associated binary images codes, which arise from the equivalence classes of Hadamard matrices.

4 Properties of the Quaternary Codes

Quaternary codes can be described by generator matrices similarly to binary codes. Any quaternary code C is *permutation-equivalent* – differing only by a permutation of the coordinates of the codewords, to a code whose generator matrix has the standard form

$$\begin{pmatrix} I_{k_1} & A & B \\ 0 & 2I_{k_2} & 2C \end{pmatrix}.$$

Where A and B are matrices with entries from Z_4 and C is a binary matrix. Since $\mathcal{C}$ is an elementary abelian group, it is a direct product ${Z_4}^{k_1} \times {Z_2}^{k_2}$ of cyclic groups Z_4 and Z_2. An invariant of $\mathcal{C}$, called the *type*, is defined as $4^{k_1}2^{k_2}$. In which case $\mathcal{C}$ contains $2^{2k_1+k_2}$ codewords. (See for example [3], [5].)

Theorem 2. *Let H_n be a representative Hadamard matrix of $\{H_n\}$ (cf. Theorem 8), then the codewords of the quaternary code $\mathcal{C}(H_n)$ are either even or odd. The coordinates of the even codewords satisfy $c_i \equiv 0$, or $2 \pmod 4$. The coordinates of the odd codewords satisfy $c_i \equiv 1$, or $3 \pmod 4$. The even codewords $\mathcal{E}_n$ comprise the torsion sub-code of $\mathcal{C}(H_n)$. The all-ones vector belongs to $\mathcal{C}(H_n)$.*

Proof. Obviously $\mathcal{E}_n$ is closed under the addition of vectors. Let $\sum \lambda_i h_i \in \mathcal{E}_n$, then the coordinates of this codeword satisfy $c_i \equiv 0, 2 \pmod 4$. Hence for any $\lambda \in Z_4$, $\lambda(\ldots, c_i, \ldots) = (\ldots, \lambda c_i, \ldots)$, and again the scaled coordinates satisfy $\lambda c_i \equiv 0, 2 \pmod 4$.

By Theorem 8, we can define $\mathcal{C}(H_n)$ by a normalized representative. Hence the all-ones vector belongs to $\mathcal{C}(H_n)$.

Theorem 3. *If $n \equiv 0 \pmod 4$ then $\mathcal{C}(H_n)$ has type $4^1 2^{k_2}$, and the quotient group $\mathcal{C}(H_n)/\mathcal{E}_n$ has order two.*

Proof. Any $\mathcal{C}(H_n)$ contains a codeword x whose coordinates are ± 1. So the translation map $c \to c + x$ swaps the even and odd codewords. Hence $|\mathcal{C}(H_n)/\mathcal{E}_n| = 2$.

Regarding $\mathcal{C}(H_n)$ as an additive group, the non-zero codewords of $\mathcal{E}_n$ have order 2, and any odd codeword has order 4. Since $|\mathcal{C}(H_n)/\mathcal{E}_n| = 2$ there is just one non-zero residue class, whose representative can be any odd codeword. This implies that the value of the k_1-index in $4^{k_1}2^{k_2}$ is $k_1 = 1$. (If $k_1 > 1$, then the k_1 generators of ${Z_4}^{k_1}$ would correspond to k_1 different elements of the quotient group.)

By Theorem 3 the quaternary codes $\mathcal{C}(H_n)$ have the following coset decomposition

$$\mathcal{C}(H_n) = \mathcal{E}_n \cup \{\mathcal{E}_n + v\}, \tag{1}$$

where v is any odd order element of $\mathcal{C}(H_n)$.

For certain classes of Hadamard matrices we can determine the k_2-index in the type of $\mathcal{C}(H_n)$. We require the following results about the *Smith normal form* of Hadamard matrices.

Recall that the Smith normal form of an $n \times n$ integer matrix A of rank r is the diagonal matrix $S(A) = (s_{ij}) = VAU$ for $V, U \in \mathrm{GL}(n, Z)$; i.e. $S(A)$ and A are Z-equivalent. The only non-zero entries of $S(A)$ are the r diagonal elements

s_{ii}, which satisfy the divisibility condition: $s_{11}|s_{22}|\cdots|s_{rr}$. The s_{ii} are uniquely determined. The following result was obtained by Newman, Spence and Wallis [16].

Theorem 4 (Newman et al. [16]). *Any Hadamard matrix of order $4m$, where m is square-free, is Z-equivalent to*

$$diag(1, \underbrace{2,\ldots,2}_{2m-1\ times}, \underbrace{2m,\ldots,2m}_{2m-1\ times}, 4m). \tag{2}$$

Theorem 4 was originally proved for Hadamard matrices of order $4m$, where m is odd by J. S. Wallis and W. Wallis; see [16].

A skew-Hadamard matrix H_n satisfies $H_n + H_n^T = 2I$. Michael and Wallis [10] have recently determined the Smith normal form of skew-Hadamard matrices. According to Theorem 2 of [10] a skew-Hadamard matrix of order $n = 4m$ for any m has the diagonal matrix (2) as its Smith normal form.

Theorem 5. *If H_n has order $n = 4m$ and m is odd, then $C(H_n)$ has type $4^1 2^{4m-2}$. If m is even and square-free then $C(H_n)$ has type $4^1 2^{2m-1}$. The same conclusion holds if H_n is a skew-Hadamard matrix of order $n = 4m$ and m is even.*

Proof. Let H_n be a particular representative of $\{H_n\}$. We regard the entries of H_n as rational integers, and reduce this matrix to a Z-equivalent matrix which has the Smith normal form (2). So that for some $V, U \in \mathrm{GL}(n, Z)$, $UH_nV = diag(1, \underbrace{2,\ldots,2}_{2m-1\ times}, \underbrace{2m,\ldots,2m}_{2m-1\ times}, 4m)$.

Let $\overline{H_n}$ denote the induced mod 4 reduction of H_n, then $\overline{UH_nV} = \overline{U}\ \overline{H_n}\ \overline{V}$. Where $\overline{U}$ and $\overline{V}$ denote the induced mod 4 row and column operations. To determine the type of $C(H_n)$, it thus suffices to reduce mod 4, the Smith normal form of H_n.

If m is odd, then reducing mod 4 the Smith normal form (2) gives the diagonal matrix $diag(1, \underbrace{2,\ldots,2}_{2m-1\ times}, \underbrace{2,\ldots,2}_{2m-1\ times})$. If m is even and square-free, we obtain $diag(1, \underbrace{2,\ldots,2}_{2m-1\ times})$. Hence the type of $C(H_n)$ is either: $4^1 2^{4m-2}$ or $4^1 2^{2m-1}$ respectively.

If H_n is skew-Hadamard and m is even without any further assumption, the above argument and Theorem 2 of [10] implies that the type is $4^1 2^{2m-1}$.

In the previous argument we used a particular representative of $\{H_n\}$. But if two Hadamard matrices are H-equivalent, they are Z-equivalent; (cf. Lemma 10.4 [16]). Hence any other representative of $\{H_n\}$ is equivalent to the same diagonal form. This proves our result.

The next theorem answers Patrick Solé's question [14].

Theorem 6. *If H_n has order $n = 4m$ and m is odd, then $C(H_n)$ is a self-dual quaternary code.*

Proof. By Theorem 5 there are $4^1 2^{4m-2}$ codewords in $\mathcal{C}(H_n)$. Since the k_1-index in the type of $\mathcal{C}(H_n)$ is 1, the dual code $\mathcal{C}(H_n)^\perp$ contains $4^{n-1-k_2}2^{k_2}$ codewords. (See p. 34 of [3].) But $k_2 = 4m-2$ and $n = 4m$, which gives that $|\mathcal{C}(H_n)^\perp| = |\mathcal{C}(H_n)|$. As $\mathcal{C}(H_n) \subseteq \mathcal{C}(H_n)^\perp$, our assertion is established.

Self-orthogonal quaternary codes which are generated by tetrads – vectors with four coordinates congruent to 1 or 3 mod 4 and the rest congruent to 0 mod 4 – were classified by Conway and Sloane [3]. By Theorem 2, the codes $\mathcal{C}(H_n)$ contain tetrads only when $n = 4$. Indeed, if the generator matrix of $\mathcal{C}(H_4)$

$$H_4 = \begin{pmatrix} 1 & 1 & 1 & 1 \\ 1 & 3 & 1 & 3 \\ 1 & 1 & 3 & 3 \\ 1 & 3 & 3 & 1 \end{pmatrix},$$

is row-reduced mod 4, the generator matrix of the code $\mathcal{D}_4^{\oplus}$ is obtained.

The Gray images: $\phi(2,0,0,2), \phi(2,2,0,0), \phi(2,0,2,0)$ and $\phi(3,3,3,3)$ comprise the e_8 *tetrad system*; (cf. [3]). Hence $\phi(\mathcal{C}(H_4))$ is equivalent to the $[8,4,4]$ extended binary Hamming code.

After this work was complete, Patrick Solé observed that the self-dual codes codes $\mathcal{C}(H_n)$ obtained by Theorem 6, are a subclass of the Klemm codes $\mathcal{K}_{4m}$; cf. [3]. I am grateful to him for allowing his proof of this fact to be included here.

Theorem 7 (Solé [15]). *If H_n has order $n = 4m$ and m is odd, then the self-dual quaternary code $\mathcal{C}(H_n)$ is equivalent Klemm's code $\mathcal{K}_n$.*

Proof. By eq. (1) we can write $\mathcal{C}(H_n) = \mathcal{E}_n \cup \{\mathcal{E}_n + 1\}$ with the all-ones vector. This is equivalent to writing $\mathcal{C}(H_n)$ as $\mathcal{R}_n + \mathcal{E}_n$ with $\mathcal{R}_n = \{0,1\}$. But $\mathcal{E}_n$ is the torsion code therefore $\mathcal{E}_n = 2\mathcal{R}_n^\perp$ by self-duality. But this is the defining equation of Klemm's codes.

5 H-equivalence

Two quaternary codes of the same length are equivalent, if they differ by a permutation or a sign-change of their coordinates. Two binary codes of the same length are equivalent if one is obtained from the other by a permutation of their coordinates.

Theorem 8. *If $H_n^{(1)}$ and $H_n^{(2)}$ are H-equivalent Hadamard matrices then the corresponding quaternary codes $\mathcal{C}(H_n^{(1)})$ and $\mathcal{C}(H_n^{(2)})$ are equivalent.*

Proof. Suppose first that $H_n^{(2)}$ is obtained from $H_n^{(1)}$ by type (i) operations. Swapping pairs of rows does not affect the Z_4-spans. Swapping a pair of columns in $H_n^{(1)}$ induces a transposition on the coordinates of $\mathcal{C}(H_n^{(2)})$; i.e. the codes are equivalent. So we only have to consider type (ii) operations.

We have identified: $-1 \equiv 3 \pmod 4$, so multiplying the rows of a Hadamard matrix by -1 does not affect the Z_4-spans of the rows. Consider now the column operations. By Theorem 2 the codewords belonging to $\mathcal{C}(H_n^{(1)})$ and $\mathcal{C}(H_n^{(2)})$ are either even or odd. Changing the signs of some of the columns of $H_n^{(1)}$ has no affect on the torsion sub-code $\mathcal{E}_n$ of $\mathcal{C}(H_n^{(1)})$. Hence $\mathcal{E}_n$ is a submodule of $\mathcal{C}(H_n^{(2)})$ and $\mathcal{C}(H_n^{(1)})$. This gives the coset decompositions: $\mathcal{C}(H_n^{(1)}) = \mathcal{E}_n \cup \{\mathcal{E}_n + v\}$ and $\mathcal{C}(H_n^{(2)}) = \mathcal{E}_n \cup \{\mathcal{E}_n + v'\}$, for $v \in \mathcal{C}(H_n^{(1)})$ and $v' \in \mathcal{C}(H_n^{(2)})$; cf. eq. (1). The coordinates of v' differ from v by a sign-change of coordinates. Hence the codewords of $\mathcal{C}(H_n^{(2)})$ can be obtained from the codewords of $\mathcal{C}(H_n^{(1)})$ by changing the signs of those coordinates in which v' differs from v.

Corollary 1. *Let $H_n^{(1)}$ and $H_n^{(2)}$ be H-equivalent Hadamard matrices, and $\mathcal{C}(H_n^{(1)})$ and $\mathcal{C}(H_n^{(2)})$ their corresponding quaternary codes. Then the binary images $\phi(\mathcal{C}(H_n^{(1)}))$ and $\phi(\mathcal{C}(H_n^{(2)}))$ are equivalent.*

Proof. By Theorem 8, $\mathcal{C}(H_n^{(1)})$ and $\mathcal{C}(H_n^{(2)})$ are equivalent quaternary codes. Hence the codewords of $\mathcal{C}(H_n^{(1)})$ and $\mathcal{C}(H_n^{(2)})$ differ by either a sign change or a permutation of their coordinates. If the sign of coordinate c_i is changed, then the coordinates of the image vector $\phi(c_i)$ are either swapped or fixed. Hence the binary images of the quaternary codes are equivalent binary codes. Similarly, a permutation $c_i \to c_j$, induces the permutation $\phi(c_i) \to \phi(c_j)$ on pairs of coordinates of the codewords of $\phi(\mathcal{C}(H_n^{(2)}))$. Hence $\phi(\mathcal{C}(H_n^{(1)}))$ and $\phi(\mathcal{C}(H_n^{(2)}))$ are equivalent.

6 Weight Enumerators

We summarize the weight enumerators which are commonly used to describe quaternary codes. (See [12] for more details.)

The *symmetrized weight enumerator* is an invariant of the equivalence classes of quaternary codes. This polynomial is defined by a four variable polynomial – called the *complete weight enumerator*

$$\mathrm{cwe}_{\mathcal{C}}(A,B,C,D) = \sum_{u\in\mathcal{C}} A^{n_0(u)} B^{n_1(u)} C^{n_2(u)} D^{n_3(u)}.$$

Where $n_i(u)$ is the number of coordinates of u which are congruent to i mod 4. Identifying the variables B and D in $\mathrm{cwe}_{\mathcal{C}}(A,B,C,D)$, gives a three variable polynomial – the symmetrized weight enumerator

$$\mathrm{swe}_{\mathcal{C}}(A,B,C) = \mathrm{cwe}_{\mathcal{C}}(A,B,C,B).$$

Finally, the *Hamming weight enumerator* of $\mathcal{C}$ is

$$\mathrm{W}_{\mathcal{C}}(A,B) = \mathrm{cwe}_{\mathcal{C}}(A,B,B,B).$$

For any $a \in \mathcal{C}$, the Hamming weight of a is the number of non-zero coordinates. The *minimum Hamming weight* d_H of $\mathcal{C}$, is the least Hamming weight of all the non-zero codewords of $\mathcal{C}$.

The Lee weight $wt_L(a)$ of a quaternary vector $a \in Z_4^n$ is the rational sum of the Lee weights of its coordinates. (The Lee weights of $0, 1, 2, 3 \in Z_4$ are respectively: $0, 1, 2, 1$.) The *Lee weight enumerator* of quaternary codes C is $\text{Lee}_C(W, X) = \text{swe}_C(W^2, WX, Y^2)$. The *minimum Lee weight* d_L is defined analogously to d_H. The analogues of the MacWilliams relations for the Lee weight and symmetrized weight enumerators were established by Klemm [7].

The Euclidean weights of the elements 0,1,2 and 3 of Z_4 are: 0,1,4 and 1 respectively. The Euclidean weight of a quaternary codeword is the rational sum of the Euclidean weights of the coordinates of the codeword. Obviously, if C_1 and C_2 are equivalent quaternary codes, then the Euclidean weights of $v \in C_1$ and the corresponding vector $v' \in C_2$ are the same. The *minimum* Euclidean weight d_E of a quaternary code C is defined as the least Euclidean weight of all the non-zero codewords of C.

The Lee weight function defines a metric on Z_4^n. The Gray map is an *isometry* from Z_4^n with respect to the Lee distance on Z_2^{2n} and the Hamming distance; see [5]. In other words, any pair of codewords $a, b \in Z_4^n$ satisfy: $d(\phi(a), \phi(b)) = d_L(a, b)$, and $wt(\phi(a)) = wt_L(a)$.

The second condition gives immediately

Theorem 9. *The Lee weight enumerator of the quaternary codes $C(H_n)$ is completely determined by the Hamming weight enumerator of the binary image code $\phi(C(H_n))$.*

To illustrate Theorem 9 consider the code $C(H_4)$. It was shown previously that $\phi(C(H_4))$ is equivalent to the $[8, 4, 4]$ Hamming code. Hence the Lee weight enumerator of $C(H_4)$ is

$$\text{Lee}_{C(H_4)}(W, X) = W^8 + 14W^4X^4 + X^8,$$

also

$$\text{swe}_{C(H_4)} = A^4 + 6A^2C^2 + C^4 + 8B^4.$$

7 Euclidean Weights

The Euclidean weights of our quaternary codes are constrained as follows.

Theorem 10. *The Euclidean weights of the quaternary codes $C(H_n)$ satisfy the following divisibility conditions. If $n \equiv 0 \pmod 8$ the Euclidean weights are divisible by 8; if $n \equiv 4 \pmod 8$ the weights are divisible by 4.*

Proof. If $n \equiv 0 \pmod 8$ the Euclidean weights of the odd codewords of $C(H_n)$ are divisible by 8. As $C(H_n)$ is self-orthogonal, and contains the all-ones vector, the inner product of the all-ones vector and any even codeword $(\ldots, c_i, \ldots)$ is 0 mod 4. Hence any non-zero even codeword has an even number of coordinates c_i which assume the value $c_i = 2$. The Euclidean weights of these coordinates are 4, hence the Euclidean weights of the even codewords are divisible by 8.

A similar argument settles the case $n \equiv 4 \pmod 8$; but now the conclusion is the Euclidean weights of the even codewords are divisible by 8. Which implies that the Euclidean weights of the odd codewords are divisible by 4. Hence the Euclidean weights of all the codewords of $\mathcal{C}(H_n)$ are divisible by 4.

We illustrate Theorem 10 by $\mathcal{C}(H_8)$, the code defined by the unique H-class of Hadamard matrices of order 8. $\mathcal{C}(H_8)$ contains $4^1 2^3$ codewords. The torsion sub-code E_{16} contains 16 codewords. The first four coordinates of the codewords of E_{16} are either: $(0,0,0,0|\ldots)$, $(2,2,2,2|\ldots)$, or all 6 combinations of $(2,2,0,0|\ldots)$. The coordinates of a codeword following the vertical bar are either the same as the first four coordinates, or are obtained by replacing a 2 by 0; e.g. $(2,2,0,0|\ldots)$ corresponds to $(2,2,0,0|2,2,0,0)$ and $(2,2,0,0|0,0,2,2)$. The minimum Euclidean distance is $d_E = 8$, and

$$\mathrm{swe}_{\mathcal{C}(H_8)} = A^8 + 16A^4C^4 + C^8 + 16B^8.$$

8 Type IV Quaternary Codes

A *Type* IV code is a quaternary self-dual code, in which all the codewords have even Hamming weights. (See [4].)

Theorem 11. *Suppose that $n \equiv 0 \pmod 4$, then the Hamming weights of all codewords belonging to $\mathcal{C}(H_n)$ are even.*

Proof. By Theorem 2 only the codewords of the torsion sub-code $\mathcal{E}_n$ can have zero coordinates. An inner product argument, similar to the one used in Theorem 10, and fact that $n \equiv 0 \pmod 4$, establishes that the Hamming weights of the codewords of $\mathcal{E}_n$ are even.

In view of Theorem 7, we include the following result only for the sake of completeness.

Corollary 2. *If H_n has order $n = 4m$ and m is odd, then $\mathcal{C}(H_n)$ is a Type IV code.*

Proof. By Theorem 6 the quaternary codes $\mathcal{C}(H_n)$ are self-dual. By Theorem 11 the Hamming weights of all the codewords of $\mathcal{C}(H_n)$ are even.

9 Binary Images

Theorem 12. *The binary image $\phi(\mathcal{C}(H_n))$ of the quaternary code $\mathcal{C}(H_n)$ is a binary linear code.*

Proof. By Theorem 2 the codewords of $\mathcal{C}(H_n)$ are either even or odd. Let a and b be any two codewords of $\mathcal{C}(H_n)$. Without loss of generality the possible combinations of parity for the pairs $\{a,b\}$ are: $\{\text{even},\text{even}\}$, $\{\text{odd},\text{odd}\}$, or $\{\text{even},\text{odd}\}$. Applying the map α (cf. Section 2), to a and b gives a vector which belongs to $\mathcal{C}(H_n)$ in all three cases. Hence by Theorem 1 $\phi(\mathcal{C}(H_n))$ is a binary linear code.

Theorem 13. *Suppose that* $n \equiv 0 \pmod 4$ *and* $\phi(C(H_n))$ *is the binary image of a quaternary code defined by a Hadamard matrix* H_n *of order* n. *Then* $\phi(C(H_n))$ *is a self-orthogonal binary code.*

Proof. Let $\phi(a)$ and $\phi(b)$ be two codewords of $\phi(C(H_n))$. The following relation is well known (cf. p. 9 [9]):

$$wt(\phi(a) + \phi(b)) = wt(\phi(a)) + wt(\phi(b)) - 2(\phi(a), \phi(b)).$$

Where $(\phi(a), \phi(b))$ denotes the standard rational valued scalar product. Theorem 12 and Corollary 9 imply that $wt_L(c) = wt_L(a) + wt_L(b) - 2(\phi(a), \phi(b))$. Where the quaternary vector c corresponds to the binary vector $\phi(a) + \phi(b)$ under the Gray map. ($\phi(C(H_n))$ is a linear code.) We conclude that $2(\phi(a), \phi(b)) = 2n - wt_L(c)$. Now Theorem 2 gives:

$$2n - wt_L(c) = \begin{cases} 2n - n, \text{ or} \\ 2n - 2n. \end{cases}$$

Hence in both cases $(\phi(a), \phi(b)) = n/2$, or 0. By the hypothesis $n/2$ is divisible by 2. So reducing $(\phi(a), \phi(b)) \mod 2$, gives 0 in both cases. Hence the codewords belonging to $\phi(C(H_n))$ are mutually orthogonal. For binary codes this condition is equivalent to $\phi(C(H_n)) \subseteq \phi(C(H_n))^\perp$.

We also obtain self-dual codes as binary images.

Theorem 14. *If a Hadamard matrix* H_n *has order* $n = 4m$ *and* m *is odd, then* $\phi(C(H_n))$ *is a self-dual binary code.*

Proof. By Theorem 13 $\phi(C(H_n)) \subseteq \phi(C(H_n))^\perp$. By Theorem 6 the quaternary code $C(H_n)$ contains 2^{4m} codewords, and so does its binary image. Since $\phi(C(H_n))$ is a binary code of length $8m$ and dimension $4m$ our assertion follows.

Theorem 15. *Let* $\phi(C(H_n))$ *be the linear binary code defined by a Hadamard matrix* H_n *of order* $n > 2$. *Then the Hamming weights of the codewords of* $\phi(C(H_n))$ *are divisible by* 4. *If* H_n *has order* $n = 4m$ *and* m *is odd, then* $\phi(C(H_n))$ *is a doubly even, or Type II code. (Cf. The Gleason, Pierce and Turyn classification; see [9].)*

Proof. By definition the codewords of $\phi(C(H_n))$ are the Gray-map images of the codewords of $C(H_n)$. By Theorem 2 these are either even or odd. An odd codeword a has one of the following possible shapes: (1^n), (3^n) or $(1^r, 3^s)$, where $r + s = n$. The Hamming weights of the binary vectors $\phi(a)$ corresponding to the codewords of these shapes are either: n in the first two cases, and $r + s = n$ for the last. Since $n \equiv 0 \pmod 4$, the Hamming weights are divisible by 4 in all cases.

The argument is similar for the even codewords. But now the possible shapes of $b \in \mathcal{E}_n$ are: (0^n), (2^n), or $(0^t, 2^u)$ where $t+u = n$. Hence there are three possible Hamming weights for $\phi(b)$: 0, $2n$, and $2u$. Since the all-ones vector is orthogonal to $\mathcal{E}_n$, it must be the case that $u \equiv 0 \pmod 4$.

The last assertion now follows from Theorem 14.

The hypothesis that $n = 4m$ and m is odd, is equivalent to $n \equiv 4 \pmod 8$. Ozeki [11] has constructed binary Type II codes from equivalence classes of Hadamard matrices satisfying this condition. By Theorem 7 and the definition of Klemm's codes, it follows that the minimum Hamming weight of the self-dual binary codes $\phi(\mathcal{C}(H_n))$ is always 4. (We have included an independent proof of Theorem 14.)

10 Examples of Codes

In this Section we survey all the quaternary and binary codes which arise from the H-equivalence classes of Hadamard matrices H_n of orders up to 20. The required calculations were performed with the Guava package of GAP [13].

$\{H_n\}$: $n \leq 12$.
For each admissible order $n \leq 12$ there is a unique equivalence class $\{H_n\}$. The codes arising from these equivalence classes have already been discussed.

$\{H_n\}$: $n = 16$.
There are five equivalence classes of $\{H_{16}\}$. They are Hall's classes: I, II, III, IV, V. Let A, B, C, and D be representatives of the first four classes. The representative matrix of class V is the transposed matrix D^T; see [16]. The quaternary codes $\mathcal{C}$(A), $\mathcal{C}$(B), and $\mathcal{C}$(C), have types: $4^1 2^4$, $4^1 2^5$, and $4^1 2^6$. $\mathcal{C}$(A) is equivalent to the code $ZRM(1,4)$; cf. [5].

D and D^T generate two inequivalent quaternary codes of length 16 and type $4^1 2^7$. The generator matrices of these codes are:

$$\mathcal{C}(\mathrm{D}) := \begin{pmatrix} 1&1&1&1&1&1&1&1&1&1&1&1&1&1&1&1 \\ 0&2&0&2&0&2&0&2&0&2&0&2&0&2&0&2 \\ 0&0&2&2&0&0&0&0&0&0&0&0&0&0&2&2 \\ 0&0&0&0&2&2&0&0&0&0&0&0&0&0&2&2 \\ 0&0&0&0&0&0&2&2&0&0&0&0&0&0&2&2 \\ 0&0&0&0&0&0&0&0&2&2&0&0&0&0&2&2 \\ 0&0&0&0&0&0&0&0&0&0&2&2&0&0&2&2 \\ 0&0&0&0&0&0&0&0&0&0&0&0&2&2&2&2 \end{pmatrix},$$

and

$$\mathcal{C}(\mathrm{D^T}) := \begin{pmatrix} 1&1&1&1&1&1&1&1&1&1&1&1&1&1&1&1 \\ 0&2&0&0&0&2&2&2&0&0&0&0&0&0&0&0 \\ 0&0&2&0&2&0&2&2&0&0&0&0&0&0&0&0 \\ 0&0&0&2&2&2&0&2&0&0&0&0&0&0&0&0 \\ 0&0&0&0&0&0&0&0&2&0&0&0&0&2&2&2 \\ 0&0&0&0&0&0&0&0&0&2&0&0&2&0&2&2 \\ 0&0&0&0&0&0&0&0&0&0&2&0&2&2&0&2 \\ 0&0&0&0&0&0&0&0&0&0&0&2&2&2&2&0 \end{pmatrix}.$$

The binary image codes: $n = 16$

The binary image $\phi(\mathcal{C}(\mathrm{A}))$ is the first order Reed-Muller code $R(1,5)$; cf. [5]. Let $\phi(\mathcal{C}(\mathrm{B}))$, $\phi(\mathcal{C}(\mathrm{C}))$, $\phi(\mathcal{C}(\mathrm{D}))$, and $\phi(\mathcal{C}(\mathrm{D}^{\mathrm{T}}))$ be the self-orthogonal binary images of the previous quaternary codes. The first three codes have parameters: $[32,7,8]$, $[32,8,8]$, $[32,9,8]$. Their dual codes have parameters: $[32,25,4]$, $[32,24,4]$, $[32,23,4]$. These are optimal codes (having the maximal possible minimal Hamming weight in each dimension). The Hamming weight enumerators of these codes are given below. (The zero and the all-ones vector is omitted; A_i denotes the number of codewords of Hamming weight i.)

Code	A_4	A_6	A_8	A_{10}	A_{12}	A_{14}	A_{16}
$\phi(\mathcal{C}(\mathrm{B}))$	0	0	4	0	0	0	118
$\phi(\mathcal{C}(\mathrm{C}))$	0	0	12	0	0	0	230
$\phi(\mathcal{C}(\mathrm{D}))$	0	0	28	0	0	0	454
$\phi(\mathcal{C}(\mathrm{B}))^{\perp}$	728	1344	165596	1006208	3533544	7352576	9410246
$\phi(\mathcal{C}(\mathrm{C}))^{\perp}$	472	6272	83164	503424	1768424	3668224	4717254
$\phi(\mathcal{C}(\mathrm{D}))^{\perp}$	344	2688	41948	252032	885864	1826048	2370758

The codes $\phi(\mathcal{C}(\mathrm{D}))$ and $\phi(\mathcal{C}(\mathrm{D}^{\mathrm{T}}))$ are inequivalent, as are $\phi(\mathcal{C}(\mathrm{D}))^{\perp}$ and $\phi(\mathcal{C}(\mathrm{D}^{\mathrm{T}}))^{\perp}$. However, both pairs of codes have identical Hamming weight enumerators. The dual codes contain more codewords of Hamming weight 4 than the corresponding (optimal) codes obtained from the $|U|U+V|$ construction; cf. [9].

$\{H_n\}$: $n = 20$.

There are three classes of Hadamard matrices of order 20. But by Theorem 7, the three classes define a single Klemm code $\mathcal{K}_{20}$.

11 Conclusion

We have described a new construction of self-orthogonal quaternary and binary codes from the H-equivalence classes of Hadamard matrices. We have examined the codes which are generated by Hadamard matrices of orders up to 20. Beyond this, only the Hadamard matrices of orders 24 and 28 are completely classified into H-equivalence classes. Partial classifications are known for higher order Hadamard matrices. We are currently investigating the quaternary and binary codes generated by the classes of Hadamard matrices of orders 24 and hope to report on these elsewhere.

References

1. A. Bonnecaze, P. Solé, C. Bachoc and B. Mourrain, Type II codes over Z_4, IEEE Trans. Inform. Theory, **43**, (1997), 969-976.

2. C. Carlet, One-weight Z_4-linear codes, preprint 1998.
3. J. H. Conway and N. J. A. Sloane, Self-dual codes over the integers modulo 4, Journ. Comb. Theory A **62** (1993), 30-45.
4. S. T. Dougherty, P. Gaborit, M. Harada, A. Munemasa, and P. Solé, Type IV self-dual codes over rings, to appear 1999.
5. A. R. Hammons, P. V. Kumar, A. R. Calderbank, N. J. A. Sloane, and P. Solé, The Z_4-linearity of Kerdock, Preparata, Goethals, and related codes, IEEE Trans. Inform. Theory, **40**, (1994), 301-319.
6. M. Harada, P. Solé, and P. Gaborit, Self-Dual codes over Z_4 and unimodular lattices: A survey, Preprint 1997.
7. M. Klemm, Selbstduale Codes über dem Ring der ganzen Zahlen modulo 4, Arch. Math. **53** (1989), 201-207.
8. V. Pless, J. S. Leon, and J. Fields, All Z_4 codes of Type II and length 16 are known, Journ. Comb. Theory A **78** (1997), 32-50.
9. F. J. MacWilliams and N. J. A. Sloane, *The Theory Of Error Correcting Codes*, North-Holland, Amsterdam, 1977.
10. T. S. Michael and W. D. Wallis, Skew-Hadamard matrices and the Smith normal form, Designs, Codes and Cryptography **13** (1998), 173-176.
11. M. Ozeki, Hadamard Matrices and doubly even self-dual error-correcting codes, Journ. Comb. Theory A **44** (1987), 274-287.
12. E. M. Rains and N. J. A. Sloane, Self-Dual Codes, in *Handbook of Coding Theory*, V. Pless and W. C. Huffman, Eds. Amsterdam, Elsevier, 1998.
13. M. Schönert et al., GAP Groups, Algorithms and Programming 3.4.4, Lehrstuhl D für Mathematik, RWTH Aachen, 1997.
14. P. Solé, private communication, Nov. 1997.
15. P. Solé, private communication, Mar. 1999.
16. W. D. Wallis, A. P. Street, and J. S. Wallis, *Combinatorics: Room Squares, Sum-Free Sets, Hadamard Matrices*, Lecture Notes in Mathematics, Vol. 292, Springer-Verlag, Berlin, New York, Heidelberg, 1972.

On Constructing Balanced Correlation Immune Functions

Thomas W. Cusick

106 Diefendorf Hall, Department of Mathematics,
State University of New York at Buffalo, Buffalo, NY 14214-3093
E-mail: cusick@acsu.buffalo.edu

Abstract. The notion of correlation immunity for Boolean functions is important for cryptographic security in the design of many stream ciphers. This paper gives a new method for constructing balanced Boolean functions which are correlation immune of order k. The nonlinearity and algebraic degree of the constructed functions can be controlled.
Index terms - Cryptography, Boolean functions, stream ciphers, correlation, nonlinearity.

1 Introduction

The notion of correlation immunity has been important in the design of stream ciphers since it was introduced by Siegenthaler [7,8]. For $Z_2 = \{0,1\}$, a Boolean function $f : Z_2^n \longrightarrow Z_2$ is said to be *correlation immune of order* k if knowing the value of any k of the input variables gives no additional information on the value of the output. Equivalently, if $\mathbf{x} = (x_1, \ldots, x_n)$, then $f(\mathbf{x})$ is correlation immune of order k if for any choice of k variables $x_{i(1)}, \ldots, x_{i(k)}$, all of the 2^k probabilities

$$\Pr(x_{i(1)} = \epsilon_1, \ldots, x_{i(k)} = \epsilon_k \mid f(\mathbf{x}) = 1)$$

are 2^{-k}, where each ϵ_i is in Z_2.

Define the *Walsh-Hadamard transform* $F_f(\mathbf{w}) = F(\mathbf{w})$ of $f(\mathbf{x})$ by

$$F(\mathbf{w}) = \sum_{\mathbf{x}} (-1)^{f(\mathbf{x}) + \mathbf{x} \cdot \mathbf{w}}$$

where the sum is over Z_2^n and $\mathbf{x} \cdot \mathbf{w}$ is the ordinary dot product. Xiao and Massey [9] proved that $f(\mathbf{x})$ is correlation immune of order k if and only if

$$F(\mathbf{w}) = 0 \quad \forall \quad \mathbf{w} \quad \text{with} \quad 1 \leq wt(\mathbf{w}) \leq k,$$

where $wt(\mathbf{w})$ is the Hamming weight. In other words, $f(\mathbf{x})$ is correlation immune of order k if and only if for any choice of $t \leq k$ variables $x_{i(j)}$, $1 \leq j \leq t$, the function

$$f(\mathbf{x}) + \sum_{j=1}^{t} x_{i(j)}$$

is *balanced* (that is, takes on equal number of 0's and 1's). Note that $f(\mathbf{x})$ is balanced if and only if $F(\mathbf{0}) = 0$.

In many cryptographic applications which use Boolean functions $f(\mathbf{x})$, it is required that these functions be balanced. In this paper, we give a method for constructing balanced Boolean functions which are correlation immune of order k. For brevity, we will say that such a function is BCI(k). There is an alternative terminology (which we do not use) going back to Chor et al. [4]: A Boolean function which is BCI(k) is said to be *k-resilient.*

2 Constructing balanced correlation immune functions

Given a Boolean function $f(\mathbf{x})$ in n variables, it is convenient to write

$$f(\mathbf{x}) = f = (v(1), \ldots, v(2^n)) \tag{1}$$

where the $v(i)'s$ are the values of $f(\mathbf{x})$ when $\mathbf{x}$ runs through the vectors $(0, \ldots, 0)$, $(0, \ldots, 0, 1)$, $\ldots$, $(1, \ldots, 1)$ of Z_2^n in lexicographical order. In fact, we shall sometimes identify f with the vector of its values in (1) and we shall refer to the vectors $(v(1), \ldots, v(2^{n-1}))$ and $(v(2^{n-1}+1), \ldots, v(2^n))$ as the *first* and *second halves* of f, respectively.

We now describe a method due to Camion et al. [1] for constructing functions f which are BCI(k). We fix n and r with $1 \leq r < n$, and we define

$$m = n - r, \qquad \mathbf{u} = (x_1, \ldots, x_m), \quad \mathbf{v} = (x_{m+1}, \ldots, x_n).$$

Let $g(\mathbf{u})$ be any Boolean function on Z_2^m and let $\phi(\mathbf{u})$ be a map from Z_2^m to Z_2^r. Define

$$f(\mathbf{x}) = f(\mathbf{u}, \mathbf{v}) = \mathbf{v}.\phi(\mathbf{u}) + g(\mathbf{u}).$$

Then it is proved in [1, p. 98] that if

$$w = \min\{wt(\phi(\mathbf{u})) \; : \mathbf{u} \in Z^m\},$$

then f is BCI(k) with $k \geq w - 1$. (Note that our notation differs from that in [1].)

In cryptographic applications, the *nonlinearity* N_f of a Boolean function $f(\mathbf{x})$ is often important. This is defined as follows: Given Boolean functions f and g in n variables, the *Hamming distance* $d(f, g)$ between them is defined by

$$d(f, g) = \mid \{\mathbf{x} : \; f(\mathbf{x}) \neq g(\mathbf{x})\} \mid,$$

where $\mid S \mid$ denotes the number of elements in the set S. Now N_f is given by

$$N_f = \min_a \; d(f, a),$$

where a runs through all of the 2^{n+1} affine functions in n variables.

It is well-known that an alternative formula for N_f is

$$N_f = 2^{n-1} - \frac{1}{2}\max_{\mathbf{w}} |F(\mathbf{w})|$$

and that for any Boolean function f in n variables we have

$$N_f \leq 2^{n-1} - 2^{(\frac{n}{2})-1}.$$

The method of constructing BCI(k) functions in [1] has been further studied by Chee et al. [3] (the journal version of this paper is [2]) and Seberry et al. [6]. These later papers concentrate on finding functions which are BCI(k) and have high nonlinearity and large degree (considered as polynomials in n variables).

In order to describe and prove the correctness of our method for constructing BCI(k) functions, it is convenient to have the two lemmas below. The first one uses the special 4-bit strings defined by

$$A = 0,0,1,1; \quad B = 0,1,0,1; \quad C = 0,1,1,0; \quad D = 0,0,0,0$$

and their complements defined by

$$\bar{A} = 1,1,0,0; \quad \bar{B} = 1,0,1,0; \quad \bar{C} = 1,0,0,1; \quad \bar{D} = 1,1,1,1.$$

We define the set

$$T = \{A, B, C, D, \bar{A}, \bar{B}, \bar{C}, \bar{D}\}$$

which is made up of some of the above 4-bit strings.

Given a Boolean Function (1) and a positive integer $j \leq n$, we say f is made up of 2^{n-j} consecutive *blocks* of bits, each block having length 2^j.

Lemma 1 *Any affine function $f = (v(1), \ldots, v(2^n))$ is a string of 2^n bits made up of 4-bit blocks $I_1, \ldots, I_{2^{n-2}}$ given as follows: The first block I_1 is in T. The second block I_2 is I_1 or $\bar{I}_1$. The next two blocks I_3 and I_4 are I_1, I_2 or $\bar{I}_1, \bar{I}_2$. The next four blocks $I_5, \ldots, I_8$ are $I_1, \ldots, I_4$ or $\bar{I}_1, \ldots, \bar{I}_4$. .... The last 2^{n-3} blocks $I_{2^{n-3}+1}, \ldots, I_{2^{n-2}}$ are $I_1, \ldots, I_{2^{n-3}}$ or $\bar{I}_1, \ldots, \bar{I}_{2^{n-3}}$.*

Proof. This is part of the "folklore" of the theory of Boolean functions. ∎

Lemma 2 *Given integers $n > k \geq 0$, let $\epsilon = (\epsilon_{n-k}, \ldots, \epsilon_n)$ be any $(k+1)$-vector of bits. Then all the blocks of length 2^{k+1} in the linear function*

$$L(\epsilon) = \sum_{i=n-k}^{n} \epsilon_i x_i$$

are the same (i.e., the period length of this linear function divides 2^{k+1}).

Proof. It is easy to see (special case of Lemma 1) that the period of x_n is $0,1$; of x_{n-1} is A; of x_{n-2} is $D\bar{D}$; and of $x_{n-j}, 3 \leq j \leq k$, is $D_{2^{j-2}}\bar{D}_{2^{j-2}}$ (here D_i means a string of i $D's$). Thus each of the linear functions $x_{n-k}, \ldots, x_n$ has period length which divides 2^{k+1}, and hence so does any sum of these linear functions. ∎

Theorem 1 *Fix n and r with $1 \leq r < n$ and let $\mathbf{x} = (x_1, \ldots, x_n)$. Define $m = n - r$,*

$$\mathbf{u} = (x_1, \ldots, x_m), \quad \mathbf{u}_0 = (0, x_2, \ldots, x_m), \quad \mathbf{v} = (x_{m+1}, \ldots, x_n).$$

Let $\phi(\mathbf{u})$ be a map from Z_2^m to Z_2^r defined as follows: $\phi(\mathbf{u}_0)$ can take any 2^{m-1} values and the remaining 2^{m-1} values of $\phi(\mathbf{u})$ are determined by the permutation defined below. Let $\epsilon(\mathbf{u})$ be a map from Z_2^m to Z_2 defined as follows: $\epsilon(\mathbf{u}_0)$ can take any 2^{m-1} values and the remaining 2^{m-1} values of $\epsilon(\mathbf{u})$ are determined by the permutation defined below. Define $f(\mathbf{x}) = (v(1), \ldots, v(2^n))$ to be a string of 2^{n-r} blocks of length 2^r, chosen as follows: We always have

$$f(\mathbf{x}) = \mathbf{v} \cdot \phi(\mathbf{u}) + \epsilon(\mathbf{u}).$$

For the first half of f $(x_1 = 0)$, define

$$f(0, x_2, \ldots, x_n) = \mathbf{v} \cdot \phi(\mathbf{u}_0) + \epsilon(\mathbf{u}_0).$$

For the second half of f, define $(v(2^{n-1}+1), \ldots, v(2^n))$ to be some permutation of the complements of the 2^{n-r-1} blocks in the first half of f. This permutation completes the definitions of $\phi(\mathbf{u})$ and $\epsilon(\mathbf{u})$. If

$$w = \min\{wt(\phi(\mathbf{u})) \ : \mathbf{u} \in Z^m\},$$

then $f(\mathbf{x})$ is BCI(k) with $k \geq w$.

Proof. By the discussion in the introduction, it suffices to prove that the Walsh-Hadamard transform $F(\mathbf{w})$ is 0 for all $\mathbf{w}$ with $0 \leq wt(\mathbf{w}) \leq w$. From its definition, f is balanced, so we have $F(\mathbf{0}) = 0$.

Now consider $\mathbf{w}$ with $1 \leq wt(\mathbf{w}) \leq w$. Let $\mathbf{w} = (\mathbf{w}_L, \mathbf{w}_R)$, where $\mathbf{w}_L = (w_1, \ldots, w_m)$ and $\mathbf{w}_R = (w_{m+1}, \ldots, w_n)$, so $\mathbf{x} \cdot \mathbf{w} = \mathbf{u} \cdot \mathbf{w}_L + \mathbf{v} \cdot \mathbf{w}_R$. If $wt(\mathbf{w}) < w$ or $wt(\mathbf{w}) = w$ and $wt(\mathbf{w}_R) < w$, then $\mathbf{v} \cdot \phi(\mathbf{u}) + \epsilon(\mathbf{u}) + \mathbf{x} \cdot \mathbf{w}$ is a nonconstant affine function of $x_{m+1}, \ldots, x_n$ for each $\mathbf{u}$ and hence

$$\sum_{\mathbf{x}} (-1)^{f(\mathbf{x}) + \mathbf{x} \cdot \mathbf{w}} = \sum_{\mathbf{u}} (-1)^{\epsilon(\mathbf{u}) + \mathbf{u} \cdot \mathbf{w}_L} \sum_{\mathbf{v}} (-1)^{\mathbf{v} \cdot \phi(\mathbf{u}) + \mathbf{v} \cdot \mathbf{w}_R} = 0 \ ; \qquad (2)$$

in fact, all of the inner sums over $\mathbf{v}$ are 0.

It remains to consider the case $wt(\mathbf{w}) = wt(\mathbf{w}_R) = w$. Here it may happen that $\mathbf{v} \cdot \phi(\mathbf{u}_0) + \epsilon(\mathbf{u}_0) + (\mathbf{u}_0, \mathbf{v}) \cdot \mathbf{w}$ is a constant function of $\mathbf{v}$ for some choices of $\mathbf{u}_0$, so some of the sums

$$\sum_{\mathbf{v}} (-1)^{\mathbf{v} \cdot \phi(\mathbf{u}_0) + \mathbf{v} \cdot \mathbf{w}_R}$$

will have value 2^r instead of 0. Consider such a value of $\mathbf{u}_0$, say $\mathbf{u}_0 = (0, a_2, \ldots, a_m)$, so that the block

$$\{\mathbf{v} \cdot \phi(\mathbf{u}_0) + \epsilon(\mathbf{u}_0) + (\mathbf{u}_0, \mathbf{v}) \cdot \mathbf{w} \mid \mathbf{v} \in Z^r\} \qquad (3)$$

is made up of 2^r $0's$ or $1's$. By our hypotheses, there is some vector $\mathbf{u}^* = (1, b_2, \ldots, b_m)$ such that $\phi(\mathbf{u}^*) = \phi(\mathbf{u}_0)$ and $\epsilon(\mathbf{u}^*) + \epsilon(\mathbf{u}_0) = 1$, and because of Lemma 2 with $k = r - 1$ the block

$$\{\mathbf{v} \cdot \phi(\mathbf{u}^*) + \epsilon(\mathbf{u}^*) + (\mathbf{u}^*, \mathbf{v}) \cdot \mathbf{w} \mid \mathbf{v} \in Z^r\}$$

is the complement of the block in (3). Thus we can always pair off the inner sums over $\mathbf{v}$ in (2) in such a way that we obtain $F(\mathbf{w}) = 0$. ∎

In comparing the construction of Theorem 1 with that of [1], we see that in Theorem 1 the minimum weight w in the definition of ϕ can be taken to be one unit smaller for a given order k of correlation immunity. This gives many more BCI(k) functions. On the other hand, Theorem 1 does not allow the addition of an arbitrary function $g(\mathbf{u})$ as in [1].

3 Nonlinearity

It is easy to give good explicit bounds on the nonlinearity of the functions defined in Theorem 1.

Theorem 2 *Let $f(\mathbf{x}) = \mathbf{v} \cdot \phi(\mathbf{u}) + \epsilon(\mathbf{u})$ be one of the functions defined in Theorem 1. Suppose the most numerous block of length 2^r in the bit string $(v(1), \ldots, v(2^n))$ of values of $f(\mathbf{x})$ occurs R times. Then the nonlinearity N_f satisfies*

$$2^{n-1} - 2^{(n/2)-1} \geq N_f \geq 2^{n-1} - 2^r R.$$

Proof. The upper bound is well-known (equality holds for bent functions). By Lemma 1, any affine function $a(\mathbf{x})$ in k variables is made up of 2^{k-2} 4-bit blocks, each of which is either one of the strings A, B, C, D or its complement. We say that $a(\mathbf{x})$ is *based on* the underlying string A, B, C or D. If a_1 and a_2 are two affine functions based on different 4-bit strings, then Lemma 1 implies $d(a_1, a_2) = 2^{k-1}$, because the Hamming distance between any two strings in the set T which are based on different letters is 2. If a_1 and a_2 are two different affine functions based on the same letter and not complements of each other, then Lemma 1 again implies $d(a_1, a_2) = 2^{k-1}$. By definition, $f(\mathbf{x}) = (v(1), \ldots, v(2^n))$ is made up of 2^{n-r} blocks of length 2^r, each of which is an affine function in r variables. If P denotes the most numerous block of length 2^r, then by our remarks above the affine function $a = PP\ldots P$ has

$$d(f, a) = 2^{n-1}.$$

Note that by the construction of Theorem 1, the block $\bar{P}$ also appears R times in f. It may be that there is an affine function b made up of blocks P and $\bar{P}$ which agrees with f at all of the blocks P or $\bar{P}$. Such a function has the maximum possible agreement with f, so we would have

$$d(f, b) = 2^n - 2^{r+1}R - \frac{1}{2}(2^n - 2^{r+1}R) = 2^{n-1} - 2^r R,$$

which gives the lower bound on N_f. If b does not exist, then we get strict inequality in the lower bound. ∎

A result similar to Theorem 2 for the functions of Camion et al. [1] has been given in [3, Th.5, p.234] and [6, Th.14, p.190] (in the former paper an exact equality for the nonlinearity is achieved). Our method of proof is different, and could also be used to obtain these results.

Because Theorem 1 allows the value of w to be one smaller than the corresponding value in the construction of [1], the construction in this paper gives a bigger set of BCI(k) functions from which to choose those with large nonlinearity. This could be advantageous in some cryptographic applications.

4 Algebraic Degree

Sometimes it is desirable for cryptographic applications to avoid Boolean functions of too small degree. However, Siegenthaler [7] proved that there is the following tradeoff between the degree $d(f)$ and the order k of correlation immunity for a Boolean function f in n variables: we always have $d(f) \leq n-k$ and if f is also balanced, then $d(f) \leq n-k-1$ for $k < n-1$.

It turns out that the maximum degree possible for the functions in Theorem 1 is m. We can produce BCI(k) functions of maximum possible degree m in Theorem 1 by using the following result, which characterizes the functions in Theorem 1 which have degree m.

Theorem 3 *Let $f(\mathbf{x}) = \mathbf{v} \cdot \phi(\mathbf{u}) + \epsilon(\mathbf{u})$ be one of the functions defined in Theorem 1. Let $\phi(\mathbf{u}) = (y_{m+1}, \ldots, y_n)$. Then $d(f) \leq m$; any term of degree m which has a nonzero coefficient is of form $x_1 \ldots x_{j-1}x_{j+1} \ldots x_m x_t$ for some j and t with $1 \leq j \leq m$ and $m+1 \leq t \leq n$; and each such term has coefficient*

$$\sum_{\{\mathbf{u}:\, x_j=0 \text{ and } \phi(\mathbf{u}) \text{ has } y_t=1\}} 1$$

Thus $d(f) = m$ if and only if at least one of the above sums has an odd number of terms.

Proof. We can expand $f(\mathbf{x})$ as

$$\begin{aligned} f(\mathbf{x}) = {} & (x_1+1)(x_2+1)\ldots(x_m+1)(\phi(\mathbf{0}) \cdot \mathbf{v} + \epsilon(\mathbf{0})) \\ & + (x_1+1)(x_2+1)\ldots x_m(\phi((0,\ldots,0,1)) \cdot \mathbf{v} + \epsilon((0,\ldots,0,1))) \\ & + \ldots \\ & + x_1x_2\ldots x_m(\phi((1,\ldots,1,1)) \cdot \mathbf{v} + \epsilon((1,\ldots,1,1))). \end{aligned}$$

By the hypotheses of Theorem 1, any value of $\phi(\mathbf{u})$ appears an even number of times in this expansion, so all terms $x_1x_2\ldots x_m x_t$ $(t \geq m+1)$ have coefficient 0; hence $d(f) \leq m$. The term $x_1x_2\ldots x_m$ has coefficient $\sum \epsilon(\mathbf{u}) = 0$, so any term of degree m in the expansion has the form given in the theorem, for some j and t. Given such a term, direct computation from the expansion of $f(\mathbf{x})$ shows that its coefficient is the expression given in the theorem. This completes the proof. ∎

5 An old construction

We remark that it is very simple to construct a large class of correlation immune functions of order 1, which we can choose to be balanced, if desired. We simply define the first half of a Boolean function of form (1) arbitrarily, and then define the second half to be the first half in reverse order. This is a "folklore" construction, but it is mentioned by Mitchell [5, p. 163] in the context of counting correlation immune functions. This construction gives many BCI(1) functions, but we cannot analyze their nonlinearity or algebraic degree without imposing some restrictions. This is why more specialized constructions, such as the one in Theorem 1, are more useful for cryptographic applications.

References

1. P. Camion, C. Carlet, P. Charpin and N. Sendrier, On correlation-immune functions, *Advances in Cryptology - Crypto '91, Springer Lecture Notes Comput. Sci.* **576** (1992), pp. 86-100.
2. Seongtaek Chee, Sangjin Lee, Kwangjo Kim and Daeho Kim, Correlation immune functions with controllable nonlinearity, *ETRI J.* **19** (1997), pp. 389-401.
3. Seongtaek Chee, Sangjin Lee, Daiki Lee and Soo Hak Sung, On the correlation immune functions and their nonlinearity, *Advances in Cryptology - Asiacrypt '96, Springer Lecture Notes Comput. Sci.* **1163** (1996), pp. 232-243.
4. B. Chor, O. Goldreich, J. Hastad, J. Freidman, S. Rudich and R. Smolensky, The bit extraction problem or t-resilient functions, *Proc. 26th Sympos. Found. Comput. Sci.* (1985), pp. 396-407.
5. C.J. Mitchell, Enumerating functions of cryptographic significance, *J. Cryptology* **2** (1990), pp. 155-170.
6. J. Seberry, X. Zhang and Y. Zheng, On constructions and nonlinearity of correlation immune functions, *Advances in Cryptology - Eurocrypt '93, Springer Lecture Notes Comput. Sci.* **765** (1994), pp.181-199.
7. T. Siegenthaler, Correlation-immunity of nonlinear combining functions for cryptographic applications, *IEEE Trans. Inform. Theory* **30** (1984), pp. 776-780.
8. T. Siegenthaler, Decrypting a class of stream ciphers using ciphertext only, *IEEE Trans. Computers* **34** (1985), pp. 81-85.
9. G. Xiao and J. Massey, A spectral characterization of correlation-immune combining functions, *IEEE Trans. Inform. Theory* **34** (1988), pp. 569-571.

On Bispecial Words and Subword Complexity of D0L Sequences *

A. Frid[1] and S. V. Avgustinovich[1]

Novosibirsk State University,
Pirogova st. 2, 630090 Novosibirsk, Russia.
Email: {frid,avgust}@math.nsc.ru

Abstract. We study the *subword complexity* of a sequence which counts the number of its distinct factors of a given length. Using the technique of *bispecial words*, we find an explicit formula for the subword complexity of a wide class of D0L sequences. We also obtain the characterization of bispecial factors and *minimal forbidden words* of that class of sequences.

1 Introduction

One of the classical functions describing the complexity of a sequence is its *subword complexity* $f(n)$ which is equal to the number of distinct factors of length n of the sequence. This function agrees with the intuition, according to which the more distinct subwords appear in a sequence, the more complicated the sequence is. Clearly, the subword complexity of a sequence on a q-letter alphabet is a non-decreasing function satisfying the inequalities $1 \leq f(n) \leq q^n$.

Two approaches of studying the subword complexity are popular: finding the complexity of given sequences and answering the question of what functions can be the subword complexity functions of sequences. The survey [1] covers mostly the first approach, and [7] illuminates the second one.

In this paper, we give an explicit formula for the subword complexity of a wide class of D0L sequences. Fixed points of morphisms, or D0L sequences, represent a class of infinite words obtained by iterating morphisms. They are helpful in many cases when an example of a sequence with given properties is needed (for recent examples, see e.g. [9] and [10]).

D0L sequences and their subword complexity have been extensively studied. In 1984 J.-J. Pansiot [16] proved that the complexity of a D0L sequence can grow only as $O(n^2)$, $O(n \log n)$, $O(n \log\log n)$, $O(n)$ or $O(1)$; he gave also a description of classes of sequences having the complexity of each type. Methods of computing the subword complexities of different classes of D0L sequences were suggested by T. Tapsoba [17] and B. Mossé [14,15]. In [8], an explicit formula for *marked uniform* D0L sequences was given: to apply it, it is sufficient to find the

* Supported in part by RFBR (grants 96-01-01800 and 97-01-01075) and Federal Aim Program "Integration" (grant 473)

subword complexity values for small lengths and substitute them to the formula as coefficients.

Here we obtain a result of a similar nature for a much wider class of D0L sequences. After finding by hand the first values of the subword complexity and some short ***bispecial factors*** of the sequence, we just substitute their parameters to a general formula.

When submitting this paper to SETA, we thought that the technique of bispecial words was our invention. Only after the paper had been accepted, we became aware of the papers of J. Cassaigne [4, 5], where he used *exactly* the same technique to compute the subword complexity of the same class of sequences. He obtained an algorithm of computing the complexity; moreover, as it was shown in [5] by a series of examples, for the binary alphabet this algorithm always leads to an explicit formula. However, passing from the algorithm to the formula is not so easy in the case of a larger alphabet: the regularity which can be easily seen in the binary case, is in general not so obvious and must be proved. This is the part we pay most attention to in this new version.

An additional point is that we give a more explicit description of bispecial factors. In particular, our description makes it possible to characterize the *minimal forbidden words* of the sequence, which turn out to be a useful tool of language theory and have a direct relationship with bispecial words [2, 6].

Thus, in comparison with the preliminary version, the paper was factually rewritten. The class of D0L words considered is widened and new considerations for this wide class are added. On the other hand, since the technique was not our invention, we rewrote its description.

In Section 2 we introduce the class of D0L sequences we work with. Then in Section 3 we describe Cassaigne's technique of bispecial words and in Section 4 show how his algorithm works. Sections 5 to 7 are concerned with full description of bispecial words and passing from the algorithm to the final formula. At last, in Section 8 we consider a famous example of a D0L word on the 3-letter alphabet and compute its subword complexity.

2 D0L Sequences

Let us consider a morphism φ on a q-letter alphabet Σ. We call φ-images of letters ***blocks***. A *D0L sequence*, or *D0L word*, is defined as a fixed point of φ, i.e., as a right infinite word w satisfying the equality

$$w = \varphi(w).$$

A word $v \in \Sigma^*$ is called a ***subword*** or *factor* of a word $u \in \Sigma^*$, if $u = s_1 v s_2$ for some words s_1 and s_2. Our goal is to find the ***subword complexity*** $f(n)$ of w, which is equal to the number of its distinct factors of length n.

Let F be the language of factors of w, and let $u \in F$. We say that (u_1, u_2) is a *synchronization point* of u (for φ) on w if $u = u_1 u_2$ and

$$\forall v_1, v_2 \in \Sigma^*, \forall s \in F \ \ \exists s_1, s_2 \in F$$

$$[v_1 u v_2 = \varphi(s) \Longrightarrow (s = s_1 s_2, v_1 u_1 = \varphi(s_1), u_2 v_2 = \varphi(s_2))]$$

(see [3]). Roughly speaking, a synchronization point marks a place in u where a demarcation between blocks lies in every occurrence of u in w.

We call a word $u \in F$ *circular* if it has at least one synchronization point on w. And a D0L sequence w is called *circular* with *synchronization delay* L if every word of length at least L which is a factor of w is circular. This notion was introduced in the papers of F. Mignosi and P. Séébold [12] and J. Cassaigne [3]; see also B. Mossé [13], [14].

In this paper, we consider a D0L sequence w having the following properties:

(A1) A block cannot be a prefix or a suffix of another block; in other terms, the set of blocks $\{\varphi(a) | a \in \Sigma\}$ is a biprefix code.
(A2) w is circular.

Given together, these conditions mean that in every occurrence of a long enough factor of w, the boundaries between blocks are disposed at the same places.

For the final formula for the subword complexity, we need also the third condition

(A3) The language F of factors is *prolongable*, i.e., each word $u \in F$ can be extended to the left (and to the right) to another element of F: $au \in F$ for some $a \in \Sigma$.

However, this condition is not really crucial, since for a non-prolongable case the resulting expression for the subword complexity just slightly changes from the prolongable case. We mention this difference in the next section, then discuss the facts valid for all the sequences satisfying **(A1)-(A2)**, and finally in Section 7 find the expression for the case of **(A1)-(A3)**.

In the next section, we consider a factorial language F; later we suppose that F is the language of factors of w.

3 Bispecial Factors, Minimal Forbidden Words, and Second Differences

A language is called *factorial* if it contains all the factors of each its element. For the sake of generality, in this section we consider a factorial language $F \subset \Sigma$; we assume also that F is *prolongable*, i.e., each its element u can be prolonged to an element of F by appending letters to the left and to the right: $aub \in F$ for some $a, b \in \Sigma$.

The set of elements of F of length n is denoted by $F(n)$. We study the subword complexity function $f(n) = \#F(n)$ and use also the auxiliary functions of first and second differences: $s(n) = f(n+1) - f(n)$ and $b(n) = s(n+1) - s(n)$.

Let us consider a word $u \in F$. We define its *left (right) speciality set* $L(u)$ ($R(u)$) as the set of letters appending which to u to the left (to the right) gives an element of F. The cardinalities of $L(u)$ and $R(u)$ are called respectively *left* and *right speciality degrees* of u and denoted by $l(u)$ and $r(u)$. The word $u \in F$ is called *right (left) special* if it can be prolonged to the left (to the right) to an element of F by at least two different letters, i.e., if $l(u) \geq 2$ ($r(u) \geq 2$). And a word is called *bispecial* if it is both left special and right special.

Note that in general, the set $F \cap \Sigma u \Sigma$ does not coincide with $L(u)uR(u)$: some words from the latter set may be not elements of F. A helpful notion serving to illustrate this fact is the bipartite *bispeciality graph* $G(u) = (L(u), R(u), E(u))$: the edges of $E(u)$ join exactly such letters $l \in L(u)$ and $r \in R(u)$ that $lur \in F$ (see Fig. 1).

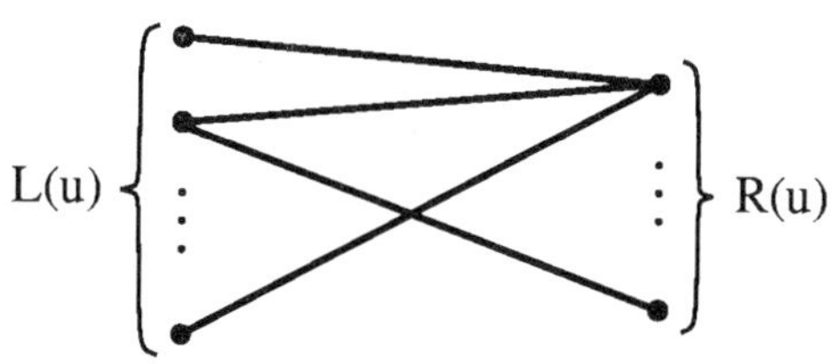

Fig. 1. A bispeciality graph

Note that the bispeciality graph of an element of a prolongable language F cannot be empty and does not contain vertices without incident edges. The number of its edges is $\#E(u) = \#(F \cap \Sigma u \Sigma)$.

The *bilateral order* $g(u)$ of a word $u \in F$ is

$$g(u) = \#E(u) - l(u) - r(u) + 1.$$

This notion was introduced by J. Cassaigne in [5]; in the same paper, the following proposition was proved:

Proposition 1. *For all* $n \geq 1$

$$b(n) = \sum_{u \in F(n)} g(u). \tag{1}$$

Another easy proposition mentioned in [5] is

Proposition 2. *If* $g(u) \neq 0$ *then* u *is bispecial; the converse is not true.*

Figure 2 classifies all the possible bispeciality graphs and bilateral orders of bispecial words on the binary alphabet.

Propositions 1 and 2 reduce computing the subword complexity of a prolongable language to finding its bispecial elements (together with their bilateral

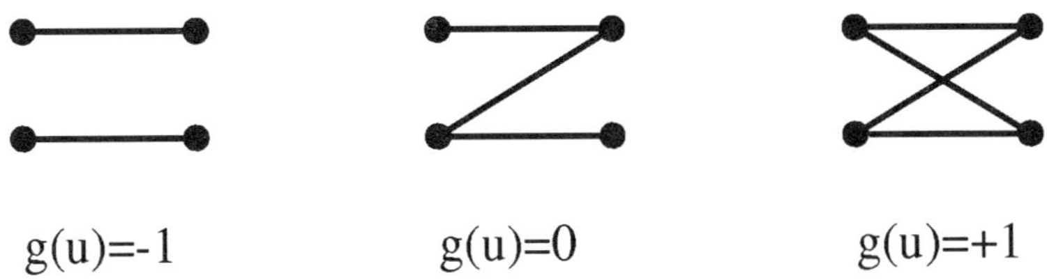

Fig. 2. Bispeciality graphs of bispecial words on the binary alphabet

orders). Moreover, Proposition 1 is valid for the non-prolongable case too, and we can use it to find the second differences for *any* D0L word satisfying **(A1)-(A2)**. The only difference of the non-prolongable case is that all the prefixes of the sequence are not bispecial, but have bilateral orders -1 and must be taken into account.

Note also the link between bispecial words and *minimal forbidden words* defined as words from $\Sigma^*\backslash F$ all whose factors are elements of F. Minimal forbidden words turn out to be very helpful [2, 6], and one of their basic properties [2] is that for a prolongable factorial language, they are exactly the words corresponding to absent edges in bispeciality graphs: a word s is minimal forbidden iff $s = lur$, where $l, r \in \Sigma$, u is bispecial, $lu, ur \in F$, but $lur \notin F$. So, since we are going to characterize all the bispecial words together with their bispeciality graphs, we shall find all the minimal forbidden words too.

In the next section, we discuss first considerations on the structure of the set of bispecial factors of a D0L sequence and describe Cassaigne's algorithm for finding the subword complexity.

4 Inheritance of Bispeciality and Cassaigne's Algorithm

Let F be the language of factors of a D0L word w satisfying **(A1)-(A2)**. Recall that by *blocks* we mean the images of letters.

Consider a circular bispecial word $u \in F$. Due to **(A1)-(A2)**, all the boundaries between blocks in it are disposed always at the same places, i.e., there exists a unique decomposition $u = \alpha\varphi(v)\beta$, where α is a suffix and β is a prefix of some blocks; α and β can be empty, but neither of them is a full block. In this case, we say that v is the *ancestor* of u and u is a *descendant* of v. Clearly, the ancestor of a circular bispecial word is unique, but in general, a word can have several descendants.

Note also that the notion of the ancestor of a (bispecial) word defined here does *not* coincide with the notion of the ancestor of an interpretation of a word defined in [12] and used in [8].

Proposition 3. *The ancestor v of a circular bispecial word $u = \alpha\varphi(v)\beta$ is bispecial with $l(v) \geq l(u)$ and $r(v) \geq r(u)$.*

From now on we always mean bispecial words when mentioning ancestors and descendants.

To use Cassaigne's algorithm with confidence, we need also another easy proposition:

Proposition 4. *Let v_i be a descendant of v_{i-1} for all $i = 1, \ldots, n$. Then v_0 can be equal to v_n only if the length of v_0 is less then the maximal length of a block.*

Let us sketch out all the circular bispecial words in a graph in which edges connect each word with all its descendants. Add to this graph the set N of uncircular bispecial words having circular descendants. Since the ancestor of a word is unique and due to Proposition 4, this graph has no cycles; so, it is a forest of infinite trees. Then, due to Proposition 3, the set of roots of trees of this forest coincide with N (see Fig. 3).

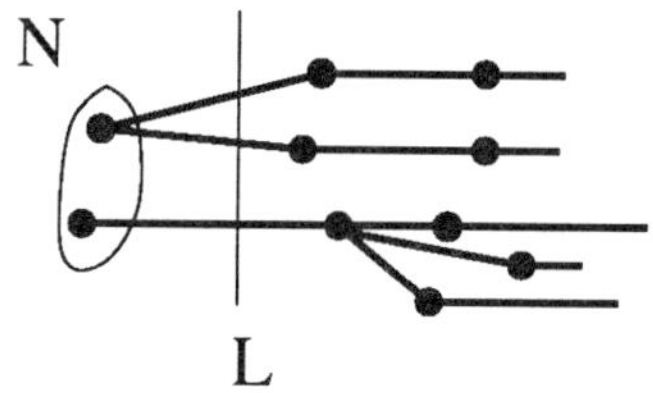

Fig. 3. The forest of bispecial words

Note that there exist only a finite number of uncircular words (and the length of each of them is at most $L - 1$). So, the set N (which is called in [5] the set of *exceptional bispecial factors*) can always be found in finite time. Then, finding in succession all the descendants of known bispecial words, we can find all the bispecial words of a given length and compute their bilateral orders; after that, we can substitute them into formula (1). This is Cassaigne's algorithm for finding the subword complexity.

Let the alphabet Σ be binary and let v be a bispecial word. Then it is not difficult to check that v has a unique descendant $u = s\varphi(v)p$, where s is the longest common suffix and p is the longest common prefix of the two blocks. Moreover, $g(u) = g(v)$. Thus, in the case of the binary alphabet each tree in the forest of bispecial words is just a chain, and as it was shown in [5] by examples, we always can find the subword complexity of a D0L word in an explicit form.

However, in the case of a larger alphabet a bispecial word can have several descendants with different bilateral orders. So, the situation *seems to be* not so easy. However, in practice starting with some length we find the same kind of regularity: each long enough bispecial word has a unique descendant; they both have equal bilateral orders and isomorphic bispeciality graphs. In what follows, we shall prove it and derive an explicit formula for the subword complexity.

5 Stability

For a set of letters $L \subseteq \Sigma$ of cardinality 2 or more, we define the word $s(L) \in \Sigma^*$ as the longest common suffix of all the blocks $\varphi(l)$, $l \in L$; analogously, we define $p(L)$ as their longest common prefix. If for all $a, b \in L$ the equality $s(\{a,b\}) = s(L)$ holds, then we say that L is *left stable*, and if $p(\{a,b\}) = p(L)$, then we say that L is *right stable*. Note that all the sets of cardinality 2 are both left stable and right stable.

Proposition 5. *Let v be the ancestor of u. Then $l(u) = l(v)$ iff $L(v)$ is left stable, and $r(u) = r(v)$ iff $R(v)$ is right stable.*

Proof. As before, we write $u = \alpha\varphi(v)\beta$. Let the strict inequality hold: $l(v) > l(u)$. It means that some word $lur \in F(n+2)$ (where $l, r \in \Sigma$) admits an ambiguity: the word $l\alpha$ is a suffix of at least two blocks $\varphi(l')$ and $\varphi(l'')$, where $l' \neq l''$, $l', l'' \in L(v)$. But since u is bispecial, there exists a third letter $l''' \in L(v)$ such that $s(\{l', l'''\}) = s(\{l'', l'''\}) = \alpha \neq s(\{l', l''\})$. Thus, $L(v)$ is not left stable.

Vice versa, if $l(u) = l(v)$, then any two blocks $\varphi(l')$ and $\varphi(l'')$ where $l', l'' \in L(v)$ have the same longest common suffix (equal to α), so that $L(v)$ is left stable. Analogous arguments are valid for $R(v)$.

For a left stable set L we define $C_l(L)$ as the set of all symbols preceding $s(L)$ in the blocks of $\varphi(L)$. By the definition of left stability, $\#C_l(L) = \#L$. Analogously for a right stable R we define $C_r(R)$ as the set of symbols following $p(R)$ in the blocks of $\varphi(R)$; by the definition of right stability, $\#C_r(R) = \#R$. Let us call a left stable set L *absolutely left (right) stable* if $L = C_l^k(L)$ $(L = C_r^k(L))$ for some $k > 0$.

We call a circular bispecial word v *stable* if $L(v)$ is left stable and $R(v)$ is right stable, and we call it *absolutely stable* if $L(v)$ and $R(v)$ are respectively absolutely left and right stable.

Proposition 6. *Let a circular bispecial word v be stable. Then*
1. *v has the unique descendant $u = s(L(v))\varphi(v)p(R(v))$.*
2. *$L(u)$ is determined by $L(v)$ and $R(u)$ is determined by $R(v)$.*
3. *The graphs $G(u)$ and $G(v)$ are isomorphic; in particular, $g(u) = g(v)$.*

Proof. 1. Clearly, $u = s(L(v))\varphi(v)p(R(v))$ is bispecial and thus is a descendant of v. On the other hand, any descendant of v must have the form $u' = \alpha\varphi(v)\beta$, where α is a suffix and β is a prefix of a block, but neither of them is a full block. From the definition of $s(L(v))$ we have that u' can be an element of F only when $s(L(v))$ is a suffix of α. But since $L(v)$ is left stable, u' can be left special only when $\alpha = s(L(v))$. Analogously u' can be right special only when $\beta = p(R(v))$, so, to be bispecial, u' must coincide with u.

2. Clearly, $L(u) = C_l(L(v))$ and $R(u) = C_r(R(v))$.

3. For every $l \in L(v)$, we find the symbol $l' = l'(l) \in L(u)$ preceding $s(L(v))$ in $\varphi(l)$; for every $r \in R(v)$, we find $r' = r'(r) \in R(u)$ following $p(R(v))$ in $\varphi(r)$. Since v is stable, all the l' are distinct, and so do all the r'. So, we have found one-to-one correspondences between $L(v)$ and $L(u)$, $R(v)$ and $R(u)$.

Furthermore, if $lvr \in F$, then $l'ur' \in F$ (since it is a subword of $\varphi(lur)$). Vice versa, due to stability, $l's(L(v))\varphi(v)$ can occur in w only as a suffix of $\varphi(lv)$, and $\varphi(v)p(R(v))r'$ can occur in w only as a prefix of $\varphi(vr)$. Thus, $l'ur' \in F$ implies $lvr \in F$. We have found a one-to-one correspondence between the sets $\Sigma v \Sigma \cap F$ and $\Sigma u \Sigma \cap F$, i.e., between the edges of $G(v)$ and the edges of $G(u)$. Consequently, $g(u) = g(v)$.

Note that the third part of Proposition 6 implies also a one-to-one correspondence between minimal forbidden words generated by v and by u.

Theorem 1. *All the bispecial words except a finite number are absolutely stable.*

Proof. In the forest of bispecial words (Fig. 3), let us consider an infinite path starting with an element of N. In going from a bispecial word to its descendant, left and right speciality degrees cannot grow (Proposition 3); moreover, an upper bound for speciality degrees is the cardinality of the alphabet, and the lower bound is 2 (since the words remain bispecial). Thus, in the path, there exists only a finite number of words where bispeciality degrees decrease, i.e., unstable words (Proposition 5).

Moreover, the length of the initial subpath containing all the unstable words is bounded by a constant depending only on the cardinality of the alphabet. Indeed, either all the left speciality or all the right speciality sets of stable words in the path between two successive unstable words must be distinct: otherwise, due to Proposition 6, both left and right speciality sets begin to repeat periodically to infinity, and the second unstable word can never occur. But there exists only a finite number of distinct left or right stable sets of given cardinality; thus, the length of the path between two successive unstable words is bounded, and so is the length of the path including all the unstable words. We denote its maximal possible length by m.

But in the forest of bispecial words, there exists only a finite number of path of length m starting with an element of N. They include only a finite number of words, and outside of them, all the bispecial words are stable.

There exists only a finite number of left (or right) stable sets which are not absolutely left (or right) stable, and each of them can occur in a path of stable words only once. As son as an absolutely stable word occurs in such a path, all the words in it become absolutely stable. So, all the bispecial words except a finite number are absolutely stable.

6 Mapping Ψ

As we proved in the previous section, each long enough bispecial word v has a unique descendant which can be expressed as $s(L(v))\varphi(v)p(R(v))$ and preserves the properties of v. From now on we also use the notation

$$\Psi(v) = s(L(v))\varphi(v)p(R(v)).$$

Proposition 7. *Without loss of generality we can assume that for any absolutely stable bispecial word v*

$$L(\Psi(v)) = L(v) \text{ and } R(\Psi(v)) = R(v).$$

Proof. First of all, note that $L(\Psi(v)) = C_l(L(v))$ and $R(\Psi(v)) = C_r(R(v))$. Denote the set of absolutely left stable sets by A_l and the set of absolutely right stable sets by A_r. Let $k_l = \text{lcm}_{L \in A_l}\{\min k | C_l^k(L) = L\}$, and $k_r = \text{lcm}_{R \in A_r}\{\min k | C_r^k(R) = R\}$. Let $k = \text{lcm}\{k_l, k_r\}$. Clearly, for all absolutely stable v, $L(v) = L(\Psi^k(v))$ and $R(v) = R(\Psi^k(v))$.

The sequence w is a fixed point of φ. But as well we can consider it as a fixed point of φ^k. After that we have to redefine the mappings Ψ, C_l and C_r (we denote the mappings obtained while considering w as a fixed point of φ^i by Ψ_i, $C_{l,i}$ and $C_{r,i}$ respectively). Note that a sequence that is circular for φ remains circular for φ^k, and that words absolutely stable for φ remain absolutely stable for φ^k. It is not difficult to prove by induction that for any absolutely stable bispecial v,

$$\Psi_k(v) = \Psi_1^k(v).$$

Thus, $L(\Psi_k(v)) = C_{l,k}(L(v)) = C_{l,1}^k(L(v)) = L(v)$; analogously, $R(\Psi_k(v)) = R(v)$.

To prove the proposition, we have to pass from φ to some its power φ^k and respectively redefine all the parameters and functions which depend on φ. However, from now on we omit the power k and just assume that $k = 1$.

Factually, we have proved the following:

1. There exists M such that each bispecial word having a descendant of length at least M is absolutely stable.
2. Each bispecial word u of length larger than or equal to M is a unique descendant of some bispecial word v and can be expressed as $u = \Psi(v)$. Moreover, $L(u) = L(v)$, $R(u) = R(v)$, and $g(u) = g(v)$.

We have found a regularity in the forest of bispecial words: starting with some length M, it has not more additional branches, and each word inherits the basic properties of its ancestor (see Fig. 4).

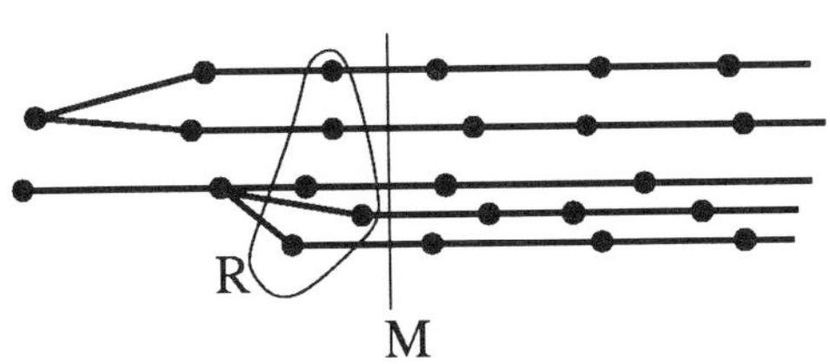

Fig. 4. The forest of bispecial words: a regularity

Now we are going to use this information to find the subword complexity of w. First of all, we rewrite formula (1) in terms of the mapping Ψ. To do it, we introduce the *root set* R of bispecial words which are shorter than M but have descendants of length at least M (see Fig. 4):

$$R = \{v | v \text{ is bispecial}, |v| < M, |\Psi(v)| \geq M\}.$$

Note that if we are interested just in complexity, then to simplify the process, we can include into R only words giving non-zero addends into (1), i.e., having non-zero bilateral orders.

As it follows from Property 2, each bispecial word of length M or more can be decomposed as $u = \Psi^k(v)$, where $v \in R$ and $k > 0$ (and such v and k are unique). Since in this case $g(u) = g(v)$, the formula (1) for $n \geq M$ can be rewritten as

$$b(n) = \sum_{v \in R} \sum_{k: |\Psi^k(v)| = n} g(v). \tag{2}$$

7 Final Formula

In this section, we transform (2) to an explicit formula for the subword complexity.

By the definitions of the first and the second differences and formula (2) it follows that for all $i \geq M$

$$s(i) = s(M) + \sum_{j=M}^{i-1} b(j) = s(M) + \sum_{j=M}^{i-1} \sum_{u \in R} \sum_{l: |\Psi^l(u)| = j} g(u),$$

and, consequently, for all $n \geq M$

$$\begin{aligned} f(n+1) &= f(M) + \sum_{i=M}^{n} s(i) \\ &= f(M) + (n - M + 1)s(M) + \sum_{i=M}^{n} \sum_{j=M}^{i-1} \sum_{u \in R} \sum_{l: |\Psi^l(u)| = j} g(u). \end{aligned}$$

For each $u \in \Sigma^*$ and $n > 0$, we define

$$l_u(n) = \max\{l \,|\, |\Psi^l(u)| \leq n - 1\}.$$

Let a word $u \in R$. In the sum $\sum_{j=M}^{i-1} \sum_{u \in R} \sum_{l: |\Psi^l(u)| = j} g(u)$, it corresponds to $l_u(i)$ non-zero addends (associated with lengths $|\Psi^l(u)|$), each of them equal to $g(u)$. Thus,

$$\sum_{j=M}^{i-1} \sum_{u \in R} \sum_{l: |\Psi^l(u)| = j} g(u) = \sum_{u \in R} g(u) l_u(i).$$

Furthermore, $l_u(n)$ is equal to l if and only if $|\Psi^l(u)| < i \le |\Psi^{l+1}(u)|$. So,

$$\sum_{i=M}^{n} \sum_{u\in R} g(u) l_u(i) =$$

$$\sum_{u\in R} g(u) \left[\sum_{l=0}^{l_u(n)-1} l(|\Psi^{l+1}(u)| - |\varphi^l(u)|) + l_u(n)(n - |\Psi^{l_u(n)}(u)|) \right].$$

Recall that we denote the cardinality of the alphabet Σ by q. The lengths $|\Psi^l(u)|$ can be computed using the $q \times q$ *composition matrix* $K = (k_{ij})$ of the morphism φ, each entry k_{ij} of which is equal to the number of occurrences of a_i in $\varphi(a_j)$. Let $\lambda_1, \ldots, \lambda_s$ be the eigenvalues of K, λ_i having order k_i. It can be shown that for a word $u \in \Sigma^*$ and for $l \ge 0$

$$|\Psi^l(u)| = \sum_{i=1}^{s} \sum_{k=0}^{k_i-1} c_u^{i,k} l^k \lambda_i^l + c_u, \tag{3}$$

where the constants $c_u^{i,k}$ and c_u are determined by u.

We omit the proof of Formula (3), but we mention, first, that it leans upon Proposition 7, and then, that for each $u \in R$ it uses the $(q+1) \times (q+1)$ matrix $\begin{pmatrix} K & p(u) \\ 0 & 1 \end{pmatrix}$, where $p(u)$ denotes the sum of the Parikh vectors of $s(L(u))$ and $p(R(u))$.

Using Formula (3), we obtain that

$$\sum_{l=0}^{l_u(n)-1} l(|\Psi^{l+1}(u)| - |\Psi^l(u)|) + l_u(n)(n - |\Psi^{l_u(n)}(u)|) =$$

$$(n - c_u) l_u(n) + \sum_{i=1}^{s} \sum_{k=1}^{k_s} c_u^{i,k-1} \sum_{l=1}^{l_u(n)} \lambda^l((l-1)^k - l^k).$$

Thus, the following theorem holds:

Theorem 2. *For all* $n \ge M$

$$f(n+1) = f(M) + (n - M + 1)s(M) +$$

$$+ \sum_{u\in R} g(u) \left[(n - c_u) l_u(n) + \sum_{i=1}^{s} \sum_{k=1}^{k_s} c_u^{i,k-1} P(l_u(n), k, \lambda_i) \right], \tag{4}$$

where $P(r, k, \lambda) = \sum_{l=1}^{r} \lambda^l((l-1)^k - l^k)$.

Note, in particular, that $P(r, 1, \lambda) = \dfrac{\lambda^{r+1} - \lambda}{1 - \lambda}$ for all $\lambda \ne 1$, and that $P(r, 1, 1) = -r$. Recall that Formula (4) is valid only for D0L sequences whose language of factors is prolongable.

8 Example

Let us find the subword complexity of a well-known [11] square-free word: the fixed point $w = abcacbabcbacabcacbacabcb\ldots$ of the morphism φ_0:

$$\begin{cases} \varphi_0(a) = abc \\ \varphi_0(b) = ac \\ \varphi_0(c) = b. \end{cases}$$

It can be easily seen that the language of factors of w is prolongable, and w is circular for φ_0. Now we must find all the absolutely stable sets of letters and find a power of φ_0 such that Proposition 7 will be valid. Since the words aa, bb, and cc can never occur in w, the left and right speciality sets of bispecial words always have cardinality 2. Let us check them. We see that

$$C_l(\{a,b\}) = \{a,b\}, \qquad C_r(\{a,b\}) = \{b,c\},$$

$$C_l(\{a,c\}) = \{b,c\}, \qquad C_r(\{a,c\}) = \{a,b\},$$

$$C_l(\{b,c\}) = \{b,c\}, \qquad C_r(\{b,c\}) = \{a,b\}.$$

We see that $\{a,b\}$ and $\{b,c\}$ are absolutely left and right stable, but $\{a,c\}$ is not. Moreover, since $C_l^2(\{a,b\}) = C_l(\{b,c\}) = \{a,b\}$, we have to pass from φ_0 to its square $\varphi = \varphi_0^2$:

$$\begin{cases} \varphi(a) = abcacb \\ \varphi(b) = abcb \\ \varphi(c) = ac. \end{cases}$$

In new notations, $\{a,b\}$ and $\{b,c\}$ are still absolutely left and right stable, and $\{a,c\}$ is not. Now $C_l(\{a,b\}) = C_r(\{a,b\}) = \{a,b\}$, and $C_l(\{b,c\} = C_r(\{b,c\} = \{b,c\}$; thus, Proposition 7 is true. Counting short bispecial words, we see that the only bispecial word that is not absolutely stable is b; all the other bispecials of length ≤ 5 are

$$a, abc, c, cba \text{ (with bilateral order } g = 1) \text{ and}$$

$$ac, ca, abcba, cbabc \text{ (with bilateral order } g = -1).$$

The lengths of all the descendants of these words are greater than 5; thus, the constant M can be assumed to be equal to 6, and the root set can be assumed to be equal to

$$R = \{a, c, ac, ca, abc, cba, abcba, cbabc\}.$$

The matrix of φ is

$$\begin{pmatrix} 2\,2\,2 \\ 1\,2\,1 \\ 1\,0\,1 \end{pmatrix},$$

its eigenvalues are 4, 1 and 0. Thus, for any $u \in R$

$$|\Psi^k(u)| = a_u \cdot 4^k + b_u.$$

Finding the constants a_u and b_u, we obtain that

$$\begin{aligned} |\Psi^k(a)| = |\Psi^k(c)| &= 2\cdot 4^k - 1, \\ |\Psi^k(ac)| = |\Psi^k(ca)| &= 3\cdot 4^k - 1, \\ |\Psi^k(abc)| = |\Psi^k(cba)| &= 4^{k+1} - 1, \text{ and} \\ |\Psi^k(abcba)| = |\Psi^k(cbabc)| &= 6\cdot 4^k - 1. \end{aligned}$$

Thus,we have to consider only four cases:

- $6\cdot 4^{k-1} - 1 < n \le 2\cdot 4^k - 1$. In this case for every $u \in R$ the equality $l_u(n) = k$ holds. It can be computed also that $f(6) = 20$, $s(6) = 4$. Substituting these values to (4),we obtain that
$$f(n+1) = 2n + 2 + 2\cdot 4^k.$$
- $2\cdot 4^k - 1 < n \le 3\cdot 4^k - 1$. Then $l_a(n) = l_c(n) = k$ and $l_u(n) = k-1$ for any other $u \in R$; thus,
$$f(n+1) = 4n + 4 - 2\cdot 4^k.$$
- $3\cdot 4^k - 1 < n \le 4\cdot 4^k - 1$. Then
$$f(n+1) = 2n + 2 + 4^{k+1}.$$
- $4\cdot 4^k - 1 < n \le 6\cdot 4^k - 1$. Then
$$f(n+1) = 4n + 4 - 4^{k+1}.$$

We see that these four cases can be reduced to two: for any $n > 6$,

$$f(n) = \begin{cases} 2n + 2^{k+2} & \text{if} \quad 3\cdot 2^k < n \le 2^{k+2}, \text{ and} \\ 4n - 2^{k+2} & \text{if} \quad 2^{k+2} < n \le 3\cdot 2^{k+1}. \end{cases}$$

We have found the subword complexity of w.

9 Acknowledgments

The results of the paper were refined during the visit of the first author to LIAFA laboratory of Université Paris 7 in October and November of 1998. I wish to thank Michel Morvan, Daniel Krob, and all the LIAFA members for their hospitality.

We are also deeply grateful to Julien Cassaigne for his inestimable moral support and fruitful discussions and to Jean-Paul Allouche who presented this paper at SETA.

References

1. J.-P. Allouche, Sur la complexité des suites infinies, *Bull. Belg. Math. Soc.* **1** (1994), 133–143.
2. M.-P. Béal, F. Mignosi, and A. Restivo, Minimal Forbidden Words and Symbolic Dynamics, *STACS'96*, Lect. Notes Comp. Sci. **1046** (1996) 555–566.
3. J. Cassaigne, An algorithm to test if a given circular HD0L-language avoids a pattern, *IFIP World Computer Congress'94*, Elsevier (North-Holland) **1** (1994), 459–464.
4. J. Cassaigne, Motifs évitables et régularités dans les mots, *Thèse de Doctorat*, Université Paris 6 (1994). Rapport LITP TH 94–04.
5. J. Cassaigne, Complexité et facteurs spéciaux, *Bull. Belg. Math. Soc.* **4** (1997), 67–88.
6. M. Crochemore, F. Mignosi, and A. Restivo, Minimal Forbidden Words and Factor Automata, *MFCS'98*, Lect. Notes Comp. Sci. **1450** (1998), 665–673.
7. S. Ferenczi, Complexity of Sequences and Dynamical Systems, *Discrete Math.* **206** (1999), to appear.
8. A. Frid, On Uniform D0L Words, *STACS'98*, Lect. Notes Comp. Sci. **1373** (1998), 544–554.
9. V. Keränen, Abelian Squares are Avoidable on 4 Letters, *ICALP'92*, Lect. Notes Comp. Sci. **623** (1992), 41–52.
10. R. Kolpakov, G. Kucherov, Minimal Letter Frequency in n-th Power-free Binary Words. *MFCS'97*, Lect. Notes Comp. Sci. **1295** (1997) 347–357.
11. M. Lothaire, "Combinatorics on Words", Addison-Wesley, Reading, Mass., 1983.
12. F. Mignosi and P. Séébold, If a D0L language is k-power-free then it is circular, *ICALP'93*, Lect. Notes Comp. Sci., **700** (1993), 507–518.
13. B. Mossé, Puissances de mots et reconnaissabilité des points fixes d'une substitution, *Theoret. Comput. Sci.* **99** (1992), 327–334.
14. B. Mossé, Reconnaissabilité des substitutions et complexité des suites automatiques, *Bull. Soc. Math. France* **124** (1996), 329–346.
15. B. Mossé, On the complexity function of fixed points of substitutions (1996), Unpublished.
16. J.-J. Pansiot, Complexité des facteurs des mots infinis engendrés par morphismes itérés, *ICALP'84*, Lect. Notes Comp. Sci. **172** (1984), 380–389.
17. T. Tapsoba, Automates calculant la complexité des suites automatiques, *Journal de Théorie des nombres de Bordeaux* **6** (1994), 127–134.

ON THE DISTRIBUTION OF THE RSA GENERATOR

John B. Friedlander,[1] Daniel Lieman[2] and Igor E. Shparlinski[3]

[1] Department of Mathematics, University of Toronto
Toronto, Ontario M5S 3G3, Canada
frdlndr@math.toronto.edu

[2] Department of Mathematics, University of Missouri
Columbia, MO 65211, USA
lieman@math.missouri.edu

[3] School of MPCE, Macquarie University
Sydney, NSW 2109, Australia
igor@mpce.mq.edu.au

1 Introduction

Let ϑ, m and e be integers such that $\gcd(\vartheta, m) = 1$. Then one can define the sequence (u_n) by the recurrence relation

$$u_n \equiv u_{n-1}^e \pmod m, \quad 0 \le u_n \le m-1, \qquad n = 1, 2, \dots, \tag{1}$$

with the *initial value* $u_0 = \vartheta$.

This sequence is known as the *power generator* of pseudo-random numbers and has many applications to cryptography, see [3, 9, 13, 16].

In two special cases $\gcd(e, \varphi(m)) = 1$, where $\varphi(m)$ is the Euler function, and $e = 2$ this sequence is known as the *RSA generator* and as the *Blum–Blum–Shub generator*, respectively.

Although various properties have been studied in a number of works, see [3, 4, 8–16], very few unconditional results are known. One such result is due to [8]. It is shown in that paper that the rightmost bit of the Blum–Blum–Shub generator takes values 0 and 1 almost equally often, provided that the period is large enough. Lower bounds on the linear complexity of the power generator have been given in [11, 15].

Here we prove that if the period of the RSA generator is large enough then the elements of this sequence are uniformly distributed modulo m and a positive proportion of the rightmost and leftmost bits is uniformly distributed.

In this paper we prove the result in the most important case for applications when $m = pl$ where p and l are distinct primes. Such numbers are called *Blum integers* (sometimes given with certain additional conditions such as that $p \equiv l \equiv 3 \pmod 4$ and that p and l are of the same order). However, the same results hold for m prime, as is easy to see from our proof, and similar, but somewhat weaker, results can be obtained for arbitrary composite moduli as well. In fact, our proof is based on the bound for exponential sums from [5] with prime denominators. In order to cover non-squarefree moduli one should extend the results of [6, 5] to prime power denominators. This, although it requires some additional work and adjusting the technique, can be done without requiring any essentially new ideas.

It is obvious that the sequence (1) eventually becomes periodic with some period t. Moreover, it is easy to see that if $\gcd(e, \varphi(m)) = 1$ then this sequence is *purely periodic*.

2 Exponential sums

We define exponential sums

$$S_a = \sum_{n=1}^{t} \mathbf{e}_m(au_n),$$

where

$$\mathbf{e}_d(z) = \exp(2\pi i z/d).$$

We obtain a non-trivial upper bound on sums S_a and derive (see Theorem 3) the uniformity of distribution modulo m of the elements u_n, $n = 1, \dots, t$, provided that $t > m^{23/24+\delta}$ with some fixed positive δ.

These exponential sums enter into our problem by means of the following well-known basic identity.

Lemma 1. *For any integer u*

$$\sum_{\lambda=0}^{m-1} \mathbf{e}_m(\lambda u) = \begin{cases} 0, & \text{if } u \not\equiv 0 \pmod m; \\ m, & \text{if } u \equiv 0 \pmod m. \end{cases}$$

We also need the following well known estimate (see Problem 11.c to Chapter 3 of [17]).

Lemma 2. *For any integer $H \geq 0$,*

$$\sum_{a=1}^{m-1} \left| \sum_{u=0}^{H} \mathbf{e}_m(au) \right| = O(m \log m).$$

The following lemma shows how to reduce the general case to bounds of exponential sums with prime power denominators (see e.g. Problem 12.d to Chapter 3 of [17]).

Lemma 3. *Let* $m = m_1 m_2$ *where* $m_1 \geq m_2 \geq 2$ *and* $\gcd(m_1, m_2) = 1$ *and let* k_1, k_2 *be such that*

$$k_1 m_2 \equiv 1 \pmod{m_1} \qquad \textit{and} \qquad k_2 m_1 \equiv 1 \pmod{m_2}.$$

Then for any polynomial $f(x)$ *with integer coefficients*

$$\sum_{\substack{x=0 \\ \gcd(x,m)=1}}^{m-1} \mathbf{e}_m(f(x)) = \sum_{\substack{x_1=0 \\ \gcd(x_1,m_1)=1}}^{m_1-1} \mathbf{e}_{m_1}(k_1 f(x_1)) \sum_{\substack{x_2=0 \\ \gcd(x_2,m_2)=1}}^{m_2-1} \mathbf{e}_{m_2}(k_2 f(x_2)).$$

Indeed, this statement follows from Problem 12.d to Chapter 3 of [17] if one remarks that

$$k_1 m_2 + k_2 m_1 \equiv 1 \pmod{m}.$$

We reduce the problem of estimating sums S_a to certain double sums which have been estimated in [5].

Lemma 4. *For any prime number* p *and any integers* a, b *with* $\gcd(ab, p) = 1$, *the bound*

$$\sum_{y=1}^{p-1} \left| \sum_{x=1}^{p-1} \mathbf{e}_p(ax + bx^y) \right|^4 = O(p^{14/3})$$

holds.

Proof. Let g be a primitive root modulo p. Then

$$\sum_{y=1}^{p-1} \left| \sum_{x=1}^{p-1} \mathbf{e}_p(ax + bx^y) \right|^4 = \sum_{y=1}^{p-1} \left| \sum_{x=1}^{p-1} \mathbf{e}_p(ag^x + bg^{xy}) \right|^4.$$

The last sum is estimated as $O(p^{14/3})$ in the proof of Theorem 8 of [5]. □

Now we are prepared to formulate our main estimate.

Theorem 1. *Let* $m = pl$ *where* p *and* l *are two distinct primes. If the sequence* (u_n) *is given by (1) with* $\gcd(e, \varphi(m)) = 1$ *then the bound*

$$\max_{\gcd(a,m)=1} |S_a| = O\left(m^{23/24}\right)$$

holds.

Proof. Because the sequence (u_n) is purely periodic with period t, for any $k = 1, \ldots, t$ we have

$$S_a = \sum_{n=1}^{t} \mathbf{e}_m(au_{n+k}).$$

Therefore,

$$t|S_a|^2 = \sum_{k=1}^{t} \left| \sum_{n=1}^{t} \mathbf{e}_m\left(au_{n+k}\right) \right|^2 = \sum_{k=1}^{t} \left| \sum_{n=1}^{t} \mathbf{e}_m\left(a\vartheta^{e^{n+k}}\right) \right|^2 .$$

Hence we obtain

$$\begin{aligned} t|S_a|^2 &\le \sum_{\substack{x=0 \\ \gcd(x,m)=1}}^{m-1} \left| \sum_{n=1}^{t} \mathbf{e}_m\left(ax^{e^n}\right) \right|^2 \\ &= \sum_{\substack{x=0 \\ \gcd(x,m)=1}}^{m-1} \sum_{n,k=1}^{t} \mathbf{e}_m\left(a\left(x^{e^n} - x^{e^k}\right)\right) \\ &\le \sum_{n,k=1}^{t} \left| \sum_{\substack{x=0 \\ \gcd(x,m)=1}}^{m-1} \mathbf{e}_m\left(a\left(x^{e^n} - x^{e^k}\right)\right) \right| \\ &\le \sum_{k=1}^{t} \sum_{\substack{y=1 \\ \gcd(y,\varphi(m))=1}}^{\varphi(m)} \left| \sum_{\substack{x=0 \\ \gcd(x,m)=1}}^{m-1} \mathbf{e}_m\left(a\left(x^{ye^k} - x^{e^k}\right)\right) \right| . \end{aligned}$$

It is easy to see that because $\gcd(e, \varphi(m)) = 1$ each internal sum does not depend on k. Therefore,

$$|S_a|^2 \le \sum_{\substack{y=1 \\ \gcd(y,\varphi(m))=1}}^{\varphi(m)} \left| \sum_{\substack{x=0 \\ \gcd(x,m)=1}}^{m-1} \mathbf{e}_m\left(a\left(x - x^y\right)\right) \right| .$$

Lemma 3 implies that there exist some integers k_p and k_l with $\gcd(p, k_p) = \gcd(l, k_l) = 1$ and such that

$$\begin{aligned} &\sum_{\substack{x=0 \\ \gcd(x,m)=1}}^{m-1} \mathbf{e}_m\left(a\left(x - x^y\right)\right) \\ &= \sum_{x_1=1}^{p-1} \mathbf{e}_p\left(ak_p\left(x_1 - x_1^y\right)\right) \sum_{x_2=1}^{l-1} \mathbf{e}_l\left(ak_l\left(x_2 - x_2^y\right)\right) . \end{aligned}$$

Taking into account that $\varphi(m) = (p-1)(l-1)$, from the previous equation and the Cauchy inequality we derive

$$|S_a|^2 \le \sum_{\substack{y=1 \\ \gcd(y,\varphi(m))=1}}^{\varphi(m)} \left| \sum_{x_1=1}^{p-1} \mathbf{e}_p\left(ak_p\left(x_1 - x_1^y\right)\right) \right| \left| \sum_{x_2=1}^{l-1} \mathbf{e}_l\left(ak_l\left(x_2 - x_2^y\right)\right) \right|$$

$$\le \varphi(m)^{1/2} \left(\sum_{\substack{y=1 \\ \gcd(y,\varphi(m))=1}}^{\varphi(m)} \left| \sum_{x=1}^{p-1} \mathbf{e}_p \left(ak_p \left(x - x^y \right) \right) \right|^4 \right)^{1/4}$$

$$\times \left(\sum_{\substack{y=1 \\ \gcd(y,\varphi(m))=1}}^{\varphi(m)} \left| \sum_{x=1}^{l-1} \mathbf{e}_l \left(ak_l \left(x - x^y \right) \right) \right|^4 \right)^{1/4}$$

$$\le \varphi(m)^{1/2} \left(\frac{\varphi(m)}{p-1} \sum_{y=1}^{p-1} \left| \sum_{x=1}^{p-1} \mathbf{e}_p \left(ak_p \left(x - x^y \right) \right) \right|^4 \right)^{1/4}$$

$$\times \left(\frac{\varphi(m)}{l-1} \sum_{y=1}^{l-1} \left| \sum_{x=1}^{l-1} \mathbf{e}_l \left(ak_l \left(x - x^y \right) \right) \right|^4 \right)^{1/4} .$$

Using the bound of Lemma 4, we obtain the desired result. □

3 Leftmost and Rightmost Bits

Here we show that leftmost and rightmost bits of (u_n) are uniformly distributed.

Let σ be a binary string of length k. Denote by $L(\sigma)$ the number of $n = 1, \ldots, t$, such that σ is the string of the k least significant bits of u_n.

Theorem 2. *Let $m = pl$ where p and l are two distinct odd primes. If the sequence (u_n) is given by (1) with $\gcd(e, \varphi(m)) = 1$ then, for any binary string σ of length k, the bound*

$$\left| L(\sigma) - t2^{-k} \right| = O\left(m^{23/24} \log m \right)$$

holds.

Proof. We denote by s the integer whose binary representation coincides with σ and put $K = 2^k$, $H = \lfloor (m-1-s)/K \rfloor$.

We remark that $L(\sigma)$ is equal to the number W of solutions of the following congruence

$$u_n \equiv Kx + s \pmod m, \qquad 1 \le n \le t,\ 0 \le x \le H.$$

Thus, using Lemma 1 we write

$$W = \frac{1}{m} \sum_{n=1}^{t} \sum_{x=0}^{H} \sum_{a=0}^{m-1} \mathbf{e}_m \left(a \left(u_n - Kx - s \right) \right)$$

$$= \frac{1}{m} \sum_{a=0}^{m-1} \mathbf{e}_m \left(-as \right) S_a \sum_{x=0}^{H} \mathbf{e}_m(-aKx).$$

The term corresponding to $a = 0$ equals

$$t(H+1)m^{-1} = t2^{-k} + O(1)$$

which contributes to the main term of the desired formula, with an admissible error. To estimate the contribution R of the remaining terms we apply Theorem 1 getting

$$R \ll m^{-1/24} \sum_{\substack{a=1 \\ \gcd(a,m)=1}}^{m-1} \left| \sum_{x=0}^{H} \mathbf{e}_m(-aKx) \right|$$

$$+\frac{1}{m}\sum_{a=1}^{l-1} |S_{ap}| \left| \sum_{x=0}^{H} \mathbf{e}_l(-aKx) \right| + \frac{1}{m}\sum_{a=1}^{p-1} |S_{al}| \left| \sum_{x=0}^{H} \mathbf{e}_p(-aKx) \right|.$$

Let us estimate the sums $|S_{ap}|$ and $|S_{al}|$. Let t_p be the period of the sequence (u_n) modulo p. It is easy to see that $t \le t_p(l-1)$. We now have

$$S_{al} = \frac{t}{t_p} \sum_{n=1}^{t_p} \mathbf{e}_p\left(au_n\right) = O(lp^{23/24})$$

if $\gcd(a,p) = 1$ (obviously Theorem 1 applies to the case when m is prime as well). A similar estimate also holds for S_{ap}.

Remarking that $\gcd(K,m) = 1$, we see that $-aK$ can be replaced with just a. Therefore, applying Lemma 2, after some simple calculations we obtain the desired estimate. □

Virtually the same proof yields the same result for the most significant bits. One simply replaces $Kx + s$ by $x + Ms$ for suitable M in the above congruence for u_n.

Since the most significant bits of a number are the ones most responsible for locating it as a point on the line this case may also be formulated somewhat differently. For an interval $\mathcal{I} = [r, r+h-1]$, where $0 \le r \le r+h-1 \le m-1$, of length $\#\mathcal{I} = h$ we denote by $N(\mathcal{I})$ the number of $n = 1, \ldots, t$ for which u_n belongs to the interval $\mathcal{I}$.

In this form there is an alternative well-known way of deriving the result from the exponential sum bound (see for example [7]) which does not however apply to the least significant bits. The result is

Theorem 3. *Let $m = pl$ where p and l are two distinct primes. If the sequence (u_n) is given by (1) with* $\gcd(e, \varphi(m)) = 1$ *then the bound*

$$\sup_{\mathcal{I} \subset [0, m-1]} \left| N(\mathcal{I}) - \frac{t}{m}\#\mathcal{I} \right| = O\left(m^{23/24} \log m\right)$$

holds.

Of course, the weaker fact that the left side is $o(m)$ for $t \ge m^{23/24+\delta}$ is precisely the statement of uniform distribution.

4 Remarks

As already mentioned, results similar to those above can be obtained for the case of a general modulus m. On the other hand, it is not clear how to study the joint distribution of s-tuples $(u_n, \dots, u_{n+s-1})$. Even the case of pairs (u_n, u_{n+1}) is of interest.

It would also be interesting to extend the results of this paper to the case of the *exponential generator*

$$v_n \equiv g^{v_{n-1}} \pmod m, \quad 0 \le v_n \le m-1, \qquad n = 1, 2, \dots,$$

which also has numerous cryptographic applications [9, 13].

Finally, studying the *Pollard generator*

$$w_n \equiv w_{n-1}^2 + a \pmod m, \quad 0 \le w_n \le m-1, \qquad n = 1, 2, \dots,$$

with some integer a is an interesting challenge as well. It is well known that this generator has important applications to integer factorization [2]. However, so far only one rigorously proved result about this generator is known [1].

References

1. E. Bach, 'Toward a theory of Pollard's rho method', *Inform. and Comp.*, **90** (1991), 139–155.
2. E. Bach and J. Shallit, *Algorithmic Number Theory*, MIT Press, 1996.
3. L. Blum, M. Blum and M. Shub, 'A simple unpredictable pseudo-random number generator', *SIAM J. Comp.*, **15** (1986), 364–383.
4. J. J. Brennan and B. Geist, 'Analysis of iterated modular exponentiation: The orbit of $x^\alpha \bmod N$', *Designs, Codes and Cryptography*, **13** (1998), 229–245.
5. R. Canetti, J. B. Friedlander, S. Konyagin, M. Larsen, D. Lieman and I. E. Shparlinski, 'On the statistical properties of the Diffie–Hellman distribution', *Israel J. Math.*, (to appear).
6. R. Canetti, J. B. Friedlander and I. E. Shparlinski, 'On certain exponential sums and the distribution of Diffie–Hellman triples', *J. London Math. Soc.*, (to appear).
7. J. H. H. Chalk, 'The Vinogradov–Mordell–Tietäväinen inequalities', *Indag. Math.*, **42** (1980), 367–374.
8. T. W. Cusick, 'Properties of the $x^2 \bmod N$ pseudorandom number generator', *IEEE Trans. Inform. Theory*, **41** (1995), 1155–1159.
9. T. W. Cusick, C. Ding and A. Renvall, *Stream Ciphers and Number Theory*, Elsevier, Amsterdam, 1998.
10. R. Fischlin and C. P. Schnorr, 'Stronger security proofs for RSA and Rabin bits', *Lect. Notes in Comp. Sci.*, Springer-Verlag, Berlin, **1233** (1997), 267–279.
11. F. Griffin and I. E. Shparlinski, 'On the linear complexity profile of the power generator', *Preprint*, 1998, 1–11.
12. J. Håstad and M. Näslund, 'The security of individual RSA bits', *Proc 39th IEEE Symp. on Foundations of Comp. Sci.*, 1998, 510–519.

13. J. C. Lagarias, 'Pseudorandom number generators in cryptography and number theory', *Proc. Symp. in Appl. Math.*, Amer. Math. Soc., Providence, RI, **42** (1990), 115–143.
14. A. J. Menezes, P. C. van Oorschot and S. A. Vanstone, *Handbook of Cryptography*, CRC Press, Boca Raton, FL, 1996.
15. I. E. Shparlinski, 'On the linear complexity of the power generator', *Tech. Report. C/TR 98-12*, Dept. of Computing, Macquarie Uni., Sydney, 1998, 1–8.
16. D. R. Stinson, *Cryptography: Theory and Practice*, CRC Press, Boca Raton, FL, 1995.
17. I. M. Vinogradov, *Elements of Number Theory*, Dover Publ., NY, 1954.

Edit Probability Correlation Attack on the Alternating Step Generator

Jovan Dj. Golić[1] and Renato Menicocci[2]

[1] School of Electrical Engineering, University of Belgrade
Bulevar Revolucije 73, 11001 Belgrade, Yugoslavia
Email: golic@galeb.etf.bg.ac.yu
[2] Fondazione Ugo Bordoni
Via B. Castiglione 59, 00142 Roma, Italy
Email: rmenic@fub.it

Abstract. Given an edit transformation defined by the stop/go clocking in the alternating step generator, an edit probability for two binary strings of appropriate lengths is proposed. corresponds alternating An efficient recursive algorithm for the edit probability computation is derived. It is pointed out how this edit probability can be used to mount a statistically optimal correlation attack on each of the clock-controlled shift registers individually. By estimating the underlying false alarm probability, it is shown that the minimum output sequence length required to be known for a successful attack is linear in the length of the respective shift register. This is illustrated by experimental attacks on relatively short shift registers.

Key words. Stream ciphers, alternating step generator, edit probability, correlation attack.

1 Introduction

Keystream generators for stream cipher applications consisting of a small number of clock-controlled shift registers combined by a suitable function can produce sequences with long period, high linear complexity, and good statistical properties (e.g., see [4]). The stop-and-go clocking is interesting particularly for high speed applications. At any time, a stop/go shift register is clocked once if the clock-control input bit is equal to 1 (or 0) and is not clocked at all otherwise.

The alternating step generator (ASG) [5] is a combination of three binary LFSRs, two of which, LFSR_1 and LFSR_2, are stop/go clocked in a special way by the third one, LFSR_3, which is regularly clocked. Instead of LFSR_3, one may also use any binary keystream generator. More precisely, if the clock-control bit is equal to 1, then LFSR_1 is clocked and LFSR_2 is not, and if the clock-control bit is equal to 0, then LFSR_2 is clocked and LFSR_1 is not. The output sequence is formed as the bitwise sum of the two stop/go clocked LFSR sequences. Let r_i denote the length of LFSR_i, $1 \leq i \leq 3$. It is shown in [5] that the initial state of the clock-control $LFSR_3$ can be recovered by a divide-and-conquer attack which can be regarded as a special kind of the linear consistency attack [7], which appeared later. Namely, if and only if the guess about the initial state of $LFSR_3$ is correct, then the first binary derivative of the output sequence gives rise to the first binary derivatives of both regularly clocked $LFSR_1$ and $LFSR_2$ sequences which are then easily tested for low linear complexity by the well-known Berlekamp-Massey algorithm. The complexity of the attack is thus $O(2^{r_3+2\log_2 \min(r_1,r_2)})$.

A correlation attack on $LFSR_1$ and $LFSR_2$ combined which is based on a specific edit distance is proposed in [3]. The edit distance between two given binary input strings and one given binary output string of appropriate lengths is defined as the minimum possible number of effective substitutions needed in the combination string produced from the two input strings by the stop/go clocking in the ASG manner to obtain the output string, where the minimum is taken over all binary clock-control strings. The minimum output sequence length required for success is both theoretically and experimentally demonstrated to be about four total lengths of $LFSR_1$ and $LFSR_2$ on average. The complexity of this attack is hence $O(2^{r_1+r_2+2\log_2(r_1+r_2)})$.

The objective of this paper is to investigate whether a divide-and-conquer attack on individual $LFSR_1$ or $LFSR_2$ is possible. The complexity of recovering the initial states of $LFSR_1$ and $LFSR_2$ with such an attack would then be close to $O(2^{r_1})$ and $O(2^{r_2})$, respectively. This would be a considerable improvement (at least for large r_1 and r_2) over the edit distance correlation attack [3], especially if r_1 and r_2 are approximately equal, as is typically suggested in the literature (e.g., see [6]). The fact that the first binary derivative of the ASG output sequence is bitwise correlated, with the correlation coefficient 1/2, to the first binary derivative of the output sequence of each of the stop/go clocked $LFSR_1$ and $LFSR_2$ suggests that such a divide-and-conquer attack may be possible. So, the question is essentially whether the stop/go clocking according to a random clock-control sequence and the bitwise addition of a random nonbalanced binary sequence with the probability of one equal to 1/4 to the first binary derivative of the considered stop/go clocked LFSR sequence is not sufficient to disguise the same, but regularly clocked LFSR sequence. Interestingly, it turns out that the edit distance reflecting the corresponding edit transformation does not ensure the desired statistical distinction between the case when the guessed LFSR sequence is correct and the case when it is random.

However, although for a different type of clocking considered, the work [2] shows that at least in principle the edit probability approach [1] may sometimes work when the edit distance one cannot. Accordingly, the problem is approached by defining an appropriate edit probability reflecting the stop/go clocking in the alternating step generator. The edit probability is defined for two binary strings: an input string corresponding to the output sequence of one of the clock-controlled shift registers when regularly clocked and an output string corresponding to the first binary derivative of the output sequence of the ASG. The (random) edit transformation then consists of the alternating clocking of the input string (the chosen LFSR sequence to be reconstructed) and a purely random string (the other LFSR sequence) by a purely random clock-control string, of the bitwise addition of the two stop/go clocked strings, and of taking the first binary derivative of the resulting string. The edit probability is then the probability that a given input string yields a given output string by the described random edit transformation.

An efficient recursive algorithm for the edit probability computation is first derived. It is then pointed out how this edit probability can be used to mount a statistically optimal correlation attack on each of the clock-controlled shift registers. By systematic computation of the underlying false alarm probability, it is shown that the minimum output sequence length required to be known for a successful attack is linear in the length of the respective shift register. The initial state of the other LFSR is then reconstructed in a similar way by using the same edit probability for one input string or, more efficiently, by using a related edit probability for two input strings (one of which is known). The complexity of this attack is $O(2^{\max(r_1,r_2)+2\log_2 \max(r_1,r_2)})$. Successful experimental attacks on relatively short shift registers are also conducted.

In Section 2, a more detailed description of the ASG is provided. The edit probability for one input string and the recursive algorithm for its efficient computation are presented in Section 3, whereas Section 4 contains similar concepts and results for the related edit probability for two input strings. The underlying false alarm probability is estimated in Section 5 and the corresponding correlation attack is explained and experimentally illustrated in Sections 6 and 7, respectively. Conclusions are given in Section 8 and a number of tables displaying the statistics of the edit probability for one input string are presented in the Appendix.

2 Description of Alternating Step Generator

As shown in Fig. 1, the output of the alternating step generator (ASG) [5] is obtained by bitwise addition (modulo 2) of the output sequences of two binary linear feedback shift registers, LFSR_1 and LFSR_2, whose stop/go clocking is defined by a binary clock-control generator (CCG), which is typically another, but regularly clocked LFSR, denoted as LFSR_3. It is assumed that LFSR_1 and LFSR_2 have different irreducible feedback polynomials of respective degrees r_1 and r_2 and coprime periods. At every step, only one LFSR is stepped and the output bit is assumed to be produced in the step-then-add manner. Let c_t denote the output bit of the CCG at step $t \geq 1$. Then, in order to obtain the ASG output bit u_t at step t, we first step LFSR_1 or LFSR_2 depending on whether $c_t = 1$ or $c_t = 0$, respectively, and then we add modulo 2 the output bits of LFSR_1 and LFSR_2. In [5], good standard cryptographic properties of the ASG, such as a long period, a high linear complexity, and approximately uniform relative frequency of short output patterns on a period are established, under the assumption that the clock-control sequence is a de Bruijn sequence and that the feedback polynomials of LFSR_1 and LFSR_2 are primitive. It is expected that similar results also hold if the CCG is a LFSR ($LFSR_3$) with a primitive feedback polynomial whose period is coprime to the periods of $LFSR_1$ and $LFSR_2$.

It is here assumed that the feedback polynomials of all three LFSRs are known to the cryptanalyst, so that the objective of the cryptanalytic attacks considered is to reconstruct the secret-key-controlled LFSR initial states from a known segment of the keystream sequence in the known plaintext scenario.

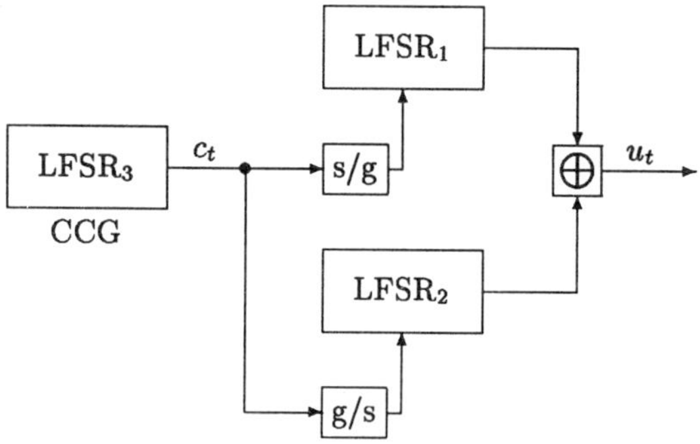

Fig. 1. The alternating step generator.

3 Edit Probability for One Input String

Let $X^{n+2} = x_1, x_2, \ldots, x_{n+2}$ and $Y^{n+2} = y_1, y_2, \ldots, y_{n+2}$ denote two binary input strings and let $C^{n+1} = c_1, c_2, \ldots, c_{n+1}$ denote a binary clock-control string. Let $U^{n+1}(X^{n+2}, Y^{n+2}, C^{n+1}) = u_1, u_2, \ldots u_{n+1}$ denote the combination string of length $n+1$ produced from X^{n+2} and Y^{n+2} by the step-then-add alternating stepping according to C^{n+1} (X^{n+2} and Y^{n+2} correspond to the regularly clocked $LFSR_1$ and $LFSR_2$ sequences of length $n+2$, respectively). Accordingly, we initially have $u_1 = x_1 \oplus y_2$ if $c_1 = 0$ and $u_1 = x_2 \oplus y_1$ if $c_1 = 1$, whereas for any $1 \leq s \leq n$ and $0 \leq l \leq s$, if l denotes the number of ones in C^s, then $u_s = x_{l+1} \oplus y_{s+1-l}$ and we have $u_{s+1} = x_{l+1} \oplus y_{s+2-l}$ if $c_{s+1} = 0$ and $u_{s+1} = x_{l+2} \oplus y_{s+1-l}$ if $c_{s+1} = 1$ ($\oplus$ denotes the binary addition (modulo 2)).

Further, the first binary derivative of a binary string $A^{n+1} = a_1, a_2, \ldots, a_{n+1}$ is defined as $\dot{A}^n = \dot{a}_1, \ldots \dot{a}_n$ where $\dot{a}_i = a_i \oplus a_{i+1}$, $1 \leq i \leq n$. Then, let $\dot{U}^n(X^{n+2}, Y^{n+2}, C^{n+1}) = \dot{u}_1, \ldots, \dot{u}_n$ denote the first binary derivative of $U^{n+1}(X^{n+2}, Y^{n+2}, C^{n+1})$. The process of producing $\dot{U}^n(X^{n+2}, Y^{n+2}, C^{n+1})$ from a given X^{n+2} is called the ASG *edit transformation* of X^{n+2} into $\dot{U}^n$ according to Y^{n+2} and C^{n+1}.

Now, assume a probabilistic model where the strings Y^{n+2} and C^{n+1} are mutually independent strings of uniformly distributed and independent binary random variables. Such strings are called here *purely random*. They define a random edit transformation of X^{n+2} into $\dot{U}^n(X^{n+2}, Y^{n+2}, C^{n+1})$.

Let $Z^n = z_1, z_2, \ldots, z_n$ denote a binary output string. The *edit probability* for a given input string X^{n+2} and a given output string Z^n, denoted as

$P(X^{n+2}, Z^n)$, is then defined as the probability that a given X^{n+2} is transformed into a given Z^n by a random edit transformation according to random Y^{n+2} and C^{n+1}. Formally, the edit probability is defined as

$$P(X^{n+2}, Z^n) \ = \ \Pr\{\dot{U}^n(X^{n+2}, Y^{n+2}, C^{n+1}) = Z^n\}. \tag{1}$$

In fact, in the probabilistic model where X^{n+2} is assumed to be purely random, the statistically optimal edit probability (minimizing the error probability when deciding on X^{n+2} given Z^n) is defined as $\Pr\{X^{n+2}, \dot{U}^n(X^{n+2}, Y^{n+2}, C^{n+1}) = Z^n\}$. It is equal to $\Pr\{\dot{U}^n(X^{n+2}, Y^{n+2}, C^{n+1}) = Z^n | X^{n+2}\} \cdot \Pr\{X^{n+2}\}$, which is further equal to $P(X^{n+2}, Z^n)\cdot \Pr\{X^{n+2}\}$, as in this model the edit probability (1) is conditioned on X^{n+2}. Since $\Pr\{X^{n+2}\} = 2^{-n-2}$ for every given X^{n+2}, we have that the edit probability (1) is also statistically optimal.

If the input strings X^{n+2} and Y^{n+2} change places, the resulting edit probability $P'(Y^{n+2}, Z^n)$ is not formally the same as the alternating stepping is not symmetric with respect to X^{n+2} and Y^{n+2} for a given clock-control string C^{n+1}. However, since $\dot{U}^n(X^{n+2}, Y^{n+2}, C^{n+1}) = \dot{U}^n(Y^{n+2}, X^{n+2}, \bar{C}^{n+1})$, where $\bar{C}^{n+1}$ denotes the bitwise binary complement of C^{n+1}, it follows that $P' = P$.

Our basic objective is to examine whether the defined edit probability can be computed efficiently by an algorithm whose computational complexity is significantly smaller than 2^{2n+3}, which corresponds to the computation of (1) by the summation of the elementary probability 2^{-2n-3} over all Y^{n+2} and C^{n+1} such that $\dot{U}^n(X^{n+2}, Y^{n+2}, C^{n+1}) = Z^n$. To this end, for any $1 \le s \le n$ and $0 \le l \le s+1$, we define the *partial edit probability* $P(l, s)$ as the joint probability

$$P(l, s) \ = \ \Pr\{\dot{U}^s(X^{s+2}, Y^{s+2}, C^{s+1}) = Z^s, w(C^{s+1}) = l\} \tag{2}$$

where $w(C^{s+1})$ is the Hamming weight of C^{s+1}. If $w(C^{s+1}) = l$, then $u_{s+1} = x_{l+1} \oplus y_{s+2-l}$, so that the edit transformation in (2) involves the prefixes X^{l+1} and Y^{s+2-l} only.

The next theorem enables the efficient computation of the edit probability based on a recursive property of the partial edit probability.

Theorem 1. *For any given X^{n+2} and Z^n, we have*

$$P(X^{n+2}, Z^n) \ = \ \sum_{l=0}^{n+1} P(l, n) \tag{3}$$

where the partial edit probability $P(l, n)$ is computed recursively by

$$P(l, s) \ = \ \frac{1}{4}P(l, s-1) + \frac{1}{2}(1 \oplus \dot{x}_l \oplus z_s)P(l-1, s-1) \tag{4}$$

for $1 \le s \le n$ and $0 \le l \le s+1$, with the initial values $P(0, 0) = P(1, 0) = 1/2$ and $P(-1, s-1) = P(s+1, s-1) = 0$ (the first and the second terms in (4) are thus computed only for $l < s+1$ and $l > 0$, respectively).

Proof First observe that (3) is an immediate consequence of the definition of the partial edit probability. Second, for $s = 1$, (2) implies that $P(0,1) = 1/8$, $P(1,1) = 1/8 + (1 \oplus \dot{x}_1 \oplus z_1)/4$, and $P(2,1) = (1 \oplus \dot{x}_2 \oplus z_1)/4$. The same values can also be obtained by (4) from the given initial values.

Now, assume that $s > 1$. We partition all clock-control strings C^{s+1} into two subsets depending on whether $c_{s+1} = 0$ or $c_{s+1} = 1$. Then (2) can be put into the form

$$P(l,s) = \Pr\{\dot{U}^s(X^{s+2}, Y^{s+2}, C^{s+1}) = Z^s, w(C^{s+1}) = l | c_{s+1} = 0\} \cdot \frac{1}{2} + \Pr\{\dot{U}^s(X^{s+2}, Y^{s+2}, C^{s+1}) = Z^s, w(C^{s+1}) = l | c_{s+1} = 1\} \cdot \frac{1}{2}. \quad (5)$$

If $c_{s+1} = 0$, then $\dot{u}_s = \dot{y}_{s+1-l}$, and if $c_{s+1} = 1$, then $\dot{u}_s = \dot{x}_l$. Consequently,

$$\begin{aligned} &\Pr\{\dot{U}^s(X^{s+2}, Y^{s+2}, C^{s+1}) = Z^s, w(C^{s+1}) = l | c_{s+1} = 0\} \\ &= \Pr\{\dot{y}_{s+1-l} = z_s | \dot{U}^{s-1}(X^{s+1}, Y^{s+1}, C^s) = Z^{s-1}, w(C^s) = l\} \\ &\quad \cdot \Pr\{\dot{U}^{s-1}(X^{s+1}, Y^{s+1}, C^s) = Z^{s-1}, w(C^s) = l\} &(6) \\ &= \Pr\{y_{s+2-l} = y_{s+1-l} \oplus z_s | \dot{U}^{s-1}(X^{s+1}, Y^{s+1}, C^s) = Z^{s-1}, w(C^s) = l\} \\ &\quad \cdot P(l, s-1) &(7) \\ &= \frac{1}{2} \cdot P(l, s-1). &(8) \end{aligned}$$

Equation (8) follows from (7) because y_{s+2-l} remains to be independent of y_{s+1-l} when conditioned on $\dot{U}^{s-1}(X^{s+1}, Y^{s+1}, C^s) = Z^{s-1}$ and $w(C^s) = l$, as this condition involves only Y^{s+1-l} (not y_{s+2-l}).

Similarly,

$$\begin{aligned} &\Pr\{\dot{U}^s(X^{s+2}, Y^{s+2}, C^{s+1}) = Z^s, w(C^{s+1}) = l | c_{s+1} = 1\} \\ &= \Pr\{\dot{x}_l = z_s | \dot{U}^{s-1}(X^{s+1}, Y^{s+1}, C^s) = Z^{s-1}, w(C^s) = l-1\} \\ &\quad \cdot \Pr\{\dot{U}^{s-1}(X^{s+1}, Y^{s+1}, C^s) = Z^{s-1}, w(C^s) = l-1\} &(9) \\ &= (1 \oplus \dot{x}_l \oplus z_s) \cdot P(l-1, s-1). &(10) \end{aligned}$$

Note that the given initial values eliminate the terms given by (8) and (10) when $l = s+1$ and $c_{s+1} = 0$ and when $l = 0$ and $c_{s+1} = 1$, respectively. Equation (4) directly follows from (5), (8), and (10). □

The time and space complexities of the recursive algorithm corresponding to Theorem 1 are clearly $O(n^2)$ and $O(n)$, respectively, since only the values of the partial edit probability for the current and the preceding value of s have to be stored at a time. The algorithm is thus feasible even if n is very large. The only computational problem is that the edit probability appears to be exponentially small in the string length, so that one has to deal with very small numbers. The following normalization turned out to be convenient in the experiments conducted:

$$\bar{P}(X^{n+2}, Z^n) = 2^{n+1} P(X^{n+2}, Z^n). \quad (11)$$

It can be computed recursively by

$$\bar{P}(X^{n+2}, Z^n) = \sum_{l=0}^{n+1} \bar{P}(l, n) \tag{12}$$

where

$$\bar{P}(l, s) = \frac{1}{2}\bar{P}(l, s-1) + (1 \oplus \dot{x}_l \oplus z_s)\bar{P}(l-1, s-1) \tag{13}$$

for $1 \leq s \leq n$ and $0 \leq l \leq s+1$, with the initial values $\bar{P}(0,0) = \bar{P}(1,0) = 1$ and $\bar{P}(-1, s-1) = \bar{P}(s+1, s-1) = 0$.

4 Edit Probability for Two Input Strings

The *edit probability* for two given input strings X^{n+2} and Y^{n+2} and a given output string Z^n, denoted as $P(X^{n+2}, Y^{n+2}, Z^n)$, is defined as the probability that given strings X^{n+2} and Y^{n+2} are transformed into a given Z^n by a random edit transformation defined by random C^{n+1}. The edit transformation is here defined as the process of producing $\dot{U}^n(X^{n+2}, Y^{n+2}, C^{n+1})$ from given X^{n+2} and Y^{n+2}. More precisely, this edit probability is defined as

$$P(X^{n+2}, Y^{n+2}, Z^n) = \Pr\{\dot{U}^n(X^{n+2}, Y^{n+2}, C^{n+1}) = Z^n\} \tag{14}$$

(in contrast with (1), Y^{n+2} is here fixed rather than random). By a similar argument as for $P(X^{n+2}, Z^n)$, this edit probability is statistically optimal in the probabilistic model where X^{n+2} is purely random. It is also statistically optimal in the probabilistic model where X^{n+2} and Y^{n+2} are both purely random (when deciding on both X^{n+2} and Y^{n+2} given Z^n). This edit probability is symmetric with respect to the input strings X^{n+2} and Y^{n+2}, because $\dot{U}^n(X^{n+2}, Y^{n+2}, C^{n+1}) = \dot{U}^n(Y^{n+2}, X^{n+2}, \bar{C}^{n+1})$.

In a similar way as for the edit probability for one input string, one can then prove the following theorem.

Theorem 2. *For any given X^{n+2}, Y^{n+2}, and Z^n, we have*

$$P(X^{n+2}, Y^{n+2}, Z^n) = \sum_{l=0}^{n+1} P(l, n) \tag{15}$$

where the partial edit probability $P(l, n)$ is computed recursively by

$$P(l, s) = \frac{1}{2}(1 \oplus \dot{y}_{s+1-l} \oplus z_s)P(l, s-1) + \frac{1}{2}(1 \oplus \dot{x}_l \oplus z_s)P(l-1, s-1) \tag{16}$$

for $1 \leq s \leq n$ and $0 \leq l \leq s+1$, with the initial values $P(0,0) = P(1,0) = 1/2$ and $P(-1, s-1) = P(s+1, s-1) = 0$.

The time and space complexities of the recursive computation are also $O(n^2)$ and $O(n)$, respectively, and the normalization can be performed in the same way as for the edit probability for one input string.

The edit probability for one input string may be used to obtain a relatively small number of candidates for the initial state of each of the stop/go clocked LFSRs from a given output sequence. However, given a number of candidate initial states for one of the LFSRs, it is more efficient to recover the initial state of the other one by using the edit probability for two input strings.

5 False Alarm Probability

In this section, a statistical model for the correlation attack to be described in Section 6 is introduced and analyzed. Suppose that a given output string Z^n is produced from an input string X^{n+2} by the ASG edit transformation. Suppose also that instead of knowing X^{n+2} exactly, we know that it belongs to a given finite set of possible input strings. Our objective is to find the correct input string from a given output string by using the edit probability (for one input string) defined in Section 3, hoping that this edit probability is sufficiently larger in the case when our guess about X^{n+2} is correct than in the opposite case. To this end, we first compute the edit probability for every possible input string and then decide on one with the maximum edit probability. However, it is not at all intuitively clear whether this strategy can work or not.

Consequently, consider the probability distribution of the edit probability $P(X^{n+2}, Z^n)$ in the following two probabilistic models (*hypotheses*):

- $\mathbf{H_0}$ (*correlated case*): X^{n+2}, Y^{n+2}, and C^{n+1} are purely random and independent and $Z^n = \dot{U}^n(X^{n+2}, Y^{n+2}, C^{n+1})$.
- $\mathbf{H_1}$ (*independent case*): X^{n+2} and Z^n are purely random and independent.

For the maximum edit probability decision rule to work, it is necessary that the separation between the probability distributions in the correlated and independent cases increases with the string length n, and the faster the increase, the smaller the string length required for successful decision making is. As a theoretical analysis of the separation between the two probability distributions seems to be very difficult, one may measure the separation experimentally. We conducted systematic experiments for the normalized edit probability and produced histograms of the two distributions for each $n = 100, (10), 1000$ on random samples of 10000 pairs (X^{n+2}, Z^n) generated according to $\mathbf{H_0}$ and $\mathbf{H_1}$, respectively. They show that although the distributions cannot be well approximated by the normal distributions, the separation becomes significant even for relatively small values of n. It thus turns out that the normalized edit probability is much larger in the correlated than in the independent case. For illustration, Tables 1 and 2 given in the Appendix display the observed minimum, maximum, mean, and median values along with the standard deviation of the normalized edit probability for each $n = 100, (200), 900$ for the independent and correlated case, respectively.

As we deal with a decision making problem, the separation between the two distributions is measured by the false alarm probability (derived from $\mathbf{H_1}$) when the missing event probability (derived from $\mathbf{H_0}$) is fixed. Since the number of correct input strings is only one, the missing event probability can be fixed to a value which need not be very small (e.g., $p_m = 0.1$, $p_m = 0.01$, or $p_m = 0.001$). Therefore, in the statistical hypothesis testing considered, a threshold is set according to p_m and a tested input string is classified under $\mathbf{H_0}$ or $\mathbf{H_1}$ depending on whether the normalized edit probability is bigger or smaller than the threshold. In this case, the false alarm probability p_f becomes a function of n, and if and only if this function is decreasing, then the separation between the two distributions increases with n, as desired. The value of n should be chosen so as to make p_f very small, that is, inversely proportional to the number of incorrect input strings. For $p_m = 0.1$, $p_m = 0.01$, and $p_m = 0.001$, Table 3 given in the Appendix displays the estimated threshold, $\bar{P}_{th}$, and p_f for each $n = 100, (200), 900$.

Indeed, for each considered p_m, the estimated false alarm probability decreases with n. Moreover, this decrease appears to be consistent with the exponential form $a\,b^n$, $b < 1$, for large n. As a consequence, the minimum string length n required for the expected number of false input string candidates to fall below one is linear in the logarithm (to the base two) of the total number of tested input strings (see Section 6). The corresponding estimates of the parameters a and b were obtained by the least mean square approximation method applied to the logarithms to the base two of the false alarm probabilistic estimates for $n = 100, (10), 1000$ and are presented in Table 4 given in the Appendix. The parameters a and b were estimated on the first 20 points for $p_m = 0.1$, on the first 20 and 40 points for $p_m = 0.01$, and on the first 20, 40, and 60 points for $p_m = 0.001$, respectively. The most reliable estimates were obtained for the first 20 points in all the cases. To be on the conservative side, the false alarm probabilities $p_f^{0.1}(n)$, $p_f^{0.01}(n)$, and $p_f^{0.001}(n)$ are approximated for large n by using the maximum values of b as

$$p_f^{0.1}(n) \approx 0.78 \cdot 0.98^n, \quad p_f^{0.01}(n) \approx 1.29 \cdot 0.988^n, \quad p_f^{0.001}(n) \approx 1.35 \cdot 0.993^n. \tag{17}$$

6 Correlation Attack

It is assumed that the LFSR feedback polynomials and a sufficiently long segment of the output sequence, in the known plaintext scenario, are known to the cryptanalyst. The objective of cryptanalysis is then to reconstruct the secret-key-dependent LFSR initial states by a method faster than the exhaustive search. More precisely, the first objective of the correlation attack described in this section is to recover the initial state of LFSR_1 or LFSR_2 individually, by using the (normalized) edit probability for one input string defined in Section 3. As argued in Section 3, such an attack is statistically optimal. Suppose a number of candidates for the LFSR_1 initial state is obtained in this way. Then the LFSR_2 initial state candidates can be obtained either by using the edit probability for one input string or the edit probability for two input strings defined in Section 4. In the final stage, the correct initial states of LFSR_1 and LFSR_2 along with the LFSR_3 initial state are all reconstructed. Note that in view of the structure of the ASG generator, one may expect that different LFSR initial state triples necessarily give rise to different output sequences, so that the solution for the LFSR initial states is unique.

We first produce n bits of the first binary derivative, Z^n, of the first $n+1$ successive bits of the known output sequence. The string Z^n comes from the (unknown) output strings of the regularly clocked LFSR_1 and LFSR_2 of maximum length $n+2$ (the actual lengths are random and depend on the unknown clock-control string of length $n+1$). In the correlation attack on LFSR_i, $i = 1, 2$, for any assumed LFSR_i initial state, we first determine an input string X^{n+2} as the first $n+2$ output bits generated by its linear recursion and then compute the normalized edit probability $\bar{P}(X^{n+2}, Z^n)$ by the recursive algorithm derived in Section 3. This is repeated for every possible initial state of LFSR_i, altogether 2^{r_i} of them. Roughly speaking, the candidates for the correct LFSR_i initial state are obtained as the ones with the computed normalized edit probability close to being maximal. More precisely, a threshold is set according to the missing event probability (for the correct hypothesis $\mathbf{H_0}$) which is fixed to a value which need not be very small (e.g., $p_m = 0.1$ or $p_m = 0.01$), and a tested initial state is classified as a candidate if the corresponding normalized edit probability is bigger than or equal to the threshold. This threshold is obtained empirically by the statistical method given in Section 5.

Ideally, if n is large enough, then there will remain only one candidate for the initial state of LFSR_i. This can happen if and only if the false alarm probability (for the alternative hypothesis $\mathbf{H_1}$) $p_f(n)$ defined in Section 5 is sufficiently small. Namely, since the expected number of false candidates for an average Z^n is $(2^{r_i} - 1)p_f(n)$, the correlation attack is deemed successful if and only if, approximately,

$$2^{r_i}\, p_f(n) \ \le\ 1 \tag{18}$$

(see [3]). If $p_f(n)$ has the exponential form $a\, b^n$, where $b < 1$, then (18) reduces to

$$n \ \ge\ \frac{r_i + \log_2 a}{-\log_2 b} \tag{19}$$

which means that the required output segment length is essentially linear in the length of LFSR_i. In view of (17), the required output segment length is then approximately given as

$$n \ge 34.3\, r_i - 12.3 \tag{20}$$

$$n \ge 57.4\, r_i - 21.1 \tag{21}$$

$$n \ge 98.7\, r_i - 42.7 \tag{22}$$

for $p_m = 0.1$, $p_m = 0.01$, and $p_m = 0.001$, respectively.

The number of obtained candidate initial states for LFSR_i cannot be reduced to just one by increasing the length n if the threshold is computed according to a given p_m, because the initial states generating relatively small (positive or negative) phase shifts of the original LFSR_i sequence give rise to the normalized edit probability values that are also close to being maximal. This can be explained by the recursion (13), which changes only in the beginning for a positive phase shift or at the end for a negative phase shift, and remains the same for the best part of the input string. Therefore, relatively small phase shifts of the original sequence are not well modeled by the hypothesis $\mathbf{H_1}$ (independent case). So, we may always expect multiple candidates, unless we further increase the threshold (i.e., increase the missing event probability) for any given n. The missing event probability is thus chosen to be equal to 0.1 and multiple candidates are ranked in order of decreasing normalized edit probabilities. It is natural to expect that the correct initial state gives rise to the normalized edit probability that is maximal or very close to being maximal, provided n is chosen according to (20).

Candidates for the initial state of the other shift register LFSR_j, $j \neq i$, can be produced in the same way by using the edit probability for one input string, regardless of the obtained initial state candidates for LFSR_i. Alternatively, one may use the edit probability for two input strings defined in Section 4 (the way the input strings X^{n+2} and Y^{n+2} are assigned to LFSR_i and LFSR_j is not important due to the symmetry). Namely, for each obtained initial state candidate for LFSR_i and each assumed initial state for LFSR_j, $j \neq i$, compute the normalized edit probability for two input strings and then apply a similar statistical hypothesis testing procedure as above to obtain the initial state candidates for LFSR_j. Then

- $\mathbf{H_0}$ (*correlated case*): X^{n+2}, Y^{n+2}, and C^{n+1} are purely random and independent and $Z^n = \dot{U}^n(X^{n+2}, Y^{n+2}, C^{n+1})$.
- $\mathbf{H_1}$ (*independent case*): X^{n+2}, $\hat{X}^{n+2}$, Y^{n+2}, and C^{n+1} are purely random and independent and $Z^n = \dot{U}^n(\hat{X}^{n+2}, Y^{n+2}, C^{n+1})$.

As the edit probability for two input strings makes use of the additional information about one of the input strings, the required output segment length (and the complexity) for the same number of multiple candidates is in this case smaller than for the edit probability for one input string. Note that the search through the LFSR_j initial states is reduced if a prefix of length smaller than r_j of the considered input string is found such that the resulting edit probability is equal to zero. In addition, this method also reduces the number of candidate initial states for the first shift register LFSR_i.

Accordingly, we obtain a relatively small number of candidate initial state pairs for LFSR_1 and LFSR_2 in time $O(2^{\max(r_1,r_2)+2\log_2 \max(r_1,r_2)})$.

Both component candidate initial states are ranked in order of decreasing normalized edit probabilities. The (unique) correct pair along with the correct LFSR_3 initial state are then all reconstructed by the edit distance method from [3]. The candidate pairs are tested according to decreasing normalized edit probabilities (gradually increasing the number of component candidate initial states examined). For each candidate pair, one stores the matrix of the corresponding partial edit distances and then by backtracking recovers all possible clock-control strings C^n that together with the corresponding input strings result in a given output string (not its binary derivative). The difference is that such clock-control strings need not exist (the edit distance need not be equal to zero) if a candidate pair is obtained by using the edit probability for one input string for both LF-SRs. The average number of such clock-control strings of length n per candidate pair can be approximated as (see [3])

$$m_n \approx 1.2 \cdot 2^{0.27\,n}. \tag{23}$$

Alternatively, if the candidate initial states for the other shift register LFSR_j are obtained by the edit probability for two input strings, all possible clock-control strings for each candidate initial state pair for LFSR_1 and LFSR_2 can also be obtained by backtracking through the matrix of positive partial edit probabilities.

So, pick $n = r_3$ and then test each obtained initial state triple for all the LFSRs by generating the corresponding ASG output sequence and comparing it with the given one. In this final stage, the unique solution for the initial states of all the LFSRs is thus found in $O(2^{0.27\,r_3})$ time and space.

7 Experimental Correlation Attacks

A number of computer simulations were conducted to show that the above edit probability correlation attack can work in practice. Only primitive feedback polynomials were used for all the LFSRs. The correlation attack was performed in the way explained in Section 6 by using the edit probability for one input string to recover the initial state candidates for both LFSR_1 and LFSR_2. To reduce the number of multiple candidates, the missing event probability was chosen to be equal to 0.1. The feedback polynomials were assumed to be known and the objective was to reconstruct the initial states of LFSR_1 and LFSR_2 along with the initial state of LFSR_3 from a sufficiently long segment of the ASG output sequence.

Some examples of the experimental results obtained are shown in Table 5. In each experiment described by the shift register lengths (r_1, r_2, r_3), for any chosen n, N_i denotes the number of candidate initial states for LFSR_i, $i = 1, 2$, $N_{1,2}$ denotes the number of candidate initial state pairs (among $N_1 N_2$ of them) that passed the zero edit distance test, k_i denotes the rank of the normalized edit probability corresponding to the correct initial state of LFSR_i, $i = 1, 2$, and N_3 stands for the number of clock-control strings of length r_3 that had to be tested per correct initial state pair. In each experiment, a unique solution for the LFSR initial states was obtained. Notice that although a relatively small number of multiple candidates for the initial states of LSFR_1 and LFSR_2 did appear, the correct initial states always ranked the best or very close to the best provided n was sufficiently large. As indicated by (20), it was observed that N_i was approximately minimized by using $n \approx 40 r_i$, $i = 1, 2$. However, in practice, it appears that k_i reduces to one or to a very small integer even if $n \approx 20r_i$. In this case, just a few candidates have to be tested in order to find the correct initial state pair.

Table 5. Experimental results.

	$n = 200$	$n = 400$	$n = 600$	(r_1, r_2, r_3)
N_1, N_2	1380, 463	106, 37	22, 25	(15, 14, 20)
$N_{1,2}, N_3$	6, 40	6, 40	6, 40	
k_1, k_2	43, 7	1, 2	1, 2	
N_1, N_2	1125, 1168	36, 33	28, 31	(15, 16, 24)
$N_{1,2}, N_3$	3, 8100	3, 8100	3, 8100	
k_1, k_2	1, 1	1, 1	1, 1	
N_1, N_2	479, 289	28, 33	23, 23	(16, 16, 28)
$N_{1,2}, N_3$	4, 3168	4, 3168	4, 3168	
k_1, k_2	7, 1	6, 1	1, 1	
N_1, N_2	4353, 2243	125, 46	17, 20	(18, 17, 30)
$N_{1,2}, N_3$	2, 2484	2, 2484	2, 2484	
k_1, k_2	3, 12	1, 37	1, 2	

8 Conclusions

It is pointed out that the stop/go clocking in the alternating step generator can be viewed as a random edit transformation of one input into one output binary string. The input string corresponds to the output sequence of one of the clock-controlled LFSRs when regularly clocked and the output string corresponds to the first binary derivative of the output sequence of the alternating step generator. The output sequences of the other clock-controlled LFSR and the clock-control LFSR are both assumed to be purely random. A related edit probability for two binary strings of appropriate lengths is then introduced. An efficient recursive algorithm for the edit probability computation is derived.

It is shown how this edit probability can be used to mount a statistically optimal correlation attack on each of the two clock-controlled LFSRs individually. The correlation attack requires the computation of the edit probability for every possible initial state of the considered LFSR. Alternatively, when the initial state of one of the LFSRs is reconstructed, the correlation attack on the other one can also be based on a more efficient edit probability for two input strings and one output string.

For the underlying statistical hypothesis testing problem, the false alarm probability is estimated by systematic experiments. It is shown that this probability can be well approximated by an exponentially decreasing function of the string length. This implies that the minimum output sequence length required to be known for a successful attack is linear in the length of the respective shift register. More precisely, about forty shift register lengths are sufficient for success.

The time complexity of recovering the initial states of the two clock-controlled LFSRs is thus $O(2^{\max(r_1,r_2)+2\log_2 \max(r_1,r_2)})$. The initial state of the clock-control LFSR can be recovered from consistent clock-control strings obtained by the edit distance method as in [3] which requires $O(2^{0.27\, r_3})$ time and space.

For illustration, successful experimental attacks on relatively short shift registers are conducted. The results demonstrate that in order to prevent the initial state reconstruction of each of the clock-controlled LFSRs individually, the length of each of them should be sufficiently long.

More generally, the methodology developed in this paper shows that the security against correlation attacks of keystream generators involving clock-controlled shift registers can be analyzed by using special edit probabilities, adapted to the structure considered.

References

1. J. Dj. Golić and S. Petrović, "A generalized correlation attack with a probabilistic constrained edit distance," Advances in Cryptology - EUROCRYPT '92, *Lecture Notes in Computer Science*, vol. 658, R. A. Rueppel ed., Springer-Verlag, pp. 472-476, 1993.
2. J. Dj. Golić and L. O'Connor, "Embedding and probabilistic correlation attacks on clock-controlled shift registers," Advances in Cryptology - EUROCRYPT '94, *Lecture Notes in Computer Science*, vol. 950, A. De Santis ed., Springer-Verlag, pp. 230-243, 1995.
3. J. Dj. Golić and R. Menicocci, "Edit distance correlation attack on the alternating step generator," Advances in Cryptology - CRYPTO '97, *Lecture Notes in Computer Science*, vol. 1294, B. Kaliski ed., Springer-Verlag, pp. 499-512, 1997.
4. D. Gollmann and W. G. Chambers, "Clock-controlled shift registers: A review," *IEEE Journal on Selected Areas in Communications*, vol. 7, pp. 525-533, May 1989.
5. C. G. Günther, "Alternating step generators controlled by de Bruijn sequences," Advances in Cryptology - EUROCRYPT '87, *Lecture Notes in Computer Science*, vol. 304, D. Chaum and W. L. Price eds., Springer-Verlag, pp. 5-14, 1988.
6. A. Menezes, P. van Oorschot, and S. Vanstone, *Handbook of Applied Cryptography*. New York: CRC Press, 1997.
7. K. Zeng, C. H. Yang, and T. R. N. Rao, "On the linear consistency test (LCT) in cryptanalysis with applications," Advances in Cryptology - CRYPTO '89, *Lecture Notes in Computer Science*, vol. 435, G. Brassard ed., Springer-Verlag, pp. 164-174, 1990.

Appendix

Table 1. Statistics of $\bar{P}$ on 10000 independent pairs (X^{n+2}, Z^n).

n	Min	Max	Mean	Median	Std Dev
100	6.059E-10	7.749E2	1.825E0	5.051E-2	1.406E1
300	2.514E-14	2.320E3	6.438E-1	1.696E-4	2.832E1
500	2.066E-19	4.170E2	2.194E-1	1.121E-6	6.085E0
700	1.535E-23	5.366E1	1.477E-2	8.478E-9	6.352E-1
900	1.842E-24	1.773E1	3.241E-3	7.086E-11	2.038E-1

Table 2. Statistics of $\bar{P}$ on 10000 correlated pairs (X^{n+2}, Z^n).

n	Min	Max	Mean	Median	Std Dev
100	1.162E-3	1.024E6	9.378E2	4.773E1	1.247E4
300	6.552E-3	1.4E11	3.625E7	1.092E4	1.532E9
500	1.006E-3	7.071E15	1.697E12	2.483E6	9.125E13
700	1.170E-3	2.514E20	2.542E16	5.479E8	2.514E18
900	5.981E-3	6.170E22	1.186E19	1.354E11	7.292E20

Table 3. Estimation of thresholds and false alarm probabilities.

n	$\bar{P}_{th}^{0.1}$	$\bar{P}_{th}^{0.01}$	$\bar{P}_{th}^{0.001}$	$p_f^{0.1}$	$p_f^{0.01}$	$p_f^{0.001}$
100	2.193E0	1.680E-1	2.005E-2	1.015E-1	3.491E-1	6.127E-1
300	5.684E1	1.093E0	6.713E-2	1.0E-3	2.29E-2	9.48E-2
500	2.729E3	9.323E0	1.061E-1	0	2.9E-3	1.95E-2
700	2.150E5	3.290E2	1.398E0	0	0	1.2E-3
900	1.282E7	6.752E3	2.673E1	0	0	0

Table 4. Estimation of a and b on 20*, 40**, and 60*** points.

p_m	a^*	a^{**}	a^{***}	b^*	b^{**}	b^{***}
0.1	.7815	-	-	.9799	-	-
0.01	1.288	1.922	-	.9879	.9858	-
0.001	1.346	2.184	2.897	.9926	.9901	.9891

Automaticity of Solutions of Mahler Equations

Fritz von Haeseler[1] and Wibke Jürgensen*[2]

[1] CeVis, University of Bremen
Universitätsallee 29, D-28359 Bremen, Germany
fritz@cevis.uni-bremen.de

[2] Department of Mathematics and Computer Science, University of Bremen
Postfach 330 440, D-28334 Bremen, Germany
wibke@math.uni-bremen.de

Abstract. We give sufficient criteria for the m-automaticity of the coefficients of formal power series satisfying special Mahler equations over finite commutative rings as well as over finite fields of any characteristic. The main tool for our investigation will be the m-equation-kernel of a rational function. We show that there is a close relation between the finiteness of an m-equation-kernel and the finiteness of the m-kernel of a solution.

Introduction

In 1979, G. Christol, T. Kamae, M. Mendès France, and G. Rauzy showed that the theory of automatic sequences (e.g., [1] for an introduction and further references) can be regarded as a study of solutions of Mahler equations with coefficients in a finite field, see [4] and [5].

Given a prime number p and a power series $F(X) = \sum_{j=0}^{\infty} f_j X^j$ with coefficients in the Galois field $\mathbb{F}_{p^\alpha}$, they proved that the sequence $(f_j)_{j\in\mathbb{N}}$ is p-automatic, i.e., the p-kernel $\{(f_{p^k n+r})_{n\geq 0} \mid k \in \mathbb{N}, 0 \leq r < p^k\}$ is a finite set, if and only if F is algebraic over the field $\mathbb{F}_{p^\alpha}(X)$ of rational functions, cf. [4, 5] and also [14] for the generalization to higher dimensions. In particular, F satisfies a Mahler equation

$$\sum_{j=0}^{d} P_j(X) F(X^{p^j}) = Q(X),$$

where P_j, $j = 0, \ldots, d$, and Q are polynomials with coefficients in $\mathbb{F}_{p^\alpha}$. Note that the 'inner' powers of F in the above equation are powers of the characteristic of $\mathbb{F}_{p^\alpha}$.

More generally, [3, 7] and [13] proved that for any integer $m \geq 2$ the power series $F(X) = \sum_{j=0}^{\infty} f_j X^j$ of an m-automatic sequence $(f_j)_{j\in\mathbb{N}}$ with values in a

* Supported by DFG-Graduiertenkolleg "Komplexe Dynamische Systeme"

finite field $\mathbb{F}_{p^\alpha}$ satisfies a Mahler equation of the form

$$\sum_{j=0}^{D} P_j(X)F(X^{m^j}) = Q(X),$$

where P_j, $j = 0,\dots,D$, and Q are polynomials with coefficients in the field under consideration. (Strictly speaking, they study m-regular sequences, a more generalized concept of m-automatic sequences.) A similar result concerning m-automatic sequences is also given in [9] (Theorem 8.5). Furthermore A. J. van der Poorten established an analogous result concerning the characteristic function of an m-automatic sequence, cf. [12].

It is therefore of some interest to study the converse, namely whether a solution of an equation of the above type is m-automatic or not. As an example in [9] shows, a non-trivial solution is not necessarily m-automatic.

In this note, we do not deal with the general question, we rather restrict our attention to a study of equations of the type

$$F(X) = R(X)F(X^m),$$

where R is a rational function with coefficients in a finite ring or in a finite field.

After some preliminaries in the first section we study the automaticity of solutions of the above equation over a finite commutative ring with identity. To this end, we introduce the notion of the m-equation-kernel of a rational function. Its finiteness ensures the finiteness of the m-kernel of an assumed solution. The main result of Sect. 2 is a sufficient criterion for the m-automaticity of a solution. It states that any solution F is m-automatic if there exist two polynomials G and H such that $R(X)G(X^m) = H(X)G(X)$ holds.

In the third section, we study a map $\Psi_m : \mathcal{R}[X] \to \mathcal{R}[X]$ which is closely related to the decimations of a series and study some of its properties for the case that the ring $\mathcal{R}$ is a finite field and Ψ_m is restricted to polynomials.

Finally, in the last chapter, we study the automaticity properties of the solutions of our Mahler equations with coefficients in a finite field. It will turn out that the order of the denominator of the rational function R plays a crucial rôle for the automaticity properties of the solutions.

Acknowledgments We are grateful to G. Skordev (University of Bremen) and J.-P. Allouche (Paris) for several helpful discussions and hints to the relevant literature. The work on this paper was completed while the first author was a guest researcher at the DFG-Graduiertenkolleg "Strukturbildungsprozesse" (Universität Bielefeld). He gratefully acknowledges the financial support of the DFG.

1 Preliminaries

Let $\mathcal{R}$ be a non-trivial finite commutative ring with identity. Denote by $\mathcal{R}[X]$ the ring of polynomials, and by $\mathcal{R}[[X]]$ the ring of Laurent series with coefficients in $\mathcal{R}$ such that for every $F(X) = \sum_{j\in\mathbb{Z}} f_j X^j$ there exists a $k \in \mathbb{Z}$ such that $f_i = 0$ for all $i < k$. The largest k with this property is called the *left degree*, $\deg_l(F)$. The ring of rational functions $\mathcal{R}(X)$ consists of all fractions P/Q with polynomials $P, Q \in \mathcal{R}[X]$ such that the coefficient of $X^{\deg_l(Q)}$ of Q is a unit in $\mathcal{R}$.

We begin with the definition of an automatic sequence. To this end, we introduce decimation operators $\partial_t^{(m)}$, see [1,6,8], and [15].

Definition 1. *Let $m \geq 2$ be a natural number and let $t \in \{0, 1, \dots, m-1\}$. Then the map $\partial_t^{(m)} : \mathcal{R}[[X]] \to \mathcal{R}[[X]]$ defined by*

$$\sum_{j\in\mathbb{Z}} f_j X^j \mapsto \sum_{j\in\mathbb{Z}} f_{mj+t} X^j$$

is called t-decimation w.r.t. m.

From now on we agree to omit the superscript m if there is no risk of confusion and we also speak of t-decimation rather than of t-decimation w.r.t. m.

For a polynomial $P \in \mathcal{R}[X]$ one can easily show that $\deg(\partial_t(P)) \leq m^{-1}(\deg(P) - t)$ for every natural number t with $0 \leq t < m$.

Definition 2. *A series $F \in \mathcal{R}[[X]]$ is called* m-automatic *if the* m-kernel, *i.e., the set*

$$\ker_m(F) = \{\partial_{t_k} \circ \cdots \circ \partial_{t_1}(F) \mid t_k, \dots, t_1 \in \{0, \dots, m-1\}, k \in \mathbb{N}\}$$

is finite.

For a further study of m-automatic series, we introduce a matrix which is built from the decimations.

Let G be a series in $\mathcal{R}[[X]]$. Then the $m \times m$ matrix $\mathcal{A}_m$ is defined by

$$\mathcal{A}_m(G) = (a_{ij})_{ij}$$

with

$$a_{ij} = \begin{cases} \partial_{i-j}(G) & \text{for } i \geq j \\ X\partial_{m+i-j}(G) & \text{for } i < j \end{cases}.$$

We call $\mathcal{A}_m$ the *m-decimation matrix*.

Example 1. Consider the polynomial $1 + X^2 + X^4 + X^5 + X^6$ over $\mathcal{R}$. Then the 3-decimation matrix of this polynomial is given by

$$\mathcal{A}_3(1 + X^2 + X^4 + X^5 + X^6) = \begin{bmatrix} 1+X^2 & X+X^2 & X^2 \\ X & 1+X^2 & X+X^2 \\ 1+X & X & 1+X^2 \end{bmatrix}.$$

Lemma 1. *Let $G, H \in \mathcal{R}[[X]]$. Then $\mathcal{A}_m(GH) = \mathcal{A}_m(G)\mathcal{A}_m(H)$.*

Proof. Each series $K \in \mathcal{R}[[X]]$ can be written as $K(X) = \sum_{t=0}^{m-1} X^t \partial_t(K)(X^m)$. Furthermore, we have for $j = 0, \ldots, m-1$ that $\partial_t(X^j K(X^m)) = K(X)$ for $j = t$ and $\partial_t(X^j K(X^m)) = 0$ for $j \neq t$.

By considering the t-decimation of

$$G(X)H(X) = \sum_{k=0}^{(m-1)^2} X^k \sum_{i+j=k} (\partial_i(G)\partial_j(H))(X^m),$$

we obtain

$$\partial_t(GH)(X) = \sum_{i+j=t} (\partial_i(G)\partial_j(H))(X) + X \sum_{i+j=m+t} (\partial_i(G)\partial_j(H))(X),$$

which gives the desired result. □

If $G \in \mathcal{R}[[X]]$ has a multiplicative inverse in the ring $\mathcal{R}[[X]]$ then, as a consequence of Lemma 1, we have that $\mathcal{A}_m(G^{-1}) = \mathcal{A}_m^{-1}(G) = (\det \mathcal{A}_m(G))^{-1} \mathcal{A}_m^{\mathrm{ad}}(G)$, where $\mathcal{A}_m^{\mathrm{ad}}(G)$ is the adjoint matrix of $\mathcal{A}_m(G)$. It will turn out that the determinant of the m-decimation matrix is of special importance for our following considerations.

Definition 3. *The map* $\Psi_m : \mathcal{R}[[X]] \to \mathcal{R}[[X]]$ *defined as*

$$\Psi_m(G) = \det \mathcal{A}_m(G)$$

is called m-determinant.

Some properties of Ψ_m are obvious from the definition. By the rule of multiplication of determinants, we obtain $\Psi_m(GH) = \Psi_m(G)\Psi_m(H)$ for series $G, H \in \mathcal{R}[[X]]$. In addition, the matrix $\mathcal{A}_m(G(X^m))$ is of diagonal form with diagonal entries G. Hence $\Psi_m(G(X^m)) = G(X)^m$.

The first column of the matrix $\mathcal{A}_m(G)$ contains the t-decimations of $G \in \mathcal{R}[[X]]$. Thus, we define the *m-decimation-vector* of G by

$$\mathcal{D}_m(G) = \mathcal{A}_m(G)(1, 0, \ldots, 0)^T = \begin{pmatrix} \partial_0(G) \\ \vdots \\ \partial_{m-1}(G) \end{pmatrix}.$$

With this notation, we can summarize the properties of t-decimations in the following way.

Lemma 2. *Let* $a, b \in \mathcal{R}$, *let* G, H *be series in* $\mathcal{R}[[X]]$, *and let* $R = P/Q \in \mathcal{R}(X)$ *be a rational function with* $\Psi_m(Q) \neq 0$. *Then the following is true*

1. $\mathcal{D}_m(aG + bH) = a\mathcal{D}_m(G) + b\mathcal{D}_m(H)$;
2. $\mathcal{D}_m(GH) = \mathcal{A}_m(G)\mathcal{D}_m(H)$;
3. $\mathcal{D}_m(G(X^m)H(X)) = G(X)\mathcal{D}_m(H)(X)$;
4. $\mathcal{D}_m(R) = (\Psi_m(Q))^{-1} \mathcal{A}_m^{\mathrm{ad}}(Q)\mathcal{D}_m(P)$.

2 Mahler equations over finite commutative rings

We consider Mahler equations of the form $F(X) = R(X)F(X^m)$ with a rational function $R \in \mathcal{R}(X)$ and present some results on the m-automaticity of non-trivial solutions $F \in \mathcal{R}[[X]]$. Calculating the m-kernel of a solution F, we obtain $\partial_{t_1}(F) = \partial_{t_1}(R)F$ in a first step. Further decimation leads to $\partial_{t_2} \circ \partial_{t_1}(F) = \partial_{t_2}(\partial_{t_1}(R)R)F$.

That observation leads to the definition of the m-equation kernel, which we define using modified t-decimations (w.r.t. m). For every $R \in \mathcal{R}[[X]]$ the modified t-decimation $\Delta_{t,R} : \mathcal{R}[[X]] \to \mathcal{R}[[X]]$ is defined as

$$\Delta_{t,R}(G) = \partial_t(RG).$$

Definition 4. *Let $R \in \mathcal{R}[[X]]$. Then the set*

$$K_m(R) = \{\Delta_{t_k,R} \circ \cdots \circ \Delta_{t_1,R}(1) \mid t_k, \dots, t_1 \in \{0, \dots, m-1\}, k \in \mathbb{N}\}$$

is called the m-equation-kernel *of R.*

In [9] it is shown for any polynomial P that a solution F of the Mahler equation $F(X) = P(X)F(X^m)$ has a finite m-kernel. Analogously, the m-equation-kernel of a polynomial is finite.

Proposition 1. *The m-equation-kernel of a polynomial in $\mathcal{R}[X]$ is finite.*

In addition to the m-equation-kernel of polynomials, the next proposition deals with the m-equation-kernel of special rational functions. The result reveals the close relation between the m-kernel of F and the m-equation-kernel of R, pointed out at the beginning of this section.

Proposition 2. *Let $G \in \mathcal{R}[[X]]$ be an invertible series. The m-equation-kernel $K_m(G(X)/G(X^m))$ is finite if and only if the m-kernel of G is finite.*

Proof. Let $R(X) = G(X)/G(X^m)$. From Lemma 2 we have for every $t \in \{0, \dots, m-1\}$ and every $H \in \mathcal{R}[[X]]$ that

$$\Delta_{t,R}(H) = \frac{1}{G}\partial_t(GH).$$

Consequently, the elements of the m-equation-kernel of R are given by

$$\Delta_{t_k,R} \circ \cdots \circ \Delta_{t_1,R}(1) = \frac{\partial_{t_k} \circ \cdots \circ \partial_{t_1}(G)}{G}$$

and we get the following close relation between the m-equation-kernel and the m-kernel:

$$G(X)K_m(G(X)/G(X^m)) = \ker_m(G(X)).$$

Since the m-kernel of G is finite, the m-equation-kernel $K_m(G(X)/G(X^m))$ is also a finite set and vice versa. □

Indeed, given a Mahler equation $F(X) = R(X)F(X^m)$ we have for each m-automatic solution which is multiplicatively invertible in $\mathcal{R}[[X]]$ that $K_m(R) = K_m(F(X)/F(X^m))$ is finite. Consequently, either all solutions of a Mahler equation are m-automatic or the trivial solution, i.e., $F \equiv 0$, is the only m-automatic solution.

It is a well-known fact that the sum as well as the product of two series with finite m-kernels are again m-automatic, cf. [9]. Considering m-equation-kernels, we prove an analogous statement for products of rational functions. However, the sum of two rational functions with finite m-equation-kernels does not necessarily have a finite m-equation-kernel.

Proposition 3. *Let $R, S \in \mathcal{R}(X)$ be rational functions with finite m-equation-kernel. Then $K_m(RS)$ is finite as well.*

Proof. We consider the set $M = \{H_0 + XH_1 + X^2H_2 \mid H_j \in L, j = 0, 1, 2\}$, where L is defined as

$$L = \left\{ \sum_{i=1}^{n} R_i S_i \;\middle|\; R_i \in K_m(R), S_i \in K_m(S), n \in \mathbb{N} \right\}.$$

The sets L and M are finite because the m-equation-kernels $K_m(R)$ and $K_m(S)$ are finite and $\mathcal{R}$ is a finite ring.

We prove that $K_m(RS) \subset M$. It follows from Lemma 2 (*2.*) that

$$\Delta_{t,RS}(1) = \partial_t(RS) = \sum_{i+j=t} \partial_i(R)\partial_j(S) + X \sum_{i+j=t+m} \partial_i(R)\partial_j(S)$$

such that $\Delta_{t,RS}(1) \in M$ for any natural number $t < m$. Furthermore, some tedious manipulation yields that for a rational function $T \in M$ the modified t-decimation $\Delta_{t,RS}(T)$ is also in M for all $t < m$. Hence M is invariant with respect to the modified decimations and $K_m(RS) \subset M$ is finite. □

We are now able to combine the last three propositions to the following theorem.

Theorem 1. *Let R be a rational function in $\mathcal{R}(X)$. The m-equation-kernel $K_m(R)$ is finite if there exist two polynomials $G, H \in \mathcal{R}[X]$, G invertible, such that*

$$R(X)G(X^m) = H(X)G(X).$$

Proof. From Proposition 2 we know that the m-equation-kernel of $G(X)/G(X^m)$ is finite. Then it follows from Propositions 1 and 3 that

$$K_m(R(X)) = K_m\left(H(X)\frac{G(X)}{G(X^m)}\right)$$

is finite as well. □

A simple consequence of Theorem 1 is obtained by choosing $G(X) = X^i$ with $i(m-1) \leq n$ and $H(X) = X^{i(m-1)-n}$ for any integer n.

Corollary 1. *$K_m(X^n)$ is finite for each integer n.*

Summing up our results, we showed in this section that a solution $F \in \mathcal{R}[[X]]$ of the Mahler equation

$$F(X) = R(X)F(X^m)$$

with a rational function $R \in \mathcal{R}(X)$ is m-automatic or, equivalently, has a finite m-kernel in case that

- $R(X) = X^n P(X)$ for $P \in \mathcal{R}[X]$ and $n \in \mathbb{Z}$ or
- there exist two polynomials $G, H \in \mathcal{R}[X]$, G invertible, such that $R(X)G(X^m) = H(X)G(X)$.

Furthermore, Proposition 3 yields that F has also a finite m-kernel if R is a product of rational functions such that every factor satisfies one of these conditions.

3 On the images of Ψ_m

So far, our approach to the m-automaticity of solutions of Mahler equations was based on finite commutative rings. For a further investigation of Mahler equations over finite fields, we have to consider the map Ψ_m introduced in Definition 3 as the determinant of the matrix $\mathcal{A}_m$.

For a prime number p, we denote by $\mathbb{F}_p$ the Galois field with p elements and the finite field with $q = p^\alpha$ elements is denoted by $\mathbb{F}_q$.

Lemma 3. *Let $P \in \mathbb{F}_q[X]$ be a polynomial of degree d with leading coefficient $c \in \mathbb{F}_q$ and roots $\xi_1, \ldots, \xi_d$ in the splitting field. Then*

$$\Psi_m(P)(X) = (-1)^{d(m-1)} c^m \prod_{j=0}^{d-1} (X - \xi_j^m).$$

Proof. We consider $\Psi_m(X - \xi_j)$ for $0 \leq j < d$ over $\mathbb{F}_{q^d}$. It is obvious from Definition 3 that

$$\Psi_m(X - \xi_j) = \det \begin{bmatrix} -\xi_j & 0 & \cdots & 0 & X \\ 1 & -\xi_j & 0 & \cdots & 0 \\ 0 & 1 & \ddots & \ddots & \vdots \\ \vdots & \ddots & \ddots & -\xi_j & 0 \\ 0 & \cdots & 0 & 1 & -\xi_j \end{bmatrix} = (-1)^{m-1}(X - \xi_j^m).$$

Furthermore, $\Psi_m(c) = c^m$ and hence

$$\Psi_m(P)(X) = \Psi_m\left(c \prod_{j=1}^{d} (X - \xi_j)\right) = (-1)^{d(m-1)} c^m \prod_{j=1}^{d} (X - \xi_j^m).$$

□

By Lemma 3, we see that the m-determinant of a polynomial P allows us to compute a polynomial Q which has the roots ξ_j^m, where $P(\xi_j) = 0$ in a splitting field, without knowing the roots of P.

In [2], the authors consider shift register sequences and their decimations to prove the p-automaticity of solutions of Mahler equations over the ring of integers modulo p^α. The characteristic polynomial of the decimation of a shift register sequence coincides there with the image of the characteristic polynomial of the non-decimated shift register sequence under Ψ_p, see also [11].

Corollary 2. *Let $P \in \mathbb{F}_q[X]$ be a polynomial. Then*

1. *the degree of P is invariant under Ψ_m;*
2. *$P(X)$ divides $\Psi_m(P)(X^m)$;*
3. *$\Psi_m(P) = P$ for $m = q$.*

Proof. The invariance of the degree of P under Ψ_m is a direct consequence of Lemma 3 since $c^m \neq 0$ over $\mathbb{F}_q$.

To prove *2.* and *3.*, we may assume that P is irreducible of degree d. It can be written as $P(X) = c\prod_{j=0}^{d-1}(X - \xi_j)$ in the splitting field $\mathbb{F}_{q^d}$. By Lemma 3, we have that

$$R(X) = \frac{\Psi_m(P)(X^m)}{P(X)} = (-1)^{d(m-1)}c^{m-1}\prod_{j=0}^{d-1}(X^{m-1} + X^{m-2}\xi_j + \ldots + \xi_j^{m-1})$$

is a rational function in $\mathbb{F}_q(X)$ as well as a polynomial with coefficients in $\mathbb{F}_{q^d}$. Therefore, R is a polynomial in $\mathbb{F}_q[X]$.

It remains to show that $\Psi_q(P) = P$. Let $\xi \in \mathbb{F}_{q^d}$ be a root of P. Then all roots of P are given by the d distinct elements $\xi_j = \xi^{q^j}$ $(j = 0, \ldots, d-1)$ in $\mathbb{F}_{q^d}$. Furthermore, the roots of $\Psi_q(P)$ are given by ξ_j^q which gives $\xi_j^q = \xi_{j+1}$ for $j < d-1$ and $\xi_{d-1}^q = \xi_0$. Since $c^q = c$ for every $c \in \mathbb{F}_q$, we obtain $\Psi_q(P) = P$. □

Remark 1. Results equivalent to Lemma 3 and Corollary 2 (*2.* and *3.*) can be obtained for polynomials over finite commutative rings, considering a suitable extension. Analogous to 2 (*1.*), we obtain that $\deg(\Psi_m(P)) \leq \deg(P)$.

We will see that the order of a polynomial over a finite field is an important integer in our studies of Mahler equations. According to [10], the order of a non-zero polynomial $P \in \mathbb{F}_q[X]$ with $P(0) \neq 0$ is the least positive integer e for which $P(X)$ divides $(X^e - 1)$. If $P(0) = 0$, then $P(X) = X^h G(X)$, where $h \in \mathbb{N}$ and $G \in \mathbb{F}_q[X]$ with $G(0) \neq 0$ are uniquely determined, and $\mathrm{ord}(P)$ is defined as $\mathrm{ord}(G)$.

Let $e \in \mathbb{N} \setminus \{0\}$. If there exists a natural number k with $e = km$ it follows that $\Psi_m(X^e - 1) = (X^k - 1)^m$. Otherwise m does not divide e and we get $\Psi_m(X^e - 1) = X^e - 1$ for m odd. Consequently, we have the following result on the order of Ψ_m of a polynomial.

Lemma 4. *Let $P \in \mathbb{F}_q[X]$ be an irreducible polynomial.*

1. *If there exists* $k \in \mathbb{N}$ *with* $\operatorname{ord}(P) = km$ *and if* $Q \in \mathbb{F}_q[X]$ *is an irreducible factor of* $\Psi_m(P)$ *then* $\operatorname{ord}(Q) = k$.
2. *If* $m \not| \operatorname{ord}(P)$ *then* $\Psi_m(P)$ *is irreducible and* $\operatorname{ord}(\Psi_m(P)) = \operatorname{ord}(P)$.

Obviously, the dynamic of Ψ_m is interesting. The images are either irreducible of order e/m or they are powers of an irreducible polynomial with order e/m. Moreover, different polynomials can have the same image.

Figure 1 displays the dynamic of Ψ_3 on irreducible polynomials of degree 4 over $\mathbb{F}_2$. The order is given in small letters. The two polynomials on the left side are mapped to $1 + X + X^2 + X^3 + X^4$ whereas this one is a fixed point of Ψ_3 because it is the only irreducible polynomial of degree 4 whose order is 5.

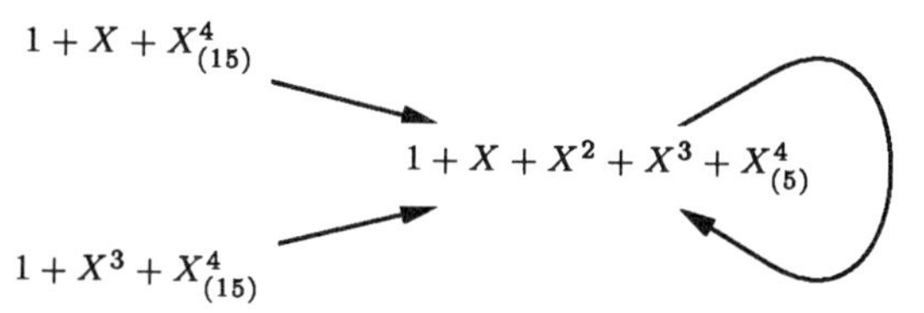

Fig. 1. Images of irreducible polynomials of degree 4 under Ψ_3 over $\mathbb{F}_2$

Irreducible polynomials whose orders are not divisible by m are mapped on irreducible polynomials of the same degree and the same order. Therefore, the images of the iterations of Ψ_m form a circle as shown in Fig. 2.

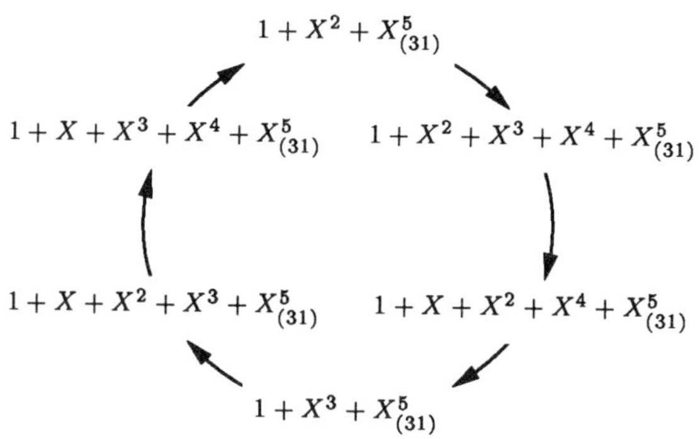

Fig. 2. Images of irreducible polynomials of degree 5 under Ψ_3 over $\mathbb{F}_2$

Apart from our interest, the map Ψ_m can be used to obtain new irreducible polynomials from known ones. Instead of going into the particulars of Ψ_m, we want to return to the automaticity of solutions of Mahler equations over finite fields.

4 Mahler equations over finite fields

In this section, we study Mahler equations over finite fields for a further investigation of the rational functions described in Theorem 1. We first consider $m = q$. In this case, the theorem of G. Christol, T. Kamae, M. Mendès France and G. Rauzy already yields the m-automaticity of all solutions [4, 5].

Let $R = P/Q$ be a rational function with polynomials $P, Q \in \mathbb{F}_q[X]$. Choosing $G = Q$ and $H = PQ^{q-2}$, we obtain that

$$R(X)G(X^q) = R(X)G(X)^q = H(X)G(X).$$

Using Theorem 1, we have:

The q-equation-kernel of a rational function in $\mathbb{F}_q(X)$ is finite.

Remark 2. Since the q-equation kernel of a rational function in $\mathbb{F}_q(X)$ is finite, it follows from arguments in [6] that the p-equation kernel ($q = p^\alpha$) is finite too.

Apart from this special case, the finiteness of the m-equation-kernel of a rational function $P/Q \in \mathbb{F}_q(X)$ depends on the order of Q.

Theorem 2. *Let Q be an irreducible polynomial with coefficients in $\mathbb{F}_q$. If $m|\mathrm{ord}(Q)$ then the m-equation-kernel $K_m(Q^{-1})$ is finite.*

Proof. We want to apply Theorem 1 again. Since $m|\mathrm{ord}(Q)$, there exists an integer $k > 0$ such that $Q(X)$ divides $(X^{mk} - 1)$. Now let $G(X) = X^k - 1$ then

$$H(X) = \frac{(X^{mk} - 1)}{(X^k - 1)} \frac{1}{Q(X)}$$

is a polynomial because $Q(X)$ and $(X^k - 1)$ are coprime polynomials which divide $(X^{mk} - 1)$.

Applying Theorem 1, the m-equation-kernel $K_m(Q^{-1})$ is finite. □

Example 2. Consider the polynomial $Q(X) = 1 + X + X^4$ over $\mathbb{F}_2$. Since Q is irreducible with $\mathrm{ord}(Q) = 15$, the 3-equation-kernel of Q is finite. Computing $K_3(Q^{-1})$ shows that

$$K_3\left(\frac{1}{Q}\right) = \left\{ \frac{P(X)}{1 + X + X^2 + X^3 + X^4} \,\middle|\, \begin{array}{l} P \in \mathbb{F}_2[X], \deg(P) \leq 2, \\ P(X) \notin \{X^2, X + X^2\} \end{array} \right\}.$$

$K_m(Q^{-1})$ consists of 6 elements.

Indeed, the above theorem is also applicable to every rational function, whose denominator factors in irreducible polynomials Q_i with $m|\mathrm{ord}(Q_i)$ for all i.

But what about the m-equation-kernels of rational functions P/Q, where m does not divide the order of Q? Provided that Q is mapped to P under an iteration of Ψ_m, the next theorem gives an answer.

Theorem 3. *Let P/Q be a rational function in $\mathbb{F}_q(X)$. If there exists a natural number n_0 such that*

$$\Psi_m^{n_0}(Q) = P,$$

the m-equation-kernel $K_m(P/Q)$ is finite.

Proof. To apply Theorem 1, we have to construct two polynomials $G, H \in \mathbb{F}_q[X]$ such that $(P(X)/Q(X))G(X^m) = H(X)G(X)$. Therefore, we consider the polynomial $E \in \mathbb{F}_q[X]$ defined by

$$E = Q\Psi_m(Q)\cdots\Psi_m^{n_0-1}(Q).$$

Using Corollary 2, we obtain that there exists a polynomial $H \in \mathbb{F}_q[X]$ with

$$\Psi_m(E)(X^m) = H(X)E(X).$$

Now let $G = \Psi_m(E)$. Then $G(X^m) = \Psi_m(E)(X^m)$ and

$$\frac{P(X)}{Q(X)}G(X^m) = P(X)H(X)\Psi_m(Q(X))\ldots\Psi_m^{n_0-1}(Q(X)) = H(X)G(X),$$

since $\Psi_m^{n_0}(Q) = P$. □

Example 3. We consider the irreducible polynomials $P(X) = 1 + X^2 + X^3 + X^4 + X^5$ and $Q(X) = 1 + X^2 + X^5$ over $\mathbb{F}_2$ with $\mathrm{ord}(P) = \mathrm{ord}(Q) = 31$. As pictured in Fig. 2, we have $\Psi_3(Q) = P$ and $K_3(P/Q)$ is finite by Corollary 3. Computing the 3-equation-kernel shows that

$$K_3\left(\frac{P}{Q}\right) \subset \left\{ \frac{G(X)}{P(X)} \;\middle|\; G(X) \in \mathbb{F}_2[X], \deg(G) \leq 5 \right\}.$$

Moreover, the number of elements of the 3-equation-kernel of P/Q is 47.

Summing up our results on Mahler equations over finite fields, we proved that a solution $F \in \mathbb{F}_q[[X]]$ of

$$F(X) = R(X)F(X^m)$$

with a rational function $R \in \mathbb{F}_q(X)$ is m-automatic resp. has a finite m-kernel in case that

- m is a power of the characteristic p of $\mathbb{F}_q$;
- $R(X) = X^n P(X)$ for $P \in \mathbb{F}_q[X]$ and $n \in \mathbb{Z}$;
- $R = Q^{-1}$ for $Q \in \mathbb{F}_q[X]$ irreducible with $m|\mathrm{ord}(Q)$;
- $R = P/Q$ for $P, Q \in \mathbb{F}_q[X]$ with $\Psi_m^{n_0}(Q) = P$ for a natural number n_0;
- R can be decomposed into a product of rational functions such that every factor satisfies one of these conditions.

By introducing the m-equation-kernel for rational functions, we get some results on the m-automaticity of solutions of simple Mahler equations over finite commutative rings as well as over finite fields. Provided that m and q are relatively prime, one can show that the last four items of this section are equivalent to the conditions mentioned in Theorem 1.

Some questions remain. Does Theorem 1 yield not only a sufficient but also a necessary criterion for finite m-equation-kernels? What about possible generalizations of the results for Mahler equations of the form

$$\sum_{i=0}^{n} P_i(X)F(X^{p^i}) = Q(X)$$

with polynomials Q and P_i for $i = 0, 1, \ldots, n$?

References

1. J.-P. Allouche. Automates finis en théorie des nombres. *Expo. Math*, 5:239–266, 1987.
2. A. Barbé, H.-O. Peitgen, and G. Skordev. Automaticity of coarse-graining invariant orbits of one-dimensional linear cellular automata. *International Journal of Bifurcation and Chaos*, 9(1), 1999. To appear.
3. P.-G. Becker. k-regular power series and Mahler-type functional equations. *J. Number Theory*, 49:269–286, 1994.
4. G. Christol. Ensembles presque périodiques k-reconnaissables. *Theoret. Comput. Sci.*, 9:141–145, 1979.
5. G. Christol, T. Kamae, M. Mendès France, and G. Rauzy. Suites algébriques, automates et substitutions. *Bull. Soc. Math. France*, 108:401–419, 1980.
6. A. Cobham. Uniform tag sequences. *Math. Systems Theory*, 6:164–192, 1972.
7. P. Dumas. Algebraic aspects of B-regular series. In A. Lingas, R. Karlsson, and S. Carlsson, editors, *Automata, Languages and Programming*, volume 700 of *Lect. Notes Comput. Sci.*, pages 457–468. Springer-Verlag, 1993. Proceedings of the 20th International Colloquium ICALP 93, Lund, Sweden.
8. S. W. Golomb. *Shift Register Sequences*. Holden-Day, San Francisco, 1967.
9. F. von Haeseler. *On algebraic properties of sequences generated by substitutions over a group*. Habilitationsschrift, University of Bremen, 1995.
10. R. Lidl and H. Niederreiter. *Finite Fields*, volume 20 of *Encyclopedia of Mathematics and its Applications*. Addison-Wesley, 1983.
11. H Niederreiter. A simple and general approach to the decimation of feedback shift-register sequences. *Problems Control Inform. Theory*, 17:327–331, 1988.
12. A. J. van der Poorten. Substitution automata, functional equations and "functions algebraic over a finite field". *Contemp. Math.*, 9:307–312, 1982.
13. B. Randé. Équations fonctionelles de Mahler et fonctions élémentaires. Groupe de travail en Théorie Analytique des Nombres, Paris, 1991. see also Thèse, Université de Bordeaux, 1992.
14. O. Salon. Suites automatique à multi-indices. *Séminaire de Théorie des Nombres de Bordeaux*, 4:4.01–4.27, 1986-1987. Followed by an Appendix by J. Shallit, 4.29A-4.36A.
15. N. Zierler. Linear recurring sequences. *J. Soc. Indust. Appl. Math.*, 7:31–48, 1959.

Correlation Distribution of the Quaternary Kasami Sequences*

T. Helleseth[1], P. V. Kumar[2], H. M. Martinsen[1], O. N. Vassbakk[1]

[1] Department of Informatics, University of Bergen
Høyteknologisenteret, N-5020 Bergen, Norway.
Email: {torh, halvard, odd}@ii.uib.no
[2] Communication Sciences Institute, University of Southern California
Los Angeles, CA 90089-2565, USA.
Email: kumar@aditya.usc.edu

Abstract. A new construction of quaternary sequences, similar to the small set of binary Kasami sequences, is presented. The correlation distribution of the code generated by these sequences is determined. As a byproduct, the Lee weight distribution of the code is given.

1 Introduction

Code-division multiple-access (CDMA) allows several users simultaneous access to a common channel by assigning a distinct *code* or *signature* sequence to each user, to enable him to distinguish his signal from those of the other users. It is desirable to have large sets of sequences possessing low values of C_{max}, the maximum magnitude of the periodic crosscorrelation and out-of-phase autocorrelation, to minimize interference as well as to help synchronization. Optimal binary sequences have been studied extensively and used in binary phase shift keying (BPSK) modulated CDMA communication systems. Such sequences are optimal in terms of having the minimum possible value of C_{max} for the given family size and symbol alphabet.

In recent years, quadriphase sequences for use in QPSK modulated CDMA communication systems have been studied, see [4–6]. Motivated by these results and by code constructions over Z_4, see for example [7], we find the correlation distribution of the Kasami code over Z_4. From the correlation distribution of the quaternary Kasami code, the weight distribution also follows.

2 Preliminaries and basic results

Let Z_n denote the integers modulo n and let $p(t)$ and $s(t)$ be two sequences with symbols in Z_n, $n \geq 2$, of period L. We say that $p(t)$ and $s(t)$ are cyclically distinct if there is no integer τ for which

$$s(t+\tau) = p(t)$$

* The work was supported in part by the Norwegian Research Council.

for all $t \geq 0$. Let $\mathcal{F} = \{s_i(t) \,|\, 1 \leq i \leq M\}$ be a family of sequences that are pairwise cyclically distinct and of period L. Let ω be a primitive n-th root of unity. The correlation between $s_i(t)$ and $s_j(t)$ at shift τ is given by

$$c_{i,j}(\tau) = \sum_{t=0}^{L-1} \omega^{s_i(t+\tau)-s_j(t)}, \ 1 \leq i,j \leq M, \ 0 \leq \tau \leq L-1.$$

If $i \neq j$, this is the crosscorrelation function. Otherwise, it reduces to the autocorrelation function.

Let $tr_m(\cdot)$ denote the trace map from the Galois field $GF(2^m)$ to Z_2. Let $m \geq 2$ and α be a primitive element in $GF(2^{2m})$. Then the small set of Kasami sequences is defined as

$$\{tr_{2m}(\alpha^t) + tr_m(b\alpha^{(2^m+1)t}) \,|\, b \in GF(2^m)\}.$$

For simplicity, in the remaining let $q = 2^m$. The spectrum of correlation values is given by $\{-1, -1-q, -1+q\}$, and the family is therefore optimal, see [3].

The Galois ring $R_m = GR(4,m)$ is an extension of Z_4 of degree m. R_m is a local ring having a unique maximal ideal $2R_m$ and the quotient $R_m \backslash 2R_m$ is isomorphic to $GF(q)$. As a multiplicative group, the set R_m^* of units of R_m contains a cyclic subgroup $\cong Z_{q-1}$. Let $\beta \in R_m^*$ be a generator of this group and let $\mathcal{T}_m = \{0, 1, \beta, \beta^2, \ldots, \beta^{q-2}\}$. It can be shown that every element in $z \in R_m$ can be expressed uniquely as

$$z = a + 2b, \ a, b \in \mathcal{T}_m,$$

and that $\alpha = \beta \pmod 2$ is a primitive element in $GF(q)$. The Frobenius map σ from R_m to R_m is defined by

$$\sigma(z) = a^2 + 2b^2$$

and the trace map from R_m to R_k, where $m = rk$ is defined by

$$T_k^m(z) = \sum_{j=0}^{r-1} \sigma^{jk}(z).$$

For convenience, in the case of $k = 1$, the trace function is denoted by $T_m(\cdot)$.

Similar to the binary Kasami sequences, it is natural to select the set of cyclically distinct quaternary Kasami sequences to be

$$\{T_{2m}((1+2\delta)\beta^t) + 2T_m(b\beta^{(q+1)t}) \,|\, \delta \in \mathcal{T}_{2m}, \ b \in \mathcal{T}_m\}$$

and

$$\{2T_{2m}(\beta^t) + 2T_m(b\beta^{(q+1)t}) \,|\, b \in \mathcal{T}_m\}.$$

This is a set of q^3+q sequences of period q^2-1. We are interested in finding the correlation distribution of these sequences. Typically we want to calculate the following,

$$\begin{aligned} c_{i,j}(\tau) &= \sum_{t=0}^{q^2-2} \omega^{T_{2m}((1+2\delta)\beta^{t+\tau})+2T_m(b\beta^{(q+1)(t+\tau)})-T_{2m}((1+2\delta^*)\beta^t)+2T_m(b^*\beta^{(q+1)t})} \\ &= \sum_{t=0}^{q^2-2} \omega^{T_{2m}((\gamma'+2\delta')\beta^t)+2T_m(b'\beta^{(q+1)t})} \\ &= \sum_{t=0}^{q^2-2} \omega^{T_{2m}(a\beta^t)+2T_m(b\beta^{(q+1)t})}, \qquad (1) \end{aligned}$$

where ω is the complex 4-th root of unity, and $a \in R_{2m}$ and $b \in \mathcal{T}_m$. It is necessary to find the values of equation (1) for all $a \in R_{2m}$ and $b \in \mathcal{T}_m$. The set consisting of the sequences $\{T_{2m}(a\beta^t)+2T_m(b\beta^{(q+1)t}) \,|\, a \in R_{2m}, b \in \mathcal{T}_m\}$ is a quaternary linear code. We will therefore find the correlation distribution of this quaternary code. In the analysis of the correlation distribution of the code it is helpful to let the codewords be of length q^2. That is, instead of summation over all non-zero elements in $\mathcal{T}_{2m}$, let the summation run through the zero element as well. For $a \in R_{2m}$ and $b \in \mathcal{T}_m$, define $c(a,b)$ as

$$c(a,b) = (0, T_{2m}(a)+2T_m(b), \ldots, T_{2m}(a\beta^{q^2-2})+2T_m(b\beta^{(q+1)(q^2-2)})).$$

The quaternary Kasami code, $\mathcal{K}_4$ is defined by

$$\mathcal{K}_4 = \{c(a,b) \,|\, a \in R_{2m},\ b \in \mathcal{T}_m\}.$$

The Lee weights of the elements $0,1,2,3$ of Z_4 are $0,1,2,1$, respectively. For a vector $c \in Z_4^n$, the Lee weight $w_L(c)$, is defined as the sum of the Lee weight of its components. Any codeword $c(a,b)$ can be represented as a complex number $\sigma_{a,b}$,

$$\sigma_{a,b} = \sum_{x \in \mathcal{T}_{2m}} \omega^{T_{2m}(ax)+2T_m(bx^{q+1})},$$

The Lee weight is then given as

$$w_L(c) = n - Re(\sum_{x \in \mathcal{T}_{2m}} \omega^{T_{2m}(ax)+2T_m(bx^{q+1})}). \qquad (2)$$

It follows that determining the real part of the different correlation values of $\mathcal{K}_4$ is equivalent to determining the Lee weight distribution of the code. So from now on, we will determine the different complex numbers representing the codewords. The weight distribution of

$$\{u + c(a,b) \,|\, u \in Z_4,\ c(a,b) \in \mathcal{K}_4\}$$

together with some of the results in this paper can be found in [8]. In the remaining of this section, we will state some basic observations and results.

Lemma 1. *[3] Let $\Gamma(a) = \sum_{x \in \mathcal{T}_m} \omega^{T_m(ax)}$ for $a \in R_{4^m}$. The distribution of the set $\{\Gamma(a) \mid a \in R_{4^m}\}$ is as follows*

$$\Gamma(a) = \begin{cases} 2^m & \text{once}, \\ 0 & q-1 \text{ times}, \\ \Gamma(1+2\delta),\ \delta \in \mathcal{T}_m & q-1 \text{ times}. \end{cases}$$

The distribution of $\Gamma(1+2\delta)$ is

m odd		m even	
$\frac{Value}{\sqrt{q}}$	Frequency	$\frac{Value}{\sqrt{q}}$	Frequency
ϵ	$q/4+\sqrt{q/8}$	1	$q/4+\sqrt{q/4}$
$\epsilon\omega$	$q/4-\sqrt{q/8}$	ω	$q/4$
$\epsilon\omega^2$	$q/4-\sqrt{q/8}$	ω^2	$q/4-\sqrt{q/4}$
$\epsilon\omega^3$	$q/4+\sqrt{q/8}$	ω^3	$q/4$

where $\epsilon = e^{\frac{2\pi i}{8}}$ is a primitive 8-th root of unity.

Lemma 2. *Let x, y be elements in $\mathcal{T}_{2m}$. Then the following holds:*

1. *Let $b \in GF(q)$. The equation,*
$$bx^q + x + 1 = 0$$
has either no solutions in $GF(q^2)$ or the same number of solutions as $bx^q + x = 0$.
2. *Let $\overline{x}$ denote the modulo 2 reduction of x. Then*
$$\sum_{x \in \mathcal{T}_{2m}} \omega^{2T_{2m}(ux)} = \sum_{\overline{x} \in GF(2^{2m})} (-1)^{tr_{2m}(\overline{ux})} = \begin{cases} q^2 & \text{if } \overline{u} = 0, \\ 0 & \text{if } \overline{u} \neq 0. \end{cases}$$
3. *If $x, y \in \mathcal{T}_{2m}$ then $z = x + y + 2\sqrt{xy} \in \mathcal{T}_{2m}$.*
4. *$T_{2m}(x) = T_1^m(T_m^{2m}(x))$.*

The result can be found in [3] with the exception of the first equation which easily can be proved. The weight distribution of the binary Kasami code is known and can be found in [1].

Lemma 3. *The Hamming weight distribution of the binary Kasami code is*

$$A_i = \begin{cases} 1 & \text{if } i = 0, \\ q^3/2 - q/2 & \text{if } i = q^2/2 - q/2, \\ q^2 - 1 & \text{if } i = q^2/2, \\ q^3/2 - q^2 + q/2 & \text{if } i = q^2/2 + q/2. \end{cases}$$

The corresponding codewords over Z_4 will have Lee weight twice the Hamming weight.

3 Weight distribution of a subcode of $\mathcal{K}_4$

This section is concerned about the correlation distribution of the cyclically distinct codewords. We will first find the complex numbers which represent the codewords, and thereafter decide the number of codewords each number represents. To determine the numbers, will it be helpful to have the values of $\sigma_{a,b}^2$ and $|\sigma_{a,b}|^2$, which will be studied in the following subsections respectively. The following observations will be used.

If a is a non-unit, then $\sigma_{a,b}$ will represent a codeword of the binary Kasami code. Therefore, in the following let a be a unit. We can assume that a is of the form $a = 1 + 2\delta$. In any case, a is of the form $\gamma + 2\delta$, $\gamma, \delta \in \mathcal{T}_{2m}$. Since a is a unit, $\gamma \neq 0$. Hence,

$$\frac{a}{\gamma} = 1 + 2\frac{\delta}{\gamma}.$$

Substitution of $y = \gamma x$ gives

$$\begin{aligned}\sigma_{a,b} &= \sum_{x \in \mathcal{T}_{2m}} \omega^{T_{2m}(ax) + 2T_m(bx^{q+1})} \\ &= \sum_{y \in \mathcal{T}_{2m}} \omega^{T_{2m}(\frac{a}{\gamma}y) + 2T_m(b\frac{y^{q+1}}{\gamma^{q+1}})} \\ &= \sum_{y \in \mathcal{T}_{2m}} \omega^{T_{2m}(a^* y) + 2T_m(b^* y^{q+1})} \\ &= \sigma_{a^*, b^*},\end{aligned}$$

where $\delta^* = \frac{\delta}{\gamma}$, $a^* = \frac{a}{\gamma}$ and $b^* = \frac{b}{\gamma^{q+1}}$. It follows that $a^* = 1 + 2\delta^*$.

Let $x, y \in \mathcal{T}_m$ and let $z = x + y + 2\sqrt{xy}$. Hence, $z = x + y \pmod 2$. Rearranging the sums gives us the following equations

$$x + y = z + 2\sqrt{x(x+z)} = z + 2x + 2\sqrt{xz} \tag{3}$$

and

$$x - y = z + 2y + 2\sqrt{xy} = 3z + 2\sqrt{xz}. \tag{4}$$

That x and y determine z is easy to see, but we also note that x and z determine y. This follows from the fact that $z = x + y + 2\sqrt{x(x+z)} = 3x + 2\sqrt{xz} + y$. If $x, y \in \mathcal{T}_{2m}$ and $b \in \mathcal{T}_m$, note that

$$\begin{aligned}2T_m(b(x^{q+1} + y^{q+1})) &= 2T_m(b(x^{q+1} + (x+z)^{q+1})) \\ &= 2T_m(bxz^q + bx^q z + bz^{q+1}) \\ &= 2T_m(bz^{q+1}) + 2T_{2m}(bxz^q).\end{aligned} \tag{5}$$

3.1 The value of $\sigma_{a,b}^2$

The value of $\sigma_{a,b}^2$ is given by

$$\sigma_{a,b}^2 = \sum_{x,y \in \mathcal{T}_{2m}} \omega^{T_{2m}(a(x+y))+2T_m(b(x^{q+1}+y^{q+1}))}.$$

Let $z = x + y + 2\sqrt{xy}$, and use equations (3) and (5) in the above expression. Then

$$\begin{aligned}
\sigma_{a,b}^2 &= \sum_{x,z \in \mathcal{T}_{2m}} \omega^{T_{2m}(az+2a(x+\sqrt{xz}))+2T_m(bxz^q+bx^qz+bz^{q+1})} \\
&= \sum_{x,z \in \mathcal{T}_{2m}} \omega^{T_{2m}(az)+2T_m(bz^{q+1})+2T_{2m}(ax+a^2xz+bxz^q)} \\
&= \sum_{z \in \mathcal{T}_{2m}} \omega^{T_{2m}(az)+2T_m(bz^{q+1})} \sum_{x \in \mathcal{T}_{2m}} \omega^{2T_{2m}(x(a+a^2z+bz^q))}.
\end{aligned}$$

Apply Lemma 2 on $\sum_{x \in \mathcal{T}_{2m}} \omega^{2T_{2m}(x(a+a^2z+bz^q))}$. Then

$$\sigma_{a,b}^2 = q^2 \sum_{\substack{z \in \mathcal{T}_{2m} \\ A(z)=0 \pmod 2}} \omega^{T_{2m}(az)+2T_m(bz^{q+1})}, \tag{6}$$

where $A(z) = bz^q + a^2z + a$.

3.2 The value of $|\sigma_{a,b}|^2$

The value of $|\sigma_{a,b}|^2$ is given by

$$|\sigma_{a,b}|^2 = \sum_{x,y \in \mathcal{T}_{2m}} \omega^{T_{2m}(a(x-y))+2T_m(b(x^{q+1}+y^{q+1}))}.$$

Let $z = x + y + 2\sqrt{xy}$, and use equations (4) and (5) together with Lemma 2 in the above expression. Then

$$\begin{aligned}
|\sigma_{a,b}|^2 &= \sum_{x,z \in \mathcal{T}_{2m}} \omega^{T_{2m}(a(3z+2\sqrt{xz}))+2T_m(bxz^q+bx^qz+bz^{q+1})} \\
&= \sum_{x,z \in \mathcal{T}_{2m}} \omega^{T_{2m}(-az)+2T_m(bz^{q+1})+2T_{2m}(a^2xz+bxz^q)} \\
&= \sum_{z \in \mathcal{T}_{2m}} \omega^{T_{2m}(-az)+2T_m(bz^{q+1})} \sum_{x \in \mathcal{T}_{2m}} \omega^{2T_{2m}(x(a^2z+bz^q))} \\
&= q^2 \sum_{\substack{z \in \mathcal{T}_{2m} \\ L(z)=0 \pmod 2}} \omega^{T_{2m}(-az)+2T_m(bz^{q+1})},
\end{aligned}$$

where $L(z) = bz^q + a^2z$. In the case $a = 1 + 2\delta$ we have $a^2 = 1$. Hence, $L(z) = bz^q + z$. We want to find the zeros of $L(z) \pmod 2$. We divide this into two distinct cases, whenever $b \neq 1$ and $b = 1$. The value of $|\sigma_{a,b}|^2$ is given by the next result.

Lemma 4. *Let $a = 1 + 2\delta$, $\delta \in \mathcal{T}_{2m}$, then*

$$|\sigma_{a,b}|^2 = \begin{cases} q^2 & \text{if } b \neq 1, \\ q^3 & \text{if } b = 1,\ \delta \in \mathcal{T}_m, \\ 0 & \text{if } b = 1,\ \delta \in \mathcal{T}_{2m} \setminus \mathcal{T}_m. \end{cases}$$

Proof: In the case of $b \neq 1$, one solution of $L(z) = 0 \pmod 2$ is $z = 0$. When $b = 0$, this is the only solution. Suppose that there exists a solution $z \neq 0$. Since $b \in \mathcal{T}_m$ and $z \in \mathcal{T}_{2m}$, we have $(bz^q)^{q+1} = z^{q+1}$. Dividing each side by z^{q+1} gives $b^{q+1} = 1$. By Euclid's algorithm, $b^{\gcd(q+1,q-1)} = 1$, which implies that $b = 1$. But we are assuming that $b \neq 1$, hence $z = 0$ is the only solution of $bz^q + z = 0 \pmod 2$ when $b \neq 1$.

In the case of $b = 1$, $|\sigma_{a,1}|^2$ is given as

$$|\sigma_{a,1}|^2 = q^2 \sum_{\substack{z \in \mathcal{T}_{2m} \\ z^q + z = 0 \pmod 2}} \omega^{T_{2m}(-az) + 2T_m(z^{q+1})}.$$

The only solutions of $z^q = z$ are for $z \in \mathcal{T}_m$. Hence,

$$|\sigma_{a,1}|^2 = q^2 \sum_{z \in \mathcal{T}_m} \omega^{T_{2m}(-az) + 2T_m(z^2)}.$$

Since $a = 1 + 2\delta$, $\delta \in \mathcal{T}_{2m}$ and $T_m^{2m}(1 + 2\delta) = 1 + 2\delta + 1 + 2\delta^q$ it follows that

$$\begin{aligned} |\sigma_{a,1}|^2 &= q^2 \sum_{z \in \mathcal{T}_m} \omega^{-T_m(z(1+2\delta+1+2\delta^q)+2z^2)} \\ &= q^2 \sum_{z \in \mathcal{T}_m} \omega^{-2T_m((1+\delta+\delta^q+1)z)}, \end{aligned}$$

since $2T_m(x^2)$ equals $2T_m(x)$. Therefore we obtain

$$|\sigma_{a,1}|^2 = q^2 \sum_{z \in GF(q)} (-1)^{tr_m(z(\overline{\delta} + \overline{\delta^q}))}.$$

If $\overline{\delta} = \overline{\delta^q}$, then $\sum_{z \in GF(q)} (-1)^{tr_m(0)} = q$. If $\overline{\delta} \neq \overline{\delta^q}$, the numbers of zeros equals the number of ones of $tr_m(\cdot)$. Hence, the sum becomes zero. □

3.3 The points corresponding to codewords in $\mathcal{K}_4$

A complex number can be represented as a point in the Cartesian plane. We know that the points, which represents the codewords in $\mathcal{K}_4$, is either the origin or on a circle of radius q or $q\sqrt{q}$ around the origin. Denote these circles of radius q and $q\sqrt{q}$ for the inner and outer circle, respectively. Using the result of the weight distribution of the binary Kasami code, together with Lemma 4 we obtain the following result.

Lemma 5. *Let*

$$\sigma_{\gamma+2\delta,b} = \sum_{x\in\mathcal{T}_{2m}} \omega^{T_{2m}((\gamma+2\delta)x)+2T_m(bx^{q+1})}.$$

Then the distribution of points, except the point corresponding to the all-zero codeword, are as follows:

Case	*Conditions*	*Circle*	*Number of Codewords*
1	$\gamma\neq 0,\ \frac{\delta}{\gamma}\in\mathcal{T}_m, b=\gamma^{q+1}$	*Outer*	$q(q^2-1)$
2	$\gamma\neq 0,\ \frac{\delta}{\gamma}\in\mathcal{T}_{2m}\backslash\mathcal{T}_m,\ b=\gamma^{q+1}$	*Origin*	$(q^2-1)(q^2-q)$
3	$\gamma\neq 0,\ \frac{\delta}{\gamma}\in\mathcal{T}_{2m},\ b\neq\gamma^{q+1}$	*Inner*	$(q^2-1)(q-1)q^2$
4	$\gamma=0,\ \delta\in\mathcal{T}_{2m},\ b\in\mathcal{T}_m$	*Lemma 3*	q^3

Note that we have counted all the words in $\mathcal{K}_4$, not only the cyclically distinct ones.

We have determined the distance from the points to the origin, but not the points themselves. The points which represent the codewords must have integer imaginary and real part. Each point on the inner circle is represented by $\sigma^2_{a,b} = q^2\omega^x$, where $x \in Z_4$. This gives us four different points on the circle of radius q^2. For $x = 0$, the value of $\sigma^2_{a,b}$ is q^2. Each time we increase x by 1 modulo 4, the corresponding point $q^2\omega^x$ shifts 90 degree. The square root of these 4 points gives 8 different points. But as pointed out earlier, the points have integer imaginary and real part. Hence the points are as stated in Figure 1. The same argumentation also holds for the outer circle. But in this case we have to distinguish between m odd or even. In the case of m odd, the four points are on the circle of radius $q\sqrt{q}$ with imaginary and real part in absolute value equal to $\sqrt{q^3/2}$, see Figure 1. If m is even then the four points are $(\pm\sqrt{q^3}, 0)$ and $(0, \pm\sqrt{q^3})$. These are not given in the figure.

3.4 The numbers of codewords mapped to each point

It is easy to see that Case 4 in Lemma 5 correspond to the binary Kasami code. In Case 2, the exact number of codewords mapped to the origin is stated. The remaining cases are 1 and 3. These will be studied in more detail in the following.

Case 1: Distribution in the outer circle. We consider the case where $a = \gamma + 2\delta$, $\gamma \neq 0$, $\frac{\delta}{\gamma} \in \mathcal{T}_m$, $b = \gamma^{q+1}$. Let β be any element in $\mathcal{T}_{2m}$. When x runs through $\mathcal{T}_{2m}$ so does $x + \beta + 2\sqrt{\beta x}$. Define $S(a) = \sigma_{a,1}$. Then,

$$\begin{aligned} S(a) &= \sum_{x\in\mathcal{T}_{2m}} \omega^{T_{2m}(ax+a\beta+2a\sqrt{x\beta})+2T_m((x+\beta+2\sqrt{x\beta})^{q+1})} \\ &= \omega^{T_{2m}(a\beta)+2T_m(\beta^{q+1})} \cdot \sum_{x\in\mathcal{T}_{2m}} \omega^{T_{2m}(x(a+2a^2\beta+2\beta^q))+2T_m(x^{q+1})} \\ &= \omega^{T_{2m}(a\beta)+2T_m(\beta^{q+1})} \cdot S(a+2a^2\beta+2\beta^q). \end{aligned}$$

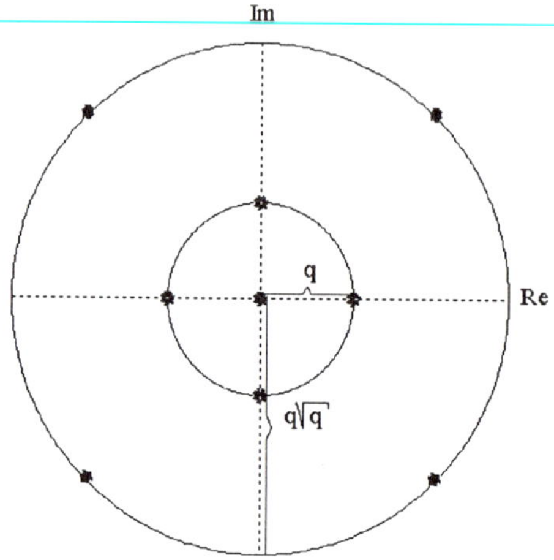

Fig. 1. Distribution of $\sigma_{a,b}$ for m odd.

Let $a = 1$, then

$$S(1) = \omega^{T_{2m}(\beta)+2T_m(\beta^{q+1})} S(1 + 2(\beta + \beta^q)).$$

Given a $\delta \in \mathcal{T}_m$, then there is always a $\beta \in \mathcal{T}_{2m}$ such that $\overline{\beta} + \overline{\beta}^q = \overline{\delta}$. It follows that $\beta + \beta^q + 2\sqrt{\beta^{q+1}} \in \mathcal{T}_m$ and $\beta + \beta^q + 2\sqrt{\beta^{q+1}} = \delta$. Hence,

$$\begin{aligned} T_{2m}(\beta) + 2T_m(\beta^{q+1}) &= T_m(T_m^{2m}(\beta) + 2\beta^{q+1})) \\ &= T_m(\beta + \beta^q + 2\beta^{q+1}) \\ &= T_m(\beta + \beta^q + 2\sqrt{\beta^{q+1}}) \\ &= T_m(\delta). \end{aligned}$$

We can then express $S(1)$ as $\omega^{T_m(\delta)} S(1+2\delta)$. The relation between $S(a)$ and $S(1)$ is stated in the next lemma.

Lemma 6. *Let* $\delta \in \mathcal{T}_m$, *then*

$$S(1 + 2\delta) = \omega^{-T_m(\delta)} S(1).$$

If we can find an expression for $S(1)$, then $S(a)$ is determined for all $a = 1 + 2\delta$. Clearly, also the following equation holds,

$$\sum_{\delta \in \mathcal{T}_m} S(1 + 2\delta) = \sum_{\delta \in \mathcal{T}_m} \omega^{-T_m(\delta)} S(1). \tag{7}$$

To help finding an expression for $S(1)$, note that

$$\begin{aligned}\sum_{\delta\in\mathcal{T}_{2m}} S(1+2\delta) &= \sum_{\delta\in\mathcal{T}_{2m}}\sum_{x\in\mathcal{T}_{2m}} \omega^{T_{2m}((1+2\delta)x)+2T_m(x^{q+1})} \\ &= \sum_{x\in\mathcal{T}_{2m}} \omega^{T_{2m}(x)+T_m(2x^{q+1})} \sum_{\bar{\delta}\in F_{2m}} (-1)^{tr_{2m}(\bar{\delta}\bar{x})} \\ &= q^2,\end{aligned}$$

since $x=0$ is the only contribution by Lemma 2.
By Lemma 4, $S(1+2\delta)=0$, when $\delta\in\mathcal{T}_{2m}\setminus\mathcal{T}_m$. Hence,

$$\sum_{\delta\in\mathcal{T}_m} S(1+2\delta) = q^2.$$

Let $\Gamma(a)=\sum_{x\in\mathcal{T}_m}\omega^{T_m(ax)}$ be given as in Lemma 1. If we solve equation (7) with respect to $S(1)$, we obtain

$$\begin{aligned}S(1) &= \frac{q^2}{\Gamma(-1)} \\ &= \frac{q^2}{\overline{\Gamma(1)}} \\ &= \frac{q^2}{(-1)^{m+1}\sqrt{q}\cdot\bar{\epsilon}^m} \\ &= \frac{q^2}{(-1)^{m+1}(1-w)^m}\cdot\frac{(1+w)^m}{(1+w)^m} \\ &= (-1)^{m+1}q(1+w)^m, \qquad (8)\end{aligned}$$

since it is shown in [2] that $\Gamma(1)=(-1)^{m+1}(1+w)^m$, where $\epsilon=(1+\omega)/\sqrt{2}$.
By [2], $\omega^{-T_m(\delta)}$ is given as

$$\omega^{-T_m(\delta)} = \frac{\Gamma(1+2\delta)}{\Gamma(1)},$$

where $\Gamma(1+2\delta)$ is given in Lemma 1. Using this in combination with Lemma 6 and equation (8), the points are given by

$$\begin{aligned}S(1+2\delta) &= \frac{\Gamma(1+2\delta)}{\Gamma(1)}(-1)^{m+1}q(1+\omega)^m \\ &= \frac{\Gamma(1+2\delta)}{(-1)^{m+1}\epsilon^m\sqrt{q}}(-1)^{m+1}q\epsilon^m\sqrt{q} \\ &= q\Gamma(1+2\delta).\end{aligned}$$

Hence, the points $S(1+2\delta)$ on the outer circle is distributed as $\Gamma(1+2\delta)$ in Lemma 1. The distribution also holds for all values of b and γ such that $b\neq 0$ and $\gamma^{q+1}=b$. There are q^2-1 choices of γ. By Lemma 1 we obtain the following lemma, which gives the distribution corresponding to Case 1.

Lemma 7. *The correlation distribution of the codewords in the subcode represented by the numbers* $\{\sigma_{\gamma+2\delta,b} \mid \gamma \neq 0,\, \gamma^{q+1} = b,\, \delta/\gamma \in \mathcal{T}_m\}$ *is given by*

m odd		*m even*	
Value $q\sqrt{q}$	Frequency (q^2-1)	Value $q\sqrt{q}$	Frequency (q^2-1)
ϵ	$q/4+\sqrt{q/8}$	1	$q/4+\sqrt{q/4}$
$\epsilon\omega$	$q/4-\sqrt{q/8}$	ω	$q/4$
$\epsilon\omega^2$	$q/4-\sqrt{q/8}$	ω^2	$q/4-\sqrt{q/4}$
$\epsilon\omega^3$	$q/4+\sqrt{q/8}$	ω^3	$q/4$

Case 3: Distribution in the inner circle. For any $\gamma \neq 0$ and $b \neq \gamma^{q+1}$, we will find the distribution of the set

$$\{\sigma_{\gamma+2\delta,b} \mid \delta \in \mathcal{T}_{2m}\}.$$

The values in the set are $q, -q, q\omega$ and $-q\omega$. Let N_1, N_2, N_3 and N_4 be the number of times each value occurs, respectively.

Lemma 8. *Let* $\gamma, \delta \in \mathcal{T}_{2m}$ *and* $b \in \mathcal{T}_m$, *then*

$$\sum_{\delta \in \mathcal{T}_{2m}} \sigma_{\gamma+2\delta,b} = q^2.$$

Proof: We have

$$\begin{aligned}\sum_{\delta \in \mathcal{T}_{2m}} \sigma_{\gamma+2\delta,b} &= \sum_{\delta \in \mathcal{T}_{2m}} \sum_{x \in \mathcal{T}_{2m}} \omega^{T_{2m}((\gamma+2\delta)x)+2T_m(bx^{q+1})} \\ &= \sum_{x \in \mathcal{T}_{2m}} \omega^{T_{2m}(\gamma x)+2T_m(bx^{q+1})} \sum_{\delta \in \mathcal{T}_{2m}} \omega^{2T_{2m}(\delta x)} \\ &= q^2.\end{aligned}$$

□

By this result we have the following equations:

$$\begin{aligned}N_1 + N_2 + N_3 + N_4 &= q^2 \\ N_1 - N_2 + \omega(N_3 - N_4) &= q,\end{aligned}$$

where the second equation leads to $N_3 = N_4$. Hence, there are 3 unknowns and only two equations. To find a new equation, the quadratic sum of $\sigma_{\gamma+2\delta,b}$ is considered. By equation (6),

$$\sigma_{a,b}^2 = q^2 \sum_{\substack{z \in \mathcal{T}_{2m} \\ A(z)=0 \pmod 2}} \omega^{T_{2m}(az)+2T_m(bz^{q+1})},$$

where $A(z) = bz^q + a^2z + a$. We may assume that $a = 1+2\delta$ and $b \neq 1$, as shown earlier. In this case, $A(z) = bz^q + z + 1$. By Lemma 2, $A(z) = 0$ has either one or no solution, since $bz^q = z$ has one solution $z = 0$. But $|\sigma_{a,b}|^2 = q^2$ when $b \neq 1$, so $A(z)$ must have one solution. Let $z_0 \neq 0$ be the zero of $A(z)$. Then

$$\sigma_{1+2\delta,b}^2 = q^2 \omega^{T_{2m}((1+2\delta)z_0)+2T_m(bz_0^{q+1})}, \tag{9}$$

where $b \neq 1$.

Lemma 9. *For $b \neq 1$ we have*

$$\sum_{\delta \in \mathcal{T}_{2m}} \sigma_{1+2\delta,b}^2 = 0.$$

Proof: By equation (9),

$$\begin{aligned}
\sum_{\delta \in \mathcal{T}_{2m}} \sigma_{1+2\delta,b}^2 &= \sum_{\delta \in \mathcal{T}_{2m}} q^2 \omega^{T_{2m}((1+2\delta)z_0)+2T_m(bz_0^{q+1})} \\
&= q^2 \omega^{T_{2m}(z_0)+2T_m(bz_0^{q+1})} \sum_{\delta \in \mathcal{T}_{2m}} \omega^{2T_{2m}(\delta z_0)} \\
&= 0.
\end{aligned}$$

□

That is, $N_1 + N_2 = N_3 + N_4$, and we now have three unknowns and three equations. Solving these three equations gives the following solution, $N_1 = q(q+2)/4, N_2 = q(q-2)/4$ and $N_3 = N_4 = q^2/4$. These values holds for every $\gamma \neq 0$ and every $b \neq \gamma^{q+1}$. This gives the following result, which gives the distribution corresponding to Case 3.

Lemma 10. *The correlation distribution of the codewords in the subcode represented by the numbers $\{\sigma_{\gamma+2\delta,b} \,|\, \gamma \neq 0, \gamma^{q+1} \neq b,\ \delta \in \mathcal{T}_{2m}\}$, is given by*

$$c_{\gamma+2\delta,b} = \begin{cases} q & (q^2-1)(q-1)q(q+2)/4 \textit{ times}, \\ -q & (q^2-1)(q-1)q(q-2)/4 \textit{ times}, \\ q\omega & (q^2-1)(q-1)q^2/4 \textit{ times}, \\ -q\omega & (q^2-1)(q-1)q^2/4 \textit{ times}, \end{cases}$$

4 The complete correlation distribution

The correlation distribution of the quaternary Kasami code is stated in the following result.

Theorem 1. *The correlation distribution of the non-zero sequences in the set*

$$\{\sum_{t=0}^{2^{2m}-2} T_{2m}(a\beta^t) + 2T_m(b\beta^{(q+1)t}) \,|\, a \in R_{2m}, b \in \mathcal{T}_m\}$$

is given by

$$c_{a,b}(\tau) = \begin{cases} -1 & q^2-1+(q^2-1)(q^2-q) \text{ times}, \\ -1+q & (q^3-q)/2 \\ & +(q^2-1)(q-1)q(q+2)/4 \text{ times}, \\ -1-q & (q^3-2q^2+q)/2 \\ & +(q^2-1)(q-1)q(q-2)/4 \text{ times}, \\ -1+q\omega & q^2/4(q^2-1)(q-1) \text{ times}, \\ -1-q\omega & q^2/4(q^2-1)(q-1) \text{ times}, \\ -1+\sqrt{q^3/2}+\sqrt{q^3/2}\omega & (q^2-1)(q/4+\sqrt{q/8}) \text{ times}, \\ -1-\sqrt{q^3/2}+\sqrt{q^3/2}\omega & (q^2-1)(q/4-\sqrt{q/8}) \text{ times}, \\ -1-\sqrt{q^3/2}-\sqrt{q^3/2}\omega & (q^2-1)(q/4-\sqrt{q/8}) \text{ times}, \\ -1+\sqrt{q^3/2}-\sqrt{q^3/2}\omega & (q^2-1)(q/4+\sqrt{q/8}) \text{ times} \end{cases}$$

if m is odd. If m is even, then

$$c_{a,b}(\tau) = \begin{cases} -1 & q^2-1+(q^2-1)(q^2-q) \text{ times}, \\ -1+q & (q^3-q)/2 \\ & +(q^2-1)(q-1)q(q+2)/4 \text{ times}, \\ -1-q & (q^3-2q^2+q)/2 \\ & +(q^2-1)(q-1)q(q-2)/4 \text{ times}, \\ -1+q\omega & q^2/4(q^2-1)(q-1) \text{ times}, \\ -1-q\omega & q^2/4(q^2-1)(q-1) \text{ times}, \\ -1+\sqrt{q^3} & (q^2-1)(q/4+\sqrt{q/4}) \text{ times}, \\ -1-\sqrt{q^3} & (q^2-1)(q/4-\sqrt{q/4}) \text{ times}, \\ -1+\sqrt{q^3}\omega & (q^2-1)q/4 \text{ times}, \\ -1-\sqrt{q^3}\omega & (q^2-1)q/4 \text{ times}. \end{cases}$$

Proof: Since we are not doing summation over the zero element in the sequence approach, we have to subtract 1 from the previous results to get the correct correlation values.

If $a = 2a'$, that is, if a is not a unit then $c_{2a',b}(\tau)$ reduces to

$$c_{a,b}(\tau) = \sum_{t=0}^{q^2-2} (-1)^{tr_{2m}(\overline{a'}\alpha^t)+tr_m(\bar{b}\alpha^{(q+1)t})}.$$

It is easy to see that $c_{a,b} \in Z$. By Lemma 3, the correlation values from these sequences are $-1-q$ and $-1+q$. If a is a unit, then the number of correlation values equal to -1 is, by Lemma 5, easily seen to be $(q^2-1)(q^2-q)$. The distribution of values on the outer circle is stated in Lemma 7. The distribution of values on the inner circle is given by Lemma 10. □

The value C_{max}, the maximum magnitude of the periodic cross-correlation and out-of-phase auto-correlation, is given by $C_{max} = \sqrt{q^3}+1$ if m is even. For m odd is $C_{max} = \sqrt{q^3+\sqrt{2q^3}+1} < \sqrt{q^3}+1$. Compared to other families of approximately same size and period, this maximum value seems large.

By equation (2) and Theorem 1, the weight distribution of the quaternary Kasami code is given by the next result.

Corollary 1. *The weight distribution of the quaternary Kasami code*

$$\mathcal{K}_4 = \{c(a,b) \mid a \in R_{2m},\, b \in \mathcal{T}_m\}.$$

for $m \geq 3$, is given by

$$C_i = \begin{cases} 1 & \text{if } i = 0, \\ (q^3 - q)/2 & \\ +(q^2-1)(q-1)q(q+2)/4 & \text{if } i = q^2 - q, \\ (q^2-1) + (q^2-1)(q^2-q) & \\ +(q^2-1)(q-1)q^2/2 & \text{if } i = q^2, \\ q^3/2 - q^2 + q/2 & \\ +(q^2-1)(q-1)q(q-2)/4 & \text{if } i = q^2 + q, \end{cases}$$

added with A_i,

$$A_i = \begin{cases} (q^2-1)(q/4 + \sqrt{q/4}) & \text{if } i = q^2 - \sqrt{q^3}, \\ (q^2-1)q/2 & \text{if } i = q^2, \\ (q^2-1)(q/4 - \sqrt{q/4}) & \text{if } i = q^2 + \sqrt{q^3}, \end{cases}$$

if m is even, or by B_i,

$$B_i = \begin{cases} (q^2-1)(q/2 + \sqrt{q/2}) & \text{if } i = q^2 - \sqrt{q^3/2}, \\ (q^2-1)(q/2 - \sqrt{q/2}) & \text{if } i = q^2 + \sqrt{q^3/2}, \end{cases}$$

if m is odd.

References

1. T. Helleseth and P. V. Kumar, The weight hierarchy of the Kasami codes, *Discrete Mathematics*, **45** (1995), 133–143.
2. K. Yang, T. Helleseth, P. V. Kumar and A. Shanbhag, On the weight hierarchy of Kerdock codes over Z_4, *IEEE Trans. Inform. Theory*, **42** (1996), 1587–1594.
3. T. Helleseth and P. V. Kumar, "Sequences with low correlation," in Handbook of Coding Theory (Eds. V. Pless and W. C. Huffman), Elsevier, 1998.
4. P. Kumar, T. Helleseth, A. Calderbank and A. Hammons, Jr., Large families of quaternary sequences with low correlation, *IEEE Trans. Inform. Theory*, **42** (1996), 579–593.
5. P. Udaya and M. Siddiqi, Large linear complexity sequences over Z_4 for quadriphase modulated communication systems having good correlation properties, *Proc. of IEEE International Symposium Inform. Theory*, (1991) 386.
6. S. Boztas, R. Hammons and P. V. Kumar, 4-phase sequences with near-optimum correlation properties, *IEEE Trans. Inform. Theory*, **38** (1992) , 1101–1113.
7. A. R. Hammons, Jr., P. V. Kumar, A. R. Calderbank, N. J. A. Sloane and P. Solè, The Z_4-linearity of Kerdoc, Preparata, Goethals, and related codes, *IEEE Trans. of Inform. Theory*, **40** (1994), 301–319.
8. O. N. Vassbakk, "Vektfordelingen og vekthiearkiet til noen Z_4-koder", Master thesis, Department of Informatics, University of Bergen, Norway, 1998.

Multicovering Radii of Reed-Muller Codes and the Existence of Secure Stream Ciphers (Extended Abstract)

Iiro Honkala[1] and Andrew Klapper[2]

[1] Department of Mathematics
University of Turku
20014 Turku, Finland
E-mail: honkala@utu.fi
[2] Department of Computer Science
University of Kentucky
Lexington, KY, USA 40506-0046
E-mail: klapper@cs.uky.edu
Project sponsored by the National Science Foundation under grant number NCR-9706078.

Abstract. The *multi-covering radii* of a code are recent generalizations of the covering radius of a code. In this paper upper bounds are found for the multicovering radii of first order Reed-Muller codes. They are exact in some cases. These bounds are then used to prove the existence of secure families of keystreams against a general class of cryptanalytic attacks.

Keywords: Error correcting code, stream cipher, covering radius, Reed-Muller code.

1 Introduction

In this paper we derive upper bounds on the multicovering radii of first order Reed-Muller codes. In some cases the bounds are shown to be exact. We then use these bounds to strengthen an earlier result concerning the existence of families of stream ciphers that are asymptotically secure against a certain very general class of cryptanalytic attack.

Multicovering radii, which were introduced recently [7], are defined as follows. We let $F_2 = \{0,1\}$ denote the field with two elements. We also denote the Hamming distance between two vectors u and v by $\text{dist}(u, v)$ and the complement of a vector v by $\bar{v}$.

Definition 1. *Let C be a binary code of length n. Let m be a positive integer. The m-*covering radius *of C, denoted by $t_m(C)$, is the smallest integer t such every m-tuple of vectors in F_2^n is contained in a ball of radius t around at least one codeword in C. That is, $\forall v_1, \cdots, v_m \in F_2^n : \exists c \in C : \forall i = 1, \cdots, m :$ $\text{dist}(c, v_i) \leq t$.*

This is a natural generalization of the classical notion of covering radius, which is exactly the case when $m = 1$ [2, 5]. When $m > 1$, it is known that there is no code of length m, dimension n, and m-covering radius t if

$$t < \left\lceil \frac{n + \lfloor \log_2(m) \rfloor - 1}{2} \right\rceil .$$

This bound is not in general tight, but no general tight bound is known.

The notion of multi-covering radii first arose in an investigation of the existence of stream ciphers secure against a large class of attacks. The security of a stream cipher depends on the inability of an adversary to infer the key sequence from a subsequence (generally with some constraints on the resources of the adversary).

Many attacks on stream ciphers have the general form: input a prefix of a sequence; find a sequence generator (from a specific class of generators) whose output agrees with the prefix. A model for this general type of attack was recently considered by Klapper [8]. Attacks can be deterministic – meaning the output from generator found must agree with the original sequence exactly – or probabilistic – meaning the output from generator found must agree with the original sequence in significantly more than half the bits. A class of sequences can be secure against all attacks *infinitely often* – meaning that for each attack, there are infinitely many sequences in the class that resist the attack – or *almost everywhere* – meaning that for each attack all but finitely many sequences in the family resist the attack. Klapper showed that there are families of sequences that are secure against all deterministic attacks almost everywhere and there are families that are secure against probabilistic attacks infinitely often, but the existence of families that are secure against all probabilistic attacks almost everywhere was left open. It was this question that led to the invention of the notion of multicovering radii.

In Section 2 we find upper bounds on the multicovering radii of strength two codes. First order Reed-Muller codes are strength two codes, so these bounds apply to them. In Section 3 we show that in some cases these bounds are tight. In Section 4 we use the bounds from Section 2 to prove the existence of families of efficiently generated sequences that are secure in the strong sense discussed above.

2 Norse Bound for Multi-Covering Radius

Let m be a positive integer. Let $v \in F_2^r$ and $b \in F_2$, and let $c^{v,b}$ be the vector indexed by F_2^r whose uth coordinate is

$$c_u^{v,b} = b + \sum_{i=1}^{m} u_i v_i = b + v \cdot u.$$

Then $RM(1,r)$, the rth first order Reed-Muller code, consists of all vectors $c^{v,b}, v \in F_2^r, b \in F_2$. Let $n = 2^r$, so $RM(1,r)$ is a $[n, r+1, n/2]$ code [9]. $RM(1,r)$

is an example of a *strength 2 code*. That is, every pair of coordinates in $RM(1,r)$ takes on any fixed pair of values for exactly one quarter of the codewords. Since $c^{v,0}$ and $c^{v,1}$ are complementary, $RM(1,r)$ is closed under complement as well.

It has long been known [6] that the ordinary covering radius t of $RM(1,r)$ is bounded by

$$t_1(RM(1,r)) \le \left\lfloor \frac{2^r - 2^{r/2}}{2} \right\rfloor .$$

This bound is known as the *Norse Bound* and it is this that we generalize. It is further known that equality holds if r is even [10].

Theorem 1. *Let C be any strength 2 code of length n. Suppose C is closed under complement. Then*

$$t_m(C) \le \left\lfloor \frac{n + \sqrt{mn/2}}{2} \right\rfloor .$$

Proof Sketch: For any $x^1, \cdots, x^m \in F_2^n$,

$$\sum_{c\in C} \sum_{i=1}^{m} (2\mathrm{dist}(x^i, c) - n)^2 = mn|C|.$$

Therefore there is at least one $c \in C$ such that

$$\sum_{i=1}^{m} (2d_i - n)^2 \le mn,$$

where $d_i = \mathrm{dist}(x^i, c)$. This can be used to show that for every i, $(2d_i - n)^2 \le mn/2$. Hence $2d_i - n \le \sqrt{mn/2}$, and the theorem follows. □

Corollary 1. *For any $r > 0$ we have*

$$t_m(RM(1,r)) \le 2^{r-1} + \left\lfloor \sqrt{m2^{r-3}} \right\rfloor .$$

Proof: The first order Reed-Muller codes are well known to be strength 2 codes.

3 Exact Values

In this section we show that the bound in Corollary 1 is tight if m is an odd power of 2 and r is even; and if m is an even power of 2 and r is odd; and for $m = 2$ and $r \in \{1, 3, 5\}$. These results lead to general lower bounds, although a significant gap between the lower and upper bounds remains.

Theorem 2. *Assume that C is a binary self-complementary code with strength two. If C has covering radius $\frac{1}{2}(n - \sqrt{n})$ then $t_2(C) = \frac{1}{2}(n + \sqrt{n})$.*

Proof Sketch: If the covering radius of C equals $\frac{1}{2}(n-\sqrt{n})$, we know that there exists a point $x \in F_2^n$ such that

$$\operatorname{dist}(x,c) \geq \frac{1}{2}(n-\sqrt{n}) \tag{1}$$

for all $c \in C$. This implies that in fact for all $c \in C$,

$$\operatorname{dist}(x,c) \in \left\{\frac{1}{2}(n-\sqrt{n}), \frac{1}{2}(n+\sqrt{n})\right\}.$$

Then for all $c \in C$ we have $\operatorname{dist}(c,x) = \frac{1}{2}(n+\sqrt{n})$ or $\operatorname{dist}(c,\bar{x}) = \frac{1}{2}(n+\sqrt{n})$, proving that $t_2(C) \geq \frac{1}{2}(n+\sqrt{n})$. □

Corollary 2. *For all s,*

$$t_2(RM(1,2s)) = 2^{2s-1} + 2^{s-1}.$$

Consider now larger values of m.

Theorem 3. *If C is self-complementary, then $t_m(C) = n$ for all $m \geq |C|$.*

For the first-order Reed-Muller codes this means that $t_m(RM(1,r)) = n$ whenever $m \geq 2^{r+1}$.

Theorem 4. *Assume that $m = 2^{2t+1}$. Then for all $s \geq t \geq 0$,*

$$t_m(RM(1,2s)) = \frac{n+\sqrt{mn/2}}{2} = 2^{2s-1} + 2^{s+t-1}.$$

Proof Sketch: We keep m fixed and prove the result by induction on s. The idea is to prove the stronger result that also there exists a *self-complementary* set T of vectors $y^1, \ldots, y^m$ such that

$$\max_{1 \leq i \leq m} \operatorname{dist}(y_i, c) \geq t_m(RM(1,2s))$$

for every $c \in RM(1,2s)$. The base of the induction is the case $s = t$.

In general, by using the recursive property

$$RM(1,r+1) = \{(u,u+v) \mid u \in RM(1,r), v \in RM(0,r)\}$$

twice, we see that all codewords $c \in RM(1,2(s+1))$ are of one of the types (u,u,u,u), $(u,\overline{u},u,\overline{u})$, $(u,u,\overline{u},\overline{u})$, $(u,\overline{u},\overline{u},u)$ where $u \in RM(1,2s)$. We now take

$$x^i = (y^i, y^i, y^i, \overline{y^i})$$

for $i = 1,\ldots,m$ and consider the various types of *c*s on a case by case basis to complete the induction. □

We obtain the following result for the codes $RM(1,2s+1)$ in exactly the same way.

Theorem 5. *Assume that $m = 2^{2t}$ with $t > 0$. Then for all $s \geq t-1$*

$$t_m(RM(1, 2s+1)) = \frac{n + \sqrt{mn/2}}{2} = 2^{2s} + 2^{s+t-1}.$$

□

Finally, we consider odd values of r in the case $m = 2$. For $r = 1$, $RM(1, r)$ is the set of all length 2 vectors. This code is known to have 2-covering radius equal to 1 [7, Proposition 3].

In general, $RM(r-2, r)$ is the extended Hamming code $\hat{\mathcal{H}}_r$. In particular, $RM(1,3) = \hat{\mathcal{H}}_3$. The unextended Hamming codes with $r \geq 3$ are known to have 2-covering radius equal to 2^{r-1} [7, Proposition 5]. Furthermore, extending a code by adding an overall parity check increases the m-covering radius by 1 [7, Corollary 1]. Therefore

$$\begin{aligned} t_2(\hat{\mathcal{H}}_r) &= t_2(RM(r-2, r)) \\ &= 2^{r-1} + 1. \end{aligned}$$

In particular, $t_2(RM(1,3) = 5$.

Theorem 6.

$$t_2(RM(1,5)) = 18.$$

Proof Sketch: By Theorem 1, $t_2(RM(1,5)) \leq 18$. We show that $t_2(RM(1,5)) \geq 18$.

Let z^1 be any point at distance 6 from $RM(1,4)$ and z^2 its complement. Without loss of generality, z^1 has weight 6 and z^2 has weight 10. Now take $x^1 = (z^1, \mathbf{0})$ and $x^2 = (z^2, \mathbf{1})$, where $\mathbf{0}$ and $\mathbf{1}$ are the all-zero and all-one words of length 16. We then use the fact that $RM(1,5) = \{(c,c), (c, \bar{c}) \mid c \in RM(1,4)\}$ to show the lower bound holds for this pair. □

In exactly the same way we can prove the following more general result. For $s = 1$ and $s = 2$ the lower and upper bounds coincide, but already for $s = 3$ we only get $68 \leq t_2(RM(1,7)) \leq 69$.

Theorem 7.

$$2^{2s} + 2^{s-1} \leq t_2(RM(1, 2s+1)) \leq 2^{2s} + \sqrt{2} \cdot 2^{s-1}.$$

The previous results and the monotonicity of $t_m(C)$ in m can be used to give lower and upper bounds for $t_m(RM(1,r))$ in the general case. Together Theorems 1, 4, and 5 give the following immediate corollary.

Theorem 8. *For every $m \geq 2$ and $r \geq \lfloor \log_2(m) \rfloor - 2$,*

$$\begin{aligned} 2^{r-1} + \sqrt{m2^{r-5}} = \frac{n + \sqrt{mn/8}}{2} & \\ &\leq t_m(RM(1,r)) \\ &\leq \frac{n + \sqrt{mn/2}}{2} = 2^{r-1} + \sqrt{m2^{r-3}}. \end{aligned}$$

4 Security of Stream Ciphers

We recall some definitions used previously to study the existence of secure stream ciphers [8].

Definition 2. *We define a* keystream generator, *or simply a generator, to be a 4-tuple* (S, F, g, s_0) *such that*

1. S *is a finite set (the* states*);*
2. $F : S \to S$ *is a function (the* state change function*);*
3. $g : S \to \{0,1\}$ *is a function (the* output function*); and*
4. s_0 *is an element of* S *(the* initial state*).*

A keystream generator outputs an infinite eventually periodic binary sequence by iterating the output and state change operations:

$$g(s_0), g(F(s_0)), g(F(F(s_0))), \cdots.$$

The *length* of a generator is $\lceil \log(|S|) \rceil$, the number of bits required to represent the states. We often use generators whose state space S is a set of n bit vectors $\bar{x} = (x_0, \cdots, x_{n-1})$ for some n, whence the length is n. We further generally use generators whose output functions are of the form $g(\bar{x}) = x_0$. In this case the generator is completely determined by F and s_0, and we often abuse the notation by identifying the generator with F. Any generator can be replaced by one of this form by possibly increasing the length by one.

The state change and output functions of our generators are described as circuits using binary gates. Such circuits can be encoded as binary strings [1]. The *size* of a generator F is the minimum number of gates in a circuit that computes the function F. This corresponds to evaluation time in a software implementation. The *depth* is the depth of the minimum depth circuit that computes F. This corresponds to evaluation time in a hardware implementation.

A *family of (keystream) generators,* $\mathcal{F}$, is an infinite collection of keystream generators. If B is an infinite eventually periodic binary sequence with eventual period period(B) and $0 < r \leq$ period(B), then a generator (F, s) with output sequence B' is said to *r-approximate* B if for any k,

$$|\{i, k \leq i \leq k + \text{period}(B) - 1 : b_i = b'_i\}| \geq r.$$

If $0 < r(p) \leq p$ is any function, then the $(\mathcal{F}, r)$-span of B, denoted $\lambda_{\mathcal{F},r}(B)$, is the least integer n such that B can be $r(\text{period}(B))$-approximated by a generator in $\mathcal{F}_n$ (or ∞ if there is no such n).

Let $\delta(n)$ be the maximum over all length n generators F in $\mathcal{F}$ of the depth of F. We say $\mathcal{F}$ is

1. *fast* if $\delta(n) \in \mathcal{O}(\log(n))$.
2. *short* if whenever there exists $F \in \mathcal{F}$ that generates sequence B, $\lambda_{\mathcal{F},m}(B)$ is $\mathcal{O}(\log(\text{period}(B)))$.

Definition 3. *Let T be an algorithm, let $\mathcal{F}$ be a family of generators, and let $0 < r(p) \leq p$. We say that T is an* r-effective $\mathcal{F}$-synthesizing algorithm *if*

1. *it runs in polynomial time;*
2. *when given the input $b_0, \cdots, b_{k-1}$, T outputs the encoding of a generator $(F, s) \in \mathcal{F}$ such that the first k output bits of F with initial state s are $b_0, \cdots, b_{k-1}$; and*
3. *there is a polynomial $g(n)$ such that if B is any eventually periodic sequence and $n = \lambda_{\mathcal{F},r}(B)$, then on input $b_0, \cdots, b_{k-1}$ with $k \geq g(n)$, T outputs $F \in \mathcal{F}$ of length n that r-approximates B.*

Theorem 9. *Let $h(n)$ be subexponential (in the sense that $h \in \mathcal{O}(a^n)$ for every $a > 1$), let $\epsilon > 1/2$, and let*

$$r(p) = \frac{p}{2} + p^{\epsilon}.$$

There exists a collection $\mathcal{B}$ of sequences such that

1. *$\mathcal{B}$ can be generated by a family $\mathcal{F}$ of fast short generators; and*
2. *for every r-effective register synthesizing algorithm T generating a family $\mathcal{F}'$ of registers, and for all but finitely many sequences $B \in \mathcal{B}$*

$$\lambda_{\mathcal{F}',r}(B) \geq h(\log(\text{period}(B))).$$

Proof Sketch: Let $T_1, T_2, \cdots$ be an enumeration of the r-effective register synthesizing algorithms. We construct the sequence $\mathcal{B}$ in stages. At the mth stage we construct B^m which simultaneously has large Hamming distance from every sequence generated by $T_1, \cdots, T_m$ with input a large enough prefix of B^m. We use Reed-Muller codes and our bounds on their multi-covering radii.

A *linear feedback shift register* (LFSR) of length k is a keystream generator with state set F_2^k, state change function of the form $F(x_1, \cdots, x_k) = (x_2, \cdots, x_k, f(x_1, \cdots, x_k))$ for some linear function f, and with output function $g(x_1, \cdots, x_k) = x_1$. The function f can be computed in depth $\log(k)$. For every k there are LFSRs whose output sequences (m-sequences) have period $2^k - 1$.

The output from the generators we construct consist of an m-sequence of period $2^k - 1$ followed by a $RM(n, 1)$ codeword c for some k and n. The generator consists of two parts: an LFSR that generates the m-sequence, and a generator that outputs c. The overall generator can be made to output the m-sequence, then switch to the generator of c. When c has been output, the generator switches back to the LFSR. The combined sequence has period $2^n + 2^k - 1$ and can be generated by a generator of length $n + k + 3 \leq 3\log(2^n + 2^k - 1)$ and depth $\mathcal{O}(\log(n) + \log(k)) = \mathcal{O}(\log(n + k + 3))$. Thus this is a fast short generator.

We want sequences that are far from given sequences, but in the Hamming metric, if c is close to b, then the complement c' of c is far from b: $dist(c', b) = 2^n - dist(c, b)$ if the length of the code is 2^n. The first order Reed-Muller code is closed under complementation, so by Theorem 1, there is a first order Reed-Muller codeword whose distance from any given set of m sequences of length 2^n

is at least

$$\frac{2^n}{2} - \frac{\sqrt{2mn}}{4}.$$

For each r-synthesis algorithm T_i, let $\mathcal{F}^i$ be the family of generators that is output by T_i. We assume that T^i is successful when given

$$g_i(\lambda_{\mathcal{F}^i,r}(B))$$

bits of any sequence B, with g_i a polynomial. At the mth stage of the diagonalization we want to find a fast generator F_m, as described above, with output B^m so that $\lambda_{\mathcal{F}^i,r}(B^m)$ is large for every $i = 1, \cdots, m$.

Let B^m be the sequence one of whose periods is the m-sequence of period $2^k - 1$ followed by a codeword $c \in RM(1,n)$. It can be seen that n, k, and c can be chosen so that

1. If a sequence B of period p satisfies

$$\lambda_{\mathcal{F}^i,r}(B) < h(\log(p)),$$

$i = 1, \cdots, m$, and agrees with the given m-sequence on its first $2^k - 1$ bits, then the sequence generated by the output of T^i, $i = 1, \cdots, m$, must agree with B on at least $r(p)$ bits;

2. The output of T^i on input the first $2^k - 1$ bits of B^m is correct on at most

$$2^k - 1 + \frac{2^n}{2} + \frac{(2m2^n)^{1/2}}{4} < r(2^n + 2^k - 1)$$

bits.

It follows that

$$\begin{aligned}\lambda_{\mathcal{F}^i,r}(B^m) &> h\left(\log\left(\frac{2^n + 2^k - 1}{2}\right)\right) \\ &> h(\log(\text{period}(B^m)))\end{aligned}$$

for $i = 1, \cdots, m$. □

5 Conclusions

We have given bounds on the multi-covering radii of first order Reed-Muller codes and have shown that these bounds are tight in some cases. It is apparent from our lower bounds, however, that even if the bounds are tightened our cryptographic results will not be improved. This would require reducing the $2^{n/2}$ term in the bounds. This may be possible using bounds on the multi-covering radii of the dth order Reed-Muller codes, $d > 1$. Thus we leave finding such bounds as an interesting open problem.

References

1. J. Balcázar, J. Díaz, and J. Gabarró, "Structural Complexity I," Springer-Verlag, Berlin, 1988.
2. G.D. Cohen, M.G. Karpovsky, H.F. Mattson, Jr., and J.R. Schatz, Covering radius – survey and recent results, *IEEE Trans. Info. Theory* **IT-31** (1985), 328–343.
3. G. Cohen, I. Honkala, S. Litsyn and A. Lobstein, "Covering Codes," Elsevier, Amsterdam, 1997.
4. G. Cohen and S. Litsyn, On the covering radius of Reed-Muller codes, *Discrete Mathematics*, **106-107** (1992), 147–155.
5. G.D. Cohen, S.N. Litsyn, A.C. Lobstein, and H.F. Mattson, Jr., "Covering Radius 1985-1994," Dept. Informatique, Ecole Nationale Supérieure des Télécommunications, Technical Report 94 D 025, 1994.
6. T. Helleseth, T. Kløve, and J. Mykkeltveit, On the covering radius of binary codes, *IEEE Trans. Info. Theory* **IT-24** (1978), 627–628.
7. A. Klapper, The Multicovering radii of codes, *IEEE Trans. Info. Theory* **43** (1997), 1372–1377.
8. A. Klapper, On the existence of secure keystream generators, to appear, *J. Cryptology.*
9. F.J. MacWilliams and N.J.A. Sloane, "The Theory of Error Correcting Codes," North-Holland, Amsterdam, 1977.
10. O. Rothaus, On 'bent' functions, *J. Combin. Thy., Se. A*, **20** (1976) 300–305.

Inclusion Relations of Boolean Functions Satisfying PC(l) of Order k

Tetsu IWATA and Kaoru KUROSAWA

Department of Electrical and Electronic Engineering
Faculty of Engineering
Tokyo Institute of Technology
2–12–1 O-okayama, Meguro-ku, Tokyo 152–8552, JAPAN
Email: {tez, kurosawa}@ss.titech.ac.jp
URL: http://tsk-www.ss.titech.ac.jp/~kurosawa/

Abstract. PC(l) of order k is an important cryptographic criterion of Boolean functions. Denote by $\mathcal{PC}_n(l,k)$ the set of Boolean functions satisfying PC(l) of order k with n input bits. Recently, Carlet and Hirose proved that $\mathcal{PC}_n(l_1,k_1) = \mathcal{PC}_n(l_2,k_2)$ for some class of l_1, k_1, l_2 and k_2. In this paper, we prove that $\mathcal{PC}_n(l_1,k_1) \neq \mathcal{PC}_n(l_2,k_2)$ for a large class of l_1, k_1, l_2 and k_2. Our result can be considered as a complement of their results.

1 Introduction

The security of block ciphers is often studied by viewing their S-boxes (or F functions) as a set of Boolean functions. A Boolean function $f(x)$ is said to satisfy propagation criterion of degree l (abbr., PC(l)) if $f(x) \oplus f(x \oplus \alpha)$ is balanced for any α with $1 \leq W(\alpha) \leq l$. $W(\alpha)$ denotes the Hamming weight of α. More generally, a Boolean function is said to satisfy PC(l) of order k if any function obtained from it by keeping any k input bits constant satisfies PC(l). Strict avalanche criterion (abbr., SAC) is equivalent to PC(1) and SAC(k) is equivalent to PC(1) of order k. A Boolean function is said to be a bent function if and only if it satisfies PC(n). These cryptographic criteria are directly related to the security against differential attack of block ciphers.

Quadratic Boolean functions satisfying PC(l) of order k were studied by Preneel et al. [7]. Seberry, Zhang and Zheng showed balanced Boolean functions satisfying PC($n-1$) with high nonlinearity [8]. Kurosawa and Satoh proposed a construction method for Boolean functions satisfying PC(l) of order k with large degree by using linear codes [5]. Carlet extended the Kurosawa and Satoh's method by using nonlinear codes [1].

Define

$$\mathcal{PC}_n(l,k) \triangleq \{f(x) \mid f(x) \text{ satisfies PC}(l) \text{ of order } k\ \} ,$$

where $x \in \{0,1\}^n$. Carlet proved that for every positive even $l \le n-4$ $(6 \le n)$ [1],

$$\mathcal{PC}_n(l, n-l-2) = \mathcal{PC}_n(l, n-l) \ .$$

Hirose showed that if n is even [4],

$$\mathcal{PC}_n(n,0) = \mathcal{PC}_n(n-1,0) = \mathcal{PC}_n(n-2,0) \ .$$

Now we ask, for given l_1, k_1, l_2 and k_2,

$$\mathcal{PC}_n(l_1,k_1) \setminus \mathcal{PC}_n(l_2,k_2) = \emptyset \ ?$$

In this paper, we show a negative answer to this question for a large class of parameters. This proves that some criteria PC(l) of order k are not weaker than some others. Our result can be considered as a complement of Carlet and Hirose's results.

2 Preliminaries

Let $x = (x_1, \ldots, x_n) \in \{0,1\}^n$. For a binary string α, $W(\alpha)$ denotes the Hamming weight of α.

A Boolean function $f(x)$ is a mapping from $\{0,1\}^n$ to $\{0,1\}$. We say that the Boolean function $f(x)$ is balanced if

$$\big|\{x \mid f(x)=0\}\big| = \big|\{x \mid f(x)=1\}\big| = 2^{n-1} \ .$$

2.1 PC(l) of Order k

Definition 2.1. [7] *A Boolean function $f(x)$ satisfies PC(l) if $f(x) \oplus f(x \oplus \alpha)$ is balanced for any $\alpha \in \{0,1\}^n$ such that $1 \le W(\alpha) \le l$.*

Definition 2.2. [7] *A Boolean function $f(x)$ satisfies PC(l) of order k if any function obtained from it by keeping any k input bits constant satisfies PC(l).*

Proposition 2.1. *If a Boolean function $f(x)$ satisfies PC(l) of order k, then it satisfies PC(l') of order k' for $1 \le l' \le l$ and $0 \le k' \le k$.*

Proposition 2.2. *If a Boolean function $f(x)$ does not satisfy PC(l) of order k, then it does not satisfy PC(l') of order k' for $l+1 \le l' \le n$ and $k+1 \le k' \le n$ where $l'+k' \le n$.*

Proposition 2.3. *Let $x \in \{0,1\}^s$ and $y \in \{0,1\}^t$. A Boolean function $f(x) \oplus g(y)$ satisfies PC(l) of order k if both Boolean functions $f(x)$ and $g(y)$ satisfy PC(l) of order k.*

There are some equivalent ways to define bent functions. In this paper, we define bent functions as follows.

Definition 2.3. [2,6] *A Boolean function $f(x)$ is a bent function if and only if $f(x)$ satisfies PC(n).*

2.2 Existence of PC(l) of Order k

Preneel showed the following proposition [7].

Proposition 2.4. [7] *Let*

$$s_n(x) \triangleq \bigoplus_{1 \le i < j \le n} x_i x_j \ .$$

Then, the Boolean function $s_n(x)$ satisfies PC(l) of order k if $l + k \le n - 1$ or $l + k = n$ and l is even.

Conversely, Carlet showed that the Boolean functions $s_n(x) \oplus h(x)$, where $h(x)$ is the affine function, are the only one to satisfy these criteria [1].

Seberry, Zhang and Zheng showed the following proposition [8].

Proposition 2.5. [8] *For odd n, let $x' = (x_1, \ldots, x_{n-1}) \in \{0,1\}^{n-1}$. Let $g(x')$ be a bent function. Define*

$$f(x) \triangleq g(x') \oplus x_n \ .$$

Then, the Boolean function $f(x) \oplus f(x \oplus \alpha)$ is balanced for any $\alpha \in \{0,1\}^n$ except $\alpha = (0, \ldots, 0, 1)$. Further, let A be a nonsingular matrix such that

$$(0, \ldots, 0, 1)A = (1, \ldots, 1) \ .$$

Then, the Boolean function $f(xA)$ satisfies PC($n-1$).

Definition 2.4. *We say that an $s \times t$ binary matrix Q satisfies the PC(l, k) condition if*

1. $W(Q\gamma_1^T) \ge k + 1$ *for any vector* γ_1 *with* $1 \le W(\gamma_1) \le l$.
2. $W(\gamma_2 Q) \ge k + 1$ *for any vector* γ_2 *with* $1 \le W(\gamma_2) \le l$.

Kurosawa and Satoh showed the following proposition [5].

Proposition 2.6. [5] *Let Q be an $s \times t$ binary matrix which satisfies the PC(l, k) condition (see Definition 2.4). Then, for any Boolean function $g(x)$,*

$$f(x, y) \triangleq (x_1, \ldots, x_s) Q (y_1, \ldots, y_t)^T \oplus g(x)$$

where $x = (x_1, \ldots, x_s)$ and $y = (y_1, \ldots, y_t)$, satisfies PC(l) of order k.

3 Converse of Proposition 2.6

In this section, we prove the converse of Proposition 2.6.

Theorem 3.1. *For an $s \times t$ binary matrix Q, define*

$$f(x,y) \triangleq (x_1,\ldots,x_s)Q(y_1,\ldots,y_t)^T \ , \quad (1)$$

where $x = (x_1,\ldots,x_s)$ and $y = (y_1,\ldots,y_t)$. Then, the Boolean function $f(x,y)$ satisfies PC(l) of order k "if and only if" the matrix Q satisfies the PC(l,k) condition (see Definition 2.4).

Proof. If the matrix Q satisfies the PC(l,k) condition, then, the Boolean function $f(x,y)$ satisfies PC(l) of order k from Proposition 2.6.

Suppose that the matrix Q does not satisfy the PC(l,k) condition. Then, we show that the Boolean function $f(x,y)$ does not satisfy PC(l) of order k. From Definition 2.4, we have the following two cases.

1. The PC(l,k) condition 1. is not satisfied. In this case, there exists some vector γ_1 such that

   $$W(Q\gamma_1^T) < k+1 \text{ and } 1 \le W(\gamma_1) \le l \ .$$

   Note that

   $$f(x,y) \oplus f(x,y\oplus\gamma_1) = xQy^T \oplus xQ(y\oplus\gamma_1)^T = xQ\gamma_1^T \ .$$

   Now, since $W(Q\gamma_1^T) < k+1$, the Boolean function $f(x,y) \oplus f(x,y\oplus\gamma_1)$ is written as

   $$f(x,y) \oplus f(x,y\oplus\gamma_1) = a_1x_{i_1} \oplus \cdots \oplus a_kx_{i_k}$$

   for some $x_{i_1},\ldots,x_{i_k}$ and $a_1,\ldots,a_k$. Then,

   $$f(x,y) \oplus f(x,y\oplus\gamma_1) = 0$$

   if $x_{i_1} = \cdots = x_{i_k} = 0$. This means that the Boolean function $f(x,y)$ does not satisfy PC(l) of order k.
2. The PC(l,k) condition 2. is not satisfied. In this case, we can show that the Boolean function $f(x,y)$ does not satisfy PC(l) of order k similarly to the above.

□

4 Inclusion Relations of PC(l) of Order k

Define

$$\mathcal{PC}_n(l,k) \triangleq \{f(x) \mid f(x) \text{ satisfies PC}(l) \text{ of order } k\ \} \ .$$

Let

$$A \setminus B \triangleq \{a \mid a \in A \text{ and } a \notin B\} \ .$$

We ask the following question. For given l_1, k_1, l_2 and k_2,

$$\mathcal{PC}_n(l_1,k_1) \setminus \mathcal{PC}_n(l_2,k_2) = \emptyset \ ?$$

In this section, we show a negative answer to this question for a large class of parameters.

Definition 4.1. *An $[N, m, d]$ binary linear code is a binary error correcting code of length N, dimension m and the minimum Hamming distance d.*

Definition 4.2. *The dual code $\mathcal{C}^\perp$ of a linear code $\mathcal{C}$ is defined as*

$$\mathcal{C}^\perp \triangleq \{u \mid u \cdot v = 0 \text{ for all } v \in \mathcal{C}\} ,$$

where $u \cdot v$ denotes the inner product of u and v. The dual minimum Hamming distance of $\mathcal{C}$ is the minimum Hamming distance of $\mathcal{C}^\perp$.

Theorem 4.1. *Suppose that there exists an $[N, m, d]$ binary linear code $\mathcal{C}$ with the dual minimum Hamming distance $d^\perp$ such that $d \geq 2$. Let $n = 2N$. Then*

1. $\mathcal{PC}_n(d-1, d^\perp - 1) \setminus \mathcal{PC}_n(d, 0) \neq \emptyset$.
2. $\mathcal{PC}_n(d-1, d^\perp - 1) \setminus \mathcal{PC}_n(1, d^\perp) \neq \emptyset$.

Proof. For a parity check matrix H of the code $\mathcal{C}$, let

$$Q \triangleq H^T H .$$

For this matrix Q, define the Boolean function $f(x, y)$ as shown in Equation (1). We show that

1. $f(x, y) \in \mathcal{PC}_n(d-1, d^\perp - 1) \setminus \mathcal{PC}_n(d, 0)$.
2. $f(x, y) \in \mathcal{PC}_n(d-1, d^\perp - 1) \setminus \mathcal{PC}_n(1, d^\perp)$.

First, we can see that

$$f(x, y) \in \mathcal{PC}_n(d-1, d^\perp - 1)$$

from Kurosawa and Satoh's result [5, Theorem 19].

1. Let γ_1 be a codeword of $\mathcal{C}$ such that $W(\gamma_1) = d$. Then,

   $$H\gamma_1^T = (0, \ldots, 0)^T .$$

   Hence,

   $$Q\gamma_1^T = H^T H \gamma_1^T = (0, \ldots, 0)^T .$$

   Therefore, the matrix Q does not satisfy the PC$(d, 0)$ condition. Hence,

   $$f(x, y) \notin \mathcal{PC}_n(d, 0)$$

   from Theorem 3.1.
2. We can choose the parity check matrix H such that the first column of H is $(1, 0, \ldots, 0)$ and the first row of H is β such that $W(\beta) = d^\perp$ since H is a generator matrix of the dual code $\mathcal{C}^\perp$. Now, let $\gamma_1 = (1, 0, \ldots, 0)$. Then,

   $$Q\gamma_1^T = H^T H \gamma_1^T = H^T (1, 0, \ldots, 0)^T = \beta .$$

   This means that the matrix Q does not satisfy the PC$(1, d^\perp)$ condition. Hence,

   $$f(x, y) \notin \mathcal{PC}_n(1, d^\perp)$$

   from Theorem 3.1.

□

Now we have the following corollary from Proposition 2.1 and Proposition 2.2.

Corollary 4.1. *Suppose that there exists an $[N, m, d]$ binary linear code C with the dual minimum Hamming distance $d^{\perp}$ such that $d \geq 2$. Let $n = 2N$. Then*

$$\mathcal{PC}_n(l_1, k_1) \setminus \mathcal{PC}_n(l_2, k_2) \neq \emptyset$$

if

1. $1 \leq l_1 \leq d-1$, $0 \leq k_1 \leq d^{\perp}-1$, $d \leq l_2 \leq n$ *and* $0 \leq k_2 \leq n - l_2$.
2. $1 \leq l_1 \leq d-1$, $0 \leq k_1 \leq d^{\perp}-1$, $1 \leq l_2 \leq n - k_2$ *and* $d^{\perp} \leq k_2 \leq n$.

5 Further Inclusion Relations

5.1 SAC

Definition 5.1. [3,9] *A Boolean function $f(x)$ satisfies SAC if $f(x) \oplus f(x \oplus \alpha)$ is balanced for any $\alpha \in \{0,1\}^n$ such that $W(\alpha) = 1$.*

Definition 5.2. [3,9] *A Boolean function $f(x)$ satisfies SAC(k) if any function obtained from it by keeping any k input bits constant satisfies SAC.*

Note that SAC is equivalent to PC(1) and SAC(k) is equivalent to PC(1) of order k.

The following form is called the algebraic normal form of the Boolean function $f(x)$.

$$f(x) = a_0 \oplus \bigoplus_{1 \leq i \leq n} a_i x_i \oplus \bigoplus_{1 \leq i < j \leq n} a_{i,j} x_i x_j \oplus \cdots \oplus a_{1,2,\ldots,n} x_1 x_2 \cdots x_n$$

The degree of the highest degree term in the algebraic normal form of the Boolean function $f(x)$ is denoted as $\deg(f)$.

Preneel showed the following proposition.

Proposition 5.1. [7] *Let $f(x)$ be a Boolean function with $n > 2$ and $\deg(f) = 2$. The Boolean function $f(x)$ satisfies SAC(k) if and only if every input variable occurs in at least $k+1$ second order terms of the algebraic normal form.*

5.2 Further Inclusion Relations

Let

$$\mathcal{SAC}_n(k) \stackrel{\triangle}{=} \{f(x) \mid f(x) \text{ satisfies SAC}(k)\} .$$

Theorem 5.1. *For $n \geq 3$,*

$$\mathcal{SAC}_n(n-3) \setminus \mathcal{PC}_n(2,0) \neq \emptyset .$$

Proof. Let

$$f_1(x) \stackrel{\triangle}{=} s_n(x) \oplus x_1x_2 \ ,$$

where the Boolean function $s_n(x)$ is defined in Proposition 2.4. We show that

$$f_1(x) \in \mathcal{SAC}_n(n-3) \setminus \mathcal{PC}_n(2,0) \ .$$

First, $f_1(x) \in \mathcal{SAC}_n(n-3)$ from Proposition 5.1 since x_1 occurs in $n-2$ second order terms, as well as x_2. Next, $f_1(x) \notin \mathcal{PC}_n(2,0)$ since

$$f_1(x) \oplus f_1(x \oplus \alpha) = 0$$

for $\alpha = (1,1,0,\ldots,0)$. □

Theorem 5.2. *Suppose that* $2(l+k+1) \leq n$, $1 \leq l$ *and* $0 \leq k$. *Then,*

$$\mathcal{PC}_n(l+1,k) \setminus \mathcal{PC}_n(l,k+1) \neq \emptyset \ \textit{ for odd } l \ ,$$
$$\mathcal{PC}_n(l,k+1) \setminus \mathcal{PC}_n(l+1,k) \neq \emptyset \ \textit{ for even } l \ .$$

Proof. Suppose that l is odd. Let

$$x = (x_1,\ldots,x_{l+k+1}) \text{ and } y = (y_1,\ldots,y_{n-(l+k+1)}) \ .$$

Define the Boolean function $f_2(x,y)$ as

$$f_2(x,y) \stackrel{\triangle}{=} s_{l+k+1}(x) \oplus s_{n-(l+k+1)}(y) \ .$$

We show that

$$f_2(x,y) \in \mathcal{PC}_n(l+1,k) \setminus \mathcal{PC}_n(l,k+1) \ .$$

At first, Boolean functions $s_{l+k+1}(x)$ and $s_{n-(l+k+1)}(y)$ both satisfy PC($l+1$) of order k from Proposition 2.4. Hence,

$$f_2(x,y) \in \mathcal{PC}_n(l+1,k)$$

from Proposition 2.3. Next, fix $k+1$ bits in such a way that

$$x_{l+1} = \cdots = x_{l+k+1} = 0 \ .$$

Then,

$$f_2(x,y) = s_l(x_1,\ldots,x_l) \oplus s_{n-(l+k+1)}(y) \ .$$

Now, for $\alpha = (1,\ldots,1)$ such that $W(\alpha) = l$, we have

$$f_2(x,y) \oplus f_2(x \oplus \alpha, y) = s_l(x_1,\ldots,x_l) \oplus s_l(x_1 \oplus 1,\ldots,x_l \oplus 1) = 1 \ .$$

This means that

$$f_2(x,y) \notin \mathcal{PC}_n(l,k+1) \ .$$

The proof for even l is similar. □

Theorem 5.3. *Suppose that $n \geq 3$. Then*

$$\mathcal{PC}_n(n,0) \setminus \mathcal{SAC}_n(1) \neq \emptyset \quad \textit{for even } n ,$$
$$\mathcal{PC}_n(n-2,0) \setminus \mathcal{SAC}_n(1) \neq \emptyset \quad \textit{for odd } n .$$

Proof. First for even n, let

$$f_3(x) = x_1x_2 \oplus x_3x_4 \oplus \cdots \oplus x_{n-1}x_n .$$

Then, the Boolean function $f_3(x)$ is known to be a Maiorana-type bent function [2]. Therefore,

$$f_3(x) \in \mathcal{PC}_n(n,0)$$

from Definition 2.3. Further, it is easy to see that

$$f_3(x) \notin \mathcal{SAC}_n(1)$$

for $x_2 = 0$ and $\alpha = (1,0,\ldots,0)$.

Next for odd n, let

$$f_4(x) = x_1x_2 \oplus x_3x_4 \oplus \cdots \oplus x_{n-2}x_{n-1} \oplus x_n .$$

Then, the Boolean function $f_4(x) \oplus f_4(x \oplus \alpha)$ is balanced for any α except $\alpha = (0,\ldots,0,1)$ from Proposition 2.5. Define an $n \times n$ matrix A as

$$A \triangleq \left(\begin{array}{c|c} I & O^T \\ \hline u & 1 \end{array}\right) ,$$

where $u = (0,1,\ldots,1)$, I is the $(n-1)\times(n-1)$ identity matrix and $O = (0,\ldots,0)$. Let

$$(y_1,\ldots,y_n) \triangleq (x_1,\ldots,x_n)A .$$

Then, the Boolean function $f_4(y) \oplus f_4(y \oplus \alpha)$ is balanced for any α except

$$\alpha = (0,\ldots,0,1)A = (0,1,\ldots,1) ,$$

where $f_4(y)$ is obtained by substituting $(x_1,\ldots,x_n) = (y_1,\ldots,y_n)A^{-1}$ into $f_4(x)$. Hence,

$$f_4(y) \in \mathcal{PC}_n(n-2,0) .$$

On the other hand, it is easy to see that $A^{-1} = A$. Therefore,

$$(x_1,\ldots,x_n) = (y_1,\ldots,y_n)A .$$

Hence,

$$\begin{cases} x_1 = y_1 \\ x_2 = y_2 \oplus y_n \end{cases}$$

holds. This means that the Boolean function $f_4(y)$ is written as

$$f_4(y) = y_1(y_2 \oplus y_n) \oplus g(y')$$

for some Boolean function $g(y')$, where $y' = (y_3, \ldots, y_n)$. Then,

$$f_4(y) \notin SAC_n(1)$$

for $y_1 = 0$ and $\alpha = (0, 1, 0, \ldots, 0)$. □

We have the following corollary applying Proposition 2.1 and Proposition 2.2 to Theorem 5.1, Theorem 5.2 and Theorem 5.3.

Corollary 5.1. *Suppose* $n \geq 3$, *then*

$$\mathcal{PC}_n(l_1, k_1) \setminus \mathcal{PC}_n(l_2, k_2) \neq \emptyset$$

for

1. $l_1 = 1$, $0 \leq k_1 \leq n-3$, $2 \leq l_2 \leq n$ *and* $0 \leq k_2 \leq n - l_2$.
2. *odd* l, $1 \leq l_1 \leq l+1$, $0 \leq k_1 \leq k$, $l \leq l_2 \leq n$ *and* $k+1 \leq k_2 \leq n - l_2$,
   even l, $1 \leq l_1 \leq l$, $0 \leq k_1 \leq k+1$, $l+1 \leq l_2 \leq n$ *and* $k \leq k_2 \leq n - l_2$,
   where $2(l+k+1) \leq n$, $1 \leq l$ *and* $0 \leq k$.
3. *even* n, $1 \leq l_1 \leq n$, $k_1 = 0$, $1 \leq l_2 \leq n$ *and* $1 \leq k_2 \leq n - l_2$,
   odd n, $1 \leq l_1 \leq n-2$, $k_1 = 0$, $1 \leq l_2 \leq n$ *and* $1 \leq k_2 \leq n - l_2$.

References

1. C. Carlet. On the propagation criterion of degree l and order k. In *Advances in Cryptology — EUROCRYPT '98 Proceedings, Lecture Notes in Computer Science* 1403, pages 462–474. Springer-Verlag, 1998.
2. J.F. Dillon. Elementary Hadamard difference sets. Ph. D. Thesis. Univ. of Maryland. 1974.
3. R. Forré. The strict avalanche criterion : spectral properties of Boolean functions and an extend definition. In *Advances in Cryptology — CRYPTO '88 Proceedings, Lecture Notes in Computer Science* 403, pages 450–468. Springer-Verlag, 1990.
4. S. Hirose. Boolean functions related to cryptography and their complexity. Doctoral dissertation. Kyoto Univ. Japan. February 1995.
5. K. Kurosawa and T. Satoh. Design of SAC/PC(l) of order k Boolean functions and three other cryptographic criteria. In *Advances in Cryptology — EUROCRYPT '97 Proceedings, Lecture Notes in Computer Science* 1233, pages 434–449. Springer-Verlag, 1997.
6. W. Meier and O. Staffelbach. Nonlinearity criteria for cryptographic functions. In *Advances in Cryptology — EUROCRYPT '89 Proceedings, Lecture Notes in Computer Science* 434, pages 549–562. Springer-Verlag, 1990.
7. B. Preneel, W. van Leekwijck, L. van Linden, R. Govaerts, and J. Vandewalle. Propagation characteristics of Boolean functions. In *Advances in Cryptology — EUROCRYPT '90 Proceedings, Lecture Notes in Computer Science* 473, pages 161–173. Springer-Verlag, 1991.

8. J. Seberry, X. Zhang and Y. Zheng. Nonlinearity and propagation characteristics of Boolean functions. In *Information and Computation*, vol. 119, No. 1, pages 1–13, 1995.
9. A.F. Webster and S.E. Tavares. On the design of S-boxes. In *Advances in Cryptology — CRYPTO '85 Proceedings, Lecture Notes in Computer Science* 218, pages 523–534. Springer-Verlag, 1986.

Notes on q-ary Interleaved Sequences *

JIANG Shaoquan[1], DAI Zongduo[1] and GONG Guang[2]

[1] State Key Lab. of Information Security
Dept. of Math., Graduate School
University of Science and Technology of China
100039-08, Beijing, China
Email: yangdai@mimi.cnc.ac.cn
[2] Center for Applied Cryptographic Research
Department of Combinatorics & Optimization
University of Waterloo,
Waterloo, Ontario N2L 3G1, Canada
Email: ggong@cacr.math.uwaterloo.ca

Abstract. A sequence u over a finite field F_q is called an $(f(x), m)$-interleaved sequence [1] if $f(x)$ is a common characteristic polynomial of all its m-decimated sequences. This paper demonstrates how the subjects such as the minimal polynomial, the linear span, the period and the correlation values of an $(f(x), m)$-interleaved sequence are related to its m-decimated sequences when $f(x)$ is irreducible, and makes a correction to an error made in [1] in studying the same problem. Moreover, this paper shows how to choose the m-decimated sequences to construct some sequences with the optimal auto and cross correlation values and maximal period among all the $(f(x), m)$-interleaved sequences when $f(x)$ is primitive and m is less than the period of $f(x)$.

1 Introduction

A sequence $u = \{u(i)\}_{i=0}^{\infty}$ over a finite field F_q is called an $(f(x), m)$-interleaved sequence [1] if $f(x)$ is a common characteristic polynomial [2] of its m-decimated sequences $u_j = \{u(im + j)\}_{i=0}^{\infty}, 0 \leq j < m$, which will be called its component sequences, where $f(x)$ is a polynomial over F_q and m is an integer greater than 1. When $f(x)$ is irreducible with degree n, if one fixes a root α of $f(x)$, it is clear that there exists a unique m-tuple $\underline{\gamma} = (\gamma_0, \gamma_1, \cdots, \gamma_{m-1})$ over F_{q^n} such that $u(im + j) = Tr_q^{q^n}(\gamma_j \alpha^i), \forall i, j$. (Here $Tr_q^{q^n}(x)$ is the trace function [2] from F_{q^n} to F_q.) Hence the m component sequences of u can be identified with the tuple $\underline{\gamma}$, and the sequence u can be denoted by $u_{\underline{\gamma}}$.

It is pointed out in [1] that many sequences can be considered as $(f(x), m)$-interleaved sequences with $f(x)$ being primitive and $m \geq per(f(x))$, such as multiplexed sequences [3], clock-controlled sequences [4], GMW-sequences [5],

* The work is supported by the Chinese Natural Science Foundation.

cascaded GMW-sequences [6], No sequences [7], and geometric sequences [8],[9], where $per(f(x))$ is the period [2] of $f(x)$. An interesting problem is how to construct some nice $(f(x), m)$-interleaved sequences when $f(x)$ is primitive and m is less than $per(f(x))$. A more elementary problem, which is related to the former one, is how to determine subjects such as the minimal polynomial, the linear span, the period and the correlation values of an $(f(x), m)$-interleaved sequence $u_{\underline{\gamma}} = \{u(i)\}_{i=0}^{\infty}$, which depend on its component sequences, *i.e.*, the m-tuple $\underline{\gamma}$. The latter problem is studied in [1] but not solved completely and an error is made in Theorem 1 (i) [1] which concerns on the relation between the period of the sequence $u_{\underline{\gamma}}$ and the tuple $\underline{\gamma}$. This paper begins by studying the latter problem, then makes a correction to an error made in Theorem 1 (i) [1], and finally shows how to choose the tuples $\underline{\gamma}$'s to construct some sequences with optimal auto and cross correlation values among all the $(f(x), m)$-interleaved sequences when $f(x)$ is primitive and m is less than $per(f(x))$.

2 Interleaved Sequences $u_{\underline{\gamma}}$ and Tuples $\underline{\gamma}$

In the sequel, we fix the field F_q, the integer $m(> 1)$ and the polynomial $f(x)$, and assume $f(x)$ is of degree n and irreducible over F_q. We fix a root of $f(x)$ in F_{q^n} and denote it by α. We let p be the characteristic of F_q.

In this section, we study how the behavior of an $(f(x), m)$-interleaved sequence $u_{\underline{\gamma}}$ is related to the tuple $\underline{\gamma}$, then based on our study we make a correction to an error made in Theorem 1 (i) [1]. Here by the (periodic) correlation function between two sequences $u_{\underline{\gamma}_1}$ and $u_{\underline{\gamma}_2}$ ($\underline{\gamma}_i \in F_{q^n}^m, i = 1, 2$) we mean

$$c_{\underline{\gamma}_1,\underline{\gamma}_2}(\tau) = \sum_{0 \le k < m(q^n-1)} \chi(u_{\underline{\gamma}_1}(k) - u_{\underline{\gamma}_2}(k+\tau)), \ \chi(x) = e^{i\frac{2\pi Tr_p^q(x)}{p}}.$$

When $\underline{\gamma}_1 = \underline{\gamma}_2 = \underline{\gamma}$, we write simply $c_{\underline{\gamma}_1,\underline{\gamma}_2}(\tau) = c_{\underline{\gamma}}(\tau)$, which is considered in [1]. All the other concepts are defined in [2],[10] or [11]. In this paper, we identify the m-tuple space $F_{q^n}^m$ to the F_{q^n}-module $F_{q^n}[x]/(x^m - \alpha)$ by the following map:

$$\underline{\gamma} = (\gamma_0, \gamma_1, \cdots, \gamma_{m-1}) \mapsto \underline{\gamma}(x) = \sum_{0 \le i < m} \gamma_i x^{m-i-1}. \tag{1}$$

The following lemma is basic in studying $u_{\underline{\gamma}}$.

Lemma 1. *Given $\underline{\gamma} = (\gamma_0, \gamma_1, \cdots, \gamma_{m-1}) \in F_{q^n}^m$, correspondingly define a sequence over F_{q^n} as follows:*

$$\varphi_{\underline{\gamma}} = \{\gamma_t\}_{t=0}^{\infty}, \ \gamma_{im+j} = \gamma_j \alpha^i, \ 0 \le j < m, 0 \le i \tag{2}$$

Let $\varphi_{\underline{\gamma}}(x) = \sum_{k=0}^{\infty} \gamma_t x^{-1-t}$, which is the generating function of $\varphi_{\underline{\gamma}}$. Similarly, let $u_{\underline{\gamma}}(x)$ be the generating function of $u_{\underline{\gamma}}$. Then

$$u_{\underline{\gamma}}(x) = \sum_{k=0}^{n-1} \varphi_{\underline{\gamma}}(x)^{\sigma^k}, \ \varphi_{\underline{\gamma}}(x) = \frac{\gamma(x)}{x^m - \alpha}, \ \text{and} \ \varphi_{\underline{\gamma}}^{\sigma^k}(x) = \frac{\gamma(x)^{\sigma^k}}{(x^m - \alpha)^{\sigma^k}},$$

where σ^k means the map which raises the coefficients of the power series or the polynomials to q^k-th power.

Proof. The first equation is a simple consequence of the definition of $u_{\underline{\gamma}}$ and $\varphi_{\underline{\gamma}}$:

$$\begin{aligned} u_{\underline{\gamma}}(x) &= \sum_{0\le j<m}\sum_{i\ge 0} Tr_q^{q^n}(\gamma_j\alpha^i)x^{-1-j-im} \\ &= \sum_{0\le k<n}\sum_{0\le j<m}\sum_{i\ge 0}\gamma_j\alpha^{iq^k}x^{-1-j-im} = \sum_{0\le k<n}\varphi_{\underline{\gamma}}(x)^{\sigma^k}. \end{aligned}$$

The third one is a consequence of the second one as follows:

$$\begin{aligned} \varphi_{\underline{\gamma}}(x) &= \sum_{j=0}^{m-1}\sum_{i=0}^{\infty}\gamma_j\alpha^i x^{-1-j-im} \\ &= \sum_{j=0}^{m-1}\gamma_j x^{-1-j}\frac{1}{1-\alpha x^{-m}} = \frac{\underline{\gamma}(x)}{x^m-\alpha}. \end{aligned}$$

□

The following lemma is related to the characteristic polynomial $x^m-\alpha$ of the sequence $\varphi_{\underline{\gamma}}$. Its proof is omitted, as it is a simple fact in algebra.

Lemma 2. *Let*

$$T = per(f),\ m = tD,\ \text{and}\ D = p^a D_0,$$

where t is the integer such that each of the prime factors of t is a factor of T, $\gcd(D,T)=1$, and $\gcd(D_0,p)=1$. Then the polynomial $x^m-\alpha$ has the following factorization:

$$\begin{aligned} x^m-\alpha &= a(x)^{p^a}, \\ a(x) &= \prod_{1\le d\le D_0,\ d|D_0} a_d(x),\ \text{and} \\ a_d(x) &= \prod_{0\le i<D_0}\ \prod_{0\le j<t,0<i<d,(i,d)=1}(x-\beta^{1+Tj}\theta^{\frac{D_0 i}{d}}), \end{aligned}$$

where $a(x)$ has no multiple roots, β is an element of order tT and $\beta^m=\alpha$, and θ is an element of order D_0.

In particular, $per(x^m-\alpha)=mT$, and the period of any non-constant factor of $x^m-\alpha$ is of the form $tTd, d|D$. □

Theorem 1. *1. $L^{\tau}u_{\underline{\gamma}} = u_{x^{\tau}\underline{\gamma}}$, where L is the left shift operator acting on the sequences, $x^{\tau}\underline{\gamma}$ is the tuple corresponding the polynomial $x^{\tau}\underline{\gamma}(x) \pmod{x^m-\alpha}$ of degree less than m.*

2. The minimal polynomial of $u_{\underline{\gamma}}$ is $\prod_{i=0}^{n-1}h(x)^{\sigma^i}$, where $h(x)$ is the minimal polynomial of $\varphi_{\underline{\gamma}}$.

3. *$LS(u_{\underline{\gamma}}) = nLS(\varphi_{\underline{\gamma}}) \leq mn$.*
4. *Using the notation from Lemma 2, we have*

$$per(u_{\underline{\gamma}}) = per(\varphi_{\underline{\gamma}})$$

and

$$per(u_{\underline{\gamma}}) = dtT \ if \ \underline{\gamma} \neq 0,$$

where

$$d = LCM(per(\underline{\gamma}_0), \cdots, per(\underline{\gamma}_{t-1})),$$
$$\underline{\gamma}_i = (\gamma_i, \alpha^{-e}\gamma_{i+t}, \cdots, \alpha^{-(D-1)e}\gamma_{i+(D-1)t}) \in F_{q^n}^m, 0 \leq i < t, \text{ and}$$
$$eD \equiv 1 \pmod{T}.$$

5. *When $f(x)$ is primitive, i.e., $per(f(x)) = q^n - 1$, we have*

$$c_{\underline{\gamma}_1, \underline{\gamma}_2}(\tau) = -m + (m - w_H(\underline{\gamma}_1 - x^\tau \underline{\gamma}_2))q^n$$

for any $\tau \geq 0$, where $\underline{\gamma}_1 - x^\tau \underline{\gamma}_2$ is the tuple corresponding the polynomial $\underline{\gamma}_1(x) - x^\tau \underline{\gamma}_2(x) \pmod{x^m - \alpha}$, and $w_H(\underline{\gamma})$ is the Hamming weight of $\underline{\gamma}$.

Proof. From Lemma 1, we have

$$\begin{aligned} L^\tau u_{\underline{\gamma}}(x) &= \sum_{0 \leq k < n} [(L^\tau \varphi_{\underline{\gamma}})(x)]^{\sigma^k} \\ &= \sum_{0 \leq k < n} [\frac{x^\tau \underline{\gamma}(x) \pmod{x^m - \alpha}}{x^m - \alpha}]^{\sigma^k} (\text{ see [12] }) \\ &= \sum_{0 \leq k < n} [\varphi_{x^\tau \underline{\gamma}}(x)]^{\sigma^k} = u_{x^\tau \underline{\gamma}}(x), \end{aligned}$$

which proves item 1.

Noticing that all the polynomials $(x^m - \alpha)^{\sigma^k}$ $(0 \leq k < n)$ have the same period and are coprime to each other, we get the item 2, 3, and $per(u_{\underline{\gamma}}) = per(\varphi_{\underline{\gamma}})$ follows by Lemma 1 easily. It is clear from Lemma 2 that if $\underline{\gamma} \neq 0$, then $per(\varphi_{\underline{\gamma}})$ is of the form $tTd, d|D$. In order to complete the proof of item 4, it is sufficient to prove that

$$per(\varphi_{\underline{\gamma}}) \mid dtT \longleftrightarrow per(\underline{\gamma}_i) \mid d \ \forall 0 \leq i < t. \tag{3}$$

Let $A_\epsilon = \{\gamma_{\epsilon+itT}\}_{i=0}^\infty, 0 \leq \epsilon < tT$, be the tT-decimated sequences of $\varphi_{\underline{\gamma}}$, then

$$per(A_\epsilon)|D, \tag{4}$$

since $per(\varphi_{\underline{\gamma}})|tTD$. It is clear that

$$per(\varphi_{\underline{\gamma}}) \mid dtT \longleftrightarrow per(A_\epsilon) \mid d \ \forall \ 0 \leq \epsilon < tT. \tag{5}$$

Assume $eD = 1 - bT$, and let $B_\epsilon = \{B_\epsilon(i) = \gamma_{\epsilon+ibtT}\}_{i=0}^{\infty}$ be the b-decimated sequence of A_ϵ. Then

$$per(A_\epsilon) = per(B_\epsilon), 0 \le \epsilon < tT, \tag{6}$$

since $\gcd(b, per(A_\epsilon)) = 1$. We notice that

$$B_\epsilon(i) = \gamma_{\epsilon+it(1-eD)} = \gamma_{\epsilon+it-iem} = \alpha^{-ie}\gamma_{\epsilon+it}$$

from (2), such that

$$B_\delta = (\underline{\gamma}_\delta, \underline{\gamma}_\delta, \cdots, \underline{\gamma}_\delta, \cdots), 0 \le \delta < t.$$

Hence

$$per(B_\delta) = per(\underline{\gamma}_\delta), 0 \le \delta < t. \tag{7}$$

Write $\epsilon = \delta + t\lambda$, $0 \le \delta < t$; then

$$\alpha^{-e\lambda} B_\epsilon(i) = \alpha^{-e(i+\lambda)}\gamma_{\delta+t(i+\lambda)} = B_\delta(i+\lambda),$$

such that

$$per(B_\epsilon) = per(B_\delta), \ if \ \epsilon \equiv \delta \pmod t. \tag{8}$$

The equations (5),(6),(7) and (8) lead to the desired equivalence (3).

Let $\underline{\beta} = \underline{\gamma}_1 - x^\tau \underline{\gamma}_2 = (\beta_0, \beta_1, \cdots, \beta_{m-1})$. It is clear that

$$u_{\underline{\gamma}_1} - L^\tau u_{\underline{\gamma}_2} = u_{\underline{\gamma}_1 - x^\tau \underline{\gamma}_2}$$

from Lemma 1, so we have

$$\begin{aligned} c_{\underline{\gamma}_1,\underline{\gamma}_2}(\tau) = c_{\underline{\gamma}_1 - x^\tau \underline{\gamma}_2}(0) &= \sum_{0\le j<m} \sum_{0\le i<q^n-1} \chi(Tr_q^{q^n}(\beta_j \alpha^i)) \\ &= - \sum_{0\le j<m, \beta_j \ne 0} 1 + \sum_{0\le j<m,\beta_j=0} (q^n - 1) \\ &= -w_H(\underline{\beta}) + (m - w_H(\underline{\beta}))(q^n - 1) \\ &= -m + (m - w_H(\underline{\gamma}_1 - x^\tau \underline{\gamma}_2))q^n, \end{aligned}$$

which proves item 5.

□

We get some information about the distribution of the linear spans and the periods of the $(f(x), m)$-interleaved sequences from Theorem 1 easily, as is shown in the following corollary.

Corollary 1.

1. *Let N be the total number of $(f(x), m)$-interleaved sequences whose linear spans equal mn. Then*

$$N = \prod\nolimits_{d|D_0} (q^{np^a n_d} - q^{n(p^a-1)n_d})^{\frac{t\Phi(d)}{n_d}} \ge (q^n - 1)^m.$$

Here $\Phi(\cdot)$ is the Euler function, and n_d is the smallest positive integer such that $q^{nn_d} \equiv 1 \pmod{tTd}$.

2. *Let M_r be the total number of non-zero $(f(x), m)$-interleaved sequences whose periods equal r. Then $M_r \neq 0$ if and only if $r = tTd, d \mid D$. Moreover, $M_{tT} = q^{nt} - 1$, and*

$$M_{tTd} = q^{ntd} - \sum_{1 \le i \le l} q^{\frac{dnt}{p_i}} + \sum_{1 \le i < j \le l} q^{\frac{dnt}{p_i p_j}} + \cdots + (-1)^l q^{\frac{dnt}{p_1 \cdots p_l}}$$

when $1 < d = p_1^{e_1} \cdots p_l^{e_l}$, where p_i are distinct primes, and $e_i > 0\ \forall i$.

Proof.

1. It is clear that different $\varphi_{\underline{\gamma}}$ correspond to different $u_{\underline{\gamma}}$, so N_{mn} is exactly the number of all possible sequences $\varphi_{\underline{\gamma}}$ with $LS(\varphi_{\underline{\gamma}}) = m$ from Theorem 1. With notation as in Lemma 2, it is clear that the polynomial $a_d(x)$ has a factorization over F_{q^n} of the form

$$a_d(x) = \prod_{1 \le j \le \frac{t\phi(d)}{n_d}} f_{d,j}(x),$$

where each $f_{d,j}(x)$ is irreducible of degree n_d over F_{q^n}, and $f_{d,j}(x) \neq f_{d,j'}(x)$ if $j \neq j'$. Therefore, the sequence $\varphi_{\underline{\gamma}}(x)$ has a decomposition of the form:

$$\varphi_{\underline{\gamma}}(x) = \frac{\underline{\gamma}(x)}{x^m - \alpha} = \sum_{d \mid D_0} \sum_{1 \le j \le \frac{t\phi(d)}{n_d}} \frac{h_{d,j}(x)}{f_{d,j}(x)^{p^a}}, \tag{9}$$

$$\text{where} \quad h_{d,j}(x) \in F_{q^n}[x], \text{ and } deg(h_{d,j}(x)) < p^a n_d \tag{10}$$

by Lemma 1 and Lemma 2. We see that $LS(\varphi_{\underline{\gamma}}(x)) = m$ if and only if $f_{d,j}(x)$ is not a factor of $h_{d,j}(x)$ for all $d \mid D_0, 1 \le j \le \frac{t\phi(d)}{n_d}$ by (9) and (10). Hence we get

$$\begin{aligned} N &= \prod_{d \mid D_0} (q^{nn_d p^a} - q^{nn_d p^a - nn_d})^{\frac{t\Phi(d)}{n_d}} \\ &\ge \prod_{d \mid D_0} [(q^{nn_d} - 1)^{p^a - 1}(q^{nn_d} - 1)]^{\frac{t\Phi(d)}{n_d}} \\ &= \prod_{d \mid D_0} (q^{nn_d} - 1)^{\frac{t\Phi(d) \cdot p^a}{n_d}} \ge (q^n - 1)^{tp^a \sum_{d \mid D_0} \Phi(d)} = (q^n - 1)^m. \end{aligned}$$

2. Let $\mathcal{M}_d$ be the set of all $(f(x), m)$-interleaved sequences of periods dividing tTd. From Theorem 1, we see $u_{\underline{\gamma}} \in \mathcal{M}_d$ if and only if $per(\underline{\gamma}_i) \mid d, \forall 0 \le i < t$, so

$$|\mathcal{M}_d| = q^{ntd}, \text{and } |\mathcal{M}_d| - 1 = \sum_{r \mid d} M_r.$$

Hence $M_{tT} = q^{nt} - 1$. For $d > 1$, using the Möbius inversion formula [13] and the fact that $\sum_{r|d} \mu(r) = 0$ for $d > 1$, we get

$$M_{tTd} = \sum_{r|d} (|\mathcal{M}_{\frac{d}{r}}| - 1)\mu(r) = \sum_{r|d} |\mathcal{M}_{\frac{d}{r}}|\mu(r),$$

which leads to the desired result.

□

Remark 1. From Theorem 1 we see that Theorem 1 (i) in [1], which claims that " If $w_H(\underline{\gamma}) < m$, then $per(u_{\underline{\gamma}}) = mT$, $T = per(f(x))$", is not true in general. In fact, Theorem 1 (i) in [1] should be corrected to be the following statement, using the same notation as in Lemma 2.

Statement: Assume $w_H(\underline{\gamma}) < m$. Then $per(u_{\underline{\gamma}}) = mT$ if and only if "$m = t$" or "$m = D$ is a prime."

Proof. It is easy to check the "if" part. If the "only if" part is not true, then one has either "$m = tD$ with both $t > 1$ and $D > 1$" or "$m = D$ is not a prime". In the first case, take $\underline{\gamma}(x) = \sum_{j=0}^{D-1} \alpha^{ej} x^{m-1-tj}$, whence $w_H(\underline{\gamma}) = D < m$. But, with the notation as in Theorem 1, we have $\underline{\gamma}_0 = (1, 1, \cdots, 1)$, $\underline{\gamma}_j = \underline{0}\ \forall 0 < j < t$, and $per(\underline{\gamma}_j) = 1\ \forall 0 \le j < t$. Hence $per(u_{\underline{\gamma}}) = tT < mT$ from Theorem 1, a contradiction. In the second case, we may assume $m = D = bc, b > 1, c > 1$. Take $\underline{\gamma}(x) = \sum_{0 \le j < c} \alpha^{ebj} x^{m-1-bj}$; then $w_H(\underline{\gamma}) < m$. But $\underline{\gamma}_0 = (1, 1, \cdots, 1)$, $per(\underline{\gamma}_j) = 1\ \forall 0 \le j < t = 1$, and hence $per(u_{\underline{\gamma}}) = tT < mT$ from Theorem 1, a contradiction again.

□

3 Sequences $u_{\underline{\gamma}}$ with Optimal Correlation

In this section, we assume $f(x)$ is primitive, *i.e.*, $per(f(x)) = q^n - 1$. We will show that, among all the $(f(x), m)$-interleaved sequences, the sequences with the optimal correlation must have the maximal period. Moreover, we will give a way to construct some sequences with the optimal correlation when $2m$ is less than the period of $f(x)$.

For $\underline{\gamma} \in F_{q^n}^m$, define

$$Max(c_{\underline{\gamma}}) = max\{\ |c_{\underline{\gamma}}(\tau)| \mid 0 < \tau < m(q^n - 1)\ \}.$$

For $\underline{\gamma}_i \in F_{q^n}^m (i = 1, 2)$, $\underline{\gamma}_1 \neq \underline{\gamma}_2$, define

$$Max(c_{\underline{\gamma}_1, \underline{\gamma}_2}) = max\{\ |c_{\underline{\gamma}_1, \underline{\gamma}_2}(\tau)| \mid 0 \le \tau < m(q^n - 1)\ \}.$$

Lemma 3. *1. (a)*

$$\sum_{0<\tau<m(q^n-1)} [m - w_H(\underline{\gamma} - x^\tau \underline{\gamma})] = w_H(\underline{\gamma})^2 + (m - w_H(\underline{\gamma}))^2 (q^n - 1) - m.$$

(b) *The conditions "$m - w_H(\underline{\gamma} - x^{rm}\underline{\gamma}) \leq 1 \; \forall \; 0 < r < q^n - 1$" hold if and only if $w_H(\underline{\gamma}) \geq m - 1$.*

(c) *Assume $w_H(\underline{\gamma}) \geq m - 1$. For any given s, $0 < s < m$, the conditions "$m - w_H(\underline{\gamma} - x^{rm+s}\underline{\gamma}) \leq 1 \; \forall 0 \leq r < q^n - 1$" hold if and only if*

$$\gamma_i\gamma_{i+s}^{-1} \neq \gamma_j\gamma_{j+s}^{-1} \; \forall 0 \leq i < j < m,$$

whenever $\gamma_i\gamma_{i+s}\gamma_j\gamma_{j+s} \neq 0$.

2. *Assume $\underline{\gamma}_1 \neq \underline{\gamma}_2$ and $w_2 \leq w_1$, where $w_1 = w_H(\underline{\gamma}_1), w_2 = w_H(\underline{\gamma}_2)$. We have*
(a)

$$\sum_{0 \leq \tau < m(q^n - 1)} [m - w_H(\underline{\gamma}_1 - x^\tau \underline{\gamma}_2)] = w_1 w_2 + (m - w_1)(m - w_2)(q^n - 1).$$

(b) *The conditions "$m - w_H(\underline{\gamma}_1 - x^{rm+s}\underline{\gamma}_2) \leq 1 \; \forall \; 0 \leq r < q^n - 1, \; 0 \leq s < m$" hold if and only if $w_1 = m$ and*

$$\gamma_{1,i}\gamma_{2,i+s}^{-1} \neq \gamma_{1,j}\gamma_{2,j+s}^{-1} \; \forall 0 \leq i < j < m, \; 0 \leq s < m,$$

whenever $\gamma_{2,i+s}\gamma_{2,j+s} \neq 0$, where $\underline{\gamma}_i = (\gamma_{i,0}, \gamma_{i,1}, \cdots, \gamma_{i,m-1}), i = 1, 2$.

Proof. All the parts are easy to check, and hence we omit their proof except for the part that the conditions "$m - w_H(\underline{\gamma}_1 - x^{rm+s}\underline{\gamma}_2) \leq 1 \; \forall \; 0 \leq r < q^n - 1, \; 0 \leq s < m$" imply "$w_1 = m$". In fact, if $w_1 \neq m$, then we have $w_2 \leq w_1 \leq m - 1$. There exist s_1 and s_2, such that

$$x^{s_1}\underline{\gamma}_1 = (0, \rho_1, \rho_2, \cdots, \rho_{m-1}), \rho_1 \neq 0$$

and

$$x^{s_2}\underline{\gamma}_2 = (0, \theta_1, \theta_2, \cdots, \theta_{m-1}), \theta_1 \neq 0.$$

Take r such that $\rho_1 - \alpha^r\theta_1 = 0$, and take $\tau = rm + s_2 - s_1$. Then

$$w_H(\underline{\gamma}_1 - x^\tau \underline{\gamma}_2) = w_H(x^{s_1}\underline{\gamma}_1 - \alpha^r x^{s_2}\underline{\gamma}_2) \leq m - 2,$$

a contradiction. □

Theorem 2. *Assume $q^n \geq 2m$. Then*

1. (a) *$Max(c_{\underline{\gamma}}) \geq -m + q^n$.*
(b) *$Max(c_{\underline{\gamma}}) = -m + q^n$ if and only if $w_H(\underline{\gamma}) \geq m - 1$ and*

$$\gamma_i\gamma_{i+s}^{-1} \neq \gamma_j\gamma_{j+s}^{-1} \; \forall 0 \leq i < j < m, \; 0 < s < m, \tag{11}$$

whenever $\gamma_i\gamma_{i+s}\gamma_j\gamma_{j+s} \neq 0$.

(c) *If $Max(c_{\underline{\gamma}}) = -m + q^n$, then the correlation function $c_{\underline{\gamma}}$ takes on the following two values for $0 < \tau < m(q^n - 1)$:*

$$\begin{array}{ll} -m + q^n & \text{occurs } m^2 - m \text{ times, and} \\ -m & \text{occurs } mq^n - m^2 - 1 \text{ times} \end{array}$$

when $w_H(\underline{\gamma}) = m$; *and*

$$\begin{array}{ll} -m + q^n & \textit{occurs } q^n - 2 + (m-1)(m-2) \textit{ times, and} \\ -m & \textit{occurs } (m-1)(q^n-1) - (m-1)(m-2) \textit{ times} \end{array}$$

when $w_H(\underline{\gamma}) = m - 1$.

2. *Assume* $\underline{\gamma}_1 \neq \underline{\gamma}_2$, *and assume* $w_H(\underline{\gamma}_1) \geq w_H(\underline{\gamma}_2)$. *Then*
   (a) $Max(c_{\underline{\gamma}_1,\underline{\gamma}_2}) \geq -m + q^n$
   (b) $Max(c_{\underline{\gamma}_1,\underline{\gamma}_2}) = -m + q^n$ *if and only if* $w_H(\underline{\gamma}_1) = m$, *and*

$$\gamma_{1,i}\gamma_{2,i+s}^{-1} \neq \gamma_{1,j}\gamma_{2,j+s}^{-1}, \ \forall 0 \leq i < j < m, \ 0 \leq s < m, \tag{12}$$

   whenever $\gamma_{2,i+s}\gamma_{2,j+s} \neq 0$, *where* $\underline{\gamma}_i = (\gamma_{i,0}, \gamma_{i,1}, \cdots, \gamma_{i,m-1}), i = 1, 2$.
   (c) *If* $Max(c_{\underline{\gamma}_1,\underline{\gamma}_2}) = -m + q^n$, *then the cross correlation function* $c_{\underline{\gamma}_1,\underline{\gamma}_2}$ *takes on the following two values:*

$$\begin{array}{ll} -m + q^n & \textit{occurs } m w_H(\underline{\gamma}_2) \textit{ times, and} \\ -m & \textit{occurs } m(q^n-1) - m w_H(\underline{\gamma}_2) \textit{ times.} \end{array}$$

3. (a) *If* $Max(c_{\underline{\gamma}}) = -m + q^n$, *then* $per(u_{\underline{\gamma}}) = m(q^n - 1)$.
   (b) *Assume* $\underline{\gamma}_1 \neq \underline{\gamma}_2$. *If* $Max(c_{\underline{\gamma}_1,\underline{\gamma}_2}) = -m + q^n$, *then* $u_{\underline{\gamma}_1}$ *is not cyclically equivalent to* $u_{\underline{\gamma}_2}$.

Proof. Both of items 1(a) and 1(c) are consequences of Theorem 1(4) and Lemma 3, 1(a). Item 1(b) is a consequence of Theorem 1(4), and Lemma 3, 1(b) and 1(c).

Both of items 2(a) and 2(c) are consequences of Theorem 1(4) and Lemma 3, 2(a). Item 2(b) is a consequence of Theorem 1(4) and Lemma 3, 2(b).

If item 3(a) is not true, then $per(u_{\underline{\gamma}}) = \tau < m(q^n - 1)$ for some τ, hence $c_{\underline{\gamma}}(\tau) = m(q^n - 1) > -m + q^n = Max(c_{\underline{\gamma}})$, a contradiction. If item 3(b) is not true, then $u_{\underline{\gamma}_2} = L^\tau u_{\underline{\gamma}_1}$ for some τ. Thus

$$c_{\underline{\gamma}_1,\underline{\gamma}_2}(\tau) = m(q^n - 1) > -m + q^n = Max(c_{\underline{\gamma}_1,\underline{\gamma}_2}),$$

a contradiction.

□

Theorem 3. *Suppose we construct* N *integer tuples as follows:*

$$\underline{d}_j = (d_{j,0} = 0, d_{j,1}, \cdots, d_{j,m-1}), 1 \leq j \leq N, \tag{13}$$

such that they satisfy the following conditions:

$$0 < d_{j,i} < d_{j,i'} < d_{j',i} < d_{j',i'} \ \forall 0 < i < i' < m, 1 \leq j < j' \leq N, \ \textit{and}$$
$$\sum_{0<i<m} d_{N,i} < (q^n - 1)/2.$$

For example,

$$\underline{d}_j = (0, (j-1)(m-1)+1, (j-1)(m-1)+2, \cdots, (j-1)(m-1)+m-1), 1 \le j \le N,$$

and N is the largest integer such that

$$N \le \frac{q^n - 3 - \epsilon}{2(m-1)^2} + \frac{m-2}{2(m-1)}$$

where $\epsilon = 0$ *if* $q \equiv 1 \pmod 2$, *and* $\epsilon \equiv 1$ *if* $q \equiv 0 \pmod 2$. *Correspondingly, let* $\underline{\gamma}_j \in F_{q^n}^m$ *be determined by* $\underline{d}_j$ *as follows:*

$$\underline{\gamma}_j = (\gamma_{j,0}, \gamma_{j,1}, \cdots, \gamma_{j,m-1}), \text{ and}$$
$$\gamma_{j,i} = \alpha^{a_{j,i}}, a_{j,i} = \sum_{0 \le k \le i} d_{j,k}.$$

Then

$$Max(c_{\underline{\gamma}_i}) = Max(c_{\underline{\gamma}_j, \underline{\gamma}_k}) = -m + q^n \; \forall 1 \le i \le N, \; 1 \le j \ne k \le N, \text{ and}$$

they attain the minimum among all the $(f(x), m)$*-interleaved sequences. Consequently, the constructed sequences have the optimal correlation values and the maximal period among all the* $(f(x), m)$*-interleaved sequences, moreover they are not cyclically equivalent to each other.*

Proof. Write $a_{j,k+s} = a_{j,k+s-m} + 1$ if $m \le k+s < m+s$. For any $1 \le j \le N$, it is easy to check ($T = q^n - 1$) that

$$-T/2 < a_{j,k'+s} - a_{j,k'} < a_{j,k+s} - a_{j,k} < a_{j,i+s} - a_{j,i} < a_{j,i'+s} - a_{j,i'} < T/2, \quad (14)$$

if

$$s \le i+s < i'+s < m \le k+s < k'+s < m+s, \; 0 < s < m.$$

The conditions (14) are equivalent to the conditions (11), hence $Max(c_{\underline{\gamma}_j}) = q^n - m$ from Theorem 2. For any $1 \le j < k \le N$, it is easy to check that

$$-(q^n-1)/2 < a_{k,i} - a_{j,i+s} < a_{k,l} - a_{j,l+s} < (q^n-1)/2, \quad (15)$$

if

$$0 \le i < l < m, \; 0 \le s < m.$$

The conditions (15) are equivalent to the conditions (12). Hence $Max(c_{\underline{\gamma}_j, \underline{\gamma}_k}) = q^n - m$ from Theorem 2. □

Acknowledgments The authors would like to thank Tor Helleseth and Douglas R. Stinson, who help them to improve the English writing of this paper.

References

1. G. Gong, "Theory and application of q-ary interleaved sequences," *IEEE Trans. Inform. Theory,* Vol. 41, No. 2, pp. 400–411, March, 1995.
2. R. Lidl and H. Niederreiter, "Finite Fields," Addison-Wesley, 1983.
3. S.M. Jennings, "Multiplexed sequences: some properties of minimum polynomial," in *Proc. EUROCRYPT'82,* Lecture Notes in Computer Science, Vol. 149. New York: Springer-Verlag, 1983.
4. D. Gollmann and W.G. Chambers, "Clock-controlled shift register: a review," *IEEE J. Selected Areas Commun.,* Vol. 7, No. 4, pp. 525–533, May 1989.
5. R.A. Scholtz and L.R. Welch, "GMW sequences," *IEEE Trans. Inform. Theory,* Vol. IT-30, No. 3, pp. 548–553, May 1984.
6. A.H. Chan and M. Goresky, "Cascaded GMW sequences," *IEEE Trans. Inform. Theory,* Vol. 39, pp. 177–183, Jan. 1993.
7. J.S. No and P.V. Kumar, "A new family of binary pseudorandom from sequences having optimal periodic correlation properties and large linear span," *IEEE Trans. Inform. Theory,* Vol. 35, No. 2, pp. 371-379, March 1989.
8. A.H. Chan and R.A. Games, "On the linear span of binary sequences from finite geometries, q odd," in *Proc. Crypto'86,* pp. 405–417, Springer Verlag, 1987.
9. A.H. Chan and R.A. Games, " On the linear span of binary sequences obtained from Q-ary M-sequences, Q odd," *IEEE Trans. Inform. Theory.* Vol. 36, No. 3, pp. 548–522, May 1990.
10. Z.X. Wan, "Algebra and Coding Theory" (in Chinese), Beijing: Science Press, 1979.
11. C. Ding and G. Xiao, "Stream Ciphers and Their Applications" (in Chinese), Beijing: The National Defense Industry Press, 1994.
12. Z.D. Dai, X.N. Feng, M.L. Liu and Z.X. Wan, "Nonlinear Feedforwad Sequences of m-Sequences I," *Discrete Mathematics* 123 (1993), pp.17–34.
13. L. G. Hua, "Introduction of Number Theory" (in Chinese), Beijing: Science Press, 1975.

A New Algorithm for the k-Error Linear Complexity of Sequences over $GF(p^m)$ with Period p^n

Takayasu Kaida[1], Satoshi Uehara[2], and Kyoki Imamura[2]

[1] Department of Information and Electronics Engineering,
Yatsushiro National College of Technology
Yatsushiro, Kumamoto, 866-8501 JAPAN. Email: kaida@as.yatsushiro-nct.ac.jp
[2] Department of Computer Science and Electronics, Kyushu Institute of Technology
Iizuka, Fukuoka, 820-8502 JAPAN. Email: {uehara, imamura}@cse.kyutech.ac.jp

Abstract. A new algorithm is given for the k-error linear complexity of sequences over $GF(p^m)$ with period p^n, p a prime. The algorithm is different from the previous one recently given by the authors in the following two points and can be regarded as generalization of the Stamp-Martin algorithm for the k-LC of binary sequences with period 2^n. First the value of k decreases at each iteration. Secondly the error vector for the k-LC can be determined at the same time when the k-LC is obtained. The key ideas of the algorithm are "shift" and "offset" of the cost matrix which are introduced by the authors to derive the Stamp-Martin algorithm for binary sequences from our previous algorithm.

1 Introduction

The linear complexity (LC) has been used as a useful measure for the randomness of sequences for reasons of the existence of good algorithms for computing the LC [1–3] and the simple model for generating the sequence. However the LC has such an instable property as an extreme increase or decrease by only one-symbol change within one period [4–6].

Stamp and Martin[7] defined the k-error LC (k-LC) of a periodic sequence as the smallest LC that can be obtained when any k or fewer symbols are changed within one period, which is effective to reduce the instability of the LC caused by symbol substitutions. The sphere complexity studied by Ding et.al.[8] is essentially the same as the k-LC. Although the introduction of the sphere complexity is earlier than that of the k-LC, we will use the k-LC because of the following two reasons; our preference for using the term LC and the existence of the algorithm for the k-LC of binary sequences with period 2^n (Stamp-Martin algorithm[7]).

Recently the authors[9–11] gave an algorithm for the k-LC of sequences over $GF(p^m)$ with period p^n, p a prime. This algorithm can find not only the k-LC but also an error vector which gives the k-LC. It uses the generalized version of the Games-Chan algorithm for computing the LC of sequences over $GF(p^m)$ with period p^n [8, 12, 13], and consists of n iterations for sequences with period p^n. In our previous algorithm[9–11] the value of k is fixed throughout all the iterations, and after finding the k-LC the determination of an error vector can be performed by reading the stored data in the reverse order from the last iteration to the first iteration.

In this paper we will give another algorithm for the k-LC of sequences over $GF(p^m)$ with period p^n. Our new algorithm can be said to be the generalization of the Stamp-Martin algorithm[7] for the binary sequences with period 2^n, since the value of k decreases as the proceeding of the iterations. Another feature of our new algorithm is that an error vector can be obtained at the same time when the k-LC is obtained. The key ideas of our new algorithm are "shift" and "offset" of the cost matrix which were introduced by the authors[10] to derive the Stamp-Martin algorithm from our previous algorithm in case of binary sequences.

2 Games-Chan Algorithm for the LC

In this paper we consider sequences over $GF(q)$ with period p^n, where $q = p^m$, p is a prime and n, m are positive integers. One period of a sequence $\{a_i\} = \{a_0, a_1, \cdots\}$ is written as $\boldsymbol{a}^{(N)} = (a_0^{(N)}, a_1^{(N)}, \cdots, a_{N-1}^{(N)})$, where $N = p^n$ is the period of $\{a_i\}$. In the case of a sequence $\{a_i\}$ with period $N = p^n$, the LC of $\{a_i\}$ denoted as $LC(\{a_i\})$ is defined as $LC(\{a_i\}) = L$ if

$$\sum_{i \geq 0} a_i x^{-(i+1)} = \frac{a(x)^{(N)}}{x^N - 1} = \frac{g(x)}{(x-1)^L}, \tag{1}$$

where $a(x)^{(N)} = a_0^{(N)} x^{N-1} + a_1^{(N)} x^{N-2} + \cdots + a_{N-1}^{(N)}$ and $g(1) \neq 0$. Since $x^N - 1 = (x-1)^N$, we can calculate $LC(\{a_i\})$ by determining the order Z of zero of $a(x)^{(N)}$ at $x = 1$. From (1), we have $LC(\{a_i\}) = N - Z$. Then the generalized Games-Chan algorithm for the LC of a sequence $\{a_i\}$ over $GF(p^m)$ with period p^n determines the order $Z = z_0 + z_1 p + \cdots + z_{n-1} p^{n-1}$ of zero at $x = 1$ in the order from z_{n-1} to z_0, where $z_i \in \{0, 1, \cdots, p-1\}$ for $i = 0, 1, \cdots, n-1$. Let $M = N/p$ and $\boldsymbol{a}(j)^{(N)} = (a_{jM}^{(N)}, a_{jM+1}^{(N)}, \cdots, a_{(j+1)M-1}^{(N)})$ be an M-tuple for $j = 0, 1, \cdots, p-1$. We will rewrite $\boldsymbol{a}^{(N)} = (\boldsymbol{a}(0)^{(N)}, \boldsymbol{a}(1)^{(N)}, \cdots, \boldsymbol{a}(p-1)^{(N)})$. We show the generalized Games-Chan algorithm as follows[8, 13].

[Generalized Games-Chan Algorithm]

(i) Initial values: $N = pM = p^n$, $LC = 0$, $\boldsymbol{a}^{(N)} = (a_0^{(N)}, a_1^{(N)}, \cdots, a_{N-1}^{(N)})$.

(ii) Repeat the following 0)∼2) until $M = 1$.

0). From a given pM-tuple $\boldsymbol{a}^{(pM)} = (\boldsymbol{a}(0)^{(pM)}, \cdots, \boldsymbol{a}(p-1)^{(pM)})$ over $GF(q)$, compute the following M-tuples over $GF(q)$,

$$\begin{aligned} \boldsymbol{b}(u)^{(M)} &= F_u(\boldsymbol{a}(0)^{(pM)}, \cdots, \boldsymbol{a}(p-1)^{(pM)}) \\ &= \sum_{j=0}^{p-u-1} c_{u,j} \boldsymbol{a}(j)^{(pM)} \end{aligned} \tag{2}$$

for $u = 0, 1, \cdots, p-2$, where

$$F_u(x_0, \cdots, x_{p-1}) = \sum_{j=0}^{p-u-1} c_{u,j} x_j \tag{3}$$

and $c_{u,j} = \binom{p-j-1}{u}$ is the binomial coefficient.

1). Choose one of the following p cases.

$$\begin{aligned} &\textbf{Case } 1 : \boldsymbol{b}(0)^{(M)} = \cdots = \boldsymbol{b}(p-2)^{(M)} = \boldsymbol{0}. \\ &\textbf{Case } w : \boldsymbol{b}(0)^{(M)} = \cdots = \boldsymbol{b}(p-w-1)^{(M)} = \boldsymbol{0}, \\ &\qquad \boldsymbol{b}(p-w)^{(M)} \neq \boldsymbol{0} \text{ for } w = 2, 3, \cdots, p-1. \\ &\textbf{Case } p : \boldsymbol{b}(0)^{(M)} \neq \boldsymbol{0}. \end{aligned} \tag{4}$$

2). If **Case** w is chosen
then $\boldsymbol{a}^{(M)} \leftarrow F_{p-w}(\boldsymbol{a}(0)^{(pM)}, \boldsymbol{a}(1)^{(pM)}, \cdots, \boldsymbol{a}(p-1)^{(pM)})$,
$LC \leftarrow LC + (w-1)M$, $M \leftarrow M/p$ and go to 0).

(iii). If $\boldsymbol{a}^{(1)} \neq (0)$ then $LC \leftarrow LC + 1$.

The final LC is equal to $\mathrm{LC}(\{a_i\})$.

We note that the Case w at iteration M corresponds to the case where $(x^M - 1)^{p-w}$ divides $a(x)^{(pM)}$ but $(x^M - 1)^{p-w+1}$ does not. The formulas (2) and (3) are given in [14].

3 Generalization of Stamp-Martin Algorithm

The k-LC of a sequence $\{a_i\}$ over $GF(q)$ with period N is defined as

$$k\text{-}LC(\{a_i\}) = \min\{LC(\{a_i + e_i\}) | w_H(\boldsymbol{e}) \leq k\}, \tag{5}$$

where $\{e_i\}$ is a sequence over $GF(q)$ with period N and $w_H(\boldsymbol{e})$ is the Hamming weight of the first N-tuple $\boldsymbol{e} = (e_0, e_1, \cdots, e_{N-1})$ of $\{e_i\}$, i.e. $w_H(\boldsymbol{e}) = \#\{i \in \{0, 1, \cdots, N-1\} | e_i \neq 0\}$, where $\#A$ is the cardinality of a set A. We don't need that N is the minimum period, i.e. if N_0 is the minimum period of $\{e_i\}$, N is a multiple of N_0. If a sequence $\{e_i\}$ over $GF(q)$ with period N whose Hamming weight is less than or equal to k gives k-LC$(\{a_i\})$, i.e. k-$LC(\{a_i\}) = LC(\{a_i + e_i\})$, then we call $\{e_i\}$ an error sequence and N-tuple $\boldsymbol{e} = (e_0, e_1, \cdots, e_{N-1})$ an error vector. In the case of $N = p^n$ over $GF(p^m)$, we can apply the generalized Games-Chan algorithm for the LC. However we must repeatedly apply the generalized Games-Chan algorithm at the worst case

$$\sum_{i=0}^{k} (q-1)^i \binom{N}{i} \tag{6}$$

times to the sequences $\{a_i + e_i\}$'s with different $\boldsymbol{e}$'s, if we have no effective algorithm for the k-LC. Unfortunately (6) becomes very large even for moderate N and k.

In this paper we will use the same logic and cost matrices $AC(M)$ and $BC(M)$ as those in our previous algorithms[9–11] except that the following two operations, i.e. "shift" and "offset" will be introduced in the computation of $AC(M)$.

Let $GF(q) = \{\alpha_0 = 0, \alpha_1, \cdots, \alpha_{q-1}\}$ and $\boldsymbol{b}(u)^{(M)} = (b(u)_0^{(M)}, b(u)_1^{(M)}, \cdots, b(u)_{M-1}^{(M)})$. Firstly the cost of $\boldsymbol{a}^{(M)}$ at iteration M is denoted as a $q \times M$ matrix

$$AC(M) = [A(h,i)_M], 0 \le h \le q-1; 0 \le i \le M-1, \tag{7}$$

where $A(h,i)_M$ is the minimum number of changes in the initial N-tuple $\boldsymbol{a}^{(N)}$ necessary and sufficient for changing $a_i^{(M)}$ into $a_i^{(M)} + \alpha_h$ under the condition that forcing **Case** w to happen is not altered. Secondly the cost of $\boldsymbol{b}(u)^{(M)}$ is denoted as a $(p-1) \times M$ matrix

$$BC(M) = [B(u,i)_M], 0 \le u \le p-2; 0 \le i \le M-1, \tag{8}$$

where $B(u,i)_M$ is the minimum number of changes in the initial N-tuple $\boldsymbol{a}^{(N)}$ necessary and sufficient for making $b(0)_i^{(M)} = b(1)_i^{(M)} = \cdots = b(u)_i^{(M)} = 0$. We define the total cost of $\boldsymbol{b}(u)^{(M)}$'s for $u = 0, 1, \cdots, p-2$ as

$$TB(u)_M = \sum_{i=0}^{M-1} B(u,i)_M \tag{9}$$

which means the minimum number of changes in the initial N-tuple $\boldsymbol{a}^{(N)}$ necessary and sufficient for making $\boldsymbol{b}(0)^{(M)} = \boldsymbol{b}(1)^{(M)} = \cdots = \boldsymbol{b}(u)^{(M)} = \boldsymbol{0}$. In iteration (ii) of the generalized Games-Chan algorithm we can force **Case** w to happen:

$$\begin{aligned}
&\textbf{Case } 1 \Leftrightarrow TB(p-2)_M \le k, \\
&\textbf{Case } w \Leftrightarrow TB(p-w-1)_M \le k < TB(p-w)_M \quad \text{for } w = 2, 3, \cdots, p-1, \\
&\textbf{Case } p \Leftrightarrow k < TB(0)_M.
\end{aligned} \tag{10}$$

The initial value of the cost of $\boldsymbol{a}^{(N)}$ denoted as $AC(N) = [A(h,i)_N]$, where $N = p^n$ and

$$A(h,i)_N = \begin{cases} 0 & \text{if} \quad h = 0, \\ 1 & \text{if} \quad h \ne 0 \end{cases} \tag{11}$$

is obvious from the definition. We can compute $BC(M)$ and $AC(M)$ from $AC(pM)$ in the following way for $M = p^{n-1}, p^{n-2}, \cdots, p^0$ in the order. The cost of $\boldsymbol{b}(u)^{(M)}$ denoted as $BC(M) = [B(u,i)_M]$ can be computed from the previous cost $AC(pM) = [A(h,i)_{pM}]$ by

$$B(u,i)_M = \min \left\{ \sum_{j=0}^{p-1} A(e_j, jM+i)_{pM} \,\middle|\, \boldsymbol{e} \in D(u,i)_M \right\}, \tag{12}$$

where $\boldsymbol{e} = (e_0, e_1, \cdots, e_{p-1}) \in [GF(q)]^p$ and

$$D(u,i)_M = \{\boldsymbol{e} | F_s(e_0, e_1, \cdots, e_{p-1}) + b(s)_i^{(M)} = 0, \quad s = 0, 1, \cdots, u\} \tag{13}$$

is the set of all the $\mathbf{e}$'s which can make

$$b(0)_i^{(M)} = b(1)_i^{(M)} = \cdots = b(u)_i^{(M)} = 0. \tag{14}$$

The computation of $AC(M)$ from $AC(pM)$ depends on the case chosen at iteration (ii) of the generalized Games-Chan algorithm. Here we will introduce "shift" and "offset" in computing $AC(M)$. The shift is used to make

$$A(0,i)_M \leq A(h,i)_M \quad \text{for } h = 1,2,\cdots,q-1; i = 0,1,\cdots,M-1. \tag{15}$$

The offset defined as

$$A(h,i)_M \leftarrow A(h,i)_M - A(0,i)_M \tag{16}$$

for $h = 0,1,\cdots,q-1$; $i = 0,1,\cdots,M-1$ is used to make $A(0,i)_M = 0$ for $i = 0,1,\cdots,M-1$ after the shift.

If we can force **Case** w to happen, we record the change vector of $\mathbf{a}^{(M)}$ as pM-tuple

$$\mathbf{c}_0(M) = (C(0,0)_M, C(0,1)_M, \cdots, C(0,pM-1)_M), \tag{17}$$

such that

$$\sum_{j=0}^{p-1} A(C(0,jM+i)_M, jM+i)_{pM} = B(p-w-1,i)_M \tag{18}$$

if $w = 1,2,\cdots,p-1$ and

$$\sum_{j=0}^{p-1} A(C(0,jM+i)_M, jM+i)_{pM} = \min\left\{ \sum_{j=0}^{p-1} A(e_j, jM+i)_{pM} \,\middle|\, \mathbf{e} \in [GF(q)]^p \right\} \tag{19}$$

if $w = p$, i.e., $C(0,jM+i)_M$ gives a change value of $a_{jM+i}^{(pM)}$ with the minimum number of changes in the initial N-tuple $\mathbf{a}^{(N)}$ under the condition that forcing **Case** w to happen is not altered at iteration (ii) of the generalized Games-Chan algorithm. We also compute the change matrix $C(M) = [C(h,i)_M]$ such that

$$A(h,i)_M = \sum_{j=0}^{p-1} A(C(h,jM+i)_M, jM+i)_{pM}. \tag{20}$$

First the shift is performed by changing the possible region $D(h,i)_M^w$ of $\mathbf{e}$ used in computing

$$A(h,i)_M = \min\left\{ \sum_{j=0}^{p-1} A(e_j, jM+i)_{pM} \,\middle|\, \mathbf{e} \in \hat{D}(h,i)_M^w \right\} \tag{21}$$

by changing as

$$\hat{D}(h,i)_M^w = \left\{ \boldsymbol{e} \;\middle|\; \begin{array}{l} F_s(e_0,\cdots,e_{p-1}) + b(s)_i^{(M)} = 0, \quad s = 0,1,\cdots,p-w-1 \\ F_{p-w}(e_0,\cdots,e_{p-1}) \\ \quad = F_{p-w}(C(0,i)_M,\cdots,C(0,(p-1)M+i)_M) + \alpha_h \end{array} \right\} \tag{22}$$

if $w = 1,2,\cdots,p-1$ and

$$\hat{D}(h,i)_M^p = \{\boldsymbol{e}|F_0(e_0,\cdots,e_{p-1}) = F_0(C(0,i)_M,\cdots,C(0,(p-1)M+i)_M) + \alpha_h\} \tag{23}$$

if $w = p$ and

$$\boldsymbol{a}^{(M)} \leftarrow F_{p-w}(\boldsymbol{a}(0)^{(pM)} + \boldsymbol{c}(0)^{(M)},\cdots,\boldsymbol{a}(p-1)^{(pM)} + \boldsymbol{c}(p-1)^{(M)}), \tag{24}$$

where $\boldsymbol{c}(j)^{(M)} = (C(0,jM)_M, C(0,jM+1)_M,\cdots,C(0,(j+1)M-1)_M)$ for $j = 0,1,\cdots,p-1$. Note that in the last relation of the definition of $\hat{D}(h,i)_M^w$, $\alpha_h + F_{p-w}(\boldsymbol{c}_0(M))$ is used instead of α_h. Since we have $\boldsymbol{c}_0(M) \in \hat{D}(0,i)_M^w$ and $\boldsymbol{c}_0(M) \notin \hat{D}(h,i)_M^w$ for $h \neq 0$, we have (15) from (18),(19) and (12),(13).

In the case of $w \neq p$ we have shown $B(p-w-1,i)_M = A(0,i)_M$ from (12) and (22). Therefore

$$\sum_{i=0}^{M-1} A(0,i)_M = TB(p-w-1)_M \leq k \tag{25}$$

from (9) and (10). Therefore in order to keep the condition until the iteration M, we need at least (25) changes in the initial N-tuple $\boldsymbol{a}^{(N)}$. After the offset of $AC(M)$, we replace k by $k - TB(p-w-1)_M$.

Finally in order to determine an error vector we compute the error matrix at $\boldsymbol{a}^{(M)}$ as

$$E(M) = [E(h,i)_M], 0 \leq h \leq q-1; 0 \leq i \leq N-1 = p^n - 1. \tag{26}$$

The definition of $E(h,i)_M$ is as follows. In order to change $a_i^{(M)}$ $(0 \leq i \leq M-1)$ by α_h under the condition that the happening of **Case** w at iteration M is not altered, we change N/M elements $a_{jM+i}^{(N)}$ $(0 \leq j \leq \frac{N}{M}-1)$ by $E(h,jM+i)_M$ in the original sequence $\boldsymbol{a}^{(N)}$ of length $N = p^n$.

Since such a change of $a_i^{(M)}$ by α_h can be obtained by changing $a_{jM+i}^{(pM)}$ $(0 \leq j \leq p-1)$ by $C(h,jM+i)_M$ from (20), we have

$$E(h,p\ell M + jM + i)_M = E(C(h,jM+i)_M, p\ell M + jM + i)_{pM} \tag{27}$$

for $h = 0,1,\cdots,q-1; j = 0,1,\cdots,p-1; \ell = 0,1,\cdots,\frac{N}{pM}-1; i = 0,1,\cdots,M-1$. The initial value of the error matrix $E(N) = [E(h,i)_N]$, where $E(h,i)_N = \alpha_h$ for $i = 0,1,\cdots,N-1$ is obvious from the definition.

4 Proposed Algorithm

After above preparation, we propose the generalized Stamp-Martin algorithm with computing an error vector.

[**Generalized Stamp-Martin Algorithm**]

(i) Initial values: $N = pM = p^n$, $k\text{-}LC = 0$, $\boldsymbol{a}^{(N)} = (a_0^{(N)}, a_1^{(N)}, \cdots, a_{N-1}^{(N)})$,

$$AC(N) = \begin{bmatrix} 0\,0\cdots 0 \\ 1\,1\cdots 1 \\ \vdots\;\vdots\quad\;\vdots \\ 1\,1\cdots 1 \end{bmatrix}, \quad E(N) = \begin{bmatrix} \alpha_0 & \alpha_0 & \cdots & \alpha_0 \\ \alpha_1 & \alpha_1 & \cdots & \alpha_1 \\ \vdots & \vdots & & \vdots \\ \alpha_{q-1} & \alpha_{q-1} & \cdots & \alpha_{q-1} \end{bmatrix}.$$

(ii) Repeat the following 0)∼2) until $M = 1$.

0). Compute $BC(M)$ and $TB(0)_M, \cdots, TB(p-2)_M$ by (2),(3),(8), (9) and (12).

1). Choose one of the following p cases.

Case 1: if $TB(p-2)_M \le k$,
Case w: if $TB(p-w-1)_M \le k < TB(p-w)_M$ for $w = 2, 3, \cdots, p-1$,
Case p: if $k < TB(0)_M$, respectively.

2). If **Case** w is chosen,
then compute $\boldsymbol{a}^{(M)}$ by (24) and $k\text{-}LC \leftarrow k\text{-}LC + (w-1)M$.
If $w \neq p$ then $k \leftarrow k - TB(p-w-1)_M$.
First consider the shift operation by computing $AC(M)$ and $C(M)$ from (21)-(23) and (20).
Next consider the offset operation (16) by computing $AC(M) = [A'(h,i)_M]$, which is the final $AC(M)$ at iteration M.
Compute $E(M)$ by (27) from $E(pM)$ and $C(M)$.
If $M \neq 1$ then $M \leftarrow M/p$ and go to 0).

(iii). Let $\boldsymbol{a}^{(1)} = (a_0^{(1)}) = (\alpha_s)$, $\alpha_t = -\alpha_s$.
If $A(t,0)_1 > k$ then $k\text{-}LC \leftarrow k\text{-}LC + 1$ and
$\boldsymbol{e} = (E(0,0)^{(1)}, E(0,1)^{(1)}, \cdots, E(0, N-1)^{(1)})$,
else $\boldsymbol{e} = (E(t,0)^{(1)}, E(t,1)^{(1)}, \cdots, E(t, N-1)^{(1)})$.

The final $k\text{-}LC$ is equal to k-LC$(\{a_i\})$ and $\boldsymbol{e}$ is an error vector.

Our proposed algorithm with shift and offset operations of cost matrices needs only one path for the computation of the k-LC and an error vector, and does not need to keep the data of $AC(M)$, $BC(M)$, $C(M)$ and $E(M)$ after the calculation of $AC(M/p)$, $BC(M/p)$, $C(M/p)$ and $E(M/p)$. It is not necessary to keep the first row of $AC(M)$.

5 Complexity Analysis of the Algorithms

In this section we will briefly compare the time-complexity and the space-complexity between our two algorithms; the previous one[11] called the generalized k-LC algorithm and the new one called the generalized Stamp-Martin algorithm presented in this paper. Our previous algorithm can be summarized as follows.

[**Generalized k-LC algorithm**]

(i) Initial Values: $N = pM = p^n$, k-LC$= 0$, $\boldsymbol{a}^{(N)} = (a_0^{(N)}, a_1^{(N)}, \cdots, a_{N-1}^{(N)})$,

$$AC(N) = [A(h,i)_N] = \begin{bmatrix} 0\,0 \cdots 0 \\ 1\,1 \cdots 1 \\ \vdots \;\vdots \;\vdots \;\vdots \\ 1\,1 \cdots 1 \end{bmatrix}$$

(ii) Repeat (0)–(6) until $M = 1$.

(0) Compute $BC(M) = [B(u,i)_M]$ for $(0 \le u \le p-2,\ 0 \le i \le M-1)$ by

$$B(u,i)_M = \min\left\{ \sum_{j=0}^{p-1} A(e_j, i+jM)_{pM} \,\middle|\, \boldsymbol{e} \in D(u,i)_M \right\}, \tag{28}$$

where $\boldsymbol{e} = (e_0, e_1, \cdots, e_{p-1}) \in [GF(q)]^p$,

$$D(u,i)_M = \left\{ \boldsymbol{e} \mid F_j(e_0, e_1, \cdots, e_{p-1}) + b_{j,i}^{(M)} = 0, \quad (0 \le j \le u) \right\}, \tag{29}$$

$$\begin{aligned} \boldsymbol{b}(u)^{(M)} &= (b_{u,0}\ (M), b_{u,1}\ (M), \cdots, b_{u,M-1}\ (M)) \\ &= F_u(\boldsymbol{a}(0)^{(pM)}, \boldsymbol{a}(1)^{(pM)}, \cdots, \boldsymbol{a}(p-1)^{(pM)}) \\ &= \sum_{j=0}^{p-u-1} c_{u,j} \boldsymbol{a}(j)^{(pM)}. \end{aligned} \tag{30}$$

(1) Compute $TB(0)_M, TB(1)_M, \cdots, TB(p-2)_M$ from

$$TB(u)_M = \sum_{i=0}^{M-1} B(u,i)_M. \tag{31}$$

(2) Here one of the following p cases should be chosen:

$$\begin{cases} \text{Case } 1 \Leftarrow TB(p-2)_M \le k \\ \text{Case } w \Leftarrow TB(p-w-1)_M \le k < TB(p-w)_M \\ \text{Case } p \Leftarrow k < TB(0)_M \end{cases} \tag{32}$$

(3) If Case w is chosen, then compute $\boldsymbol{a}^M$ and k-LC from

$$\boldsymbol{a}^{(M)} = F_{p-w}(\boldsymbol{a}(0)^{(pM)}, \boldsymbol{a}(1)^{(pM)}, \cdots, \boldsymbol{a}(p-1)^{(pM)}) \tag{33}$$

and

$$k\text{-}LC \leftarrow k\text{-}LC + (w-1)M. \tag{34}$$

(4) compute $AC(M) = [AC(h,i)_M]$ for $0 \le h \le q-1,\ 0 \le i \le M-1$ from

$$A(h,i)_M = \min\left\{ \sum_{j=0}^{p-1} A(e_j, i+jM)_{pM} \,\middle|\, \boldsymbol{e} \in \hat{D}(h,i)_M^w \right\}, \tag{35}$$

where

$$\hat{D}^w_M = \left\{ e \,\middle|\, \begin{array}{l} F_j(e_0, e_1, \cdots, e_{p-1}) + b^{(M)}_{j,i} = 0 \\ (0 \le j \le p-2) \\ F_{p-w}(e_0, e_1, \cdots, e_{p-1}) - \alpha_h = 0 \end{array} \right\} \quad (36)$$

if $1 \le w \le p-1$ or

$$\hat{D}^p_M = \{ e \mid F_0(e_0, e_1, \cdots, e_{p-1}) - \alpha_h = 0 \} \quad (37)$$

if $w = p$.

(5) Let $e(h,i)_M$ in $E(M) = [e(h,i)_M]$ for $0 \le h \le q-1$, $0 \le i \le M-1$ be e in right hand side of (35) when $A(h,i)_M$ is calculated at the iteration (4).

(6) If $M \ne 1$, then let $M \leftarrow M/p$ and go to (0).

(iii) If $A(-a_0^{(1)}, 0)_1 > k$, then $k\text{-}LC \leftarrow k\text{-}LC + 1$

(iv) Let

$$\begin{cases} A(-a_0^{(1)}, 0)_1 \le k \Rightarrow e(1) = e(-a_0^{(1)}, 0)_1, \\ A(-a_0^{(1)}, 0)_1 > k \Rightarrow e(1) = e(s, 0)_1, \end{cases}$$

where s is such that $A(s,0)_1 = \min\{A(h,0)_1 \mid 0 \le h \le q-1\}$.

(v) Compute $e(pM) = (e_0^{(pM)}, e_1^{(pM)}, \cdots, e_{pM-1}^{(pM)})$ from $e(M) = (e_0^{(M)}, e_1^{(M)}, \cdots, e_{M-1}^{(M)})$ and $E(pM)$ by

$$e_{i+jM}^{(pM)} = e(e_{i,j}, i+jM)_{pM} \quad (0 \le j \le p-1,\ 0 \le i \le M-1), \quad (38)$$

where $e_i^{(M)} = (e_{i,0}, e_{i,1}, \cdots, e_{i,p-1})$.

(vi) Let $M \leftarrow pM$.
If $M \ne p^{n-1}$, then go to (v).

(vii) Compute $e = (e_0, e_1, \cdots, e_{N-1})$ by

$$e_{i+jM} = e_{i,j} \quad (0 \le i \le M-1,\ 0 \le j \le p-1). \quad (39)$$

The iterations from (i) to (iii) give k-LC and the iterations from (iv) to (vii) give an error vector at above algorithm.

5.1 Evaluation of the Time-complexity

We will count the number of additions and subtractions in the worst case in order to evaluate the time-complexity. Between the two algorithms the number of computations is the same for $BC(M)$ and $TB(u)_M$ ($u = 0, 1, \cdots, p-2$); the former is $pM \times \dfrac{p^{m(p-1)} - 1}{p^m - 1}$ and the latter $(p-1)M$. The number of computations for $AC(M)$ is $p^{m+1}M$ in the generalized k-LC algorithm and $p^{(n-1)m+1}M$ in the generalized Stamp-Martin algorithm. The generalized Stamp-Martin algorithm needs $p^n M$ offsets and n changes of value of k. The summation of the above values for $M = p^{n-1}, p^{n-2}, \cdots, p^0$ gives the time-complexity of the order $O(p^{(p-1)m+n})$ for both of the two algorithms. More precisely the number of the computations in the generalized Stamp-Martin algorithm is about half of that in the generalized k-LC algorithm.

5.2 Evaluation of the Space-complexity

We will count the number of necessary memories in order to evaluate the space-complexity. The computation of $AC(M)$ needs $p^m(p+1)p^{n-1}$ memories in both of the two algorithms. The generalized k-LC algorithm needs p^{m+1+i} memories for the computation of $E(p^i)$ $(i = 0, 1, \cdots, p-1)$, which results in $p^{M+1} \times \dfrac{p^n - 1}{p-1}$ memories in total. The generalized Stamp-Martin algorithm needs $2p^{m+n}$ memories for the computation of $E(p^n)$ and p^{m+n} memories for the computation of $C(M)$. Therefore both of the two algorithms have the space-complexity of the same order $O(p^{m+n})$. More precisely the number of the necessary memories for the generalized k-LC algorithm is about half of that for the generalized Stamp-Martin algorithm.

6 Example

In this section we show a small example for 3-LC, i.e. $k = 3$, of a sequence over $GF(3^2)$ with period 3^3. Let $\{a_i\}$ be a sequence over

$$GF(3^2) = \{\alpha_0 = \alpha^* = 0, \alpha_1 = \alpha^0 = 1, \alpha_2 = \alpha^1, \cdots, \alpha_8 = \alpha^7\} \tag{40}$$

with period $3^3 = 27$, where α is a primitive element of $GF(9)$ such that $\alpha^2 = 2\alpha + 1$. We will denote α^*, α^i as $*$, i, respectively for simplicity. Let one period of $\{a_i\}$ be

$$\boldsymbol{a}^{(27)} = (012\ 345\ 670\ 012\ 345\ 671\ 012\ 345\ 672). \tag{41}$$

We will compute 3-LC $(k = 3)$ and an error vector as follows.

[A Numerical Example]

(initial values)
$\boldsymbol{a}^{(27)} = (012\ 345\ 670\ 012\ 345\ 671\ 012\ 345\ 672)$

$$AC(27) = \begin{bmatrix} 000\ 000\ 000\ 000\ 000\ 000\ 000\ 000\ 000 \\ 111\ 111\ 111\ 111\ 111\ 111\ 111\ 111\ 111 \\ 111\ 111\ 111\ 111\ 111\ 111\ 111\ 111\ 111 \\ 111\ 111\ 111\ 111\ 111\ 111\ 111\ 111\ 111 \\ 111\ 111\ 111\ 111\ 111\ 111\ 111\ 111\ 111 \\ 111\ 111\ 111\ 111\ 111\ 111\ 111\ 111\ 111 \\ 111\ 111\ 111\ 111\ 111\ 111\ 111\ 111\ 111 \\ 111\ 111\ 111\ 111\ 111\ 111\ 111\ 111\ 111 \\ 111\ 111\ 111\ 111\ 111\ 111\ 111\ 111\ 111 \end{bmatrix}$$

$$E(27) = \begin{bmatrix} *** & *** & *** & *** & *** & *** & *** & *** & *** \\ 000 & 000 & 000 & 000 & 000 & 000 & 000 & 000 & 000 \\ 111 & 111 & 111 & 111 & 111 & 111 & 111 & 111 & 111 \\ 222 & 222 & 222 & 222 & 222 & 222 & 222 & 222 & 222 \\ 333 & 333 & 333 & 333 & 333 & 333 & 333 & 333 & 333 \\ 444 & 444 & 444 & 444 & 444 & 444 & 444 & 444 & 444 \\ 555 & 555 & 555 & 555 & 555 & 555 & 555 & 555 & 555 \\ 666 & 666 & 666 & 666 & 666 & 666 & 666 & 666 & 666 \\ 777 & 777 & 777 & 777 & 777 & 777 & 777 & 777 & 777 \end{bmatrix}$$

$(M = 9)$

$\boldsymbol{b}(0)^{(9)} = (*** \; *** \; **4) \qquad \boldsymbol{b}(1)^{(9)} = (*** \; *** \; **6)$

$$BC(9) = \begin{bmatrix} 000\,000\,001 \\ 000\,000\,002 \end{bmatrix} \quad \begin{matrix} TB(0)_9 = 1 \\ TB(1)_9 = 2 \end{matrix}$$

We choose **Case 1**, since $TB(1)_9 = 2 \le k = 3$.

$\boldsymbol{a}^{(9)} = (012\ 345\ 670)$

$k\text{-}LC \leftarrow 0 \quad k \leftarrow 3 - 2 = 1$

$$AC(9) = \begin{bmatrix} 000\,000\,000 \\ 333\,333\,331 \\ 333\,333\,331 \\ 333\,333\,331 \\ 333\,333\,331 \\ 333\,333\,331 \\ 333\,333\,330 \\ 333\,333\,330 \\ 333\,333\,331 \end{bmatrix}$$

$$E(9) = \begin{bmatrix} *** & *** & *** & *** & *** & **2 & *** & *** & **1 \\ 000 & 000 & 000 & 000 & 000 & 003 & 000 & 000 & 007 \\ 111 & 111 & 111 & 111 & 111 & 110 & 111 & 111 & 115 \\ 222 & 222 & 222 & 222 & 222 & 226 & 222 & 222 & 220 \\ 333 & 333 & 333 & 333 & 333 & 331 & 333 & 333 & 334 \\ 444 & 444 & 444 & 444 & 444 & 445 & 444 & 444 & 446 \\ 555 & 555 & 555 & 555 & 555 & 557 & 555 & 555 & 55* \\ 666 & 666 & 666 & 666 & 666 & 66* & 666 & 666 & 663 \\ 777 & 777 & 777 & 777 & 777 & 774 & 777 & 777 & 772 \end{bmatrix}$$

$(M = 3)$

$\boldsymbol{b}(0)^{(3)} = (4\ 5\ 6) \quad \boldsymbol{b}(1)^{(3)} = (2\ 3\ 4)$

$$BC(3) = \begin{bmatrix} 3\,3\,1 \\ 6\,6\,3 \end{bmatrix} \quad \begin{matrix} TB(0)_3 = 7 \\ TB(1)_3 = 15 \end{matrix}$$

We choose **Case 3**, since $k = 1 < TB(0)_3 = 7$.

$\boldsymbol{a}^{(3)} = (4\ 5\ 6)$

$k\text{-}LC \leftarrow 0 + 2 \times 3 = 6 \quad k \leftarrow 1$

$$AC(3) = \begin{bmatrix} 0\,0\,0 \\ 3\,3\,1 \\ 3\,3\,1 \\ 3\,3\,1 \\ 3\,3\,1 \\ 3\,3\,1 \\ 3\,3\,0 \\ 3\,3\,0 \\ 3\,3\,1 \end{bmatrix} \quad E(3) = \begin{bmatrix} \ast\ast\ast\ast\ast\ast\ast\ast\ast\ast\ast\ast\ast\ast\ast\ast\ast 2 \ast\ast\ast\ast\ast\ast\ast\ast 1 \\ \ast\ast\ast\ast\ast\ast\ 000\ \ast\ast\ast\ast\ast\ast\ 003\ \ast\ast\ast\ast\ast\ast\ 007 \\ \ast\ast\ast\ast\ast\ast\ 111\ \ast\ast\ast\ast\ast\ast\ 110\ \ast\ast\ast\ast\ast\ast\ 115 \\ \ast\ast\ast\ast\ast\ast\ 222\ \ast\ast\ast\ast\ast\ast\ 226\ \ast\ast\ast\ast\ast\ast\ 220 \\ \ast\ast\ast\ast\ast\ast\ 333\ \ast\ast\ast\ast\ast\ast\ 331\ \ast\ast\ast\ast\ast\ast\ 334 \\ \ast\ast\ast\ast\ast\ast\ 444\ \ast\ast\ast\ast\ast\ast\ 445\ \ast\ast\ast\ast\ast\ast\ 446 \\ \ast\ast\ast\ast\ast\ast\ 555\ \ast\ast\ast\ast\ast\ast\ 557\ \ast\ast\ast\ast\ast\ast\ 55\ast \\ \ast\ast\ast\ast\ast\ast\ 666\ \ast\ast\ast\ast\ast\ast\ 66\ast\ \ast\ast\ast\ast\ast\ast\ 663 \\ \ast\ast\ast\ast\ast\ast\ 777\ \ast\ast\ast\ast\ast\ast\ 774\ \ast\ast\ast\ast\ast\ast\ 772 \end{bmatrix}$$

$(M = 1)$
$b(0)^{(1)} = (0) \quad b(1)^{(1)} = (2)$

$$BC(1) = \begin{bmatrix} 1 \\ 3 \end{bmatrix} \quad \begin{matrix} TB(0)_1 = 1 \\ TB(1)_1 = 3 \end{matrix}$$

We choose **Case 2**, since $TB(0)_1 = 1 \le k = 1 < TB(1)_1 = 3$.
$a^{(1)} = (2) \quad k\text{-}LC \leftarrow 6 + 1 = 7 \quad k \leftarrow 1 - 1 = 0$

$$AC(1) = \begin{bmatrix} 0 \\ 2 \\ 2 \\ 2 \\ 3 \\ 2 \\ 2 \\ 2 \\ 3 \end{bmatrix} \quad E(1) = \begin{bmatrix} \ast\ast\ast\ast\ast\ast\ast\ast 4 \ast\ast\ast\ast\ast\ast\ast\ast 5 \ast\ast\ast\ast\ast\ast\ast\ast 6 \\ \ast\ast\ast\ast\ast\ast 4 \ast\ast\ast\ast\ast\ast\ast\ast 4 \ast 2 \ast\ast\ast\ast\ast\ast 4 \ast 1 \\ \ast\ast\ast\ast\ast\ast 5 \ast 6 \ast\ast\ast\ast\ast\ast 5 \ast\ast\ast\ast\ast\ast\ast\ast 5 \ast 3 \\ \ast\ast\ast\ast\ast\ast 6 \ast 5 \ast\ast\ast\ast\ast\ast 6 \ast 7 \ast\ast\ast\ast\ast\ast 6 \ast\ast \\ \ast\ast\ast\ast\ast\ast\ \ast 31\ \ast\ast\ast\ast\ast\ast\ \ast 30\ \ast\ast\ast\ast\ast\ast\ \ast 35 \\ \ast\ast\ast\ast\ast\ast\ \ast 4\ast\ \ast\ast\ast\ast\ast\ast\ \ast 42\ \ast\ast\ast\ast\ast\ast\ \ast 41 \\ \ast\ast\ast\ast\ast\ast\ \ast 56\ \ast\ast\ast\ast\ast\ast\ \ast 5\ast\ \ast\ast\ast\ast\ast\ast\ \ast 53 \\ \ast\ast\ast\ast\ast\ast\ \ast 65\ \ast\ast\ast\ast\ast\ast\ \ast 67\ \ast\ast\ast\ast\ast\ast\ \ast 6\ast \\ \ast\ast\ast\ast\ast\ast\ \ast 72\ \ast\ast\ast\ast\ast\ast\ \ast 76\ \ast\ast\ast\ast\ast\ast\ \ast 70 \end{bmatrix}$$

Since $a^{(1)} = (\alpha^2)$, $-\alpha^2 = \alpha^7$ and $A(8,0)_1 = 3 > k = 0$, then $k\text{-}LC \leftarrow 7+1 = 8$ and an error vector e is the first row of $E(1)$. Hence $k\text{-}LC(\{a_i\}) = 8$ and

$$e = (\ast\ast\ast\ \ast\ast\ast\ \ast\ast 4\ \ast\ast\ast\ \ast\ast\ast\ \ast\ast 5\ \ast\ast\ast\ \ast\ast\ast\ \ast\ast 6). \tag{42}$$

7 Conclusion

The new algorithm using shift and offset of the cost matrix $AC(M)$ in this paper is a generalization of the Stamp-Martin algorithm[7], since in case of binary sequences we can show that the part of computing the k-LC in our algorithm is the same as the Stamp-Martin algorithm.

We have interest in the extension of our algorithm to sequences with an arbitrary period not restricted to the power of a prime. For such a purpose the algorithm for the LC given by Blackburn[15] seems to be useful, since it computes the LC using the generalized Games-Chan algorithm and the discrete Fourier transform.

References

1. J.L. Massey, Shift register synthesis and BCH decoding, *IEEE Trans. Inform. Theory* **IT-15** (1969), 122–127.

2. W.H. Mills, Continued fractions and linear recurrences, *Math. of Computation*, **29** (1975), 173–180.
3. J.L. Massey and T. Schaub, Linear complexity in coding theory, *Lecture Notes in Computer Science* **331** (1986), 19-32.
4. Z. Dai and K. Imamura, Linear complexity for one-symbol substitution of a periodic sequence over $GF(q)$, *IEEE Trans. Inform. Theory* **44** (1998), 1328–1331.
5. S. Uehara and K. Imamura, Linear complexity of periodic sequences obtained from $GF(q)$ sequences with period $q^n - 1$ by one-symbol insertion, *IEICE Trans.* **E79-A** (1996), 1739–1740.
6. S. Uehara and K. Imamura, Linear complexity of periodic sequences obtained from a sequence over $GF(p)$ with period $p^n - 1$ by one-symbol deletion, *IEICE Trans.* **E80-A** (1997), 1164-1166.
7. M. Stamp and C.F. Martin, An algorithm for the k-error linear complexity of binary sequences with period 2^n, *IEEE Trans. Inform. Theory* **39** (1993), 1398–1401.
8. C. Ding, G. Xiao and W. Shan, "The Stability Theory of Stream Ciphers", Lecture Notes in Computer Science, **561**, Springer-Verlag, 1991.
9. T. Kaida, S. Uehara and K. Imamura, An algorithm for the k-error linear complexity of sequences over $GF(3)$ with period 3^n, *Proc. 1996 Int'l Symp. Inform. Theory and Its Applications* (1996), 155–158.
10. T. Kaida, S. Uehara and K. Imamura, Computation of the k-error linear complexity of binary sequences with period 2^n, J. Jaffar and R.H.C. Yap (eds.), *Concurrency and Parallelism, Programming, Networking, and Security*, Lecture Notes in Computer Science **1179** (1996), 182–191.
11. T. Kaida, S. Uehara and K. Imamura, An algorithm for the k-error linear complexity of sequences over $GF(p^m)$ with period p^n, p an odd prime, submitted to *Information and Computation* (1999).
12. R.A. Games and A.H. Chan, A fast algorithm for determining the complexity of a binary sequence with period 2^n, *IEEE Trans. Inform. Theory* **IT-29** (1983), 144–146.
13. K. Imamura and T. Moriuchi, A fast algorithm for determining the linear complexity of p-ary sequence with period p^n, p a prime, *IEICE Tech. Rept.* **IT93-75** (1993), 73–78 (in Japanese).
14. S. Uehara, K. Imamura and T. Kaida, Value distribution of linear complexity for p-ary periodic sequences with period p^n, p a prime, *IEICE Trans.* **E80-A** (1997), 920–921.
15. S.R. Blackburn, A generalisation of the discrete Fourier transform: determining the minimal polynomial of a periodic sequence, *IEEE Trans. Inform. Theory* **40** (1994), 1702–1704.

Sequences of I.I.D. Binary Random Variables Using Chaotic Dynamics

Tohru Kohda

Dept. of Computer Science and Communication Engineering,
Kyushu University,
Fukuoka 812-8581 Japan,
Email:kohda@csce.kyushu-u.ac.jp

Abstract. We have recently given a simple sufficient condition for a class of ergodic maps with some symmetric properties to produce a sequence of independent and identically distributed (i.i.d.) binary random variables. This condition is shown to consist of two simple symmetric properties: the *equidistributivity property* (or briefly *EDP*) and *constant summation property* (or briefly *CSP*).

1 Introduction

A sequence of i.i.d. binary random variables [1],[2] has found significant applications in modern digital communication systems, such as in spread spectrum (SS) communication systems [3] or cryptosystems [4]. For such binary sequences, LFSR (linear feedback shift register) sequences have often been used [5].

It is well known from probability theory [1], [2] that coin tossing has always been considered as the symbol of randomness. Randomness itself, however, is very difficult to define. In fact, throwing a die may be seen as both a deterministic and a random event. Both of the dyadic map and its associated binary function, called the Rademacher function, have been considered to realize coin tossing [1],[2]. Thus, it is an important problem whether any other maps than the dyadic map can realize coin tossing or not. In addition, there is another question: "What are their associated binary functions"?

We have recently given [6] a simple sufficient condition for some class of ergodic maps to produce a sequence of i.i.d. binary random variables. In this paper, we briefly review that the map and the associated binary functions satisfy two simple symmetric properties: the *equidistributivity property* (or briefly *EDP*) and *constant summation property* (or briefly *CSP*), respectively.

2 Bernoulli Shift and Rademacher Function

Let us start with reviewing fundamental subjects from the well-known textbooks of elementary probability theory [1],[2].

Let ω be a point drawn at random from the unit interval $(0,1)$. With each ω associate its nonterminating binary expansion

$$\omega = \sum_{k=1}^{\infty} \frac{d_k(\omega)}{2^k} = 0.d_1(\omega)d_2(\omega)\cdots, \text{ where } d_i(\omega) \in \{0,1\}. \tag{1}$$

Imagine now a coin with faces labeled 1 and 0 instead of the usual heads (H) and tails (T). If ω is drawn at random, $\{d_k(\omega)\}_{k=1}^{\infty}$ behaves as if it resulted from an infinite sequence of tosses of a fair coin. To see this, consider first the event that the first n tosses give the outcomes $u_1, u_2, \cdots, u_n$ in sequence. Then the set of ω for which $d_i(\omega) = u_i, i = 1, 2, \cdots, n$ is represented by

$$[\omega : d_i(\omega) = u_i, 1 \le i \le n] = \left(\sum_{i=1}^{n} \frac{u_i}{2^i}, \sum_{i=1}^{n} \frac{u_i}{2^i} + \sum_{i=n+1}^{\infty} \frac{1}{2^i} \right] \tag{2}$$

which is referred to as the dyadic interval.

On the other hand, define the map $\tau_B(\cdot)$, referred to as the *Bernoulli shift* or *dyadic map* by

$$\tau_B(\omega) = 2\omega, \quad \bmod 1 \tag{3}$$

which gives

$$\tau_B(\omega) = 0.d_2(\omega)d_3(\omega)\cdots. \tag{4}$$

This implies that $\tau_B(\cdot)$ shifts the digits one place to the left. Namely,

$$d_k(\tau_B(\omega)) = d_{k+1}(\omega), \quad \text{for } k \ge 1 \tag{5}$$

The functions $d_k(\omega), k = 1, 2, \cdots$ furnish us with a *model* of independent tosses of a "fair" coin as follows. The argument becomes simpler if the $d_k(\omega)$ are replaced by the *Rademacher functions* [1],[2],

$$r_k(\omega) = 1 - 2d_k(\omega), \quad k = 1, 2, \cdots. \tag{6}$$

For any c_k, the function $\sum_{k=1}^{n} c_k r_k(\omega)$ satisfies the following equation [1]

$$\int_0^1 \exp[i \sum_{k=1}^{n} c_k r_k(\omega)] d\omega = \prod_{k=1}^{n} \int_0^1 \exp[i c_k r_k(\omega)] d\omega \tag{7}$$

which implies that $r_k(\omega)$ (and hence $d_k(\omega)$), $k = 1, 2, \cdots$ are "independent random variables". Figure 1 shows the binary functions $d_k(\omega)$ defined by Rademacher functions $r_k(\omega)$.

3 EDP and CSP

In the application of chaotic dynamics to cryptosystem, the dyadic map is impracticable with the help of a computer with its limited accuracy in the sense

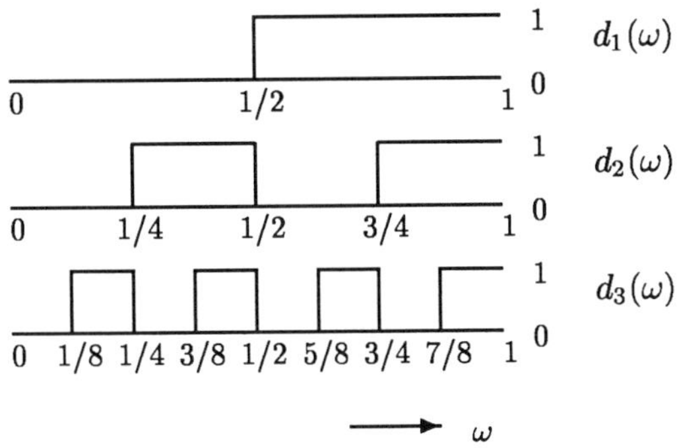

Fig. 1. Binary functions $d_k(\omega)$ defined by Rademacher functions $r_k(\omega)$.

that the period of its chaotic orbit is very short. Of crucial importance is the question whether any other maps than the dyadic map can realize coin tossing or not.

To discuss this problem, let us start with giving three simple methods [7],[8] for obtaining binary sequences from chaotic real-valued sequences $\{\omega_n\}_{n=0}^{\infty}$, generated by an ergodic map $\tau(\cdot) : I = [d, e] \to I$

$$\omega_{n+1} = \tau(\omega_n),\ \omega_n = \tau^n(\omega_0) \in I,\ n = 1, 2, \cdots. \tag{8}$$

Method-1: We define a threshold function $\Theta_t(\omega)$ as

$$\Theta_t(\omega) = \begin{cases} 0 \text{ for } \omega < t \\ 1 \text{ for } \omega \geq t \end{cases} \tag{9}$$

and define its complementary function

$$\overline{\Theta}_t(\omega) = 1 - \Theta_t(\omega). \tag{10}$$

Using these functions, we can obtain a binary sequence $\{\Theta_t(\tau^n(\omega))\}_{n=0}^{\infty}$, which is referred to as a *chaotic threshold sequence.*

Method-2: We write the value of ω ($|\omega| \leq 1$) in a binary representation:

$$|\omega| = 0.A_1(\omega)A_2(\omega)\cdots A_i(\omega)\cdots, \text{ where } A_i(\omega) \in \{0, 1\}. \tag{11}$$

The i-th bit $A_i(\omega)$ can be expressed as

$$A_i(\omega) = \sum_{r=1}^{2^i}(-1)^r \left\{\Theta_{\frac{r}{2^i}}(\omega) + \overline{\Theta}_{-\frac{r}{2^i}}(\omega)\right\}. \tag{12}$$

Thus we can obtain a binary sequence $\{A_i(\omega_n)\}_{n=0}^{\infty}$ which we call a *chaotic bit sequence.* Since $\Theta_t(\omega)$ can be regarded as a Boolean function whose variable, seed ω, is not binary but real-valued, $A_i(\omega)$ can be rewritten by

$$A_i(\omega) = \bigoplus_{r=1}^{2^i} \left\{ \Theta_{-\frac{r}{2^i}}(\omega) \oplus \Theta_{\frac{r}{2^i}}(\omega) \right\} \tag{13}$$

where $\oplus$ denotes modulo 2 addition.

Furthermore, for any map $\tau(\cdot)$ defined on the interval $[d, e]$, we can give a generalized version of the second method [6] as follows.

Method-3: We define a binary function with the set of thresholds, $T = \{t_r\}_{r=0}^{N}$, defined by

$$C_T(\omega) = \bigoplus_{r=0}^{N} \Theta_{t_r}(\omega) = \sum_{r=0}^{N} (-1)^r \Theta_{t_r}(\omega). \tag{14}$$

This function $C_T(\omega)$ gives the following two simple examples.

Example 1 of $C_T(\omega)$: When $N = 0$, $C_T(\omega) = \Theta_{t_0}(\omega)$.

Example 2 of $C_T(\omega)$: We write the value of $\dfrac{\omega - d}{e - d} \in [0, 1]$ in a binary representation:

$$\frac{\omega - d}{e - d} = 0.B_1(\omega)B_2(\omega)\cdots B_i(\omega)\cdots, \text{ where } B_i(\omega) \in \{0, 1\} \tag{15}$$

which implies that

$$C_T(\omega) = B_i(\omega), \text{ when } N = 2^i, t_r = (e - d)\frac{r}{2^i} + d. \tag{16}$$

If the interval $I = [0, 1]$, then $A_i(\omega) = B_i(\omega)$.

Thus each of $\{A_i(\tau^n(\omega))\}_{n=0}^{\infty}$ and $\{B_i(\tau^n(\omega))\}_{n=0}^{\infty}$ is referred to as a *chaotic bit sequence.* Furthermore, it is reasonable that the binary function $C_T(\omega)$ is referred to as the binary pulse function. Figure 2 shows examples of these pulse functions defined by $C_T(\omega)$. Note that $B_i(\omega) = d_i(\omega)$ when $\tau(\cdot) = \tau_B(\cdot)$.

Now let us move on to conditions for such a pulse function to generate a sequence of i.i.d. binary random variables according to the orbit $\{\tau^n(\omega)\}_{n=0}^{\infty}$.

We consider a piecewise monotonic map $\tau : [d, e] \to [d, e]$ that satisfies the following properties:

(i) There is a partition $d = d_0 < d_1 < \cdots < d_{N_\tau} = e$ of $[d, e]$ such that for each integer $i = 1, \cdots, N_\tau$ ($N_\tau \geq 2$) the restriction of τ to the interval $[d_{i-1}, d_i)$, denoted by τ_i ($1 \leq i \leq N_\tau$), is a C^2 function; as well as
(ii) $\tau((d_{i-1}, d_i)) = (d, e)$, that is, τ_i is onto;
(iii) $\tau(\omega)$ has a unique absolutely continuous invariant (or briefly ACI) measure, denoted by $f^*(\omega)d\omega$ and is mixing on I with respect to this measure.

If the interval I is given by $I = [d, e]$, then the P-F operator P_τ of the map τ is defined by [9]

$$P_\tau H(\omega) = \frac{d}{d\omega} \int_{\tau^{-1}([d,\omega])} H(y)dy. \tag{17}$$

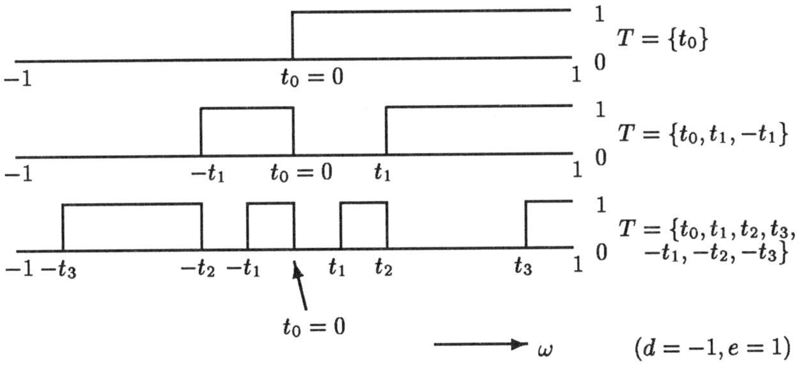

Fig. 2. *Independent* pulse functions defined by $C_T(\omega)$.

Note that the function $f^*(\omega)$ is the eigenfunction with the eigenvalue 1 of the following P-F equation

$$P_\tau f^*(\omega) = f^*(\omega). \tag{18}$$

It is known that the ensemble-average defined by

$$\langle F \rangle = \int_I F(\omega) f^*(\omega) d\omega \tag{19}$$

is useful in evaluating statistics of $\{F(\tau^n(\omega))\}_{n=0}^{\infty}$. Here define

$$p_\tau(t) = \langle \Theta_t \rangle = \int_t^e f^*(\omega) d\omega. \tag{20}$$

Note that $p_\tau(t)$ is a monotonically decreasing function of t.

Several famous ergodic maps and their ACI measures are listed as follows.

(1) R-adic map

$$\tau(\omega) = R\omega \bmod 1,\ R = 2, 3, 4, \cdots, \omega \in [0, 1], \quad f^*(\omega) d\omega = d\omega$$

(2) tent map

$$\tau(\omega) = \begin{cases} 2\omega, & \omega \in [0, \frac{1}{2}) \\ 2(1-\omega) & \omega \in [\frac{1}{2}, 1], \end{cases} \omega \in [0, 1], \quad f^*(\omega) d\omega = d\omega$$

(3) logistic map [10]

$$\tau(\omega) = 4\omega(1-\omega), \omega \in [0, 1], \quad f^*(\omega) d\omega = \frac{d\omega}{\pi\sqrt{\omega(1-\omega)}},$$

(4) Chebyshev map of degree k [11],[12]

$$\tau(\omega) = \cos(k \cos^{-1} \omega),\ k = 2, 3, 4, \cdots, = \omega \in [-1, 1], \quad f^*(\omega) d\omega = \frac{d\omega}{\pi\sqrt{1-\omega^2}}.$$

Let $G(\omega)$ and $H(\omega)$ be any two L_1 functions of bounded variation. The P-F operator P_τ has the important property:

$$\int_I G(\omega) P_\tau\{H(\omega)\} d\omega = \int_I G(\tau(\omega)) H(\omega) d\omega. \tag{21}$$

For the above class of maps, we have [9]

$$P_\tau H(\omega) = \sum_{i=1}^{N_\tau} |g_i'(\omega)| H(g_i(\omega)) \tag{22}$$

where $g_i(\omega) = \tau_i^{-1}(\omega)$.

Of major importance to the investigation of statistical properties of two sequences $\{G(\tau^n(\omega))\}_{n=0}^\infty$ and $\{H(\tau^n(\omega))\}_{n=0}^\infty$ is their 2nd-order cross-covariance function, which is defined by

$$\langle \rho^{(2)}(\ell; G, H) \rangle = \int_I (G(\omega) - \langle G \rangle)(H(\tau^\ell(\omega) - \langle H \rangle) f^*(\omega) d\omega \quad \text{where } \ell = 0, 1, 2, \cdots. \tag{23}$$

Note that when $G = H$, these denote the auto-covariance function.

Using eq. (21), we get

$$\langle \rho^{(2)}(\ell; G, H) \rangle = \int_I P_\tau^\ell \{G(\omega) f^*(\omega)\} H(\omega) d\omega - \langle G \rangle \langle H \rangle. \tag{24}$$

For several maps, such as the tent map, the logistic map, and the Chebyshev maps, the auto-covariance functions of real-valued sequences were already evaluated [13].

We now consider a class of the above piecewise monotonic maps satisfying

$$|g_i'(\omega)| f^*(g_i(\omega)) = \frac{1}{N_\tau} f^*(\omega),\ 1 \le i \le N_\tau \tag{25}$$

which is referred to as the *equidistributivity property* (or briefly *EDP*) [6]. Note that this class contains the well known maps, such as the R-adic map, the tent map, the logistic map, and the Chebyshev map of degree k, where $N_\tau = R, 2, 2, k$, respectively. Thus we give the following interesting lemma [6] which is very useful in evaluating covariance functions of chaotic threshold and bit sequences.

Lemma 1. *For the piecewise monotonic maps satisfying eq. (25), we can get*

$$P_\tau\{(\Theta_t(\omega) - p_\tau(t)) f^*(\omega)\} = \frac{s(\tau'(t))}{N_\tau} (\Theta_{\tau(t)}(\omega) - p_\tau(\tau(t))) f^*(\omega) \tag{26}$$

where $s(\omega)$ is the signum function defined by

$$s(\omega) = \begin{cases} -1 & \text{for } \omega < 0 \\ 1 & \text{for } \omega \ge 0. \end{cases} \tag{27}$$

Lemma 1 gives

$$P_\tau\{\Theta_{d_i}(\omega)f^*(\omega)\} = p_\tau(d_i)f^*(\omega). \tag{28}$$

Now, if $T = \{t_r\}_{r=0}^{2M}$ is the set of *symmetric thresholds*, defined by

$$t_r + t_{2M-r} = d + e, \quad r = 0, 1, \cdots, 2M, \tag{29}$$

then $C_T(\omega)$ is referred to as a *symmetric binary function* [6].

Next let us restrict our attention to the map satisfying

$$f^*(d + e - \omega) = f^*(\omega), \quad \omega \in [d, e], \tag{30}$$

which is referred to as a *symmetric property of the invariant measure.* Note that such a class of maps contains the well known maps, such as the R-adic map, the tent map, the logistic map, and the Chebyshev map.

Remark 1. For the maps with the symmetric property of the invariant measure eq. (30), we get

$$\langle C_T \rangle = \frac{1}{2}. \tag{31}$$

Furthermore, we consider a somewhat restricted class of piecewise monotonic maps satisfying eq. (25) which also satisfy the *symmetric property of the map*

$$\tau(d + e - \omega) = \tau(\omega), \quad \omega \in [d, e]. \tag{32}$$

Such a class includes the tent map, the logistic map, and the Chebyshev map of even degree k. The fact that τ is monotonic and onto gives

$$\tau\left(\frac{d+e}{2}\right) = d \text{ or } e. \tag{33}$$

The following lemma [6] plays an important role in estimating the covariance functions of symmetric binary sequences as shown in the Corollary that follows.

Lemma 2. *For the piecewise monotonic maps satisfying both eq. (25) and eq. (30), and their symmetric binary functions, we can get*

$$P_\tau\{C_T(\omega)f^*(\omega)\} = \langle C_T \rangle f^*(\omega). \tag{34}$$

which is a generalized version of eq. (28).

Remark 2. It is obvious from eq. (22) that if for a class of maps with the EDP, the function $C_T(\omega)$ satisfies the following relation, referred to as the *constant summation property* (or briefly *CSP*)

$$\frac{1}{N_\tau}\sum_{i=1}^{N_\tau} C_T(g_i(\omega)) = \langle C_T \rangle, \tag{35}$$

then eq. (34) automatically holds.

Corollary 1. *Consider the piecewise monotonic maps with both eq. (25) and eq. (32). Denote two different sets of symmetric thresholds by* $T = \{t_r\}_{r=0}^{2M}$ *and* $T' = \{t'_r\}_{r=0}^{2M'}$, *where*

$$d = t_0 < t_1 < \cdots < t_{2M} = e, \tag{36}$$
$$d = t'_0 < t'_1 < \cdots < t'_{2M'} = e, \tag{37}$$
$$t_r + t_{2M-r} = d + e, \quad r = 0, 1, \cdots, 2M, \tag{38}$$
$$t'_r + t'_{2M'-r} = d + e, \quad r = 0, 1, \cdots, 2M'. \tag{39}$$

Then we can obtain

$$\langle \rho^{(2)}(\ell; C_T, C_{T'}) \rangle = \begin{cases} Q^C_{TT'} & \text{for } \ell = 0 \\ 0 & \text{for } \ell \geq 1 \end{cases} \tag{40}$$

where

$$Q^C_{TT} = \langle C_T \rangle = \frac{1}{2}, \tag{41}$$
$$Q^C_{TT'} = \int_I C_T(\omega) C_{T'}(\omega) f^*(\omega) d\omega = \int_{I^C_{TT'}} d\omega, \tag{42}$$
$$I^C_{TT'} = \left(\bigcup_{r=1}^{M} I^C_T(r) \right) \bigcap \left(\bigcup_{s=1}^{M'} I^C_{T'}(s) \right), \tag{43}$$
$$I^C_T(r) = [p_r(t_{2r}), p_r(t_{2r-1})]. \tag{44}$$

Remark 3. Let $C_T = B_i$ and $C_{T'} = B_j$. Then $\langle \rho^{(2)}(0; B_i, B_j) \rangle = Q_{ij}$. We get $Q_{ij} = \frac{1}{4}$ for the maps with the uniform invariant density. On the other hand, for the maps with the nonuniform invariant densities, such as the logistic and the Chebyshev map, we can get

$$\lim_{\substack{i \to \infty \\ \text{or } j \to \infty}} Q_{ij} = \frac{1}{4} \quad \text{for } i \neq j. \tag{45}$$

Remark 4. Consider the R-adic map. When R is even, we get

$$\langle \rho^{(2)}(\ell; B_i, B_j) \rangle = \langle B_i \rangle \langle B_j \rangle \text{ for all } \ell. \tag{46}$$

Note that the symmetric binary function $C_T(\omega)$ is a generalized version of the Rademacher function for the dyadic map.

Next, let $\mathbf{U} = U_0 U_1 \cdots U_{m-1}$ be an arbitrary string of m binary digits where $U_n \in \{0, 1\}$ $(0 \leq n \leq m-1)$. Then there are 2^m possible strings. Let $\mathbf{u}^{(r)} = u_0^{(r)} u_1^{(r)} \cdots u_{m-1}^{(r)}$ be the r-th string. Introducing a binary random variable

$$\Gamma_n(\omega; G, \mathbf{u}^{(r)}) = G(\omega) u_n^{(r)} + \overline{G}(\omega) \overline{u}_n^{(r)} \tag{47}$$

for any binary function $G(\omega)$. Then the probability of the event $\boldsymbol{u}^{(r)}$ in an infinite binary sequence $\{G(\tau^n(\omega))\}_{n=0}^{\infty}$ is given by

$$\begin{aligned}\mathrm{Prob}(\boldsymbol{u}^{(r)};G) &= \int_I \left\{\prod_{n=0}^{m-1} \Gamma_n(\tau^n(\omega);G,\boldsymbol{u}^{(r)})\right\} f^*(\omega)d\omega \\ &= \langle\rho^{(m)}(\underbrace{1,1,\cdots,1}_{m-1};\Gamma_0(G,\boldsymbol{u}^{(r)}),\Gamma_1(G,\boldsymbol{u}^{(r)}),\cdots,\Gamma_{m-1}(G,\boldsymbol{u}^{(r)}))\rangle.\end{aligned} \tag{48}$$

where $\langle\rho^{(m)}(\cdot;\cdot)\rangle$ is the higher-order (m-th order) correlation function defined by

$$\begin{aligned}&\langle\rho^{(m)}(\ell_{m-1},\ell_{m-2},\cdots,\ell_1;H_m,H_{m-1},\cdots,H_1)\rangle \\ &\quad= \int_I H_m(\omega)H_{m-1}(\tau^{\ell_{m-1}}(\omega))H_{m-2}(\tau^{\ell_{m-1}+\ell_{m-2}}(\omega))\cdot \\ &\qquad\cdots H_1(\tau^{\ell_{m-1}+\ell_{m-2}+\cdots+\ell_1}(\omega))f^*(\omega)d\omega \quad \text{for all integers } \ell_i \geq 0,\end{aligned} \tag{49}$$

where each of $H_i(\omega)$ denotes an L_1 real-valued function ($i = 1,2,\cdots,m$). It is, in general, difficult to evaluate such higher-order correlation functions explicitly. However, it is simplified if the the CSP (or eq. (34)) is satisfied. In fact, we can get [6] the following theorem for any binary function $C_T(\omega)$ with the CSP (or eq. (34)).

Theorem 1. *Any binary function $C_T(\omega)$ with the CSP gives*

$$Prob(\mathbf{u}^{(r)};C_T) = \langle C_T\rangle^s(1-\langle C_T\rangle)^{m-s}, \tag{50}$$

where s is the number of 1's in $\{u_n^{(r)}\}_{n=0}^{m-1}$.

This implies that $\{C_T(\tau^n(\omega))\}_{n=0}^{\infty}$ is a sequence of i.i.d. binary random variables in the sense that it can realize a Bernoulli sequence with probability $\langle C_T\rangle$. Note that we can get a fair Bernoulli sequence when $\langle C_T\rangle = \frac{1}{2}$, that is, an m-distributed binary random sequence.

Note that eq. (34) (or CSP) can be generalized to the design for a sequence of p-ary random variables (*i.e.*, a $p(\geq 3)$-valued function $C_T(\omega)$) [14].

Figure 3 shows a stream cipher system whose running-key sequences are sequences of i.i.d. binary random variables using chaotic dynamics [15].

4 Concluding Remarks

We have reviewed a simple sufficient condition for a binary function to produce a sequence of i.i.d. binary random variables using chaotic dynamics. Furthermore, we have made it clear that any binary function with the *constant summation property* (or briefly CSP) according to the chaotic dynamics with the *equidistributivity property* (or briefly EDP) can generate a sequence of i.i.d. random variables.

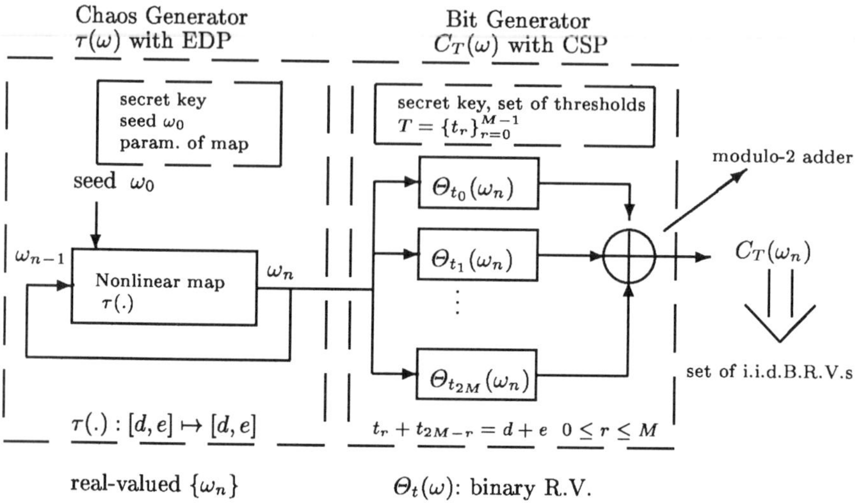

Fig. 3. Running-key sequence generator for a stream cipher system based on chaotic dynamics.

References

1. M. Kac, *Statistical Independence in Probability Analysis and Number Theory*, The Mathematical Association of America, 1959.
2. P. Billingsley, *Probability and Measure*, John Wiley & Sons, 1995.
3. M. K. Simon, J. K. Omura, R. A. Scholtz, and B.K. Levitt, *Spread Spectrum Communications Handbook*, McGraw-Hill,1994.
4. J. L. Massey, "An Introduction to Contemporary Cryptology," Proc. IEEE, **76**-5, pp. 533–549, May 1988.
5. D. V. Sarwate and M. B. Pursley, "Crosscorrelation Properties of Pseudorandom and Related Sequences," Proc. IEEE,**68**-3, pp. 593–619, 1980.
6. T. Kohda and A. Tsuneda, "Statistics of Chaotic Binary Sequences", *IEEE Trans. Information Theory*, **43**-1, pp. 104–112, Jan. 1997.
7. T. Kohda and A. Tsuneda, "Pseudonoise Sequences by Chaotic Nonlinear Maps and Their Correlation Properties," *IEICE Trans.* **E76-B**-8, pp. 855–862, 1993.
8. T. Kohda and A. Tsuneda, "Explicit Evaluations of Correlation Functions of Chebyshev Binary and Bit Sequences Based on Perron-Frobenius Operator," *IEICE., Trans.* **E77-A**-11, pp. 1794–1800, 1994.
9. A. Lasota and M. C. Mackey, *Chaos, Fractals, and Noise*, Springer-Verlag, 1994.
10. S.L. Ulam and J. von Neumann, "On combination of stochastic and deterministic processes", Bull. Math. Soc., **53**, p.1120, 1947.
11. R.L. Adler and T.J. Rivlin, "Ergodic and mixing properties of Chebyshev polynomials," Proc. Amer.Math.Soc.,**15**, pp. 794-796,1964.
12. T.J. Rivlin, *Chebyshev polynomials- From Approximation Theory to Algebra and Number Theory*, A Wiley-Interscience Publication, (1990).
13. S. Grossmann and S. Thomae, "Invariant distributions and stationary correlation functions of one-dimensional discrete processes," *Z. Naturforsch.* vol. 32a, pp. 1353–1363, 1977.

14. T. Kohda and A. Tsuneda, "Design of sequences of p-ary random variables", Proc. of 1997 IEEE Int. Sympo. on Information Theory, p.76, 1997
15. T. Kohda and A. Tsuneda, "Chaotic Bit Sequences for Stream Cipher Cryptography and Their Correlation Functions", *Proc. SPIE's Photonics East '95 Sympo., (Chaotic Circuit for Communication)*, **SPIE 2612**, pp. 86–97, 1995.

Explicit Sequence Expansions

David Kohel, San Ling, and Chaoping Xing

Department of Mathematics
National University of Singapore
Kent Ridge Crescent, Singapore 119260
Email: kohel,matlings,xingcp@math.nus.edu.sg

Abstract. Examples of d-perfect sequences are constructed based on the method in Xing *et al* [8]. In particular examples of 1-perfect sequences based on genus 0 curves over binary and the ternary fields are computed, as are 2-perfect binary sequences based on an elliptic curve. The complexity profile of certain of the 2-perfect sequences are experimentally determined to follow that of 1-perfect sequences. Based on two algebraic reformulations of the known characterization of binary 1-perfect sequences, these sequences are proved to be 1-perfect.

1 Introduction

In the previous articles [8] and [9] of Xing *et al.*, several constructions of d-perfect sequences are given based on functions on a curve over a finite field. The freedom to choose a local parameter and a function on a curve (see Theorem 1) allows many different d-perfect sequences to be derived from a given curve. In this paper, we present a method to compute these d-perfect sequences, and give examples of d-perfect sequences, for small d, by choosing functions of small degree on a curve.

The choice of the curve as well as the functions permits great flexibility in constructing d-perfect sequences. With the interest of considering curves which admit functions of small degree, we will concentrate on curves of genera 0 and 1, namely, the projective line and elliptic curves. This paper is arranged in the following way. In Section 2, some notations concerning d-perfect sequences and a result regarding the construction of d-perfect sequences in Xing *et al.* [8] will be recalled. Those 1-perfect sequences given by this construction using the curves of genus 0 over the binary and ternary fields are completely listed in Section 3. Section 4 is devoted to an example of sequences derived from an elliptic curve over $\mathbf{F}_2$. Certain 2-perfect sequences constructed in this manner are experimentally observed to be 1-perfect to a high degree of accuracy. This observation is proved based on two algebraic reformulations of the known characterization of binary 1-perfect sequences. The reformulations themselves give infinite families of binary 1-perfect sequences constructed as series expansion of functions on curves and on surfaces.

2 Background

Let us first recall some concepts regarding d-perfect sequences over a finite field $\mathbf{F}_q$ of q elements. For a sequence $\mathbf{s}$ of finite length, we denote by $\ell_{\mathbf{s}}$ the linear complexity of $\mathbf{s}$. Thus we obtain the linear complexity profile (simply denoted by *lcp*) for an infinite sequence $\mathbf{a}$, namely, the integer sequence

$$\{\ell_{\mathbf{a}}(n)\}_{n=1}^{\infty},$$

where $\ell_{\mathbf{a}}(n)$ is the linear complexity of the first n terms of the sequence $\mathbf{a}$. The sequence $\mathbf{a}$ is called d-perfect for a positive integer d if the following condition holds:

$$\frac{n+1-d}{2} \leq \ell_{\mathbf{a}}(n) \leq \frac{n+d}{2} \quad \text{for all } n.$$

Definition. Let $\mathcal{X}/\mathbf{F}_q$ be a curve and x a function in $\mathbf{F}_q(\mathcal{X})$. The *degree* of a nonconstant x is defined to be the degree of the field extension $\mathbf{F}_q(\mathcal{X})/\mathbf{F}_q(x)$, and a constant function x is defined to have degree zero.

For a function x of $\mathbf{F}_q(\mathcal{X})-\{0\}$, let $\mathrm{div}(x)$ be the principal divisor associated with x. Put

$$\mathrm{div}(x) = \mathrm{div}_0(x) - \mathrm{div}_\infty(x),$$

where $\mathrm{div}_0(x)$ and $\mathrm{div}_\infty(x)$ are two effective divisors and the supports of $\mathrm{div}_0(x)$ and $\mathrm{div}_\infty(x)$ are disjoint. Then $\mathrm{div}_0(x)$ and $\mathrm{div}_\infty(x)$ are uniquely determined by x and $\deg(\mathrm{div}_0(x))$ is equal to the degree of x. Let v_P denote the normalized discrete valuation corresponding to the point P on $\mathcal{X}/\mathbf{F}_q$.

We apply the following theorem to the explicit construction of d-perfect sequences (see Theorem 3.1 of [8]).

Theorem 1. *Let $\mathcal{X}/\mathbf{F}_q$ be a curve for which there exist an $\mathbf{F}_q$-rational point P on $\mathcal{X}$ and a degree 2 function t in $\mathbf{F}_q(\mathcal{X})$ such that $v_P(t) = 1$. Let x be a function of degree d on $\mathcal{X}$ such that $\mathbf{F}_q(\mathcal{X}) = \mathbf{F}_q(x,t)$ with $(x/t)(P) = 1$. Then there exists a unique power series expansion of the form*

$$x(t) = t + a_2t^2 + a_3t^3 + a_4t^4 + \cdots$$

for x which defines an embedding of the function field $\mathbf{F}_q(\mathcal{X})$ in $\mathbf{F}_q((t))$. The corresponding sequence $(1, a_2, a_3, a_4, \ldots)$ is d-perfect.

If the sequence of coefficients of a series $x(t)$ is d-perfect we say that the series is d-perfect. The series $x(t)$ can be rapidly computed via the effective form of Hensel's lemma (see Chapter II, Proposition 2 of Lang [1]).

Theorem 2 (Hensel's Lemma). *Let x and t be as in the previous theorem, and let x have minimal polynomial $F(X)$ over $\mathbf{F}_q(t)$. Then the power series expansion for x in $\mathbf{F}_q((t))$ can be determined by setting $x_1(t) = t$, and for all i,*

$$x_{i+1} = x_i - \frac{F(x_i)}{F'(x_i)},$$

where $F'(X)$ is the derivative of $F(X)$ with respect to X, and the sequence (x_i) satisfies $v_P(F(x_i)) \geq 2^i$.

3 Genus zero curves

3.1 General theory

To obtain 1-perfect sequences using the above construction, we require a curve $\mathcal{X}/\mathbf{F}_q$ which has a degree one function x. Such a curve has function field $\mathbf{F}_q(x)$ and it is immediately seen that it is of genus zero. Thus we begin with a pair of a genus zero curve $\mathcal{X}/\mathbf{F}_q$ and a function x generating $\mathbf{F}_q(\mathcal{X})$. Let P be the zero of x, and let t be a degree two function on $\mathcal{X}$ such that $v_P(t) = 1$ and $(x/t)(P) = 1$. The form of t is described by means of the following proposition.

Proposition 1. *Let $\mathcal{X}/\mathbf{F}_q$ be a genus zero curve with $\mathbf{F}_q(\mathcal{X}) = \mathbf{F}_q(x)$, and let P be the zero of x. A degree two function $t \in \mathbf{F}_q(x)$ satisfying $(x/t)(P) = 1$ has the form*

$$t = \frac{x + ax^2}{1 + bx + cx^2}$$

for some a, b, and c in $\mathbf{F}_q$, and where a or c is nonzero, and $\gcd(1 + ax, 1 + bx + cx^2) = 1$.

Every generator x_1 for the rational function field $\mathbf{F}_q(x)$ is of the form

$$x_1 = \frac{ax + b}{cx + d},$$

where a, b, c, and d are elements of the base field $\mathbf{F}_q$ with $ad - bc \neq 0$. Given any x_1 in $\mathbf{F}_q(x)$ and a point P on $\mathcal{X}$, we may replace it with $x_1 - x_1(P)$. Thus by scaling we may assume that

$$v_P(x_1) = 1, \quad \text{and} \quad (x_1/x)(P) = 1.$$

Then $x_1(P) = b/d = 0$, and $(x_1/x)(P) = a/d = 1$. Thus any degree one function x_1 can be normalized such that it is of the form

$$x_1 = \frac{x}{1 + cx}.$$

By Theorem 1, the corresponding power series expansion for x_1 in t is also 1-perfect, which proves the following proposition.

Proposition 2. *Suppose x_0 and x_1 are degree one functions on a curve $\mathcal{X}$ and t is a degree two uniformizing parameter at a point P. Suppose moreover that $v_P(x_0) = v_P(x_1) = 1$ and $(x_1/x_0)(P) = 1$. Then x_1 is of the form*

$$x_1 = \frac{x_0}{1 + cx_0},$$

and the series $x_0(t)$ and $x_1(t)$ are both 1-perfect.

3.2 1-Perfect sequences over $\mathbf{F}_2$

From Proposition 1, we see that there are four possible choices for the degree two local parameter t over the binary field. These are:

$$(0.0)\ t = x + x^2, \quad (1.0)\ t = \frac{x + x^2}{1 + x + x^2},$$

$$(0.1)\ t = \frac{x}{1 + x^2}, \quad (1.1)\ t = \frac{x}{1 + x + x^2}.$$

In this case Hensel's lemma construction for a power series solution $x = x(t)$ to one of the above equations is a special case of a reversion formula for power series. That is, if we take equation $(i.j)$ to define the power series $t(x)$ in $\mathbf{F}_2[x]$, then $t(x(t)) = t$ and $x(t(x)) = x$.

We note that the linear fractional transformation $A(u) = u/(1 + u)$ determines an automorphism of order two of the projective line $\mathbf{P}^1(\mathbf{F}_2)$. If we denote by $x_{ij}(t)$ the power series which is a root of equation $(i.j)$, and set $x(t) = x_{00}(t)$, then

$$x_{ij}(t) = A^j(x(A^i(t))).$$

We find the following sequences associated to the t-expansions of x in the four cases.

$$\begin{array}{ll}
(0.0) & (1,1,0,1,0,0,0,1,0,0,0,0,0,0,0,1,0,0,0,0,0,0,0,0,0,0,0,0,0,0,\ldots),\\
(0.1) & (1,0,1,0,0,0,1,0,0,0,0,0,0,0,1,0,0,0,0,0,0,0,0,0,0,0,0,0,0,1,\ldots),\\
(1.0) & (1,0,1,1,1,0,1,0,1,0,1,1,1,0,1,1,1,0,1,1,1,0,1,0,1,0,1,1,1,0,1,\ldots),\\
(1.1) & (1,1,0,0,1,1,1,1,1,1,0,0,1,1,0,0,1,1,0,0,1,1,1,1,1,1,0,0,1,1,1,\ldots).
\end{array}$$

These initial segments can be verified to follow the linear complexity of 1-perfect sequences.

The sequence (0.0) is the well-known 1-perfect sequence of the power series (see [2–5, 7]).

$$\sum_{i=0}^{\infty} t^{2^i} = t + t^2 + t^4 + t^8 + t^{16} + \cdots,$$

and the sequence (0.1) is that of the power series

$$\sum_{i=1}^{\infty} t^{2^i - 1} = t + t^3 + t^7 + t^{15} + t^{31} + \cdots.$$

By making the substitution $t \mapsto t/(1+t)$, the corresponding sequences (1.0) and (1.1) can be seen to have the forms

$$\sum_{m=1}^{\infty}\sum_{i=0}^{\infty} t^{m2^i} \quad \text{and} \quad \sum_{m=1}^{\infty}\sum_{i=0}^{\infty} t^{m(2^i-1)},$$

respectively.

3.3 1-Perfect sequences over $\mathbf{F}_3$

From Proposition 1 we see that there are 18 possible choices of degree 2 local parameters t over $\mathbf{F}_3$, which we subdivide into a block of nine functions:

$$(0.0)\ t = \frac{x}{1+x+x^2}, \qquad (1.0)\ t = \frac{x}{1-x+x^2}, \qquad (2.0)\ t = \frac{x}{1+x^2},$$

$$(0.1)\ t = x+x^2, \qquad (1.1)\ t = \frac{x+x^2}{1+x+x^2}, \qquad (2.1)\ t = \frac{x+x^2}{1-x-x^2},$$

$$(0.2)\ t = \frac{x-x^2}{1-x+x^2}, \qquad (1.2)\ t = x-x^2, \qquad (2.2)\ t = \frac{x-x^2}{1+x-x^2}.$$

and a second block of nine functions:

$$(0.0)\ t = \frac{x}{1-x^2}, \qquad (1.0)\ t = \frac{x}{1+x-x^2}, \qquad (2.0)\ t = \frac{x}{1-x-x^2},$$

$$(0.1)\ t = \frac{x+x^2}{1-x}, \qquad (1.1)\ t = \frac{x+x^2}{1+x^2}, \qquad (2.1)\ t = \frac{x+x^2}{1+x-x^2},$$

$$(0.2)\ t = \frac{x-x^2}{1+x}, \qquad (1.2)\ t = \frac{x-x^2}{1-x-x^2}, \qquad (2.2)\ t = \frac{x-x^2}{1+x^2}.$$

The linear fractional transformation $A(u) = u/(1+u)$ determines an automorphism of $\mathbf{P}^1(\mathbf{F}_3)$ of order three. As in Section 3.2, it is easy to verify that roots of any two equations in the same block can be exchanged by action of the group $\langle A \rangle$. Since A generates the group of automorphisms of $\mathbf{P}^1(\mathbf{F}_3)$ which fix the point $P = (0:1)$ and the residue of functions at P, we may think of each of the two blocks of equations as comprising an equivalence class over $\mathbf{F}_3$. The corresponding sequences for the first block are:

(0.0) $(1,1,2,1,0,0,0,1,2,1,1,2,1,0,0,0,0,0,0,0,0,0,0,0,1,2,1,1,2,1,\ldots)$,
(0.1) $(1,2,2,1,2,0,0,0,2,1,2,2,1,2,0,0,0,0,0,0,0,0,0,0,2,1,2,2,1,\ldots)$,
(0.2) $(1,0,1,1,0,0,0,0,1,1,0,1,1,0,0,0,0,0,0,0,0,0,0,0,1,1,0,1,1,\ldots)$,

(1.0) $(1,2,2,2,0,0,0,2,2,2,1,1,1,0,0,0,0,0,0,0,0,0,0,2,2,2,1,1,1,\ldots)$,
(1.1) $(1,0,1,2,0,0,0,0,1,2,0,2,1,0,0,0,0,0,0,0,0,0,0,0,1,2,0,2,1,\ldots)$,
(1.2) $(1,1,2,2,2,0,0,0,2,2,2,1,1,1,0,0,0,0,0,0,0,0,0,0,2,2,2,1,1,\ldots)$,

(2.0) $(1,0,1,0,2,0,2,0,2,0,0,0,0,0,0,0,2,0,2,0,2,0,1,0,1,0,1,0,0,0,0,\ldots)$,
(2.1) $(1,1,2,0,0,1,2,2,1,0,0,0,0,0,0,0,0,1,2,2,1,0,0,2,1,1,2,0,0,0,0,\ldots)$,
(2.2) $(1,2,2,0,0,2,2,1,1,0,0,0,0,0,0,0,0,2,2,1,1,0,0,1,1,2,2,0,0,0,0,\ldots)$,

and for the second block:

(0.0) $(1,0,2,0,2,0,1,0,2,0,0,0,0,0,0,0,2,0,1,0,2,0,2,0,1,0,2,0,0,0,0,\ldots)$,
(0.1) $(1,1,0,2,0,2,0,1,0,2,0,0,0,0,0,0,2,0,1,0,2,0,2,0,1,0,2,0,0,0,\ldots)$,
(0.2) $(1,2,0,1,0,1,0,2,0,1,0,0,0,0,0,0,1,0,2,0,1,0,1,0,2,0,1,0,0,0,\ldots)$,

(1.0) (1, 1, 0, 1, 0, 1, 2, 0, 2, 1, 1, 0, 1, 0, 1, 2, 2, 0, 2, 2, 0, 2, 0, 2, 1, 1, 0, 1, 1, 0, 1, ...),
(1.1) (1, 2, 0, 0, 1, 1, 2, 1, 1, 0, 1, 2, 2, 0, 0, 1, 0, 2, 2, 1, 0, 0, 2, 2, 1, 2, 0, 0, 1, 2, 2, ...),
(1.2) (1, 0, 2, 2, 1, 1, 0, 1, 1, 2, 1, 0, 0, 2, 2, 1, 2, 0, 0, 1, 2, 2, 0, 0, 1, 0, 2, 2, 1, 0, 0, ...),

(2.0) (1, 2, 0, 2, 0, 2, 2, 0, 2, 2, 1, 0, 1, 0, 1, 1, 2, 0, 2, 1, 0, 1, 0, 1, 1, 2, 0, 2, 1, 0, 1, ...),
(2.1) (1, 0, 2, 1, 1, 2, 0, 2, 1, 1, 1, 0, 0, 1, 2, 2, 2, 0, 0, 2, 2, 1, 0, 0, 1, 0, 2, 1, 1, 0, 0, ...),
(2.2) (1, 1, 0, 0, 1, 2, 2, 2, 1, 0, 1, 1, 2, 0, 0, 2, 0, 1, 2, 2, 0, 0, 2, 1, 1, 1, 0, 0, 1, 1, 2, ...).

4 Genus one curves

In this section we consider the sequences derived from series expansion for functions on the elliptic curve

$$y^2 + xy = x^3 + x$$

over the field $\mathbf{F}_2$ of two elements. Since every point on an elliptic curve can be translated to any other, we consider only expansions around the fixed point O at infinity.

On an elliptic curve there exist no degree one functions, so Theorem 1 provides no means of constructing sequences from functions on the curve which are provably 1-perfect. However, at the end of this section we prove that certain 2-perfect sequences obtained from functions on this curve are in fact 1-perfect. In the interest of minimizing d we consider only the series expansions of non-linearly dependent functions of degree two and local parameters for O. We classify these functions as follows.

The set of rational points on E over $\mathbf{F}_2$ consists of the four points O, $(0,0)$, $(1,0)$, and $(1,1)$. Since the 2-torsion group contains only two elements, $E(\mathbf{F}_2)$ must be isomorphic to the group $\mathbf{Z}/4\mathbf{Z}$.

The automorphism $[-1]$ on E is the map $(x, y) \mapsto (x, x+y)$, which stabilizes $(0,0)$, so we identify $(0,0)$ as the 2-torsion point. Any degree two function on E which has a zero of order one at O has exactly one other zero on E, which must be one of the rational points $(0,0)$, $(1,0)$, or $(1,1)$.

The zeros and poles of functions on E satisfy an additional relation. Let $\mathrm{div}(f)$ be the divisor of the function f, and let $\mathrm{Div}_0(E, \overline{\mathbf{F}}_2)$ be the degree zero divisors defined over an algebraic closure of $\mathbf{F}_2$. Then there exists an exact sequence

$$1 \longrightarrow \overline{\mathbf{F}}_2(E)^* \longrightarrow \mathrm{Div}_0(E, \overline{\mathbf{F}}_2) \longrightarrow E(\overline{\mathbf{F}}_2) \longrightarrow 0,$$

where the first map is to take a function to its principal divisor, and the second map is the group homomorphism which takes a point divisor $[P]$ to the point P. From the exact sequence we may classify functions in $\mathbf{F}_2(E)$ by their divisors. Precisely they correspond to divisors in the kernel of the second map which are invariant under the Galois group $\mathrm{Gal}(\overline{\mathbf{F}}_2/\mathbf{F}_2)$.

In the following table we enumerate the possible divisors of degree two functions and the function to which they correspond. In addition to the divisors of points in $E(\mathbf{F}_2)$ the two degree two divisors $\mathcal{P}$ and $\mathcal{Q}$ corresponding to the Galois invariant pairs of points in $E(\mathbf{F}_4)$ disappearing on the ideals $(x^2 + x + 1, y + 1)$ and $(x^2 + x + 1, y + x + 1)$ may appear.

First we determine the divisors of certain "building block" functions on E. We denote $\mathrm{div}(f) = \mathrm{div}_0(f) - \mathrm{div}_\infty(f)$.

f	$\mathrm{div}_0(f)$	$\mathrm{div}_\infty(f)$
x	$2[(0,0)]$	$2[O]$
$x+1$	$[(1,0)]+[(1,1)]$	$2[O]$
y	$[(0,0)]+2[(1,0)]$	$3[O]$
$y+x$	$[(0,0)]+2[(1,1)]$	$3[O]$
$y+1$	$[(1,1)]+[\mathcal{P}]$	$3[O]$
$y+x+1$	$[(1,0)]+[\mathcal{Q}]$	$3[O]$

By enumerating all possible degree two divisors which can occur as $\mathrm{div}_\infty(f)$, we obtain the classification below of all degree two functions on E over $\mathbf{F}_2$, expressed as quotients of the functions in the previous table.

	f	$\mathrm{div}_0(f)$	$\mathrm{div}_\infty(f)$
(1)	x/y	$[O]+[(0,0)]$	$2[(0,1)]$
(2)	$x/(y+x)$	$[O]+[(0,0)]$	$2[(1,1)]$
(3)	$y/(x^2+x)$	$[O]+[(1,0)]$	$[(0,0)]+[(1,1)]$
(4)	$(x+1)/(y+1)$	$[O]+[(1,0)]$	$[\mathcal{P}]$
(5)	$(y+x)/(x^2+x)$	$[O]+[(1,1)]$	$[(0,0)]+[(1,0)]$
(6)	$(x+1)/(y+x+1)$	$[O]+[(1,1)]$	$[\mathcal{Q}]$

On the following page we give the minimal polynomial for the function f over the field $\mathbf{F}_2(t)$, where f and t are one of the above six degree two functions on $E/\mathbf{F}_2$. The entry $(i.j)$ corresponds to the pair (f,t), where f is entry (i) in the above table, and t is entry (j).

The sequence of coefficients for the series expansion of f with respect to t at O is proven to be 2-perfect by Theorem 1. For a given pair, the functions may in fact be 1-perfect. We give the experimentally determined value d_0 such that the series expansion is believed to be d_0-perfect, however this value is provably only a lower bound. The functions (1) through (6) fall in three classes which are equivalent under a linear fractional transformation. For such a pair we set d_0 equal to 0 in the table, to indicate that the corresponding sequence is periodic.

	$F(X)$	d_0
(1.1)	$X + t$	0
(1.2)	$(1+t)X + t$	0
(1.3)	$(t+t^2)X^2 + (1+t)X + t$	1
(1.4)	$(1+t+t^2)X^2 + (1+t)X + t$	1
(1.5)	$tX^2 + (1+t)X + t + t^2$	1
(1.6)	$(1+t+t^2)X^2 + (1+t)X + t + t^2$	1

	$F(X)$	d_0
(2.1)	$(1+t)X + t$	0
(2.2)	$X + t$	0
(2.3)	$tX^2 + (1+t)X + t + t^2$	1
(2.4)	$(1+t+t^2)X^2 + (1+t)X + t + t^2$	1
(2.5)	$(t+t^2)X^2 + (1+t)X + t$	1
(2.6)	$(1+t+t^2)X^2 + (1+t)X + t$	1

	$F(X)$	d_0
(3.1)	$t^2X^2 + (1+t+t^2)X + t$	2
(3.2)	$X^2 + (1+t+t^2)X + t$	2
(3.3)	$X + t$	0
(3.4)	$(1+t)X + t$	0
(3.5)	$tX^2 + (1+t+t^2)X + t$	2
(3.6)	$(1+t+t^2)X^2 + (1+t+t^2)X + t$	2

	$F(X)$	d_0
(4.1)	$t^2X^2 + (1+t+t^2)X + t + t^2$	2
(4.2)	$(1+t^2)X^2 + (1+t+t^2)X + t + t^2$	2
(4.3)	$(1+t)X + t$	0
(4.4)	$X + t$	0
(4.5)	$(t+t^2)X^2 + (1+t+t^2)X + t + t^2$	2
(4.6)	$(1+t+t^2)X^2 + (1+t+t^2)X + t + t^2$	2

	$F(X)$	d_0
(5.1)	$X^2 + (1+t+t^2)X + t$	2
(5.2)	$t^2X^2 + (1+t+t^2)X + t$	2
(5.3)	$tX^2 + (1+t+t^2)X + t$	2
(5.4)	$(1+t+t^2)X^2 + (1+t+t^2)X + t$	2
(5.5)	$X + t$	0
(5.6)	$(1+t)X + t$	0

	$F(X)$	d_0
(6.1)	$(1+t^2)X^2 + (1+t+t^2)X + t + t^2$	2
(6.2)	$t^2X^2 + (1+t+t^2)X + t + t^2$	2
(6.3)	$(t+t^2)X^2 + (1+t+t^2)X + t + t^2$	2
(6.4)	$(1+t+t^2)X^2 + (1+t+t^2)X + t + t^2$	2
(6.5)	$(1+t)X + t$	0
(6.6)	$X + t$	0

We note that due to the automorphism $[-1]$ of the curve, the nontrivial minimal polynomials appear in pairs, corresponding to function pairs which are exchanged under the automorphism induced by $[-1]$.

From the above table we note that the sequence associated to the root $f(t)$ of valuation 1 in $\mathbf{F}_2((t))$ to any of the four exceptional polynomials

$$\begin{aligned}
&(1)\ F(X) = (t+t^2)X^2 + (1+t)X + t\\
&(2)\ F(X) = (1+t+t^2)X^2 + (1+t)X + t\\
&(3)\ F(X) = tX^2 + (1+t)X + t + t^2\\
&(4)\ F(X) = (1+t+t^2)X^2 + (1+t)X + t + t^2
\end{aligned}$$

over $\mathbf{F}_2(t)$, were observed to exhibit a 1-perfect complexity profile. The sequences associated with the roots of above four polynomials are:

(1) $(1,1,0,1,0,1,1,1,0,1,1,1,0,1,0,1,0,1,1,1,0,1,0,1,0,1,1,1,0,1,1,1,\ldots)$,
(2) $(1,0,1,0,0,1,0,1,1,1,1,0,1,0,0,1,0,0,1,1,0,0,1,0,0,1,0,1,1,1,1,0,\ldots)$,
(3) $(1,0,1,1,1,1,0,0,1,1,0,0,1,1,1,1,1,1,0,0,1,1,1,1,1,1,0,0,1,1,0,0,\ldots)$,
(4) $(1,1,0,0,1,0,0,0,0,1,0,1,1,1,1,1,1,1,1,0,1,0,0,1,0,0,1,1,0,0,1,1,\ldots)$.

In particular we note that the above four sequences are different from the 1-perfect binary sequences in Section 3.

The following theorem of Wang and Massey [7] (see Niederreiter [2]) characterizes all binary perfect sequences.

Theorem 3. *A sequence* $(1, a_2, a_3 \ldots)$ *over the binary field is* 1*-perfect if and only if* $a_n + a_{2n} + a_{2n+1} = 0$.

In terms of a series $f(t) = t + a_2t^2 + a_3t^3 + \cdots$ we easily check that the coefficient of t^{2n+1} in

$$t\,f(t)^2 + (1+t)f(t) + t$$

is $a_n + a_{2n} + a_{2n+1}$ for all $n \geq 1$. Taking the derivative with respect to t, gives the following corollary.

Corollary 1. *A series* $f(t) = t + a_2t^2 + a_3t^3 + \cdots$ *is* 1*-perfect if and only if it satisfies the differential equation*

$$(1+t)\frac{df}{dt}(t) = f(t)^2 + f(t) + 1.$$

By Corollary 1, it follows immediately that roots $f(t)$ of the four exceptional polynomials are 1-perfect. For the first polynomial, we have

$$F(f(t)) = (t+t^2)f(t)^2 + (1+t)f(t) + t = 0.$$

Taking the first derivative, we obtain the characterizing differential equation for binary perfect sequences. For each of the other polynomials we obtain the same derivative, proving that the corresponding series are 1-perfect as experimentally observed.

The preceeding corollary allows us to determine whether a function $f(t)$ is 1-perfect by differentiating its minimal polynomial over $\mathbf{F}_2((t))$. The following corollary gives an alternative algebraic characterization of $f(t)$.

Corollary 2. *Every binary* 1*-perfect series* $f(t) = t + a_2 t^2 + a_3 t^3 + \cdots$ *can be uniquely written in the form* $f(t) = v^2 + t\,u^2$ *where* u *lies in* $1 + t\,\mathbf{F}_2[[t]]$ *and* v *is the root of*

$$v^2 + v = 1 + u + t\,u^2,$$

lying in $t\,\mathbf{F}[[t]]$. *The series* u *and* v *are uniquely defined by* $f(t)$ *and conversely every* u *in* $1 + t\,\mathbf{F}_2[[t]]$ *gives rise to a unique solution* v *in* $t\,\mathbf{F}_2[[t]]$ *such that* $f(t) = v^2 + t\,u^2$ *is* 1*-perfect.*

Proof. Every $f(t)$ in $\mathbf{F}_2[[t]]$ can be written in the form $f(t) = v^2 + t\,u^2$. The leading term is t if and only if u lies in $1 + t\,\mathbf{F}_2[[t]]$ and v in $t\,\mathbf{F}_2[[t]]$, and one verifies that $df/dt = u^2$. By Corollary 1, the series $f(t)$ is 1-perfect if and only if

$$(1+t)\frac{df}{dt} = f^2 + f + 1.$$

After a substitution and rearrangement, this is equivalent to $(v^2 + v + 1 + u + t\,u^2)^2 = 0$. Since $\mathbf{F}_2((t))$ is a field, the results follows. □

A binary 1-perfect sequence $f(t)$ arises as the series expansion with respect to t of a function on a curve $\mathcal{X}$ if and only if u is algebraic over $\mathbf{F}_2(t)$. The function t has degree 2 on $\mathcal{X}$ as in Theorem 1 if and only if exactly one of u or v is a rational function in t. If u is transcendental over $\mathbf{F}_2(t)$ then the $\mathbf{F}_2(u,t)$ is the function field of the projective plane, and $f(t)$ gives a function on a quadratic surface covering the plane.

References

1. Lang, S. *Algebraic Number Theory*. Second Edition. Graduate Texts in Mathematics, **110**. Springer-Verlag, New York, 1994.
2. Niederreiter, H. "Continued fractions for formal power series, pseudorandom numbers, and linear complexity of sequences." in *Contributions to General Algebra 5* (Salzburg, 1986), 221–233, Hölder-Pichler-Tempsky, Vienna, 1987.
3. Niederreiter, H. "Sequences with almost perfect linear complexity profile," in *Advances in Cryptology–EUROCRYPT '87*, D. Chaum and W. L. Price, eds. Springer-Verlag, Berlin: Lecture Notes in Computer Science, vol. 304, 1988, 37–51.
4. Rueppel, R. A. *Analysis and Design of Stream Ciphers*. Communication and Control Engineering Series. Springer-Verlag, Berlin-New York, 1986.
5. Rueppel, R. A. "Stream ciphers," in *Contemporary Cryptology–The Science of Information Integrity*, G. J. Simmons, Ed., 65–134, IEEE, New York, 1992.
6. Silverman, J. H. *The arithmetic of ellitic curves*, Graduate Texts in Mathematics, **106**, Springer-Verlag, New York, 1986.
7. Wang, M.-Z. and Massey, J. L. *The characterization of all binary sequences with a perfect linear complexity profile*, Paper presented at *EUROCRYPT '86*, Linköping, 1986.
8. C. P. Xing and K. Y. Lam, *Sequences with almost perfect linear complexity profiles and curves over finite fields*, To appear in *IEEE Trans. Inform. Theory*.
9. C. P. Xing, H. Niederreiter, K. Y. Lam and C. Ding, *Constructions of sequences with almost perfect linear complexity profiles from curves over finite fields*, To appear in *Finite Fields and Their Applications*, 1999.

Counting Functions and Expected Values in the Stability Theory of Stream Ciphers

Harald Niederreiter and Heike Paschinger

Institute of Discrete Mathematics, Austrian Academy of Sciences, Sonnenfelsgasse 19, A-1010 Vienna, Austria; e-mail:{niederreiter, heike.paschinger}@oeaw.ac.at

Abstract. In the stability theory of stream ciphers, the philosophy is that a good keystream should not only have a large linear complexity, but that a small number of bit changes should not cause a significant drop in the linear complexity. This requirement leads to the concept of the k-error linear complexity $L_{n,k}(S)$ of a bit string S of length n. The main results of the paper concern the number of bit strings S of length n with a given value of $L_{n,k}(S)$ or with a given upper bound on $L_{n,k}(S)$. On the basis of these results, bounds for the expected value of $L_{n,k}(S)$ for fixed n and k and random bit strings S of length n are established.

1 Introduction

The linear complexity is a basic complexity measure in the theory of keystreams for stream ciphers (see Rueppel [7], [8]). Recall that for an ultimately periodic binary sequence S the *linear complexity* $L(S)$ is equal to 0 if S is the zero sequence, and otherwise $L(S)$ is defined to be the length of the shortest linear feedback shift register (LFSR) that can generate S. Similarly, if n is a positive integer and S a finite binary sequence of length at least n or an infinite binary sequence, then the *nth linear complexity* $L_n(S)$ is equal to 0 if the first n terms of S are all 0, and otherwise $L_n(S)$ is equal to the length of the shortest LFSR that can generate the first n terms of S. We always have $0 \leq L_n(S) \leq n$.

In the stability theory of stream ciphers as described in the book of Ding, Xiao, and Shan [2], the philosophy is that a good keystream should not only have a large (nth) linear complexity, but that a small number of bit changes should not cause a significant drop in the (nth) linear complexity. This requirement leads to the refined concept of linear complexity in Definition 1 below. We follow the terminology of Stamp and Martin [9] since our definition is just a slight variation of theirs. We write $B(S,k)$ for the Hamming ball with center $S \in \mathbf{F}_2^n$ and radius $k \geq 0$, i.e.,

$$B(S,k) = \{T \in \mathbf{F}_2^n : d(S,T) \leq k\},$$

where $d(\cdot,\cdot)$ is the Hamming distance in the space $\mathbf{F}_2^n$. For the purposes of the present paper it suffices to consider finite binary sequences (or bit strings) of fixed length n in Definition 1. This definition can of course be extended to infinite binary sequences (see [6, Definition 3]).

Definition 1. Let n and k be integers with $n \geq 1$ and $0 \leq k \leq n$. Then the (nth) *k-error linear complexity* $L_{n,k}(S)$ of the bit string $S \in \mathbf{F}_2^n$ is defined by

$$L_{n,k}(S) = \min_{T \in B(S,k)} L_n(T).$$

The k-error linear complexity $L_{n,k}(S)$ is closely connected with the k-sphere complexity $SC_k(S)$ introduced in [2, Definition 4.9]. In fact, we have

$$L_{n,k}(S) = \min(SC_k(S), L_n(S)).$$

Note that for $k = 0$ we have $L_{n,0}(S) = L_n(S)$. Stamp and Martin [9] designed an efficient algorithm for computing the k-error linear complexity of periodic binary sequences whose period length is a power of 2. A generalization of the Stamp-Martin algorithm to arbitrary finite fields was recently developed by Kaida, Uehara, and Imamura [4].

In this paper we study the distribution of values of the k-error linear complexity over bit strings of fixed length. The following two counting functions will play principal roles.

Definition 2. Let n, k, and c be integers with $n \geq 1, 0 \leq k \leq n$, and $0 \leq c \leq n$. Then $N_{n,k}(c)$, respectively $M_{n,k}(c)$, is the number of bit strings $S \in \mathbf{F}_2^n$ with $L_{n,k}(S) = c$, respectively $L_{n,k}(S) \leq c$.

It is clear that

$$M_{n,k}(c) = \sum_{r=0}^{c} N_{n,k}(r).$$

In Section 2 we present exact formulas and bounds for $N_{n,k}(c)$, whereas Section 3 is devoted to results on $M_{n,k}(c)$. It turns out that, as far as lower bounds are concerned, $M_{n,k}(c)$ is easier to handle than $N_{n,k}(c)$. In Section 3 we also establish a formula for the number of purely periodic binary sequences with a given linear complexity. The last section contains bounds for the expected value of $L_{n,k}(S)$ for fixed n and k and random bit strings S of length n. The results of this paper can be used as a basis for testing the randomness of keystreams.

2 Results on $N_{n,k}(c)$

For $k = 0$ we have the following special case of a well-known formula of Gustavson [3] (see also [7, p. 36]).

Lemma 1. *We have*

$$N_{n,0}(c) = \begin{cases} 1 & \text{for } \; c = 0, \\ 2^{\min(2c-1, 2n-2c)} & \text{for } \; 1 \leq c \leq n. \end{cases}$$

For $k \geq 1$ there are various cases for which an exact formula for $N_{n,k}(c)$ can be easily established. We collect these cases in the following theorem.

Theorem 1. *The following formulas hold:*

(i) $N_{n,k}(0) = \sum_{t=0}^{k} \binom{n}{t}$ *for* $1 \leq k \leq n$;
(ii) $N_{n,k}(1) = \sum_{t=0}^{k} \binom{n}{t} + \binom{n-1}{k}$ *for* $1 \leq k < \lfloor n/2 \rfloor$;
(iii) $N_{n,k}(1) = \sum_{t=k+1}^{n} \binom{n}{t}$ *for* $\lfloor n/2 \rfloor \leq k \leq n$;
(iv) $N_{n,k}(c) = 0$ *for* $\lfloor n/2 \rfloor \leq k \leq n$ *and* $2 \leq c \leq n$;
(v) $N_{n,k}(n) = 0$ *for* $1 \leq k \leq n$.

Proof. (i) This follows from $N_{n,k}(0) = |B(Z,k)|$ with $Z = (0,\ldots,0) \in \mathbf{F}_2^n$.
(ii) There are two bit strings in $\mathbf{F}_2^n$ with nth linear complexity equal to 1, namely $S = (1,0,\ldots,0)$ and $T = (1,1,\ldots,1)$. All $R \in B(T,k)$ satisfy $L_{n,k}(R) = 1$ since the balls $B(T,k)$ and $B(Z,k)$ are disjoint for $k < \lfloor n/2 \rfloor$, where Z is as in (i). This contributes

$$\sum_{t=0}^{k} \binom{n}{t}$$

to $N_{n,k}(1)$. Now we consider $R \in B(S,k)$. Note that $B(S,k)$ and $B(T,k)$ are disjoint since $k < \lfloor n/2 \rfloor$. We must not count those $R \in B(S,k)$ with $L_{n,k}(R) = 0$, i.e., which are also in $B(Z,k)$. This means that we count only those R that we obtain by changing exactly k bits in S and not the first bit. This contributes $\binom{n-1}{k}$ to $N_{n,k}(1)$.

(iii) If Z and T are as in (i) and (ii), respectively, then for $\lfloor n/2 \rfloor \leq k \leq n$ we have

$$B(Z,k) \cup B(T,k) = \mathbf{F}_2^n.$$

Thus, $L_{n,k}(R) \leq 1$ for all $R \in \mathbf{F}_2^n$, and so by (i),

$$N_{n,k}(1) = 2^n - N_{n,k}(0) = \sum_{t=k+1}^{n} \binom{n}{t}.$$

(iv) This follows from the proof of (iii).
(v) There is only one bit string in $\mathbf{F}_2^n$ with nth linear complexity equal to n, namely $V = (0,\ldots,0,1)$, and it satisfies $L_{n,k}(V) = 0$ for $1 \leq k \leq n$. □

For $k = 1$ and small positive values of c it is not difficult to derive formulas for $N_{n,k}(c)$. Note that for $c = 1$ we have $N_{n,1}(1) = 2n$ for $n \geq 4$ by Theorem 1(ii). We indicate the argument for the next case $c = 2$. We have verified similar formulas for $c \leq 7$ and this has led to the subsequent conjecture.

Proposition 1. *For* $k = 1$ *and* $c = 2$ *we have*

$$N_{n,1}(2) = 8n \quad \textit{for all}\, n \geq 10.$$

Proof. There are eleven binary sequences $S_i, 1 \leq i \leq 11$, of linear complexity at most 2, with S_1, S_2, S_3 in the following list having linear complexity at most 1:

S_1: $0,0,\dots$	(period 1)
S_2: $1,0,0,\dots$	(preperiod 1, period 1)
S_3: $1,1,\dots$	(period 1)
S_4: $0,1,0,0,\dots$	(preperiod 2, period 1)
S_5: $1,1,0,0,\dots$	(preperiod 2, period 1)
S_6: $0,1,0,1,\dots$	(period 2)
S_7: $1,0,1,0,\dots$	(period 2)
S_8: $0,1,1,\dots$	(preperiod 1, period 1)
S_9: $0,1,1,0,1,1,\dots$	(period 3)
S_{10}: $1,0,1,1,0,1,\dots$	(period 3)
S_{11}: $1,1,0,1,1,0,\dots$	(period 3)

Let $n \geq 10$ and put $B_i = B(S_i^{(n)}, 1)$ for $1 \leq i \leq 11$, where $S_i^{(n)}$ is the string consisting of the first n terms of S_i. Then

$$N_{n,1}(2) = |\bigcup_{i=1}^{11} B_i| - N_{n,1}(0) - N_{n,1}(1) = |\bigcup_{i=1}^{11} B_i| - 3n - 1,$$

where we used Theorem 1(i),(ii) in the last step. Since any four different balls B_i have an empty intersection, the inclusion-exclusion formula yields

$$|\bigcup_{i=1}^{11} B_i| = \sum_{i=1}^{11} |B_i| - \sum_{1 \leq i < j \leq 11} |B_i \cap B_j| + \sum_{1 \leq h < i < j \leq 11} |B_h \cap B_i \cap B_j|.$$

There are seven pairs (i,j) with $1 \leq i < j \leq 11$ for which $B_i \cap B_j$ is nonempty, namely

$$(1,2), (1,4), (1,5), (2,4), (2,5), (4,5), (3,8).$$

In each of these cases we have $|B_i \cap B_j| = 2$. There are four triples (h,i,j) with $1 \leq h < i < j \leq 11$ for which $B_h \cap B_i \cap B_j$ is nonempty, namely

$$(1,2,4), (1,2,5), (1,4,5), (2,4,5).$$

In each of these cases we have $|B_h \cap B_i \cap B_j| = 1$. Thus,

$$|\bigcup_{i=1}^{11} B_i| = 11(n+1) - 7 \cdot 2 + 4 \cdot 1 = 11n + 1,$$

and the desired formula for $N_{n,1}(2)$ follows. □

Conjecture. *Let c be a positive integer and $k = 1$. Then there exists a positive integer n_0 depending on c such that*

$$N_{n,1}(c) = 2^{2c-1} n \quad \textit{for all } n \geq n_0.$$

We have considered various other special cases for $k = 2,3,4$ and $c = 2,3$. In all cases the resulting formula has the form

$$N_{n,k}(c) = N_{n,0}(c) \binom{n}{k} + f(n)$$

for all sufficiently large n, where f is a polynomial of degree $k-2$. We include one more example. It is clear that this proof technique becomes infeasible for large values of k or c.

Proposition 2. *For $k=2$ and $c=2$ we have*

$$N_{n,2}(2) = 8\binom{n}{2} + 12 \quad \textit{for all } n \geq 16.$$

Proof. Proceed as in the proof of Proposition 1, but with balls of radius 2 instead of balls of radius 1. □

With the help of Helmut Postl (Austrian Ministry of Defense) and powerful workstations provided by DEC Austria, we have computed tables of $N_{n,k}(c)$ for all $n \leq 32$. These tables are available from the authors on request. To conclude this section, we note the following obvious upper bound for $N_{n,k}(c)$.

Proposition 3. *For any integers $n \geq 1, 0 \leq k \leq n$, and $0 \leq c \leq n$ we have*

$$N_{n,k}(c) \leq \min\left(2^n, N_{n,0}(c)\sum_{t=0}^{k}\binom{n}{t}\right).$$

3 Results on $M_{n,k}(c)$

We consider now the counting function $M_{n,k}(c)$ introduced in Definition 2. For $k=0$ it follows immediately from Lemma 1 that

$$M_{n,0}(c) = \sum_{r=0}^{c} N_{n,0}(r) = 1 + \sum_{r=1}^{c} 2^{2r-1} = \frac{1}{3}\left(2^{2c+1}+1\right) \text{ for } c \leq \lfloor n/2 \rfloor. \quad (1)$$

A similar formula can be derived for $\lfloor n/2 \rfloor < c \leq n$. We have a general upper bound analogous to that in Proposition 3.

Proposition 4. *For any integers $n \geq 1, 0 \leq k \leq n$, and $0 \leq c \leq n$ we have*

$$M_{n,k}(c) \leq \min\left(2^n, M_{n,0}(c)\sum_{t=0}^{k}\binom{n}{t}\right).$$

Proof. Let $S_1, \ldots, S_h$ be the bit strings of length n with nth linear complexity at most c, so that $h = M_{n,0}(c)$. Then

$$M_{n,k}(c) = \left|\bigcup_{i=1}^{h} B(S_i,k)\right| \leq \sum_{i=1}^{h} |B(S_i,k)| = M_{n,0}(c)\sum_{t=0}^{k}\binom{n}{t},$$

and the desired bound follows immediately. □

The main part of this section is devoted to the proof of a lower bound for $M_{n,k}(c)$. For this proof we need a result of independent interest concerning the following counting function.

Theorem 2. *The number $P(c)$ of purely periodic binary sequences S with $L(S) = c$ satisfies $P(0) = 1$ and*

$$P(c) = \frac{1}{3}\left(2^{2c} - 1\right) \quad \textit{for } c \geq 1.$$

Proof. We proceed by induction on c. The case $c = 0$ is trivial, so we assume $c \geq 1$. An ultimately periodic binary sequence S with $L(S) = c$ has a minimal polynomial $m \in \mathbf{F}_2[x]$ of degree c. We write

$$m(x) = x^r g(x)$$

with $g \in \mathbf{F}_2[x]$ and $g(0) \neq 0$. Note that r is the length of the preperiod of S and g is the minimal polynomial of the periodic part of S. If S is not purely periodic, then $1 \leq r \leq c$. If we fix r, then the periodic part of S has linear complexity $c - r$, and so there are $P(c - r)$ choices for the periodic part. For each such choice the rth term of S is uniquely determined by the condition that r is the length of the preperiod, and so there are 2^{r-1} choices for the preperiod. Thus, if $U(c)$ denotes the number of ultimately periodic, but not purely periodic binary sequences S with $L(S) = c$, then

$$U(c) = \sum_{r=1}^{c} 2^{r-1} P(c - r) = \sum_{r=0}^{c-1} 2^{c-r-1} P(r).$$

Furthermore, we have $P(c) + U(c) = 2^{2c-1}$ by Lemma 1. Thus, using the induction hypothesis we get

$$\begin{aligned} P(c) = 2^{2c-1} - U(c) &= 2^{2c-1} - \sum_{r=0}^{c-1} 2^{c-r-1} P(r) \\ &= 2^{2c-1} - 2^{c-1} - \frac{1}{3}\sum_{r=1}^{c-1} 2^{c-r-1}(2^{2r} - 1) = \frac{1}{3}(2^{2c} - 1), \end{aligned}$$

and the induction is complete. □

Corollary 1. *The number $U(c)$ of ultimately periodic, but not purely periodic binary sequences S with $L(S) = c$ satisfies $U(0) = 0$ and*

$$U(c) = \frac{1}{3}(2^{2c-1} + 1) \quad \textit{for } c \geq 1.$$

Proof. This follows from Theorem 2 and the identity $P(c) + U(c) = 2^{2c-1}$ for $c \geq 1$ noted in the proof of that theorem. □

Corollary 2. *The number $Q(c)$ of purely periodic binary sequences S with $L(S) \leq c$ is given by*

$$Q(c) = \frac{1}{9}(2^{2c+2} - 3c + 5) \quad \textit{for } c \geq 0.$$

Proof. This follows from

$$Q(c) = \sum_{r=0}^{c} P(r)$$

and Theorem 2. □

Corollary 3. *The number $V(c)$ of ultimately periodic, but not purely periodic binary sequences S with $L(S) \leq c$ is given by*

$$V(c) = \frac{1}{9}(2^{2c+1} + 3c - 2) \quad \textit{for } c \geq 0.$$

Proof. This follows from

$$V(c) = \sum_{r=0}^{c} U(r)$$

and Corollary 1. □

Before we state the main result of this section, we note the following combinatorial lemma which can be proved by induction or derived from the inclusion-exclusion formula.

Lemma 2. *For any finite sets $A_1, \ldots, A_s$ we have*

$$|\bigcup_{i=1}^{s} A_i| \geq \sum_{i=1}^{s} |A_i| - \sum_{1 \leq i < j \leq s} |A_i \cap A_j|.$$

Theorem 3. *For any integers $c \geq 1, k \geq 1$, and $n \geq (4k+3)c$ we have*

$$M_{n,k}(c) \geq M_{n,0}(c) \sum_{t=0}^{k} \binom{n}{t} - \left(\sum_{t=0}^{k-1} \binom{n}{t} + \binom{n-1}{k-1} \right) \cdot \left(\sum_{r=0}^{c-1} 2^r Q(c-r-1) \sum_{t=0}^{2k-1} \binom{r}{t} + \sum_{r=1}^{c-1} 2^{r-1} Q(c-r-1) \sum_{t=1}^{2k} \binom{r}{t} \right),$$

where Q is the counting function in Corollary 2.

Proof. Let $S_1, \ldots, S_h$ be as in the proof of Proposition 4, with $h = M_{n,0}(c)$, and put $B_i = B(S_i, k)$ for $1 \leq i \leq h$. Then using Lemma 2,

$$M_{n,k}(c) = |\bigcup_{i=1}^{h} B_i| \geq \sum_{i=1}^{h} |B_i| - \sum_{1 \leq i < j \leq h} |B_i \cap B_j|$$

$$= h \sum_{t=0}^{k} \binom{n}{t} - \sum_{1 \leq i < j \leq h} |B_i \cap B_j|.$$

By invoking [1, Theorem 2.4.8], we get that for $1 \leq i < j \leq h$ we have

$$|B_i \cap B_j| \leq |B(Z,k) \cap B(S,k)| = \sum_{t=0}^{k-1} \binom{n}{t} + \binom{n-1}{k-1},$$

where Z and S are as in the proof of Theorem 1. If $I_{n,k}(c)$ is the number of pairs (i,j) with $1 \leq i < j \leq h$ for which $B_i \cap B_j$ is nonempty, then it follows that

$$M_{n,k}(c) \geq h \sum_{t=0}^{k} \binom{n}{t} - \left(\sum_{t=0}^{k-1} \binom{n}{t} + \binom{n-1}{k-1} \right) I_{n,k}(c). \tag{2}$$

The rest of the proof is concerned with bounding $I_{n,k}(c)$ from above. Note first that $B_i \cap B_j$ is nonempty if and only if $d(S_i, S_j) \leq 2k$. Since $n \geq (4k+3)c \geq 2c$, the minimal polynomial $m_i \in \mathbf{F}_2[x]$ of each bit string S_i is uniquely determined (compare with [5, Theorem 1]). For $1 \leq i \leq h$ we write

$$m_i(x) = x^{r_i} g_i(x)$$

with $g_i \in \mathbf{F}_2[x]$ and $g_i(0) \neq 0$. Note that $\deg(m_i) \leq c$.

Consider a pair (i,j) with $g_i \neq g_j$. We compare the first $(4k+3)c$ corresponding bits of S_i and S_j in the following way. In each of the bit strings S_i and S_j split up the segment from position $c+1$ to position $(4k+3)c$ into $2k+1$ blocks of $2c$ consecutive terms. In S_i (respectively S_j) this yields the blocks $D_i^{(b)}$ with $0 \leq b \leq 2k$ (respectively $D_j^{(b)}$ with $0 \leq b \leq 2k$), reaching from position $(2b+1)c+1$ to position $(2b+3)c$. Each bit string $D_i^{(b)}$ (respectively $D_j^{(b)}$) has the unique minimal polynomial g_i (respectively g_j). Since $g_i \neq g_j$, it follows that $D_i^{(b)} \neq D_j^{(b)}$ for $0 \leq b \leq 2k$, and so $d(S_i, S_j) \geq 2k+1$. Hence B_i and B_j are disjoint.

Thus, when studying $I_{n,k}(c)$, we need to take into account only those pairs (i,j) with $g_i = g_j$. Consider such a pair and put $r = \max(r_i, r_j)$. Observe that from position $r+1$ on, any c consecutive terms of S_i (respectively S_j) uniquely determine all positions from $r+1$ to n in S_i (respectively S_j) by the same linear recursion (note that we can not only go forward in the string but also backward since $g_i(0) = g_j(0) \neq 0$). Hence by the same argument as above, but by splitting up into blocks of length c instead of $2c$, one proves that $d(S_i, S_j) \leq 2k$ can happen only if S_i and S_j agree in positions $r+1$ through n.

Thus we have shown that in the study of $I_{n,k}(c)$ it suffices to consider the pairs (i,j) for which S_i and S_j differ at most in the first $r = \max(r_i, r_j)$ terms. First we treat the case where $r_i \neq r_j$, say w.l.o.g. $r = r_i > r_j$. Then $1 \leq r \leq c$ and $\deg(g_i) \leq c - r$. For fixed r, the number of choices for S_i is determined as follows. For the part from position $r+1$ to n (the "purely periodic" part) there are $Q(c-r)$ choices. Among the first r terms of S_i (the "preperiod"), the rth term is uniquely determined by the condition that the length of the "preperiod" is r, whereas the first $r-1$ terms are free. Altogether, this yields $2^{r-1}Q(c-r)$ choices for S_i. For fixed S_i, we now consider the number of choices for S_j. As noted above, only the first r terms of S_j can still be chosen. However, the rth term of S_j must be different from that of S_i (for otherwise we would have $r_j = r$), and so only the first $r-1$ terms of S_j can still be selected. Recall that we must have $d(S_i, S_j) \leq 2k$. The Hamming distance between the strings of first $r-1$ terms of S_i and S_j can therefore be at most $2k-1$. Hence the number of choices for S_j is at most

$$\sum_{t=0}^{2k-1} \binom{r-1}{t}.$$

Thus, the contribution of the case $r_i \neq r_j$ to $I_{n,k}(c)$ is at most

$$\sum_{r=1}^{c} 2^{r-1}Q(c-r) \sum_{t=0}^{2k-1} \binom{r-1}{t} = \sum_{r=0}^{c-1} 2^{r}Q(c-r-1) \sum_{t=0}^{2k-1} \binom{r}{t}. \tag{3}$$

Now we treat the remaining case $r_i = r_j = r$. By the same analysis as above, there are $2^{r-1}Q(c-r)$ choices for S_i. For fixed S_i, again only the first r terms of S_j can still be chosen. The rth terms of S_j and S_i must now agree since $r_j = r$, and so only the first $r-1$ terms of S_j can still be selected. The Hamming distance between the strings of first $r-1$ terms of S_i and S_j must be in the interval $[1, 2k]$. Therefore the number of choices for S_j is at most

$$\sum_{t=1}^{2k} \binom{r-1}{t}.$$

Note that in the case $r_i = r_j$ the pairs (i,j) and (j,i) are both counted, hence the contribution of this case to $I_{n,k}(c)$ is at most

$$\frac{1}{2}\sum_{r=2}^{c} 2^{r-1}Q(c-r) \sum_{t=1}^{2k} \binom{r-1}{t} = \sum_{r=1}^{c-1} 2^{r-1}Q(c-r-1) \sum_{t=1}^{2k} \binom{r}{t}.$$

By combining this with (3), we obtain

$$I_{n,k}(c) \leq \sum_{r=0}^{c-1} 2^{r}Q(c-r-1) \sum_{t=0}^{2k-1} \binom{r}{t} + \sum_{r=1}^{c-1} 2^{r-1}Q(c-r-1) \sum_{t=1}^{2k} \binom{r}{t}.$$

In view of (2) this completes the proof of the theorem. □

For the range of n in Theorem 3 a formula for $M_{n,0}(c)$ was given in (1). In the case $k = 1$ the lower bound in Theorem 3 can be greatly simplified by elementary algebraic manipulations and reduces to the following form.

Corollary 4. *For any integers $c \geq 1$ and $n \geq 7c$ we have*

$$M_{n,1}(c) \geq \frac{n-1}{3}(2^{2c+1}+1)+2^{c+1}.$$

4 Expected Values

For fixed integers $n \geq 1$ and $k \geq 0$ let $E_{n,k}$ be the expected value of the nth k-error linear complexity $L_{n,k}(S)$ for random bit strings S of length n, that is,

$$E_{n,k} = \frac{1}{2^n} \sum_{S \in \mathbb{F}_2^n} L_{n,k}(S).$$

The following lemma establishes a connection between $E_{n,k}$ and the counting function $M_{n,k}$ in Definition 2.

Lemma 3. *For any integers $n \geq 1$ and $k \geq 0$ we have*

$$E_{n,k} = n - \frac{1}{2^n} \sum_{c=0}^{n-1} M_{n,k}(c).$$

Proof. We have

$$\begin{aligned} 2^n E_{n,k} &= \sum_{S \in \mathbb{F}_2^n} L_{n,k}(S) = \sum_{c=1}^{n} c N_{n,k}(c) = \sum_{c=1}^{n} c(M_{n,k}(c) - M_{n,k}(c-1)) \\ &= \sum_{c=0}^{n} c M_{n,k}(c) - \sum_{c=0}^{n-1} (c+1) M_{n,k}(c) \\ &= n M_{n,k}(n) - \sum_{c=0}^{n-1} M_{n,k}(c) = n 2^n - \sum_{c=0}^{n-1} M_{n,k}(c). \end{aligned}$$

This yields the claimed identity. □

For $k = 0$ we have the formula

$$E_{n,0} = \frac{n}{2} + \frac{4 + R_2(n)}{18} + o(1) \tag{4}$$

shown in [7, Proposition 4.2], where $R_2(n)$ is the least residue of n modulo 2. Here and in the following, $o(1)$ denotes a term that tends to 0 as n tends to ∞. Now we derive a lower bound on $E_{n,k}$ for $k \geq 1$.

Theorem 4. *For any integers $n \ge 1$ and $k \ge 1$ we have*

$$E_{n,k} \ge \frac{n}{2} - \frac{1}{2}\log_2\left(\sum_{t=0}^{k}\binom{n}{t}\right) - 1.041 + o(1),$$

where $\log_2$ *denotes the logarithm to the base 2.*

Proof. Put

$$A = \left\lfloor \frac{1}{2}\log_2 \frac{3\cdot 2^{n-1}}{\sum_{t=0}^{k}\binom{n}{t}} \right\rfloor$$

and note that $0 \le A \le \lfloor n/2 \rfloor$. For $0 \le c \le A$ we use the bound

$$M_{n,k}(c) \le \frac{1}{3}(2^{2c+1}+1)\sum_{t=0}^{k}\binom{n}{t}$$

which follows from (1) and Proposition 4. For $A < c \le n-1$ we use the trivial bound

$$M_{n,k}(c) \le 2^n.$$

This yields

$$\begin{aligned}
\frac{1}{2^n}\sum_{c=0}^{n-1} M_{n,k}(c) &\le \frac{1}{3\cdot 2^n}\left(\sum_{t=0}^{k}\binom{n}{t}\right)\sum_{c=0}^{A}(2^{2c+1}+1) + n - 1 - A \\
&\le \frac{1}{3\cdot 2^{n-1}}\left(\sum_{t=0}^{k}\binom{n}{t}\right)\sum_{c=0}^{A}4^c + n - \frac{1}{2}\log_2\frac{3\cdot 2^{n-1}}{\sum_{t=0}^{k}\binom{n}{t}} + o(1) \\
&\le \frac{4^{A+1}}{9\cdot 2^{n-1}}\sum_{t=0}^{k}\binom{n}{t} + \frac{n+1}{2} + \frac{1}{2}\log_2\left(\sum_{t=0}^{k}\binom{n}{t}\right) - \frac{1}{2}\log_2 3 + o(1) \\
&\le \frac{4}{3} + \frac{n+1}{2} + \frac{1}{2}\log_2\left(\sum_{t=0}^{k}\binom{n}{t}\right) - \frac{1}{2}\log_2 3 + o(1) \\
&\le \frac{n}{2} + \frac{1}{2}\log_2\left(\sum_{t=0}^{k}\binom{n}{t}\right) + 1.041 + o(1).
\end{aligned}$$

The desired bound is now obtained from Lemma 3. □

An upper bound on $E_{n,k}$ can be easily derived from (4), namely for any integers $n \ge 1$ and $k \ge 1$ we have

$$E_{n,k} = \frac{1}{2^n}\sum_{S\in\mathbb{F}_2^n} L_{n,k}(S) \le \frac{1}{2^n}\sum_{S\in\mathbb{F}_2^n} L_{n,0}(S) = E_{n,0} \le \frac{n}{2} + \frac{5}{18} + o(1).$$

References

[1] G. Cohen, I. Honkala, S. Litsyn, and A. Lobstein, "Covering Codes", Elsevier, Amsterdam, 1997.

[2] C. Ding, G. Xiao, and W. Shan, "The Stability Theory of Stream Ciphers", Lecture Notes in Computer Science, Vol. 561, Springer-Verlag, Berlin, 1991.

[3] F.G. Gustavson, Analysis of the Berlekamp-Massey linear feedback shift-register synthesis algorithm, *IBM J. Res. Develop.* **20** (1976), 204–212.

[4] T. Kaida, S. Uehara, and K. Imamura, A new algorithm for the k-error linear complexity of sequences over $GF(p^m)$ with period p^n, this volume.

[5] H. Niederreiter, Sequences with almost perfect linear complexity profile, in "Advances in Cryptology – EUROCRYPT '87" (D. Chaum and W.L. Price, eds.), Lecture Notes in Computer Science, Vol. 304, pp. 37–51, Springer-Verlag, Berlin, 1988.

[6] H. Niederreiter, Some computable complexity measures for binary sequences, this volume.

[7] R.A. Rueppel, "Analysis and Design of Stream Ciphers", Springer-Verlag, Berlin, 1986.

[8] R.A. Rueppel, Stream ciphers, in "Contemporary Cryptology: The Science of Information Integrity" (G.J. Simmons, ed.), pp. 65–134, IEEE Press, New York, 1992.

[9] M. Stamp and C.F. Martin, An algorithm for the k-error linear complexity of binary sequences with period 2^n, *IEEE Trans. Inform. Theory* **39** (1993), 1398–1401.

On-line Constraint-based Pattern Matching on Sequences

Vladimir A. Oleshchuk

Department of Information and Communication Technology
Agder College
and
Sentek, Agder Research Foundation
N-4876 Grimstad, Norway.
Email: vladimir.oleshchuk@hia.no

Abstract. We consider pattern matching problems where patterns are presented as sequences of constraints on input elements. Given an infinite alphabet A, a pattern P is a sequence $\langle p_{i_1}, p_{i_2}, ..., p_{i_m} \rangle$ of predicates p_{i_j} defined on A. The pattern P matches an input sequence $t \in A^*$ if $t = ux_1x_2 \cdots x_m v$ such that $p_{i_j}(x_j) = true$, $j = 1, 2, ..., m$ for some $u, v \in A^*$.
We address the following problem: given a pattern P and an input sequence t, find all positions in t where P matches t. We present an on-line algorithm that solves the problem in general and give an efficient linear-time algorithm for some classes of patterns. The proposed algorithms can be used for efficient on-line pattern matching in digitized presentation of continuous reality such as, for example, digitized images and sounds, or noisy telemetric data.

1 Introduction

The main characteristic of multimedia applications is that they manage unconventional data such as images and sounds. Such applications are characterized by the extremely large volumes of data processed. The continued decline in the cost of mass storage devices and increasing speed and capacity of the computers make multimedia applications such as, for example, multimedia databases one of the fastest growing category of applications. However, many objects of real world such as images and sounds that presented in such applications are continuous by their nature. They need to be discretized in some way to be presented in computers due to the finite presentation restriction of computer representation. Discretizing is a one-to-many process, that is, the same continuous object in the real world corresponds to many discretized presentation due to, for example, limited equipment resolution, noises or precise measurement problems. Efficient use of multimedia databases in many applications requires implementation of operations available for conventional database systems such as searching, selecting, updating data in databases, etc. Since digitized presentations are approximate presentations of continuous objects, the traditional algorithms developed for discrete objects, for example texts, cannot always be used or are inefficient. Therefore, it is important to develop new algorithms that are fast, require little memory and limited buffering and can operate in real-time on digitized presentations of continuous objects. This is especially important for multimedia data since in this case both pattern and data can be very large.

Pattern matching is the problem of locating a specific pattern inside raw data. This is an important component of many tasks, including text editing, data retrieval, data compression, data mining. Another application area is real-time monitoring and event detection in manufacturing processes by examining noisy sensor data [9]. Formally, the pattern matching consists of finding one, or more generally, all the occurrences of a pattern inside sequential raw data. Raw data can be seen as a sequence over some finite or even infinite alphabet. The patterns are usually a collection of sequences described in some formal language.

When row data are texts over finite alphabets the pattern matching problem is known as string-matching. Many solutions based on the use of automata or combinatorial properties of strings over finite alphabets have been proposed in the literature [1–5]. These problems and proposed solutions heavily depend on a finiteness property of the underlying alphabets and cannot be applied directly to solve the problems in the continuous setting where alphabets are infinite or very large and precise matching, as in digitized presentations, often is not necessary or even possible [7].

In this paper we deal with pattern matching problems applied to sequences representing unconventional data as, for example, continuous data in digitized presentation. Since under digitization both a single pattern and row data can map into one-of-many digitized patterns and digitized row sequences (due to problems mentioned above) it is unlikely that an exact match can always be achieved. We present an on-line algorithm that solves the problem in general, and then modify it to develop a more efficient linear-time version for some classes of patterns.

The paper is organized as follows. In Section 2 we provide the necessary definitions and notations and present some examples as motivation and demonstration of usability of our generalized definition of pattern matching problem. The main algorithm is presented in Section 3. The linear-time algorithm and its complexity analysis is given in Section 4. General discussion of complexity of proposed algorithms is given in Section 5. Finally, concluding remarks are made in the last section.

2 Preliminaries

The alphabet, an infinite set of elements, is denoted by A. Let A^* denote the set of all finite-length sequences of elements from A. The length of a finite sequence $t_n = a_1 a_2 \cdots a_n$ is denoted as $|t_n|$, that is, $|t_n| = n$. We assume that $t_0 = \varepsilon$, where ε denotes the empty sequence. Let A^m denote the set of all sequences of length m formed using elements from A. We say that a sequence w is a *prefix* of a sequence t if $t = wv$ for some sequence $v \in A^*$. Similarly, we say that a sequence w is a *suffix* of a sequence t if $t = uw$ for some sequence $u \in A^*$. A sequence w is a *subsequence* of a sequence t if $t = uwv$ for some $u, v \in A^*$.

Let $A = \{a_1, a_2, ...\}$ be an alphabet and $\Omega_A = \{p_1, p_2, ...\}$ be a set of predicates defined on A, that is, $p_i : A \to \{true, false\}$ for any $p_i \in \Omega_A$. A *pattern* P of length m is a sequence $\langle p_{i_1}, p_{i_2}, ..., p_{i_m} \rangle$ of predicates from Ω_A. The pattern $P = \langle p_{i_1}, p_{i_2}, ..., p_{i_m} \rangle$ represents a subset Q_P of sequences from A^m defined as

$$Q_P = \left\{ a_{i_1} a_{i_2} \cdots a_{i_m} | \bigwedge_{j=1}^{m} p_{i_j}\left(a_{i_j}\right) = true \text{ where } a_{i_j} \in A \right\}$$

We say that a pattern $P = \langle p_{i_1}, p_{i_2}, ..., p_{i_m} \rangle$ from Ω_A *matches* a sequence $t_n = x_1 x_2 \cdots x_n$ from A^*, if a sequence $y_1 y_2 \cdots y_m$ from Q_P occurs as a subsequence of t, that is, if $t = u y_1 y_2 \cdots y_m v$ for some $u, v \in A^*$. The pattern P occurs beginning at position $k+1$ in sequence $t_n = x_1 x_2 \cdots x_n$ or matches t in position $k+1$ if

$$\bigwedge_{j=k+1}^{k+m} p_{i_j}\left(x_j\right) = true \text{ where } 0 \leq k \leq n - m$$

In this paper we address the following *pattern matching problem* :

> Given a pattern P from Ω_A and an input sequence t from A^*, find all positions in t where P matches t.

Many pattern matching problems can be formulated in this form. We give some illustration of proposed formulation of pattern matching problem as a generalization of pattern matching problems that have been studied in the literature. The following examples present both pattern matching problems that have already been studied and have not yet been studied.

Example 1. Let $A = \{a_1, a_2, ..., a_t\}$ and $\Omega_A = \{p_1, p_2, ..., p_t\}$ such that

$$p_i(x) = \begin{cases} true, & \text{if } x = a_i \\ false, & \text{otherwise} \end{cases}$$

where $i = 1, 2, ..., t$. Then pattern $P = \langle p_{i_1}, p_{i_2}, ..., p_{i_k} \rangle$ corresponds a one-element set $Q_P = \{a_{i_1} a_{i_2} \cdots a_{i_k}\}$. Thus, the pattern matching based on constraints from Ω_A is a string-matching problem that has been studied intensively (see surveys [1–4]).

Example 2. Let $A = \{a_1, a_2, ..., a_t\}$ and $\Omega_A = \{p_1, p_2, ..., p_t\}$ such that

$$p_i(x) = \begin{cases} true, & \text{if } x \in A_i \\ false, & \text{otherwise} \end{cases}$$

where $A_i \subseteq A$ for $i = 1, 2, ..., t$. Then pattern $P = \langle p_{i_1}, p_{i_2}, ..., p_{i_k} \rangle$ corresponds to $Q_P = \left\{ a_{i_1} a_{i_2} \cdots a_{i_k} | a_{i_j} \in A_{i_j}, j = 1, 2, ..., k \right\}$. In this example the pattern matching based on constraints from Ω_A is an approximate string-matching problem that has been studied intensively (see surveys [1, 2]). Note that the predicate p_i such that $A_i = A$ represents the *don't care* property.

Example 3. Let A be a set of real numbers and $\Omega_A = \{p_0, p_1, ...\}$ such that

$$p_i(x) = \begin{cases} true, & \text{if } x \in A_i \\ false, & \text{otherwise} \end{cases}$$

where $A_i = \{a | l_i \le a < r_i\}$ represents a half-open interval. The pattern matching problem in such setting has been studied by Landau and Vishkin [7] in relation to algorithmic pattern matching in image processing.

Example 4. Let A be a set of elements, E be an equivalence relation on A and $E_i, i = 0, 1, ..., t$ denote equivalence classes induced by E on A. Let $\Omega_A = \{p_1, p_2, ..., p_t\}$ be defined as

$$p_i(x) = \begin{cases} true, & \text{if } x \in E_i \\ false, & \text{otherwise.} \end{cases}$$

3 Main algorithm

In this section we present a general on-line algorithm for constraint-based pattern matching.

Let t_k be a string $a_1 a_2 \cdots a_k$ from A^* and P be a pattern $\langle p_{i_1}, p_{i_2}, \cdots, p_{i_m} \rangle$ from Ω_A. Let S_k^P denote a set of lengths of all prefixes of P matching some suffix of t_k, that is, $j \in S_k^P$ if and only if

$$p_{i_1}(a_{k-j+1}) \wedge p_{i_2}(a_{k-j+2}) \wedge \cdots \wedge p_{i_j}(a_k) = true$$

Therefore, P matches t_k if and only if $m \in S_j^P$ for some $j \le k$. Thus, if we want to perform pattern matching based on this idea, we have to construct a sequence of sets $S_0^P, S_1^P, ..., S_k^P, ...$ for a given pattern $P = \langle p_{i_1}, p_{i_2}, ..., p_{i_m} \rangle$ and input sequence $t = a_1 a_2 \cdots a_k \cdots$, and check whether m is in S_k^P for some $k > 0$. Then, for any S_k^P such that $m \in S_k^P$ we report that P matches t in position $k - m + 1$.

The algorithm `MATCH`, based on the above idea, is presented in Figure 1. The output of the algorithm `MATCH`(t, P) is the set of all positions of occurrences of P in t. The algorithm uses the procedure `UPDATE`(S^P, x, P) to construct S_i^P based on S_{i-1}^P and x, where x is a new input element of sequence t.

Let us describe in details the algorithm `UPDATE`. The main purpose of `UPDATE` is to construct $S_0^P, S_1^P, ..., S_k^P, ...$ for a given pattern $P = \langle p_{i_1}, p_{i_2}, ..., p_{i_m} \rangle$ and input sequence $t = a_1 a_2 \cdots a_k \cdots$. The algorithm `UPDATE` is presented in Figure 2. According to 2, `UPDATE` calculates S_{i+1}^P based on S_i^P, the next input element x and the length of the prefix of P that has been already matched.

The difference between the algorithm `UPDATE` and a naive brute-force algorithm is that in `UPDATE` some unnecessary computations are removed from consideration. For each input element x of input t, `UPDATE` considers only those predicates of the pattern $P = \langle p_{i_1}, p_{i_2}, ..., p_{i_m} \rangle$ that still may be a part of some occurrences of P in t. When an input of length i has been analyzed, `UPDATE`

```
Algorithm MATCH(t, P)
Input: a sequence of data t and a pattern P = ⟨p_{k_1}, p_{k_2}, ..., p_{k_m}⟩
Output: begining positions of all matching patterns
S ← {0};
pos ← 0;
while input is not empty do
     x ←read next element from t;
     pos ← pos + 1;
     S ← UPDATE (S, x, P);
     if m ∈̇ S then
          pattern matching at (pos − m + 1)
          S ← S \ {m}
end.
```

Fig. 1. Algorithm MATCH

```
Algorithm UPDATE(S^P, x, P)
Input: Set S^P such that S^P ⊆ {0, 1, ..., m − 1},
          pattern P = ⟨p_{i_1}, p_{i_2}, ..., p_{i_m}⟩ from Ω_A and x from A
Output: Set S^P such that S^P ⊆ {0, 1, ..., m}
S' ← ∅
for each j from S^P do
     S^P ← S^P \ {j}
     if p_{i_{j+1}} (x) = true then S' ← S' ∪ {j + 1}
S^P ← S' ∪ {0}
return S^P
```

Fig. 2. Algorithm UPDATE

contains in S_i^P references on all predicates that need to be evaluated on the next input x, that is, $j \in S_i^P$ represents p_{i_j} from P.

However, the time complexity of UPDATE depends on properties of the pattern P, that is, on both the complexity of evaluation of predicates from P and their interdependency. The interdependency presents the knowledge of how truth values for some predicates of P can be found without their explicit evaluation but based only on truth values of some already evaluated predicates of P.

Properties of UPDATE are summarized in Lemma 1.

Lemma 1. *Let $P = \langle p_{i_1}, p_{i_2}, ..., p_{i_m} \rangle$ be a pattern. Let $S_0^P, S_1^P, ..., S_k^P, ...$ be a sequence of sets generated by algorithm* UPDATE *based on input $t = x_1 x_2 \cdots x_k \cdots$ such that $S_0^P = \{0\}$ and $S_k^P = \text{UPDATE}\left(S_{k-1}^P, x_k, P\right)$. Then S_k^P satisfies the following properties:*

(i) $S_k^P \subseteq \{0, 1, ..., m\}$ *for any* $k = 0, 1, ...$
(ii) an element j is in $S_k^P \setminus \{0\}$ if and only if j is the length of some suffix of $x_1 x_2 \cdots x_k$ such that $p_{i_1}(x_{k-j+1}) \wedge \cdots \wedge p_{i_j}(x_k) = true$
(iii) if $j = \max\left\{S_k^P\right\}$ then j is the length of the longest suffix of $x_1 x_2 \cdots x_k$ such $p_{i_1}(x_{k-j+1}) \wedge \cdots \wedge p_{i_j}(x_k) = true$

Proof. *Case (i).* Since $S_{k-1}^P \subseteq \{0, 1, ..., m-1\}$ for any call UPDATE(S_{k-1}^P, x_k, P) and any element from S_{k-1}^P can increase only in line 4, $S_k^P \subseteq \{0, 1, ..., m\}$.

Case (ii). We shall show that S_k^P enumerates all prefixes of P that matches some suffix of a given input $t_k = x_1 x_2 \cdots x_k$. We first prove that for

$$\forall\, j, \quad j \in S_k^P \text{ implies } p_{i_1}(x_{k-j+1}) \wedge \cdots \wedge p_{i_j}(x_k) = true \tag{1}$$

We prove Claim 1 by induction on k.

Claim 1 is true for $k = 0$, since $S_k^P = \{0\}$, and a zero-length prefix matches any suffix.

Suppose that the claim is true for S_{k-1}^P. We can show that it is true for S_k^P. According to UPDATE, for each j from S_{k-1}^P statement $p_{i_1}(x_{k-j+1}) \wedge \cdots \wedge p_{i_j}(x_{k-1})$ is *true*. For the next input element is x_k, the number $(j+1)$ will be included in S_k^P (in line 4) if and only if $p_{i_{j+1}}(x_k)$ is *true*. If for all j from S_{k-1}^P, $p_{i_{j+1}}(x_k)$ is *false*, then $S_k^P = \{0\}$ (line 5), and the zero-length prefix matches any suffix.

Case (iii). It follows directly from case *(ii)*. □

The following theorem shows the correctness of the algorithm MATCH.

Theorem 1. *Algorithm* MATCH *finds on-line all positions in t where P matches t.*

Proof. Assume that MATCH processes input $t = x_1 x_2 \cdots x_{pos} \cdots$. We prove correctness by induction on *pos*.

First, the algorithm MATCH works correctly for the empty sequence t_0.

Suppose that MATCH has performed correctly for all $pos = 1, 2, ..., k$, that is, MATCH has found all positions in $t = x_1 x_2 \cdots x_k$ where P matches t. As follows from the algorithm (Fig. 1) (line 9), S is equal to $S_k^P \setminus \{m\}$. Thus, according to Lemma 1, S contains lengths of all proper prefixes of P that match some suffixes of t_k.

Let us show that MATCH works correctly for $pos = k + 1$.

Assume that MATCH has been processed an input $t_k = a_1 a_2 \cdots a_k$, $k > 0$. When the next element $x = a_{k+1}$ has been read (line 4), MATCH calls UPDATE(S, x, P) in order to update S with respect to x. By updating S in line 6 we guarantee that, according to the inductive hypothesis and Lemma 1, S contains lengths of all prefixes of P that match some suffixes of t_{k+1}. Therefore, m will be in S if and only if P matches $a_{k-m+2} a_{k-m+3} \cdots a_{k+1}$.

Thus, any time $m \in S$, the pattern is found (lines 7, 8). This concludes the proof. □

The time complexity of algorithm MATCH is linear in the length of input plus the time complexity of all UPDATE's calls. We discuss the complexity of MATCH in Section 5.

4 Linear-time algorithm

In this section we consider a class of patterns for which a linear-time pattern matching algorithm can be designed. We present modified versions of MATCH and

UPDATE, called MATCHM and UPDATEM, correspondingly. We show that an average amortized cost of any UPDATEM's call is $O(1)$ and, therefore, MATCHM is a linear-time algorithm.

Let us introduce some additional definitions and notation.

Given a pattern $P = \langle p_{i_1}, p_{i_2}, ..., p_{i_m} \rangle$ from Ω_A, the *prefix relation* for P is a relation $\sigma_P : \{1, 2, ..., m\} \to \{0, 1, ..., m-1\}$, where $\sigma_P[k]$ is the length of the longest prefix $p_{i_1}, p_{i_2}, ..., p_{i_{\sigma_P[k]}}$ of P such that

$$p_{i_1}(x_1) \wedge \cdots \wedge p_{i_k}(x_k) = true \text{ implies } p_{i_1}\left(x_{k-\sigma_P[k]+1}\right) \wedge \cdots \wedge p_{i_{\sigma_P[k]}}(x_k) = true$$

for some $x_1, x_2, ..., x_k$ from A.

However, depending on properties of the predicates in P, the prefix relation σ_P can be a function. We say that σ_P is a *prefix function* for P if for all $0 < k \le m$, $|\sigma_P[k]| = 1$. It means that for all $x_1, \ldots, x_k$ from A

$$p_{i_1}(x_1) \wedge \cdots \wedge p_{i_k}(x_k) = true \text{ implies } p_{i_1}\left(x_{k-\sigma_P[k]+1}\right) \wedge \cdots \wedge p_{i_{\sigma_P[k]}}(x_k) = true$$

and there is no $x_1, \ldots, x_k$ from A such that for some $t > \sigma_P[k]$,

$$p_{i_1}(x_1) \wedge \cdots \wedge p_{i_k}(x_k) = true \text{ implies } p_{i_1}(x_{k-t+1}) \wedge \cdots \wedge p_{i_t}(x_k) = true.$$

Therefore, for any k a unique number $\sigma_P[k]$ can be found based on truth of $p_{i_1}(x_1) \wedge \cdots \wedge p_{i_k}(x_k)$. This is exactly the property that string-matching algorithms over finite alphabets satisfy [2, 3]. Similarly to the Knuth, Morris, and Pratt algorithm [6], the notion of prefix function can be used to avoid testing of useless cases in the course of pattern matching.

In this section we consider patterns such that their prefix relations are functions.

According to Lemma 1, the largest element j from S_k^P is the length of the longest prefix of P that matches the suffix of length j of a given sequence $x_1 x_2, \ldots, x_k$. Furthermore, the set S_k^P represents all prefixes of P that match some suffixes of $x_1 x_2, \ldots, x_k$. For any P such that σ_P is a function, the following proposition can be easily proved.

Proposition 1. *Let $P = \langle p_{i_1}, p_{i_2}, ..., p_{i_m} \rangle$ be a pattern and $S_0^P, S_1^P, ..., S_k^P, ...$ be a sequence of sets generated by* UPDATE *based on input $t = x_1 x_2 \cdots x_k \cdots$ such that $S_0^P = \{0\}$ and $S_k^P =$ UPDATE $\left(S_{k-1}^P, x_k, P\right)$. If σ_P is a prefix function, then the sets S_i^P and S_j^P are equal if and only if $\max\left\{S_i^P\right\} = \max\left\{S_j^P\right\}$ for any $i, j = 1, 2, ...$*

Proof. If S_i^P and S_j^P are equal, then $\max\left\{S_i^P\right\} = \max\left\{S_j^P\right\}$.

Suppose that $\max\left\{S_i^P\right\} = \max\left\{S_j^P\right\} = r$. According to Lemma 1, r is the length of the longest suffix of both $t_i = x_1 x_2 \cdots x_i$ and $t_j = x_1 x_2 \cdots x_j$ such that $p_{k_1}(x_{i-r+1}) \wedge \cdots \wedge p_{k_r}(x_i) = true$ and $p_{k_1}(x_{j-r+1}) \wedge \cdots \wedge p_{k_r}(x_j) = true$, correspondingly.

Assume that $S_i^P \neq S_j^P$ for some $S_i^P = \{e_1, ..., e_r\}$ and $S_j^P = \{e'_1, ..., e'_r\}$. Then the largest l, $0 < l < r$ such that $e_l \neq e'_l$ can be found. It follows from the

```
Algorithm UPDATEM(S^P, x, P)
Input: Set S^P such that S^P ⊆ {0,1,...,m−1},
          pattern P = ⟨p_{i_1}, p_{i_2}, ..., p_{i_m}⟩ from Ω_A and x from A
Output: Set S^P such that S^P ⊆ {0,1,...,m}
j ← max{S^P}
if p_{i_{j+1}}(x) = true then
    if B_{j+1} is defined then return B_{j+1}
    else
        find largest element k in S^P \ {j} such that p_{i_{k+1}}(x) = true
        if k is undefined then B_{j+1} ← {(j+1)}
            else B_{j+1} ← {(j+1)} ∪ B_{k+1}
        return B_{j+1}
else
    find largest element k in S^P such that p_{i_{k+1}}(x) = true
    if k is undefined then return {0}
        else return B_{k+1}
```

Fig. 3. Algorithm UPDATEM

definitions of S_k^P and σ_P that $e_l, e_l' \in \sigma_P[e_{l+1}]$. Therefore, e_l has to be equal to e_l' since σ_P is a function, that is, $|\sigma_P[e_{l+1}]| = 1$. We have a contradiction to the assumption. □

According to Proposition 1, since σ_P is a function there exist at most m different sets in the sequence $S_1^P, S_2^P, \ldots$ for a given P. Therefore, we need to construct S_k^P by calling UPDATE only ones for each possible $k = 1, 2, \ldots, m$, and then reuse them later instead of calling UPDATE. Then when S_k^P has been constructed, its updating takes constant time.

The above idea has been used in the modified version of UPDATE algorithm, called UPDATEM (see Figure 3).

Algorithm UPDATEM uses the fact that σ_P is a function. The prefix function σ_P is constructed by UPDATEM on-the-fly as a sequence of sets $B_1, B_2, \ldots, B_m$. Supposing that $B_1, B_2, \ldots, B_m$ are initially undefined, UPDATEM constructs $B_1, B_2, \ldots, B_m$ such that $j \in B_k$, $j \leq k$ if and only if for all $x_1, x_2, \ldots, x_k$ from A

$$p_{i_1}(x_1) \wedge \cdots \wedge p_{i_k}(x_k) = true \text{ implies } p_{k_1}(x_{1+k-j}) \wedge \cdots \wedge p_{k_j}(x_k) = true$$

It means that if $\langle p_{i_1}, p_{i_2}, \ldots, p_{i_k} \rangle$ matches eventually some $x_1, x_2, \ldots, x_k$ then B_k contains lengths of prefixes of $\langle p_{i_1}, p_{i_2}, \ldots, p_{i_k} \rangle$ that match suffixes of $x_1, x_2, \ldots, x_k$. Ones B_i becomes defined, it will not be changed until the pattern matching is completed.

Lemma 2 summarizes properties of UPDATEM.

Lemma 2. *Let $P = \langle p_{i_1}, p_{i_2}, \ldots, p_{i_m} \rangle$ be a pattern such that σ_P is a prefix function. Let $S_0^P, S_1^P, \ldots, S_k^P, \ldots$ be a sequence of sets generated by algorithm* UPDATEM *based on input $t = x_1 x_2 \cdots x_k \cdots$ such that $S_0^P = \{0\}$ and $S_k^P =$* UPDATEM$\left(S_{k-1}^P, x_k\right)$. *Let $B_1, B_2, \ldots, B_m$ be a sequence of sets generated inside* UPDATEM. *Then S_k^P satisfies the following properties:*

(i) $S_k^P \subseteq \{0, 1, ..., m\}$ *for* $k = 0, 1, ...$
(ii) j *is in* $S_k^P \setminus \{0\}$ *if and only if* j *is the length of some suffix of* $t_k = x_1 x_2 \cdots x_k$ *such that* $p_{i_1}(x_{k-j+1}) \wedge \cdots \wedge p_{i_j}(x_k) = true$
(iii) *if* $j = \max\{S_k^P\}$ *then* j *is the length of the longest suffix of* $t_k = x_1 x_2 \cdots x_k$ *such that* $p_{i_1}(x_{k-j+1}) \wedge \cdots \wedge p_{i_j}(x_k) = true$
(iv) *if* $j = \max\{S_k^P\}$, $j > 0$ *for some* $t_k = x_1 x_2 \cdots x_k$ *then* B_j *is defined and contains lengths of all prefixes of* P *that matches some suffixes of* t_k, $k = 1, 2, ...$
(v) *if* B_j *is defined, then* $j \in B_j$ *for any* $j = 1, 2, ..., m$

Proof. Cases (i)–(iii) can be proved similarly to cases *(i)–(iii)* in Lemma 1.

Case (iv). The proof is by induction on k. For $k = 1$ the statement is trivially verified.

Let us suppose that the statement is satisfied for all integers $k \leq j$. We show that the statement *(iv)* is true for B_j.

Assume that $B_1, B_2, ..., B_j$ are defined and B_{j+1} is undefined. Consider a call of UPDATEM(S^P, x, P) such that $\max\{S^P\} = j$ and $p_{i_{j+1}}(x) = true$. Since $p_{i_{j+1}}(x) = true$, B_{j+1} will be defined either in line 6 or line 7.

If there is no other k in $S^P \setminus \{j\}$ such that $p_{i_{k+1}}(x) = true$, then the only prefix of P that matches a suffix of the input has length $j + 1$. Therefore, $B_j = \{j + 1\}$ (line 6).

If k is the largest element in $S^P \setminus \{j\}$ such that $p_{i_{k+1}}(x) = true$, then the next longest prefix of P that matches a suffix of the input whose length is $k+1$. Since $k + 1 < j + 1$, B_{k+1} is defined. Based on the inductive hypothesis, we can see that B_{k+1} contains lengths of all prefixes of P of that matches some suffixes of the input except whose length is $j + 1$. Therefore, B_{j+1} is equal to $B_{k+1} \cup \{j + 1\}$ (line 7) and contains lengths of all prefixes of P that matches some suffixes of the input.

Case (v). The proof follows directly from *Case (iv)*. □

From Lemma 2 it follows that from MATCH's point of view algorithms UPDATEM and UPDATE proceed similarly. By replacing in MATCH all calls to UPDATE by calls to UPDATEM, we develop a new, more efficient pattern matching algorithm called MATCHM, and presented in Figure 4.

Theorem 2. *Let* t *be an input sequence,* $P = \langle p_{i_1}, p_{i_2}, ..., p_{i_m} \rangle$ *be a pattern with predicates* p_{i_j} *that can be evaluated in constant time and* σ_P *be a function. The algorithm* MATCHM *takes* $O(n)$ *time to find on-line all patterns* P *in an input sequence* t*, where* $|t| = n$.

Proof. According to Lemma 2, from MATCH's point of view algorithms UPDATEM and UPDATE are indistinguishable. Therefore, MATCHM finds on-line all patterns P in t since MATCH does it (Theorem 1).

The time complexity of algorithm MATCH is $O(n)$ plus time complexity of all UPDATEM's calls.

We will use the potential method of amortized analysis [10] to show that the average cost of an UPDATEM's call is $O(1)$, if we averages over a sequence of all calls during the matching, even though a single call might be $O(m)$.

```
Algorithm MATCHM(t, P)
Input: a sequence of data t and a pattern P = ⟨p_i1, p_i2, ..., p_im⟩
such that σ_P is a prefix function
Output: begining positions of all matching patterns
S ← {0};
pos ← 0;
while input is not empty do
    x ← read next element from t;
    pos ← pos + 1;
    S ← UPDATEM (S, x, P);
    if m ∈ S then
        pattern matching at (pos − m + 1)
        S ← S \ {m}
end.
```

Fig. 4. Algorithm MATCHM

We start with an initial data structure S_0^P on which n UPDATEM operations will be performed, where $n = |t|$. For each $i = 1, 2, ..., n$, we let c_i be the actual cost of the ith UPDATEM call and S_i^P be the data structure that results after applying the ith UPDATEM to S_{i-1}^P.

A potential function Φ maps each data structure S_i^P to number $\Phi\left(S_i^P\right)$ which is the potential associated with S_i^P and represents the prepaid work that can be released to pay for future UPDATEM's calls. The amortized cost $\widehat{c}_i$ of the ith UPDATEM's call with respect to potential function Φ is defined by $\widehat{c}_i = c_i + \Phi\left(S_i^P\right) - \Phi\left(S_{i-1}^P\right)$. Then the total amortized cost of the n UPDATEM's calls is

$$\sum_{i=1}^{n} \widehat{c}_i = \sum_{i=1}^{n} \left(c_i + \Phi\left(S_i^P\right) - \Phi\left(S_{i-1}^P\right)\right) = \sum_{i=1}^{n} c_i + \Phi\left(S_n^P\right) - \Phi\left(S_0^P\right)$$

Thus the total amortized cost $\sum_{i=1}^{n} \widehat{c}$ is an upper bound on the actual total cost $\sum_{i=1}^{n} c_i$.

We define the potential function Φ on the set S_i^P to be the largest number in the set S_i^P. Since we start with the set $S_0^P = \{0\}$, we have $\Phi\left(S_0^P\right) = 0$. Since $S_i^P \subseteq \{0, 1, ..., m\}$, $i = 1, 2, ..., m$, the data structure S_i^P that results after the ith UPDATEM's call has nonnegative potential, and thus $\Phi\left(S_i^P\right) \geq \Phi\left(S_0^P\right) = 0$. The total amortized cost of n UPDATEM's calls with respect to Φ represents an upper bound on the actual cost.

Let us compute the amortized cost of the various UPDATEM's calls.

Suppose that $j = \max\{S_{i-1}^P\}$ and for ith call UPDATEM$\left(S_{i-1}^P, x, P\right)$.

If $p_{i_{j+1}}(x) = true$ and B_{j+1} is *defined* then B_{j+1} is returned. The actual cost of this case is 1, and the potential difference is $\Phi(S_i^P) - \Phi(S_{i-1}^P) = j + 1 - j = 1$. The amortize cost of this case is

$$\widehat{c}_i = c_i + \Phi(S_i^P) - \Phi(S_{i-1}^P) = 2$$

If $p_{k_{j+1}} = true$ but B_{j+1} is ***not defined*** then additional work should be done. We will go through the set S^P which is equal to B_j to find the largest k such that $p_{i_{k+1}}(x) = true$. The actual cost c_i in this case is no bigger then $|B_j|$ or $|S^P|$ and equal r, and the potential difference is

$$\Phi(S_i^P) - \Phi(S_{i-1}^P) \leq (j - r) - j = -r$$

The amortized cost of this case is

$$\widehat{c_i} = c_i + \Phi(S_i^P) - \Phi(S_{i-1}^P) \leq r - r = 0$$

If $p_{k_{j+1}} \neq true$ then we will go through the set S^P which is equal to B_j to find the largest k such that $p_{i_{k+1}}(x) = true$. The actual cost c_i in this case is the number of cases r that must be tested no bigger then $|B_j|$, and the potential difference is

$$\Phi(S_i^P) - \Phi(S_{i-1}^P) \leq (j - r) - j = -r$$

The amortized cost of this case is

$$\widehat{c_i} = c_i + \Phi(S_i^P) - \Phi(S_{i-1}^P) \leq r - r = 0$$

The amortized cost of each `UPDATEM`'s call in all three cases is $O(1)$, thus the total amortized cost of the sequence of n `UPDATEM`'s calls is $O(n)$. Since $\Phi(S_i^P) \geq \Phi(S_0^P) = 0$ for all i, the total amortized cost of n `UPDATEM`'s calls is an upper bound on the total actual cost. So the worst-case cost of n `UPDATEM`'s calls is $O(n)$. □

5 Complexity

In this section we make some comments about the complexity of algorithm `MATCH` presented in Section 3.

The naive brute force algorithm consists of checking at all positions in the input sequence between 0 and $n-m$ whether an occurrence of the pattern begins there or not. Then, after each attempt, it shifts the pattern exactly one position to the right. If predicates can be evaluated in constant time, the complexity of such a brute force algorithm is $O(nm)$ both in the worst and average cases. However, by taking into consideration the history of matching to avoid unnecessary the actual time can be reduced, since not all predicates of a pattern need to be evaluated after each shift. When predicates $p_{i_1}, p_{i_2}, ..., p_{i_m}$ in a pattern P are interdependent, that is, truth value of one predicate implies the truth value of another, then the time complexity can be improved even further. Collecting historical data we can use such interdependency to improve performance of such algorithms. In fact there are cases, such as these presented in Section 4 or in [1, 2,4], when predicates are interdependent.

Both the algorithms `MATCH` and `MATCHM` presented in this paper collect data dynamically and avoid evaluation of useless cases. The time complexity of the algorithm `MATCH` and its modification `MATCHM` depends on properties of predicates in P. When predicates of P are interdependent with property that σ_P is a function, we have developed in Section 4 a linear-time algorithm. We say that predicates p_{i_j} and p_{i_k} are ***independent*** if $p_{i_j}(x) \neq p_{i_k}(x)$ and $p_{i_j}(x)$ does not imply $p_{i_k}(x)$ for all x in A. This means that truth values of $p_{i_k}(x)$ cannot be derived from truth value of $p_{i_j}(x)$. In the case where all predicates in P are independent we cannot make a conclusion about truth values of predicates of P without their evaluation. Also in many cases we can avoid evaluating all predicates and still have asymptotically worst-case running time $O(mn)$ for algorithm `MATCH`. However, the actual time can be much better since the algorithm does not analyze useless cases.

6 Conclusion

We have presented a new approach for the pattern matching problem over infinite alphabets where patterns are presented as sequences of constraints or predicates defined on elements of some, generally, infinite alphabet. We have presented both general and linear-time algorithms to perform such constraint-based pattern matching. Our scheme permits on-line matching on the input representing digitized continuous reality such as sounds, images or even noisy telemetric data without knowing ***a priori*** either the whole sequence or the whole pattern. Another example when our algorithms can be efficiently applied is the problem of finding the longest prefix of patterns that occurs within an input sequence. If the pattern is long, preprocessing can be costly and unnecessary if only relatively short prefixes will occur in input.

Further work should be done to find new classes of predicates with sound applications for which linear-time algorithms exist. It is interesting to make experiments to compare the actual complexity of such algorithms with other known algorithms when such exist.

References

1. A. V. Aho, Algorithms for Finding Patterns in Strings. In *Handbook of Theoretical Computer Science*, vol. **A**, J. van Leeuwen, ed., Elsevier Science Publishers, 255 – 300, 1990.
2. M. Crochemore and W. Rytter, Text Algorithms, Oxford University Press, 1994.
3. M. Crochemore and C. Hancart, Automata for Matching Patterns. In *Handbook of Formal Languages*, vol. **2**, G. Rozenberg and A. Salomaa, eds., Springer-Verlag, 399–462, 1997.
4. M. Crochemore and C. Hancart, Pattern Matching in Strings. In *Algorithms and Theory of Computation Handbook*, M. J. Atallah, ed., CRC Press, Boca Raton, 11.1–11.28, 1998.
5. M. Crochemore and T. Lecroq, Pattern Matching and Text Compression Algorithms. In *The Computer Science and Engineering Handbook*, A. B. Tucker, ed., CRC Press, 162–202, 1997.

6. D. E. Knuth, J. Morris, and V. Pratt, Fast Pattern Matching in Strings, *SIAM Journal on Computing* **6** (1977), 323 – 350.
7. G. M. Landau and U. Vishkin, Pattern Matching in a Digitized Image, *Algorithmica* **12** (1994), 375 – 408.
8. U. Manber and G. Myers, Suffix Arrays: a New Method for On-line String Searches, *SIAM Journal on Computing* **22** (1993), 935 – 948.
9. J.P. Morrill, Distributed Recognition of Patterns in Time Series Data, *Communication of the ACM,* **41** (1998), 45 – 51.
10. R. E. Tarjan, Amortized Computational Complexity, *SIAM Journal on Algebraic and Discrete Methods*, **6** (1985), 306 – 318.

On the Randomness of a $[d, k]$ Self-Decimation Stream Key Generator

Fumio Sato and Kaoru Kurosawa

Department of Electrical and Electronic Engineering
Faculty of Engineering
Tokyo Institute of Technology
2-12-1 O-okayama, Meguro-ku, Tokyo 152-8552, JAPAN
Email: {satoh,kurosawa}@ss.titech.ac.jp
URL: http://tsk-www.ss.titech.ac.jp/~kurosawa/

Abstract. In this paper, we first determine the number of cycles of a $[1, k]$ self-decimated m-sequence obtained from an LFSR of length L for $k \leq L$ and $k \geq 2^L - L$. Next, we derive a lower bound on the period of a $[1, k]$ self-decimated m-sequence for $3 \leq k \leq L$. Finally, we present the exact values of the bit pair frequencies of $[1, 2]$ self-decimated m-sequences. We prove that each bit pair occurs with probability almost 1/4.

1 Introduction

A self-decimation system is a well-known keystream generator proposed by Rueppel for stream ciphers [3]. A $[d, k]$ self-decimation system is a Linear Feedback Shift Register (LFSR) which has additional hardware such as follows (see Fig. 1):

(1) If the output of the LFSR is '0', then d clocks are given to the LFSR,
(2) If the output of the LFSR is '1', then k clocks are given to the LFSR.

This system is highly nonlinear in the sense that the state transition diagram has more than one cycle in general (therefore, it cannot be realized by an LFSR). Further, since the system is simple and the generated sequence has large linear complexity (the equivalent size of an LFSR), it is often used as a part of a more complicated keystream generator.

Let L denote the length of the LFSR of a $[d, k]$ self-decimation system such that the characteristic polynomial of the LFSR is primitive. Then it is known that a $[d, k]$ self-decimation system is equivalent to a $[1, k']$ self-decimation system if $\gcd(d, 2^L - 1) = 1$, where $k' = k/d \bmod 2^L - 1$.

Rueppel [3] presented the period of $[1, 2]$ self-decimated m-sequences. He also derived an upper bound on the length of the largest cycle of $[1, k]$ self-decimated sequences for $k \geq 3$ such that $\gcd(k - 1, 2^L - 1) = 1$. Further, he showed a lower and an upper bound on the bit pair frequencies of a $[1, 2]$ self-decimated m-sequence.

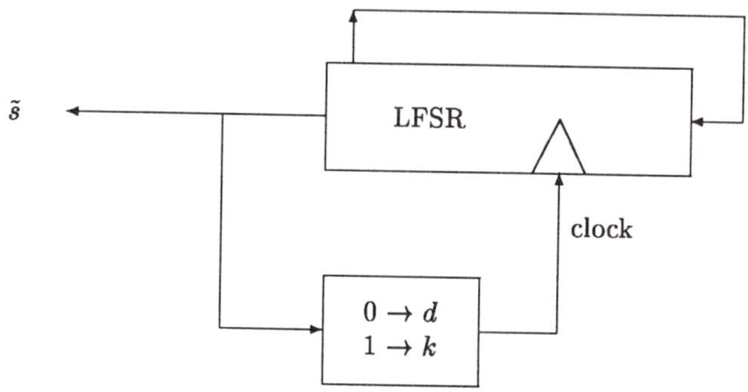

Fig. 1. $[d, k]$ self-decimation system

In this paper, we first determine the number of cycles of a $[1, k]$ self-decimated m-sequence for $k \leq L$ and $k \geq 2^L - L$.

Next, we derive a lower bound on the period of a $[1, k]$ self-decimated m-sequence for $3 \leq k \leq L$. While the upper bound of Rueppel is very loose, our lower bound is shown to be very tight by numerical simulation. Further, from a view point of users, lower bounds are more important than upper bounds on the period.

Finally, we present the exact values of the bit pair frequencies of $[1, 2]$ self-decimated m-sequences. We prove that each bit pair occurs with probability almost 1/4 and their frequencies are independent of the characteristic polynomial of the m-sequence.

2 Preliminaries

We consider a $[d, k]$ self-decimation system with an LFSR whose characteristic polynomial is primitive. Let L denote the size of the LFSR. We say that the output sequence of a $[d, k]$ self-decimation system is *a $[d, k]$ self-decimated sequence.*

Rueppel proved the following propositions [3].

2.1 Resetting Sequence

Definition 2.1. *Suppose that a symbol x always appears in a $[d, k]$ self-decimated sequence whenever a subsequence $b_0 b_1 \cdots b_l x$ appears in the original sequence of the LFSR. Then $b_0 b_1 \cdots b_l$ is called a resetting sequence.*

A state of a $[d, k]$ self-decimation system is defined as the contents of the registers of the LFSR, that is, an L bit sequence.

Proposition 2.1. *The state transition diagram of a $[d, k]$ self-decimation system has only one cycle if there exists a resetting sequence.*

2.2 Period of a $[1, k]$ Self-Decimation System

Let T_2 denote the length of a period of a $[1, 2]$ self-decimated m-sequence. Let U_k denote the length of the largest cycle of a $[1, k]$ self-decimation system.

Proposition 2.2.

$$T_2 = \left\lfloor \frac{2}{3}(2^L - 1) \right\rfloor .$$

Proposition 2.3. *If* $k \geq 3$ *and* $\gcd(k - 1, 2^L - 1) = 1$, *then*

$$U_k \leq \left\lfloor \frac{3}{4}(2^L - 1) \right\rfloor .$$

Proposition 2.3 was originally shown for $[d, d + 1]$ self-decimation systems.

3 Number of Cycles

Let C_k denote the number of cycles of the state transition diagram of a $[1, k]$ self-decimation system.

In this section, we determine C_k for $k \leq L$ and $k \geq 2^L - L$.

Lemma 3.1. 0^{k-1} *is a resetting sequence if* $k \leq L$.

Proof. Consider a subsequence $x0^{k-1}y$ of the LFSR. In a $[1, k]$ self-decimation system, 1 clock or k clocks are given to the LFSR. Therefore, an output bit of the $[1, k]$ self-decimated sequence appears somewhere in $x0^{k-1}$ since the length of $x0^{k-1}$ is k. If x is the output bit and $x = 1$, then y is the next output bit since k clocks are given. Otherwise, '0' is the output bit. In this case, it is easy to see that y is also an output bit. □

Theorem 3.1. $C_k = 1$ *if* $k \leq L$.

Proof. From Proposition 2.1 and Lemma 3.1. □

We also denote one period of a $[1, k]$ self-decimated m-sequence by a *cycle*.

Theorem 3.2. *Suppose that* $2^L - L \leq k < 2^L - 1$. *Let* $k' \stackrel{def}{=} 2^L - 1 - k$. *Then each cycle is* $0^{k'}1$. *Further,* $C_k = 2^{L-k'-1}$.

Proof. First, we prove that no cycles appear other than $0^{k'}1$. A $[1, k]$ self-decimated sequence is also a $[1, -k']$ self-decimated sequence since

$$-k' = -(2^L - 1 - k) = k \bmod 2^L - 1.$$

In the $[1, -k']$ self-decimation system, 1 clock is given or $-k'$ clocks are given to the LFSR. On the other hand, the length of $0^{k'}1$ is $k' + 1$. Therefore, an output bit of the $[1, -k']$ self-decimated sequence appears somewhere in $0^{k'}1$. Then we can see that no cycles appear other than $0^{k'}1$.

Next, from $k' = 2^L - 1 - k$, we have that

$$k' + 1 = 2^L - k \leq L,$$

since $2^L - L \leq k$. Therefore, $0^{k'}1$ appears in the original m-sequence. Further, it is easy to see that $0^{k'}1$ appears $2^{L-k'-1}$ times in the original m-sequence[2]. Therefore, $C_k = 2^{L-k'-1}$. □

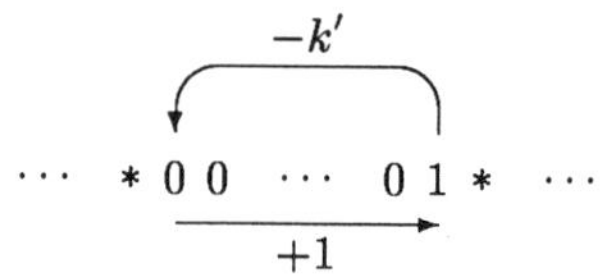

Fig. 2. Cycle

4 Period

Theorem 3.1 shows that there exists just one cycle in a $[1, k]$ self-decimation system if $k \leq L$. This means that we can define the period of $[1, k]$ self-decimated sequences for $k \leq L$.

In this section, we derive a lower bound on the length of the period of such sequences. We then show that our lower bound is very tight by numerical simulation while the upper bound of Proposition 2.3 is very loose. From a view point of users, lower bounds are more important than upper bounds generally.

Definition 4.1. *T_k denotes the length of a period of $[1, k]$ self-decimated sequences.*

Definition 4.2. *$(ab\cdots)^*$ denotes a null sequence or*

$$ab \cdots ab \cdots ab \cdots.$$

$(ab \cdots \mid cd \cdots)$ denotes $(ab \cdots)$ or $(cd \cdots)$.

(See [1] for regular expression notation.)

Theorem 4.1. *In a $[1, k]$ self-decimated sequence such that $3 \leq k \leq L$, any resetting sequence has a form 0^{k-1} or $0^{k-1}Y$ such that*

$$Y = \Big(1 \underbrace{(0 \mid 1) \cdots (0 \mid 1)}_{k-1}\Big)^* \Big((0 \mid 00 \mid \cdots \mid 0^{k-2})\big(1 \underbrace{(0 \mid 1) \cdots (0 \mid 1)}_{k-1}\big)^*\Big)^*, \quad (1)$$

where 0^{k-1} does not appear in $0^{k-2}Y$.

Proof. Suppose that x appears in a $[1, k]$ self-decimated sequence. Then 0^{k-1} appears somewhere before x in the original sequence of the LFSR. We can assume that

$$00^{k-2}Yx$$

appears in the original sequence of the LFSR, where 0^{k-1} does not appear in 0^{k-2}Y. From the definition of a $[1, k]$ self-decimation system, k clocks are given to '1' and 1 clock is given to '0'. Therefore, it is easy to see that Y is written like (1) or Y is a null sequence. □

Corollary 4.1. *A resetting sequence of a $[1, 3]$ self-decimation system is one of the following,*

(1) $00\big(1(01|10|11)\big)^u$,
(2) $00\big(1(01|10|11)\big)^u 1(0|1)10\big(1(01|10|11)\big)^w$,
(3) $00\big(1(01|10|11)\big)^u 1(0|1)10\big(1(01|10|11)\big)^{v_1} \cdots 1(0|1)10\big(1(01|10|11)\big)^{v_n}$ $1(0|1)10\big(1(01|10|11)\big)^w$.

Proof. From Theorem 4.1. □

Example 4.1. Consider a subsequence of an LFSR such that

$$001a10x,$$

where $a = 0$ or 1. Then $10x$ appears in the $[1, 3]$ self-decimated sequence because 00, $001a1$ and $001a10$ are resetting sequences.

From the above example, it is easy to see the following lemma.

Lemma 4.1. *If $k \leq L$, then*

T_k = *The number of occurrences of resetting sequences in a period of the LFSR sequence.*

Proof. Each resetting sequence contributes a bit to a period of a $[1, k]$ self-decimated sequence. □

From this lemma, we can compute T_k by counting resetting sequences.

Now we present a lower bound on T_3. Our method can be easily generalized to T_k such that $4 \leq k \leq L$.

Theorem 4.2. *If $L = 3l + 2$ ($l \geq 2$),*

$$T_3 \geq \frac{2}{5} \cdot 2^L - \frac{1}{5} \cdot 3^{l+1} - 1 + \frac{2}{25}\{2^L - 5(l-1) \cdot 2^k \cdot 3^{l-1} - 2^{k+3} \cdot 3^{l-1}\} + \sum_{n=1}^{\lfloor \frac{1}{4}(L-6) \rfloor} \sum_{i=0}^{\lfloor \frac{1}{3}(L-6-4n) \rfloor} \binom{i+n+1}{i} \cdot 3^i \cdot 2^{L-5-3n-3i}.$$

If $L = 3l$ or $L = 3L + 1$ ($l \geq 2$),

$$T_3 \geq \frac{2}{5} \cdot 2^L - \frac{1}{5} \cdot 2^{k+1} \cdot 3^l - 1 + \frac{2}{25}\{2^L - 5(l-1) \cdot 2^k \cdot 3^{l-1} - 2^{k+3} \cdot 3^{l-1}\} + \sum_{n=1}^{\lfloor \frac{1}{4}(L-6) \rfloor} \sum_{i=0}^{\lfloor \frac{1}{3}(L-6-4n) \rfloor} \binom{i+n+1}{i} \cdot 3^i \cdot 2^{L-5-3n-3i}.$$

Proof. We compute how many times resetting sequences of length L or less appear in a period of the LFSR sequence.

Let F_1 denote the number of occurrences of resetting sequences, as specified in Corollary 4.1 (1), in the leftmost shift register positions in a period of the LFSR. Similarly, let F_2 and F_3 denote such numbers of resetting sequences specified in Corollary 4.1 (2) and (3), respectively. Then $T_3 \geq F_1 + F_2 + F_3$.

F_1, F_2 and F_3 are computed as follows.

$$F_1 = \sum_{u=0}^{\lfloor \frac{1}{3}(L-2) \rfloor} 3^u \cdot 2^{L-2-3u} - 1 = \begin{cases} \frac{2}{5} \cdot 2^L - \frac{1}{5} \cdot 3^{l+1} - 1 & (k=2) \\ \frac{2}{5} \cdot 2^L - \frac{1}{5} \cdot 2^{k+1} \cdot 3^l - 1 & (\text{otherwise}) \end{cases}$$

$$F_2 = \sum_u \sum_w 3^{u+w} \cdot 2 \cdot 2^{L-6-3(u+w)} - 1 = 2^{L-5} \sum_{i=0}^{l-2} (i+1) \left(\frac{3}{8}\right)^i = \frac{2}{25}\{2^L - 5(l-1) \cdot 2^k \cdot 3^{l-1} - 2^{k+3} \cdot 3^{l-1}\}$$

$$F_3 = \sum_u \sum_w \sum_{v_1} \cdots \sum_{v_n} 3^{u+w+\sum_j v_j} \cdot 2^{n+1} \cdot 2^{L-6-4n-3(u+w+\sum_j v_j)} = \sum_{n=1}^{\lfloor \frac{1}{4}(L-6) \rfloor} \sum_{i=0}^{\lfloor \frac{1}{3}(L-6-4n) \rfloor} \binom{i+n+1}{i} \cdot 3^i \cdot 2^{L-5-3n-3i}$$

□

The following table shows the exact value of T_3, our lower bound of Theorem 4.2 and the upper bound of Proposition 2.3. (Note that T_k depends on the characteristic polynomial of the LFSR.)

Table 1. The period of self-decimated sequences

L	lower bound	period	upper bound
6	23	$29, 31, 33$	47
7	47	$61, 63, 65$	95
8	104	$125, 127, 129$	191
9	221	$251, 253, \cdots, 259$	383

5 Exact Bit Pair Frequency of a [1, 2] Self-Decimated m-sequence

Let $N_L(b_1b_2)$ denote the number of appearances of b_1b_2 in one period of a $[1,2]$ self-decimated m-sequence. Rueppel showed a lower bound and an upper bound on $N_L(b_1b_2)$ such as follows [3].

Proposition 5.1.

$$N_L^0(b_1b_2) \le N_L(b_1b_2) \le N_L^0(b_1b_2) + 2,$$

where

$$N_L^0(00) = \left\lfloor \frac{1}{6}(2^L - 1) \right\rfloor - 1,$$

$$N_L^0(01) = \left\lfloor \frac{1}{6}(2^L - 1) \right\rfloor,$$

$$N_L^0(10) = \left\lfloor \frac{1}{6}(2^L - 4) \right\rfloor \quad \textit{and}$$

$$N_L^0(11) = \left\lfloor \frac{1}{6}(2^L - 4) \right\rfloor.$$

In this section, we show the exact value of $N_L(b_1b_2)$ for any b_1b_2. Let $\mathbf{r} = \{r_i\}_{i=0}^{\infty}$ denote the m-sequence which satisfies the following equation.

$$r_{j+L} = \sum_{i=1}^{L} c_i r_{j+L-i} \qquad (j \ge 0).$$

Lemma 5.1. *If a characteristic polynomial* $C(D) = D^L + \sum_{i=1}^{L} c_i D^{L-i}$ *is irreducible, then*

$$c_L = 1 \quad \textit{and} \quad \sum_{i=1}^{L} c_i = 0 \bmod 2.$$

Proof. $C(D)$ can't be divided by D and $D+1$. Therefore, $C(0) = 1 = c_L$ and $C(1) = 1 = 1 + \sum_{i=1}^{L} c_i \bmod 2$. □

Lemma 5.2. *In one period of the m-sequence :*

(1) 01^L0 *appears once.*
(2) No 1^{L+1} *appears.*
(3) No $01^{L-1}0$ *appears.*

Proof. It is clear that 1^L appears once in one period of the m-sequence because L is the length of the LFSR. This implies (1) and (2).

Next, suppose that $r_j = 0$ and $r_{j+i} = 1$ for $1 \le i \le L-1$. Then

$$r_{j+L} = \sum_{i=1}^{L} c_i r_{j+L-i} = \sum_{i=1}^{L-1} c_i = 1$$

from Lemma 5.1. This implies (3). □

Let $\#\{\beta\}$ denote the number of how many times a subsequence β appears in a period of the LFSR sequence. It is easy to see that any resetting sequence of a [1,2] self-decimated system has a form 01^m, where $m \ge 0$ is even.

Now we first show the exact value of $N_L(b_1b_2)$ for $b_1 = 0$.

Theorem 5.1. *In a [1,2] self-decimated m-sequence,*

$$N_L(00) = \left\lfloor \frac{1}{6}(2^L + 2) \right\rfloor - 1,$$

$$N_L(01) = \left\lfloor \frac{1}{6}(2^L + 2) \right\rfloor.$$

Proof. Since 01^m is the resetting sequence ($m \ge 0$ is even), we have

$$N_L(0b) = \sum_{m=0:\text{ even}}^{L} \#\{01^m 0b\}.$$

from Lemma 5.2 (2). First, from the theory of m-sequences,

$$\#\{01^m0b\} = 2^{L-m-3} \text{ for } m = 0, 1, \ldots, L-3$$

except for

$$\#\{000\} = 2^{L-3} - 1.$$

On the other hand, from Lemma 5.2 (3),

$$\#\{01^{L-1}0b\} = 0.$$

Next, let's compute $\#\{01^L0b\}$. From Lemma 5.2, 01^L0 appears once. Suppose that $r_{j+i} = 1$ $(0 \le i \le L-2)$ and $r_{j+L-1} = 0$. Then

$$r_{j+L} = \sum_{i=1}^{L} c_i r_{j+L-i} = \sum_{i=2}^{L} c_i = c_1.$$

from Lemma 5.1. Finally, let's compute $\#\{01^{L-2}0b\}$. Suppose that $r_j = 0$, $r_{j+1} = 1$ $(1 \le i \le L-2)$ and $r_{j+L-1} = 0$. Then

$$r_{j+L} = \sum_{i=1}^{L} c_i r_{j+L-i} = \sum_{i=2}^{L-1} c_i = c_1 + 1 \bmod 2$$

from Lemma 5.1. Therefore,

$$\#\{01^L0b\} + \#\{01^{L-2}0b\} = 1.$$

From the above, we obtain that

$$N_L(00) = \begin{cases} -1 + \displaystyle\sum_{m=0,\ m:\text{even}}^{L-4} 2^{L-m-3} + 1 = \frac{1}{6}(2^L - 4) & (L \text{ is even}), \\ -1 + \displaystyle\sum_{m=0,\ m:\text{even}}^{L-3} 2^{L-m-3} = \frac{1}{6}(2^L - 8) & (L \text{ is odd}). \end{cases}$$

$$N_L(01) = \begin{cases} \displaystyle\sum_{m=0,\ m:\text{even}}^{L-4} 2^{L-m-3} + 1 = \frac{1}{6}(2^L + 2) & (L \text{ is even}), \\ \displaystyle\sum_{m=0,\ m:\text{even}}^{L-3} 2^{L-m-3} = \frac{1}{6}(2^L - 2) & (L \text{ is odd}). \end{cases}$$

□

We next present the exact value of $N_L(b_1b_2)$ for $b_1 = 1$.

Theorem 5.2. *In a [1,2] self-decimated m-sequence,*

$$N_L(10) = \left\lfloor \frac{1}{6}(2^L + 2) \right\rfloor,$$

$$N_L(11) = \left\lfloor \frac{1}{6}(2^L + 2) \right\rfloor - (-1)^L.$$

Proof. $1b$ appears in the [1,2] self-decimated sequence if and only if 01^m1ab appears in the original m-sequence because 01^m is the resetting sequence, where $m \geq 0$ is even. Therefore,

$$N_L(1b) = \sum_{m=1:\ \text{odd}}^{L} \#\{01^m ab\}$$

from lemma 5.2 (2). First, from the theory of m-sequence,

$$\#\{01^m ab\} = 2 \times 2^{L-m-3} \text{ for } m = 1, \ldots, L-3.$$

Next, let's compute $\#\{01^m ab\}$ for $m = L-2, L-1, L$. If $a = 0$, then we can show that

$$\#\{01^L 0b\} + \#\{01^{L-2} 0b\} = 1,$$
$$\#\{01^{L-1} 0b\} = 0$$

similarly to the proof of Theorem 5.1. If $a = 1$, then we have

$$\#\{01^L 1b\} = \#\{01^{L+1} b\} = 0,$$
$$\#\{01^{L-1} 1b\} = \#\{01^L b\} = \begin{cases} 1 & (b=0), \\ 0 & (b=1), \end{cases}$$
$$\#\{01^{L-2} 1b\} = \#\{01^{L-1} b\} = \begin{cases} 0 & (b=0), \\ 1 & (b=1), \end{cases}$$

from Lemma 5.2. Therefore,

$$\#\{01^L ab\} + \#\{01^{L-2} ab\} = \begin{cases} 1 & (b=0), \\ 2 & (b=1), \end{cases}$$
$$\#\{01^{L-1} ab\} = \begin{cases} 1 & (b=0), \\ 0 & (b=1). \end{cases}$$

Consequently, we obtain that

$$N_L(10) = \begin{cases} \displaystyle\sum_{m=1,\ m:\text{odd}}^{L-3} 2^{L-m-3} + 1 = \frac{1}{6}(2^L + 2) & (L \text{ is even}), \\ \displaystyle\sum_{m=1,\ m:\text{odd}}^{L-4} 2^{L-m-3} + 1 = \frac{1}{6}(2^L - 2) & (L \text{ is odd}), \end{cases}$$

$$N_L(11) = \begin{cases} \displaystyle\sum_{m=1,\ m:\text{odd}}^{L-3} 2^{L-m-3} = \frac{1}{6}(2^L - 4) & (L \text{ is even}), \\ \displaystyle\sum_{m=1,\ m:\text{odd}}^{L-4} 2^{L-m-3} + 2 = \frac{1}{6}(2^L + 4) & (L \text{ is odd}). \end{cases}$$

□

6 Conclusion

By identifying the patterns of resetting sequences, we were able to count the number of occurrences of resetting sequences in a period of an m-sequence. This result allowed us to construct lower bounds on the period of $[d, k]$ self-decimated sequences. Numerical experiments confirm that these bounds are quite tight.

Self-decimation systems should be used as a part of a more complicated stream key generators [4] because its direct use is not secure in general [3–5].

It will be a further work to study the case for $L < k < 2^L - L$.

References

1. A. V. Aho, "Algorithms for Finding Patterns in String," *Handbook of Theoretical Computer Science, Volume A: Algorithms and Complexity*, J. van Leeuwen ed., MIT Press, 1990, pp. 255–300.
2. S. W. Golomb, *Shift Register Sequences*(revised edition), Aegean Park Press, 1982.
3. R. A. Rueppel, "When Shift Registers Clock Themselves," *Advances in Cryptology – EUROCRYPT '87*, Lecture Notes in Computer Science 304, Springer-Verlag, 1988, pp. 53–64.
4. R. A. Rueppel, "Stream Ciphers," *Contemporary Cryptology: The Science of Information Integrity*, G. J. Simmons ed., IEEE Press, 1992, pp. 65-134.
5. K. Zeng, C.-H. Yang, D.-Y. Wei and T.R.N. Rao, "Pseudorandom Bit Generators in Stream-Cipher Cryptography," *IEEE Computer*, v. 24, n. 2, Feb 1991, pp. 8–17.

Sequence Families with Optimum Aperiodic Mean-Square Correlation Parameters

Hans D. Schotten

Institute of Communications Engineering,
Aachen University of Technology
Melatener Str. 23, D-52056 Aachen, Germany
Email: schotten@ient.rwth-aachen.de

Abstract. In this paper, the aperiodic mean-square correlation (AMSQC) parameters of sequence families are investigated. These parameters are widely accepted performance measures for correlation properties of sequences applied in code-division multiple-access (CDMA) systems and other technical fields. The AMSQC parameters of many known families attain the Welch-bound and result in approximately the same values for the auto- and crosscorrelation. Since in many technical applications either the auto- or the crosscorrelation properties are more relevant, the construction of sequence families which, on the one hand, attain the Welch-bound and which, on the other hand, have a wide range of auto- and crosscorrelation parameters is considered in this paper.

1 Introduction

Mean-square correlation parameters are widely accepted performance measures for correlation properties of sequences and sequence families. In this paper, the mean-square correlation properties of the aperiodic correlation are investigated. Code-division multiple-access (CDMA) systems are one of the most important applications of sequence families. The performance of these systems crucially depends on the correlation properties of the spreading-sequences assigned to the users. The average power of the multiuser interference can be expressed in terms of the aperiodic mean-square **cross**correlation parameter (interference parameter). The mean-square parameter of the **auto**correlation function describes the synchronization behavior and the spectral properties of the spreading-sequences. These spectral properties have a major effect on the system performance in case of transmission of signals over frequency-selective fading channels. In this paper, the construction and optimization of sequence families with respect to their aperiodic mean-square correlation (AMSQC) parameters are considered.
The paper is organized as follows: In Section 2, the aperiodic mean-square correlation parameters are defined. In Section 3, the technical meaning of these parameters is discussed. Bounds on the AMSQC parameters are derived in Section 4. The AMSQC parameters for various well-known sequence families are

discussed in Section 5 and 6. In Section 7, the general construction of sequence families which attain the Welch-bound and which have a wide range of mean-square auto- and crosscorrelation parameters is discussed.

2 Aperiodic Mean-Square Correlation Parameters

We consider a family $\mathcal{F} = \{s_k \,|\, 1 \leq k \leq K\}$ of K sequences where $s_k = (s_k(0), s_k(1), \ldots, s_k(N-1))$, $1 \leq k \leq K$, denotes a sequence of length N. All sequences are P-phase sequences, i.e.,

$$s_k(n) \in \left\{ \exp(\jmath \frac{2\pi}{P} l) \,|\, 0 \leq l < P \right\}, \quad 1 \leq k \leq K, \quad 0 \leq n < N,$$

where P is the alphabet size of the sequences and $\jmath = \sqrt{-1}$.
$\mathbf{S}(\mathcal{F})$ denotes the $K \times N$ matrix of which the rows are the sequences of $\mathcal{F}$. The columns of $\mathbf{S}$ are denoted by $c_i = (c_i(1), c_i(2), \ldots, c_i(K))^T$, $0 \leq i < N$:

$$\mathbf{S}(\mathcal{F}) = \begin{vmatrix} s_1(0) & s_1(1) & \ldots & s_1(N-1) \\ s_2(0) & s_2(1) & \ldots & s_2(N-1) \\ \vdots & \vdots & & \vdots \\ s_K(0) & s_K(1) & \ldots & s_K(N-1) \end{vmatrix} = \begin{vmatrix} c_0(1) & c_1(1) & \ldots & c_{N-1}(1) \\ c_0(2) & c_1(2) & \ldots & c_{N-1}(2) \\ \vdots & \vdots & & \vdots \\ c_0(K) & c_1(K) & \ldots & c_{N-1}(K) \end{vmatrix}.$$

The aperiodic correlation function of two sequences s_k and s_i of length N is defined by

$$C_{ki}(m) = \sum_{n=0}^{N-1-m} s_k(n) s_i^*(n+m) \quad \text{for} \quad 0 \leq m < N$$

and $C_{ki}(-m) = C_{ik}^*(m)$. $C_{kk}(m)$ is the autocorrelation function, and $C_{ki}(m)$, $k \neq i$, is the crosscorrelation function.
The aperiodic mean-square correlation (AMSQC) parameters can be derived from the identity [1]

$$\mu_{ki}(t) = \sum_{m=-(N-1)}^{N-1} C_{ki}(m) C_{ki}^*(m+t) = \sum_{m=-(N-1)}^{N-1} C_{kk}(m) C_{ii}^*(m+t), \quad 0 \leq t < N.$$

The sidelobe energy $\mathrm{SE}^{(k)}$ of sequence s_k is

$$\mathrm{SE}^{(k)} = \sum_{m=1}^{N-1} |C_{kk}(m)|^2.$$

With this, we define the mean-square autocorrelation parameter

$$\overline{\mu}_a = \frac{1}{K} \mu_a = \frac{2}{K} \sum_{k=1}^{K} \mathrm{SE}^{(k)} = \frac{2}{K} \sum_{k=1}^{K} \sum_{m=1}^{N-1} |C_{kk}(m)|^2.$$

The corresponding parameters for the crosscorrelation are

$$\mu_c^{(k)}(t) = \sum_{\substack{i=1 \\ i \neq k}}^{K} \mu_{ki}(t), \quad \mu_c = \sum_{k=1}^{K} \mu_c^{(k)}(t) \quad \text{and,} \quad \overline{\mu_c}(t) = \frac{\mu_c(t)}{K} = \frac{1}{K} \sum_{k=1}^{K} \mu_c^{(k)}(t).$$

To simplify the notation, the argument of all parameters is omitted if $t = 0$, e.g., $\mu_c(0) = \mu_c$.

3 CDMA Systems

We consider a direct-sequence code-division multiple-access (DS-CDMA) system with asynchronous channel access and K active users as introduced in [2, 3]. Each user is assigned a user-specific spreading-sequence s_k, $1 \leq k \leq K$, of length N. $\mathcal{F}$ denotes the set of these sequences.
At the transmitter, MPSK sequence-modulation is performed, i.e., depending on the L-ary data symbol $a^{(k)}$ one of the sequences $\omega^l s_k(n)$ where $\omega = \exp(\jmath \frac{2\pi}{L})$ and $0 \leq l < L$ is transmitted. We assume that the signal delays $\tau^{(k)}$, $1 \leq k \leq K$, are mutually independent. For the theoretical analysis considering an AWGN channel the receiver is a matched filter as depicted in Figure 1 (single-user detector). For simulations of the transmission over frequency-selective and time-variant fading channels a Rake receiver was used. A simplified discrete-time baseband model for the AWGN scenario is shown in Figure 1.

Mean-square crosscorrelation parameter μ_c (interference parameter)

In [2, 3], it is shown that for an AWGN channel the signal-to-noise-and-interference ratio SNIR at a matched filter receiver can be expressed in terms of aperiodic mean-square crosscorrelation parameters (interference parameters) and the power of the additive white Gaussian noise present in the channel. For a BPSK sequence-modulation scheme the SNIR at receiver k, denoted by $\mathrm{SNIR}^{(k)}$, is

$$\mathrm{SNIR}^{(k)} = \left(\frac{N_0}{2E_b} + \frac{1}{6N^3} r^{(k)} \right)^{-1/2}$$

where $E_b = N$ is the energy per symbol and $N_0/2$ is the noise power-spectral-density. The parameter $r^{(k)}$ depends on the considered system model. If we consider a chip-synchronous model where the signal delays $\tau^{(k)}$, $1 \leq k \leq K$, are multiples of the chip duration and equally probable in $0 \leq \tau^{(k)} < T$ where T denotes the period of a sequence, then $r^{(k)} = 3\mu_c^{(k)}$. If we assume that the delays are uniformly distributed in the interval $0 \leq \tau^{(k)} < T$, then $r^{(k)}$ depends on the sub-pulses (chips) by which the sequence elements are modulated. For rectangular pulses of length T/N, we get $r^{(k)} = 2\mu_c^{(k)} + \mu_c^{(k)}(1)$. Compared to this, for sub-pulses which are typically used in mobile communications, the impact of $\mu_c^{(k)}(1)$ on the interference parameter is much smaller and can often

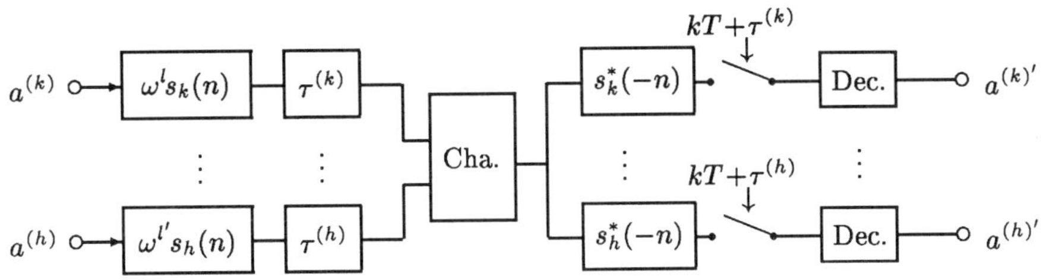

Fig. 1. Discrete-time baseband model of an asynchronous DS-CDMA system (without Rake-receiver). 'Dec'= Decision unit.

be neglected. Therefore, $\mu_c^{(k)}$ is the significant measure for the average multiuser interference at receiver k. μ_c or $\overline{\mu}_c$ are the commonly used performance measures for sequence families.

Mean-square autocorrelation parameter μ_a

Especially in mobile communications, frequency-selective and time-variant fading due to multipath propagation results in a remarkable degradation of the system performance. The fading sensitivity of the transmission depends on the flatness of the spectra of the spreading-sequences, which can quantitatively be described by the aperiodic mean-square autocorrelation parameter:

$$\mathrm{SE}^{(k)} = \sum_{m=1}^{N-1} |C_{kk}(m)|^2 = \frac{1}{2}\int_{-1/2}^{1/2} \left(|S_k(f)|^2 - N \right)^2 \mathrm{d}f$$

where $S_k(f) = \sum_n s_k(n) e^{-j2\pi n f}$ denotes the Fourier-transform of $s_k(n)$.
The performance measure for the family is μ_a. Sequences with a flat spectrum or a small sidelobe energy ensure uniform distribution of the sequence energy over the entire frequency range considered. Only for some simple channel models a quantitative description of the performance degradation due to frequency-selective and time-variant fading and the impact of the spreading-sequences on this effect is known. However, extensive simulations for various channel models have shown that for many interesting channels μ_a is a significant performance measure for the fading sensitivity [4]. Additionally, a low μ_a guarantees good synchronization properties of the sequences.

4 Bounds and 'Optimum' Families

4.1 Bounds

For the second moment of the aperiodic correlation we get

$$M^{(2)} = \sum_{k=1}^{K}\sum_{i=1}^{K}\sum_{m=-(N-1)}^{N-1} |C_{ki}(m)|^2 = \sum_{k=1}^{K}\sum_{i=1}^{K}\sum_{m=-(N-1)}^{N-1} C_{kk}(m) C_{ii}^*(m)$$

$$= K^2N^2 + 2\sum_{m=1}^{N-1} |\psi(m)|^2$$

where $\psi(m) = \sum_{k=1}^{K} C_{kk}(m)$. Expressing $M^{(2)}$ in terms of the above defined mean-square auto- and crosscorrelation parameters we get $M^{(2)} = KN^2 + \mu_a + \mu_c$ and, with this,

$$\mu_a + \mu_c = K(K-1)N^2 + 2\sum_{m=1}^{N-1} |\psi(m)|^2. \tag{1}$$

This results in the **Welch-bound** for the AMSQC [5]:

$$\mu_a + \mu_c \;\geq K(K-1)N^2 \quad \text{and} \quad \overline{\mu}_a + \overline{\mu}_c \;\geq (K-1)N^2 \quad \text{(Welch-bound)}. \tag{2}$$

Obviously, sequence-families attain this bound with equality if and only if $\psi(m) = 0$ for $1 \leq m < N$ [6]. Sets which fulfill this condition are called aperiodic complementary sets (ACS).

The Welch-bound considered in this paper corresponds to a system with asynchronous channel access where the average of the mean-square correlation for all signal delays $\tau^{(k)}$ has to be considered. In contrast to this, in systems with synchronous channel access only the correlation for $\tau^{(k)} = 0$, $1 \leq k \leq K$, has to be considered. The bound on the mean-square correlation for the synchronous case has been investigated in several papers and is often also called Welch-bound. In [7], sets of sequences attaining the Welch-bound for the synchronous case with equality are called Welch-bound equality (WBE) -sets. The definition is also used in [8] in this book. This definition of WBE-sets is equivalent to the definition of the S2-sets in the next section of this paper. WBE-sets or S2-sets form a subclass of the ACS. Sequence sets, e.g., Golay complementary sequences, exist which attain the Welch-bound for the asynchronous case (Eqn 2) with equality but which do not form WBE-sets. For this reason, the term WBE-set will not be used in this paper to avoid misunderstandings.

Besides the Welch-bound two other bounds on the AMSQC parameters which do not depend on the alphabet size of the sequences can be derived. To simply the notation, we define

$$\mu_a' = \frac{\mu_a}{K(K-1)N^2} \quad \text{and} \quad \mu_c' = \frac{\mu_c}{K(K-1)N^2}.$$

Using these abbreviations, the Welch-bound can be rewritten by

$$\mu_a' + \mu_c' \geq 1.$$

Summing up

$$\sum_{m=-(N-1)}^{N-1} |C_{kk}(m) - C_{ii}(m)|^2 \geq 0$$

for $1 \leq i \neq k \leq K$ and $1 \leq k \leq K$ results in

$$\mu_c' - (K-1)\mu_a' \leq 1 \qquad \text{(EACF-bound)},$$

where equality holds if and only if all sequences have the same autocorrelation function. These sets are called equal autcorrelation function sets (EACF sets). They are of interest in certain code-frequency-division multiple-access schemes [9].

In this paper, only polyphase sequences with $|s_k(n)| = 1$ are considered. For these sequences $|C_{kk}(m)| \leq N - m$ for $1 \leq m < N$. This results in

$$2\mathrm{SE}^{(k)} \leq \frac{1}{3}N(2N-1)(N-1), \quad 1 \leq k \leq K,$$

and

$$\mu_a' \leq \frac{(N-1)\,(2N-1)}{3N\,(K-1)} \qquad \text{(maxSE-bound)},$$

where equality holds if and only if the autocorrelation function of all sequences fulfills $|C_{kk}(m)| = N - m$ for $1 \leq m < N$.

4.2 Optimum Sequence Families

Figure 2 compares these bounds (dotted lines) to the empirically derived convex hull (dashed lines) of all parameter sets (μ_a', μ_c') for families with $K = 8$ sequences of length $N = 8$ (binary and quadriphase sequences). Obviously, besides the above mentioned general bounds additional restrictions which depend on the alphabet size of the sequences limit the range on which the parameters can vary. Sequence families for CDMA systems should have a low μ_a as well as a low μ_c. However, a simultaneous optimization of both parameters is obviously not possible. From a technical point of view, a family $\mathcal{F}_{opt}$ with the parameters $\mu_a(\mathcal{F}_{opt})$ and $\mu_c(\mathcal{F}_{opt})$ is optimum if no other family $\mathcal{F}_2$ exists with at least one of the parameters $\mu_a(\mathcal{F}_2)$ and $\mu_c(\mathcal{F}_2)$ being smaller than the corresponding parameter of $\mathcal{F}_{opt}$ (lines $E_{2,4} - A_{2,4} - B_{2,4}$ in Figure 2). Depending on the technical scenario the multiuser interference or the performance degradation due to the fading effect can result in a more severe restriction on the system performance so that the optimum choice of μ_a' and μ_c' depends on this scenario. Roughly spoken, points E_2 and E_4 mark the optimum parameter combinations for a scenario where the system performance is limited by the fading effect. Points B_2 and B_4 mark the parameter combinations of optimum sequence sets for a scenario where the system performance is mainly limited by the multiuser interference.

At least for a remarkable part of these "optimum" families (and for all families which are usually of technical interest) the Welch-bound is attained. Hereafter, we will only consider these families (lines $A_2 - B_2$, $A_4 - B_4$ in Figure 2). They will be called optimum. So, the construction of sequence sets which, on the one hand, attain the Welch-bound and which, on the other hand, have different μ_a and μ_c is of interest (lines $A_2 - B_2$, $A_4 - B_4$ in Figure 2).

In [10] and [11], polyphase sequences with a wide range of AMSQC parameters are described, and it is shown that the above described selection of sequences depending on the technical scenario and the AMSQC parameters can result in a remarkable improvement of the system performance. However, the sequences described in these papers have a quite large phase alphabet by which their application can be restricted in certain cases. Moreover, they do not always meet the Welch-bound.

5 Classification Scheme

To simply the derivation of the AMSQC parameters in Section 6 it is useful to define two parameters:

$$\mathcal{S}^{(2)}(n,n+m)=\sum_{k=1}^{K}s_k(n)s_k^*(n+m)$$

$$\mathcal{S}^{(4)}(n,l+m,l,n+m)=\sum_{k=1}^{K}s_k^*(n)s_k^*(l+m)s_k(l)s_k(n+m).$$

With this, the correlation parameters can be rewritten:

$$\psi(m)=\sum_{n=0}^{N-1-m}\mathcal{S}^{(2)}(n,n+m)\quad\text{for}\quad 1\le m<N,$$

$$M^{(2)}=K^2N^2+2\sum_{m=1}^{N-1}\left|\sum_{n=0}^{N-1-m}\mathcal{S}^{(2)}(n,n+m)\right|^2,$$

$$\mu_a=2\sum_{m=1}^{N-1}\sum_{n=0}^{N-1-m}\sum_{l=0}^{N-1-m}\mathcal{S}^{(4)}(n,l+m,l,n+m),$$

and

$$\mu_c=M^{(2)}-KN^2-\mu_a.$$

We define four classes of families [12–14]:

1. Sequence families form **ACS** iff $\psi(m)=0$ for $1\le m<N$. ACS attain the Welch-bound, i.e.,

$$\mu_a+\mu_c=K(K-1)N^2\quad\text{and}\quad 2\mathrm{SE}^{(k)}+\mu_c^{(k)}=(K-1)N^2\quad\text{for } 1\le k\le K.$$

2. Families with

$$\mathcal{S}^{(2)}(n,n+m)=\begin{cases}K & \text{if } m=0,\\ 0 & \text{else}\end{cases}$$

are called **S2-sets**. They form a subclass of the ACS and can only exist for $K\ge N$. Note that ACS with $K<N$ exist, e.g., Golay complementary sequences.

3. Families with

$$\mathcal{S}^{(4)}(n, l+m, l, n+m) = \begin{cases} K & \text{if } m = 0 \text{ or } n = l, \\ 0 & \text{else} \end{cases}$$

are called **S4-sets**. For S4-sets

$$\sum_{k=1}^{K} |C_{kk}(m)|^2 = K(N-m), \quad 1 \leq m < N, \quad \text{and} \quad \overline{\mu}_a = N(N-1).$$

4. **S2-S4-sets** are families which fulfill the S2- and the S4-property. The aperiodic mean-square correlation parameters of these families are:

$$\frac{\overline{\mu}_a}{N-1} = N \quad \text{and} \quad \frac{\overline{\mu}_c}{(K-1)N} = N - \frac{N-1}{K-1}.$$

Note that the first term describes the average energy per autocorrelation sidelobe sample. This term is obviously independent of the family size K. $\overline{\mu}_c \,/\, ((K-1)N)$ is the average interference power per interferer at a receiver. For $K \geq N$, $\overline{\mu}_c \,/\, ((K-1)N)$ is almost independent of K, too. The average energy of the auto- and crosscorrelation values is approximately the same.

The above defined family properties crucially depend on the cyclic phase of the sequences.
Similar results depending on the linear and cyclic structure of certain codes are described in [6,8].

6 AMSQC Parameters of Known Families

We have investigated the mean-square correlation parameters of various sequence families which have been proposed for CDMA systems [12–14,17].
To simplify the description of the families we define the product $\mathcal{F}_R = \mathcal{F}_S \otimes \mathcal{F}_T$ of two families $\mathcal{F}_S$ and $\mathcal{F}_T$ of size K_S and K_T, respectively:

$$\mathcal{F}_R = \{R \mid R(n) = S(n)T(n), \ S \in \mathcal{F}_S, T \in \mathcal{F}_T\}$$

with $K_R = |\mathcal{F}_R| = K_S K_T$. Note that a family is a multiset rather than a set. The sum parameters of these product families are

$$\mathcal{S}^{(k)}_{\mathcal{F}_R}(.) = \mathcal{S}^{(k)}_{\mathcal{F}_S}(.)\,\mathcal{S}^{(k)}_{\mathcal{F}_T}(.) \quad \text{for} \quad k = 2, 4.$$

Clearly, an S2- or S4-property of $\mathcal{F}_S$ or $\mathcal{F}_T$ is passed on to $\mathcal{F}_R$.
By $\mathcal{F}_R = \mathcal{F}_S \oplus \mathcal{F}_T$ we denote the union of two sets containing all sequences of $\mathcal{F}_S$ and $\mathcal{F}_T$. Since $\mathcal{F}_R$ is a multiset, we have $K_R = |\mathcal{F}_R| = K_S + K_T$ and

$$\mathcal{S}^{(k)}_R(\cdot) = \mathcal{S}^{(k)}_S(\cdot) + \mathcal{S}^{(k)}_T(\cdot) \quad \text{for} \quad k = 2, 4.$$

The sum of two S2-sets (S4-sets) forms an S2-set (S4-set). These operations allow to derive S2- or S4-sets with various parameters K and P from basic families.

Some examples are discussed hereafter. Note that not the families introduced in the original papers but versions where some sequences are omitted are described (e.g., expurgated set of Gold-sequences).
The most important example for an S2-S4-family is the m-sequence code $\mathcal{F}_m$ which contains the $N = p^r - 1$ cyclically shifted versions of a p-phase m-sequence (p prime) and the all-one sequence. The size of this family is $K_m = p^r = N + 1$. Another well known basic family is $\mathcal{F}_{exp} = \{s_k(n) = \exp(\jmath \frac{2\pi kn}{N}) \mid 1 \leq k \leq K\}$ of size $K = N$ and alphabet size $P = N$. This family forms an S2-set but no S4-set.
Another S2-set which is of technical interest is the set of Walsh-Hadamard-sequences.
The product of $\mathcal{F}_m$ with arbitrary polyphase sequences (cosets of $\mathcal{F}_m$) are S2-S4-families. Important examples are the expurgated versions (one sequence is omitted) of Gold-families, of the quadriphase family $\mathcal{A}$ [15], or of the Boztas-Kumar family [16]. These results can also be found in [6,8,7,17,12,13,18].
Larger S2-S4-families can simply be constructed by multiplying these families with arbitrary families. Some of the resulting families are expurgated versions of known large linear families, e.g., certain Kasami- or modified Gold-families, families derived from linear cyclic codes, or certain quadriphase families [15]. So, the binary large Kasami-families are S2-S4-families for $r \equiv 2 \bmod 4$. Large Kasami-families for $r \equiv 0 \bmod 4$ are S2-sets but no S4-sets.
The $\mathbb{Z}_4$-linear binary families with $K > N$ form S2-sets [14]. The small set of the $\mathbb{Z}_4$-linear binary families forms an ACS for certain cyclic phases of the sequences. These sets can not form S2-sets since $K < N$ [14].
An extensive examination of the AMSQC parameters of many sequence families derived from known codes in Galois-fields or -rings (not necessarily linear codes) shows that all of these families result in approximately the same ratio of the mean-square auto- and crosscorrelation parameters and, thus, do not allow to select sequence families optimized for different technical scenarios [13].
It is known from various numerical investigations that for many families $\mu(1)$ can be neglected in comparison to $\mu(0)$ [19]. However, some analytical results are known and will briefly be reviewed [20,13]. Confer also the more detailed comments in [8] in this book.

1. Consider a family $\mathcal{F}_1 = \{s_k \mid 1 \leq k \leq K\}$ with the AMSQC parameters μ_c, $\mu_c(1)$, and

$$r(\mathcal{F}_1) = \sum_{k=1}^{K} r^{(k)}(\mathcal{F}_1) = \sum_{k=1}^{K} (2\mu_c^{(k)} + \mu_c^{(k)}(1)) = 2\mu_c + \mu_c(1).$$

   The family $\mathcal{F}_2 = \{g_k \mid 1 \leq k \leq K\}$ where $g_k(n) = (-1)^n s_k(n)$ has the average interference parameter

$$r(\mathcal{F}_2) = \sum_{k=1}^{K} r^{(k)}(\mathcal{F}_2) = 2\mu_c - \mu_c(1).$$

   So, either $\mathcal{F}_1$ or $\mathcal{F}_2$ has an average interference parameter $r \leq 2\mu_c$.

2. For sequence families which simultaneously contain a sequence s and a sequence g where $g(n) = (-1)^n s(n)$, we get $\mu_c(1) = 0$. This class includes the $\mathbb{Z}_4$-linear binary families.
3. For sequence families which can be described as the product of $\mathcal{F}_{exp}$ and arbitrary other families, we get $\mu_c(1) = 0$.
4. Summing up
$$|C_{ki}(m) \pm C_{ki}(m+1)|^2 \geq 1$$
for binary sequences or
$$|C_{ki}(m) \pm C_{ki}(m+1)|^2 \geq 0$$
for arbitrary polyphase sequences results in
$$|\mu_{ki}(1)| \leq \mu_{ki}(0) - N \quad \text{and} \quad |\mu_{ki}(1)| \leq \mu_{ki}(0) - 1$$
for binary and polyphase sequences, respectively.
5. For ACS, we get
$$|\mu_c(1)| \leq \begin{cases} \mu_a - K & \text{for polyphase sequences,} \\ \mu_a - K(N-1) & \text{for binary sequences.} \end{cases}$$
6. For S2-S4-sets, we get
$$|\mu_c(1)| \leq \begin{cases} KN(N-2) & \text{for polyphase sequences,} \\ K(N-1)(N-2) & \text{for binary sequences.} \end{cases}$$
In both cases, this results in
$$\frac{|\mu_c(1)|}{\mu_c} \leq \frac{1}{K-2}.$$

From these results, we can conclude that the impact of $\mu_c(1)$ on the interference parameter can be neglected for many families.

7 Optimization of μ_a and μ_c of ACS

Sequence families attain the Welch-bound if and only if they form complementary sets. Up to now, no construction method is known which allows to construct all aperiodic complementary sets (ACS) for a given N and K or ACS with certain combinations of μ_a and μ_c. For this reason, we have investigated simple and efficient search methods for ACS with different mean-square correlation parameters.

A well-known and closely related problem is the minimization of the interference parameter of families with good periodic peak correlation parameters by phase selection (cf. [8]). The periodic peak correlation parameter (cf. [8]) - a commonly used design criterion for sequence families - is invariant to cyclically shifting the sequences. This allows to select a cyclic phase of the sequences which minimizes the interference parameter and simultaneously to preserve the good periodic peak correlation parameters. Since the number of cyclic phases (N^K) is too large to perform a full search, simple phase selection criteria were investigated which result in a small interference parameter. They are usually based on the autocorrelation parameters of the sequences, so that only KN cyclic phases have to be examined. A well-known criterion is the least sidelobe energy (LSE) phase where for each sequence the phase with the minimum sidelobe energy is selected. The resulting interference parameters for the LSE-phase are not much better than the results known for random sequences or S2-S4-families [19, 21, 22]. From the Welch-bound we can conclude that a large sidelobe energy allows smaller interference parameters. For this reason, in [23, 24] O'Farrell investigates cyclic phases with a maximum sidelobe energy (Maximum Sidelobe Energy (MSE) criterion). The MSE-phase usually results in a smaller interference parameter than the LSE-phase. Note that both criteria do not guarantee that the resulting families attain the Welch-bound.

In Figure 3, the AMSQC parameter combinations for all combinations of cyclic phases are compared to the Welch-bound (dashed line) for a binary family and a quadriphase family. Obviously, cyclic phases with different μ_a and μ_c exist which attain the Welch-bound. In [13], an efficient search method for these phases is described. However, for increasing length or family size the possible improvements compared to the parameters of S2-S4-families become negligible.

7.1 ACS with $K \geq N$

The basic idea of the construction of ACS with different AMSQC parameters is to apply operations on known ACS which preserve the ACS-property but which change μ_a and μ_c. For $K \geq N$, we restrict our investigations on S2-sets (a subclass of the ACS) since this allows a simple description of the invariance operations. Here and hereafter, only sequence families which form S2-sets are considered. Invariance operations of S2-sets are for example discussed in [7]. In this paper, only operations which do not change N, K, and the phase alphabet size P are investigated.

The following invariance operations result in equivalent S2-sets: permutation of columns or rows of $\mathbf{S}$ or multiplication of columns or rows by a Pth root of unity, e.g., (-1) for binary sets. For many parameters N and K more than one non-equivalent S2-set exist. For $K = N$, (binary) S2-sets are Hadamard-matrices. Invariance operations of these matrices are well-known. The sizes of the resulting equivalence classes are discussed in [26].

However, for the CDMA systems with asynchronous channel access considered in this paper the permutation of columns or rows and the multiplication of columns or rows by -1 have different technical meanings. Sequence sets are CDMA-equivalent if sequences (rows of $\mathbf{S}$) are multiplied with a Pth root of unity, e.g., -1 for binary sequences, or if sequences are permuted. CDMA-equivalent sequence sets always result in the same system performance and the same μ_a and μ_c. For this reason, we can assume $s_k(0) = 1$ for $1 \leq k \leq K$.

Permutation of columns of **S** or multiplication of columns with a Pth root of unity preserve the S2-property but can change the AMSQC parameters. These operations can be used to construct S2-sets with different parameters.
For $K = N = 4$, there is for example only one equivalence class of binary Hadamard matrices. However, there are two binary matrices which are not CDMA-equivalent and there are 10 binary ACS which are not CDMA-equivalent.
An efficient search method for S2-sets with different μ_a is based on the following formalism. The mean-square autocorrelation of the S2-set $\mathcal{F} = \{s_k \,|\, 1 \leq k \leq K\}$ is

$$\begin{aligned}\sum_{k=1}^{K} |C_{kk}(m)|^2 &= \sum_{n=0}^{N-1-m} \sum_{l=0}^{N-1-m} \sum_{k=1}^{K} s_k(n)s_k^*(l)s_k^*(n+m)s_k(l+m) \\ &= \sum_{n=0}^{N-1-m} \sum_{l=0}^{N-1-m} \sum_{k=1}^{K} c_n(k)c_l^*(k)c_{n+m}^*(k)c_{l+m}(k)\end{aligned}$$

As example, hereafter binary sequences will be considered. Let denote $c_3 = c_1c_2$ an elementwise product of c_1 and c_2, i.e., $c_3(k) = c_1(k)c_2(k)$ for $1 \leq k \leq K$. Let now consider sequence families where the product of two columns of **S** is either a column c_a of **S** or the complement $-c_a$ of a column of **S**. We define:

$$n_{a,m}^{+} = \#\{c_nc_{n+m} = c_a \mid n = 0, \ldots, N-1-m\} \quad \text{for } a = 0..N-1.$$

$$n_{a,m}^{-} = \#\{c_nc_{n+m} = -c_a \mid n = 0, \ldots, N-1-m\} \quad \text{for } a = 0..N-1.$$

Obviously, $\sum_{a=0}^{N-1}(n_{a,m}^{+} + n_{a,m}^{-}) = N - m$. With this, we get

$$\sum_{k=1}^{K} |C_{kk}(m)|^2 = K \sum_{a=0}^{N-1} (n_{a,m}^{+} - n_{a,m}^{-})^2. \tag{3}$$

To describe the effect of the invariance operations, we define a $N \times N$ matrix $\mathbf{M} = (m_{ij})$ where $m_{ij} = c_ic_j \in \{\pm c_l \,|\, 0 \leq l < N\}$. $n_{a,m}^{+}$ and $n_{a,m}^{-}$ count the c_a and $-c_a$, respectively, on the line $m_{n,n+m}$, $n = 0 \ldots N-1-m$ parallel to the diagonal. The effect of a permutation of columns or multiplication of columns by (-1) can now be described by the application of the same operations on the corresponding columns and rows of **M**. This allows to determine the columns which must be multiplied by (-1) to increase or decrease μ_a. When this operations are applied on the Walsh-Hadamard sequences only a very few number of permutations is necessary to construct S2-sets with the full possible range of μ_a.
For many families, derived from linear codes the columns of **S** are closed under the above defined multiplication, i.e., $n_{a,m}^{-} = 0$. Two important families belong to this class:
μ_a is minimum for $n_{a,m}^{+} \in \{0, 1\}$ for $a = 0, \ldots, N-1$ resulting in $\sum_{k=1}^{K} |C_{kk}(m)|^2 = K(N-m)$ and $\overline{\mu}_a = N(N-1)$. This case describes the **S2-S4-families**.

Another family where the columns of $\mathbf{S}$ are closed under the above defined multiplication are the **Walsh-Hadamard matrices** with $K = N = 2^r$. To calculate μ_a for these families we define

$$X_m^{(r)} = \frac{1}{K}\sum_{k=1}^{K} |C_{kk}(m)|^2.$$

Exploiting the symmetry properties of these matrices, we get:

$$\begin{aligned} X_{N-m}^{(r)} &= X_{N/2-m}^{(r-1)} \quad \text{for} \quad m = 1,\ldots,N/2-1, \\ X_{N/2}^{(r)} &= (N/2)^2, \\ X_m^{(r)} &= 4X_m^{(r-1)} + X_{N/2-m}^{(r-1)} \quad \text{for} \quad m = 1,\ldots,N/2-1. \end{aligned}$$

With this, the mean-square autocorrelation parameter is $\overline{\mu}_a = N^2(N^{\log_2(3)-1} - 1)$. From Figure 2, we can conclude that with increasing alphabet size the scale on which the mean-square auto- and crosscorrelation parameters can vary becomes significantly larger. The minimum $\overline{\mu}_c$ the author has observed for binary families with $K = N$ is

$$\overline{\mu}_c = N(N^2 - N^{\log_2(3)})$$

for Walsh-Hadamard sequences. For family $\mathcal{F}_{exp} = \{s_k(n) = \exp(\jmath\frac{2\pi}{N}kn) \mid 1 \leq k \leq K = N\}$ with $P = N$, we get $|C_{kk}(m)| = N - m$, $1 \leq m < N$, and, thus,

$$\overline{\mu}_c = \frac{1}{3}N(N^2 - 1) \quad \text{(point D in Figure 2).}$$

For an AWGN channel, the application of polyphase sequences results in a SNIR which can be remarkably better (some dB, depending on N) compared to the case where binary sequences are applied. Due to their bad autocorrelation properties usually neither the Walsh-Hadamard sequences nor family $\mathcal{F}_{exp}$ will be applied in CDMA systems without additional scrambling.

7.2 ACS with $K < N$

For $K < N$, S2-sets can not exist. Three construction methods for aperiodic complementary sets with $K < N$ have been investigated:

1. iterative construction methods based on the Golay complementary sets,
2. construction methods based on Reed-Muller codes [25],
3. families derived from the small family of $\mathbb{Z}_4$-linear binary sequences [14].

Operations preserving the complementarity property (based on the Welti-code notation) have been applied on all types. They allow to construct Welch-optimum families with a wider range of AMSQC parameters than all other known methods. However, it can be shown that they do not allow to construct all possible parameter combinations. The general construction of aperiodic complementary sets for $K < N$ remains an open question.

8 Conclusion

In this paper, we have investigated the aperiodic mean-square correlation parameters of families of P-phase sequences. A technical interpretation of the parameters in the context of asynchronous DS-CDMA systems is given. The mean-square autocorrelation parameter is a performance measure for the antifading capability of the sequences, whereas the mean-square crosscorrelation parameters describe the multiuser interference. Due to the Welch-bound a simultaneous optimization of both parameters is not possible. We have investigated the correlation parameters of several known families. For most of these families the Welch-bound is attained. However, they show similar ratios of the auto- and crosscorrelation parameters. For this reason, general construction methods for ACS have been investigated. They allow to construct ACS with almost all possible combinations of mean-square auto- and crosscorrelation parameters.

References

1. D.V. Sarwate and M.B. Pursley. Crosscorrelation properties of pseudorandom and related sequences. *Proc. IEEE*, 68:593–619, 1980.
2. M.B. Pursley. Performance evaluation for phase-coded spread-spectrum multiple-access communication – part I: System analysis. *IEEE Trans. Commun.*, COM-25:795–799, 1977.
3. M.B. Pursley and D.V. Sarwate. Performance evaluation for phase-coded spread-spectrum multiple-access communication – part II: Code sequence analysis. *IEEE Trans. Commun.*, COM-25:800–803, 1977.
4. H.D. Schotten, H. Elders-Boll, and A. Busboom. Optimization of spreading-sequences for DS-CDMA systems and frequency-selective fading channels. In *Fifth International Symposium on Spread Spectrum Techniques and Applications ISSSTA'98, Sun City*, pages 33 – 37, 1998.
5. L.R. Welch. Lower bounds on the maximum cross correlation of signals. *IEEE Trans. Inform. Theory*, IT-20:397 – 399, 1974.
6. D. Sarwate. Sets of complementary sequences. *Electron. Lett.*, vol. 19, pp. 711 – 712, 1983.
7. J. Massey and T. Mittelholzer, *Welch's Bound and sequence sets for code-division multiple-access systems in* Sequences II: methods in communication, security, and computer science, pages 63 – 78. Springer-Verlag, Eds. R. Capocelli, A. De Santis, and U. Vacarro, 1993.
8. D. Sarwate. Meeting the Welch bound with equality. *this volume*, 1999.
9. H.D. Schotten and J. Ruprecht. On the Optimum Spreading Sequence and Mismatched Filter Design for Cellular CTDMA. In *Fourth International Symposium on Spread Spectrum Techniques and Applications ISSSTA'96, Mainz*, pages 677 – 682, 1996.
10. I. Oppermann. Orthogonal complex-valued spreading sequences with a wide range of correlation properties. *IEEE Trans. Commun.*, COM-45:1379 – 1380, 1997.
11. I. Oppermann and B.S. Vucetic. Complex spreading sequences with a wide range of correlation properties. *IEEE Trans. Commun.*, COM-45:365–375, 1997.
12. H.D. Schotten. Mean-square correlation parameters of code-sequences in DS-CDMA systems. In *Proc. IEEE International Symposium on Information Theory ISIT'97, Ulm*, page 486, 1997.

13. H.D. Schotten. *Algebraische Konstruktion gut korrelierender Sequenzen und Analyse ihrer Leistungsfähigkeit in Codemultiplex-Systemen. (Algebraic construction of sequences with low correlation and their application in CDMA systems)* PhD Thesis, Aachen Univ. of Technology, 1997.
14. H.D. Schotten and H. Elders-Boll. Aperiodic mean-square correlation parameters of binary $\mathbb{Z}_4$-linear sequences. In *Proc. IEEE International Symposium on Information Theory ISIT'98, Cambridge*, page 402, 1998.
15. P.V. Kumar, T. Helleseth, A.R. Calderbank, and A.R. Hammons. Large families of quaternary sequences with low correlation. *IEEE Trans. Inform. Theory*, IT-42:579 – 592, 1996.
16. S. Boztas and P.V. Kumar. Binary sequences with Gold-like correlation but larger linear span. *IEEE Trans. Inform. Theory*, IT-40:532 – 537, 1994.
17. H.D. Schotten, B. Liesenfeld, and H. Elders-Boll. Large families of code-sequences for direct-sequence CDMA systems. *International Journal of Electronics and Communications (AEÜ)*, 49:143 – 150, 1995.
18. F.W. Sun and H. Leib. Optimal phases for a family of quadriphase sequences. *IEEE Trans. Inform. Theory*, IT-43:1205 – 1217, 1997.
19. K. Kärkkäinen. Mean-square cross-correlation as a performance measure for spreading code families. In *Proc. IEEE Int. Symp. on Spread Spectrum Techniques and Its Applications ISSSTA'92, Yokohama*, pages 147 – 150, 1992.
20. H.D. Schotten. Bounds on the average interference parameter of binary sequences in DS-CDMA systems. In *Proc. IEEE International Symposium on Information Theory and Its Applications ISITA'94, Sydney*, pages 1383 – 1387, 1994.
21. M.B. Pursley and D.V. Sarwate. Evaluation of correlation parameters for periodic sequences. *IEEE Trans. Inform. Theory*, IT-23:508–513, 1977.
22. M.B. Pursley and H.F.A. Roefs. Numerical evaluation of correlation parameters for optimal phases of binary shift-register sequences. *IEEE Trans. Commun.*, COM-27:1597–1604, 1979.
23. T. O'Farrell. *Code-Division Multiple-Access (CDMA) Techniques in Optical Fibre Local Area Networks.* Manchester University, PhD thesis, 1989.
24. T. O'Farrell. New signature code sequence design techniques for CDMA systems. *Electron. Lett.*, 27:371 – 373, 1991.
25. J.A. Davis and J. Jedwab. *Peak-to-mean power control in OFDM, Golay complementary sequences and Reed-Muller codes.* Technical Report HPL-97-158, Hewlett-Packard Labs., Bristol, 1997.
26. S Golomb et al. *Digital Communications with Space Applications.* Englewood Cliffs, NJ: Prentice Hall, 1964

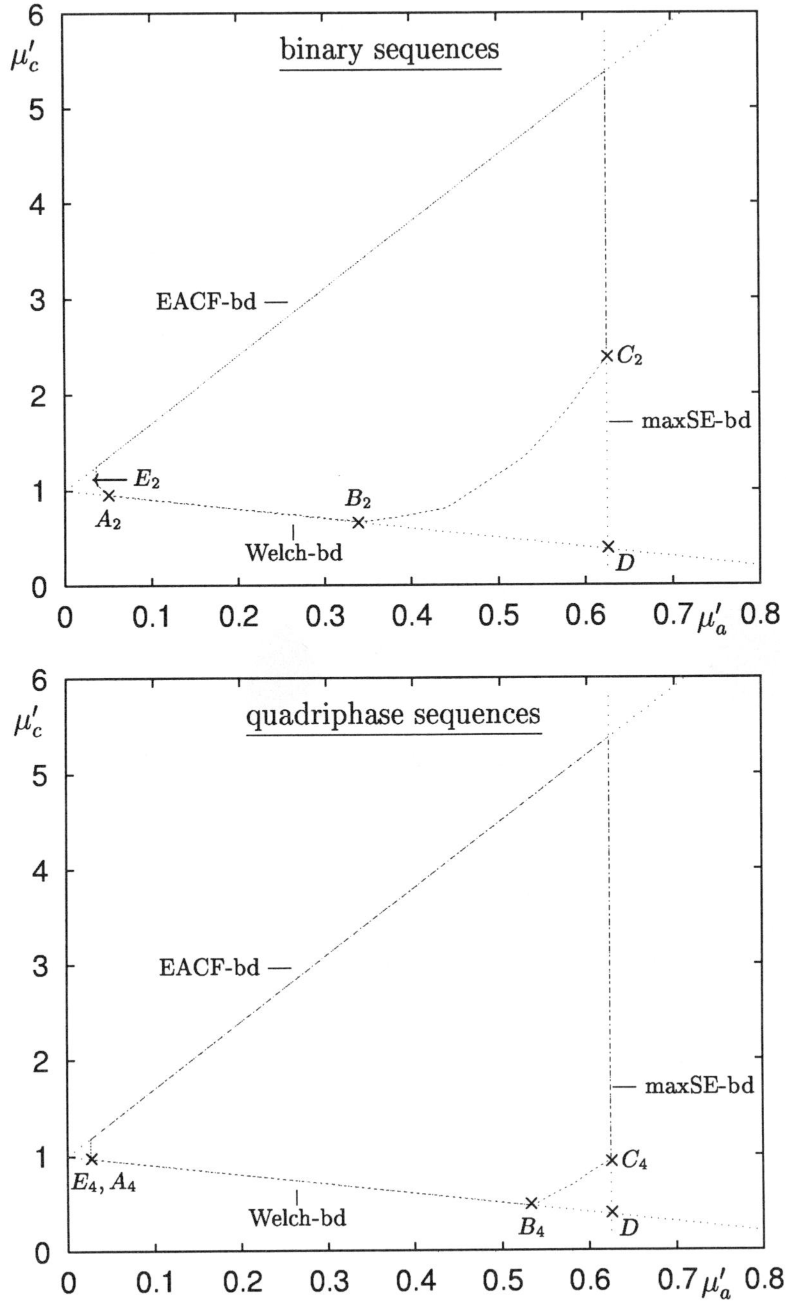

Fig. 2. Convex hull for (μ'_a, μ'_c) for sets of binary and quadriphase sequences ($N = 8$, $K = 8$) (dashed line) and general bounds on (μ'_a, μ'_c) (dotted lines).

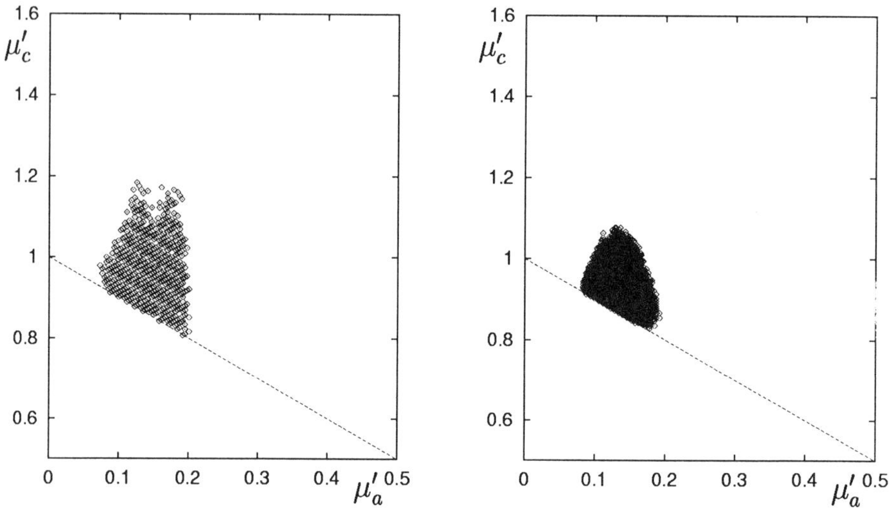

Fig. 3. AMSQC parameters (μ_a', μ_c') of an expurgated binary Gold-family (left, $N = 7$, $K = 8$) and quadriphase family $\mathcal{A}$ (right, $N = 7$, $K = 8$).

Period and Linear Complexity of Cascaded Clock-Controlled Generators

Chik How Tan

DSO National Laboratories
20 Science Park Drive
Singapore 118230.
Email: tchikhow@dso.org.sg

Abstract. In this paper, we examined the properties of Liu-Xiao clock-controlled generators [7] and two kinds of cascaded clock-controlled generators, namely, Gollmann-Chambers and Liu-Xiao r-cascaded clock-controlled generators. We derived some new results on the period, the linear complexity and the characteristic polynomials of these three clock-controlled generators under certain conditions.

1 Introduction

Pseudo random sequences are often generated by non-linear combinations of linear feedback shift registers. There are many ways of combining linear feedback shift registers, for examples, stop-and-go generators [1], alternating step generators [5] and the cascading of clock-controlled registers, etc. In the past, researchers were interested in studying cascaded clock controlled shift registers [2], [3], [4], [7], [8] and [9] which achieved high linear complexity. In [2] and [3], the author showed the period and the linear complexity of cascading stop-and-go sequences under the condition that all registers are of the same length and prime. In [4], the authors also derived the period and the linear complexity of the cascaded sequences by different stepping of registers with the same conditions as before. In [7], the authors gave a new type of cascading the registers by introducing the delay in between. They also obtained the period and the linear complexity under the conditions that all the registers are of primitive polynomials and of the same length. Although there were many results on these areas in the literature, many properties are still not known at this moment. For some applications, we are interested in sequences generated from different length registers to achieve long period and large linear complexity. In this paper, we study two kinds of cascaded clock controlled shift registers proposed by Gollmann & Chambers and Liu & Xiao in 1989 and 1992 respectively in a different setting by considering different lengths of the registers. We obtain results on the period, linear complexities and the characteristic polynomial of these two cascaded generators. Some of these results are the generalisation of that of [7].

In Section 2, we will give a general definition and list some results for clock controlled generators required for the subsequent sections. In Section 3, we first study Liu-Xiao clock-controlled generators, proposed in 1992, which is used as a basic building block for Liu-Xiao cascaded generators. We obtained its period and linear complexity as well. In Section 4 and 5, we study Gollmann-Chambers cascaded generators and Liu-Xiao cascaded generators under the different conditions of registers respectively. We show some new results on the linear complexities of two cascaded generators.

2 Preliminary

Lemma 1. *([6], Theorem 3.35) Let* $f_1(x)$, $f_2(x)$, $\ldots$, $f_N(x)$ *be all the distinct monic irreducible polynomials in* $\mathbf{F_2}[\mathbf{x}]$ *of degree* n *and order* e *and let* $T \geq 2$ *be an integer whose prime factors divide* e *but not* $(2^n - 1)/e$. *Then* $f_1(x^T)$, $f_2(x^T)$, $\ldots$, $f_N(x^T)$ *are all distinct monic irreducible polynomials in* $\mathbf{F_2}[\mathbf{x}]$ *of degree* nT *and order* eT.

Now, let us consider a basic clock-controlled generator which consists of two linear feedback shift registers (LFSR). The first register is called the control register $LFSR_1$ which controls the clocking of the second register $LFSR_2$, called the generator register. If the output sequence (called the control sequences) x_t from $LFSR_1$ is 1, then $LFSR_2$ clocks k_1 times, otherwise, $LFSR_2$ clocks k_0 times. The output sequence of $LFSR_2$ (called the generator sequence) controlled by the control sequence is then $\{y_{\sigma(t)}\}$, where $\{y_t\}$ is the output sequence from $LFSR_2$ with regular clocking and $\sigma(t) = \sum_{i=1}^{t}\{k_1 x_i + k_0(x_i \oplus 1)\}$.

Assume that $LFSR_1$ and $LFSR_2$ are associated with polynomials $f_1(x)$ and $f_2(x)$ of degree n_1 and n_2, with order T_1 and T_2 respectively.

Let

$$K = \sigma(T_1).$$

In [8], Smeets showed that the characteristic polynomial of $\{y_{\sigma(t)}\}$ divides $f_2^{(K)}(x^{T_1})$, where $f_2^{(K)}(x)$ denotes the characteristic polynomial of the K-decimated sequence of $\{y_t\}$, i.e.. $\{y_{Kt}\}$.

First, we state and prove the following well known Lemma which will be used in the proof of the subsequent Theorem.

Lemma 2. *[8]Let the control register* $LFSR_1$ *and the generator register* $LFSR_2$ *be associated with polynomials* $f_1(x)$ *and* $f_2(x)$ *of degree* n_1 *and* n_2, *with order* T_1 *and* T_2 *respectively. Let* K *be the number of clocking of* $LFSR_2$ *over the period* T_1, *then*

a. The period of $\{y_{\sigma(t)}\}$ *is* $\frac{T_1.T_2}{(K,T_2)}$,

b. Assume that T_1 *divides* $2^{n_2} - 1$, $(K, 2^{n_2} - 1) = 1$ *and the characteristic polynomial of* $LFSR_2$ *is primitive, then the characteristic polynomial of* $\{y_{\sigma(t)}\}$ *is* $f_2^{(K)}(x^{T_1})$, *which is irreducible, and its linear complexity is* n_2T_1.

Proof. The proof of (a) is trivial. We proceed the proof of part (b) as follows. Given that $(K, 2^{n_2} - 1) = 1$ and $f_2(x)$ is a primitive polynomial, then $f_2^{(K)}(x)$ is also a primitive polynomial and its order is $2^{n_2} - 1$. Given that T_1 divides $2^{n_2} - 1$, this implies that all the prime factors of T_1 divides $2^{n_2} - 1$, but they do not divide 1. By Lemma 1, $f_2^{(K)}(x^{T_1})$ is irreducible. Hence, the linear complexity of $\{y_{\sigma(t)}\}$ is $n_2 T_1$. ◇

3 Liu-Xiao clock-controlled generators

The Liu-Xiao clock-controlled generator [7] consists of two linear feedback shift registers which is quite similar to the basic clock-controlled generators, except that the previous output sequence from $LFSR_1$ is used to control the clocking of the second register and the resultant sequence is the exclusive-or of the two sequences, the control sequence and the generator sequence. At time t, if the output sequence x_{t-1} from $LFSR_1$ is 1, then $LFSR_2$ clocks k_1 times, otherwise, $LFSR_2$ clocks k_0 times. Assume that $LFSR_1$ and $LFSR_2$ are associated with primitive polynomials $f_1(x)$ and $f_2(x)$ of degree n_1 and n_2 respectively and $\{y_t\}$ be a sequence from $LFSR_2$ with regular clocking. Then, the output sequence z_t can be expressed as follows :

$$z_t = x_{t-1} \oplus y_{\sigma(t)}, \text{ where } \sigma(t) = \sum_{i=1}^{t} \{k_1 x_{i-1} + k_0(x_{i-1} \oplus 1)\} \text{ with } x_0 = 0.$$

Let

$$K = \sigma(2^{n_1} - 1).$$

Now, we state and prove the following Theorem which gives the linear complexity and the period of $\{z_t\}$.

Theorem 1. *If $LFSR_1$ and $LFSR_2$ are primitive polynomials of degrees n_1 and n_2 respectively and $(K, 2^{n_2} - 1) = 1$, then we have*

a. The period of $\{z_t\}$ is $(2^{n_1} - 1)(2^{n_2} - 1)$.

b. If n_1 divides n_2, then the linear complexity of $\{z_t\}$ is $n_2(2^{n_1} - 1) + n_1$ and the characteristic polynomial of $\{z_t\}$ is $f_1(x) f_2^{(K)}(x^{2^{n_1}-1})$.

Proof. a. Given that $(K, 2^{n_2} - 1) = 1$, by Lemma 2 (a), the period of $\{y_{\sigma(t)}\}$ is $(2^{n_1} - 1)(2^{n_2} - 1)$. Hence, the period of $\{z_t\}$ is $(2^{n_1} - 1)(2^{n_2} - 1)$.

b. By Lemma 2 (b), the characteristic polynomial of $\{y_{\sigma(t)}\}$ is $f_2^{(K)}(x^{2^{n_1}-1})$ which is irreducible, where $f_2^{(K)}(x)$ is the characteristic polynomial of K-decimated sequence of $\{y_t\}$, i.e. $\{y_{Kt}\}$. By the fact that the characteristic polynomial of the exclusive-or of two sequences is equal to LCM of the two characteristic polynomials, hence, the characteristic polynomial of $\{z_t\}$ is LCM $\{f_1(x), f_2^{(K)}(x^{2^{n_1}-1})\} = f_1(x) f_2^{(K)}(x^{2^{n_1}-1})$ and the linear complexity of $\{z_t\}$ is $n_2(2^{n_1} - 1) + n_1$. ◇

In the above Theorem part b, if we let $n_1 = n_2$ (say $= n$), then the linear complexity of $\{z_t\}$ is $n(2^n - 1) + n = n2^n$ which is the result shown by Liu and Xiao in [7]. Hence, our result of Theorem 1(b) is a generalisation of that of Liu and Xiao.

4 Gollmann-Chambers Cascaded generators

The Gollmann-Chambers r-cascaded generator [4] consists of r linear feedback shift registers $LFSR_1, LFSR_2, \ldots, LFSR_r$ and cascading them one by one. The Gollmann-Chambers 2-cascaded generator is like the basic clock-controlled generator, except that the output sequence from $LFSR_1$ is the exclusive-or with 1 to control the clocking of the second register and the resultant sequence is exclusive-or of the control sequence and generator sequence. Now, consider the Gollmann-Chambers 3-cascaded generator, the output sequences from 2-cascaded generator will control the clocking of the 3rd shift register $LFSR_3$ and the resultant output sequence is the exclusive-or of the sequences from $LFSR_3$ and the output sequences from the 2-cascaded generator. By a similar construction, the clocking of rth shift register $LFSR_r$ will be controlled by the output sequences from the $(r-1)$-cascaded generator.

Let $LFSR_1, LFSR_2, \ldots, LFSR_r$ be associated with the primitive polynomials $f_1(x), f_2(x), \ldots, f_r(x)$ of degree $n_1, n_2, \ldots, n_r$ respectively. If the control sequence $z_t^{(i)}$ is 1, then $LFSR_{i+1}$ clocks $k_{i,1}$ times, otherwise $LFSR_{i+1}$ clocks $k_{i,0}$ times. Let x_t be the output sequence from $LFSR_1$ and $y_t^{(i)}$ a sequence from $LFSR_i$ with regular clocking, for $2 \leq i \leq r$, then $y^{(i)}_{\sigma_{i-1}(t)}$ is the output from $LFSR_i$ with the clock-controlled by $(i-1)$-cascaded generator. Therefore, the intermediate output sequences $z_t^{(i)}$ are given by

$$z_t^{(i)} = z_t^{(i-1)} \oplus y^{(i)}_{\sigma_{i-1}(t)}, \quad \text{for } 2 \leq i \leq r,$$

where

$$\sigma_{i-1}(t) = \sum_{j=1}^{t} \{k_{i-1,1} z_j^{(i-1)} + k_{i-1,0}(z_j^{(i-1)} \oplus 1)\}, \sigma_0(t) = t$$

and

$$z_t^{(1)} = x_t \oplus 1.$$

Let T_i be the period of i-cascaded generator and

$$K_{i-1} = \sigma_{i-1}(T_{i-1}) \text{ for } 2 \leq i \leq r.$$

The period and its linear complexity of Gollmann-Chambers cascaded generators are given in the following theorem.

Theorem 2. *If $LFSR_1$, $LFSR_2$, $\ldots$, $LFSR_r$ are primitive polynomials of degrees $n_1, n_2, \ldots, n_r$ respectively with $n_1 \mid n_2 \mid \ldots \mid n_r$ and $(K_{i-1}, 2^{n_i} - 1) = 1$ for $2 \leq i \leq r$, then* *a. The period of* $\{z_t^{(r)}\}$ *is*

$$T_r = \prod_{j=1}^{r}(2^{n_j} - 1),$$

b. The characteristic polynomial of $\{z_t^{(r)}\}$ *is*

$$(x+1)f_1(x)\prod_{j=2}^{r} f_j^{(K_{j-1})}(x^{T_{j-1}}),$$

c. The linear complexity of $\{z_t^{(r)}\}$ *is*

$$L_r = \sum_{i=1}^{r} n_i \prod_{j=0}^{i-1}(2^{n_j} - 1) + 1, \text{ where } n_0 = 1.$$

Proof. We prove (a), (b) and (c) simultaneously by induction. When $r = 1$, the sequence of $\{z_t^{(1)}\}$ is just obtained by toggling the sequence from $LFSR_1$. Therefore, the period of $\{z_t^{(1)}\}$ is $2^{n_1} - 1$ and its characteristic polynomial is $(x+1)f_1(x)$. Hence, the linear complexity of $\{z_t^{(1)}\}$ is $n_1 + 1$. Therefore, (a), (b) and (c) is true for $r = 1$. Now, assume that it is true for $r = m > 1$, then the period and the linear complexity of Gollmann-Chambers m-cascaded generator are T_m and L_m respectively. Now, we consider Gollmann-Chambers $(m+1)$-cascaded generator. Given that $(K_m, 2^{n_{m+1}} - 1) = 1$ and $n_1 \mid n_2 \mid \ldots \mid n_{m+1}$, by Lemma 2 (b), the characteristic polynomial of $\{y_{\sigma_m(t)}^{(m+1)}\}$ is $f_{m+1}^{(K_m)}(x^{T_m})$ and is irreducible, where $f_{m+1}^{(K_m)}(x)$ is the characteristic polynomial of K_m-decimated sequence of $f_{m+1}(x)$, and the period of $\{y_{\sigma_m(t)}^{(m+1)}\}$ is $T_m(2^{n_{m+1}} - 1) = T_{m+1}$. Since $z_t^{(m+1)} = z_t^{(m)} \oplus y_{\sigma_m(t)}^{(m+1)}$, then the characteristic polynomial of $\{z_t^{(m+1)}\}$ is the LCM of the characteristic polynomials of $\{z_t^{(m)}\}$ and $\{y_{\sigma_m(t)}^{(m+1)}\}$, that is, $f_{m+1}^{(K_m)}(x^{T_m})(x+1)f_1(x)\prod_{j=2}^{m} f_j^{(K_{j-1})}(x^{T_{j-1}})$. Hence, the linear complexity is $L_m + n_{m+1}T_m = L_{m+1}$. The period of $\{z_t^{(m+1)}\}$ is LCM $\{T_m, T_{m+1}\}$ which is equal to T_{m+1} as T_m divides T_{m+1}. Therefore, (a), (b) and (c) are true for $r = m+1$. ◇

In the above Theorem, if we let $n_1 = n_2 = \ldots = n_r (= n$ say) to be a prime number with $2^n - 1$ also a prime number, then the period of this r-cascaded generator is $(2^n - 1)^r$ which is the same as that of Gollmann and Chambers. Furthermore, we also give the linear complexity and its characteristic polynomial of this cascaded generator in the above theorem.

Example 1. Consider $r = 3, n_1 = 4, n_2 = 8$ and $n_3 = 8$

a. Let $k_{1,1} = k_{2,1} = 1, k_{1,0} = k_{2,0} = 0$, then $K_1 = 7$ and $K_2 = 1913$, and we can check that $(K_1, 255) = (K_2, 255) = 1$. Hence, the results of Theorem 2 can be applied to achieve maximal period and complexity.

b. Let $k_{1,1} = k_{2,1} = 2, k_{1,0} = k_{2,0} = 1$, then $K_1 = 22$ and $K_2 = 5738$, and we can check that $(K_1, 255) = (K_2, 255) = 1$. So, the results of Theorem 2 can be applied to achieve maximal period and complexity.

From the examples given above, this shows the existence of K_i such that $(K_{i-1}, 2^{n_i} - 1) = 1$ for $2 \leq i \leq r$.

5 Liu-Xiao Cascaded generators

The Liu-Xiao r-cascaded generator [7] consists of r linear feedback shift registers $LFSR_1$, $LFSR_2$, ..., $LFSR_r$ and cascading them one by one in the form of Liu-Xiao clock-controlled generators. The Liu-Xiao 2-cascaded generator is just the Liu-Xiao clock-controlled generator. Now, consider the Liu-Xiao 3-cascaded generator, the previous output sequences from 2-cascaded generator will control the clocking of the 3rd shift register $LFSR_3$ and the resultant sequence is the exclusive-or of the sequence from $LFSR_3$ with the control sequences, that is, the previous output sequences from the 2-cascaded generator. By a similar construction, the clocking of rth shift register $LFSR_r$ will be controlled by the previous output sequences from the $(r-1)$-cascaded generator.

Let $LFSR_1, LFSR_2, \ldots, LFSR_r$ be associated with the primitive polynomials $f_1(x), f_2(x), \ldots, f_r(x)$ of degrees $n_1, n_2, \ldots, n_r$ respectively. If the control sequence, an output sequence from i-cascaded generators, $z_t^{(i)}$ is 1, then $LFSR_{i+1}$ clocks $k_{i,1}$times, otherwise $LFSR_{i+1}$ clocks $k_{i,0}$ times. Let x_t be the output sequence from $LFSR_1$ and $y_t^{(i)}$ a sequence from $LFSR_i$ with regular clocking, for $2 \leq i \leq r$, then $y^{(i)}_{\sigma_{i-1}(t)}$is the output from $LFSR_i$ with the clock-controlled by $(i-1)$-cascaded generator. Therefore, the intermediate output sequences $z_t^{(i)}$ are given by

$$z_t^{(i)} = z_{t-1}^{(i-1)} \oplus y^{(i)}_{\sigma_{i-1}(t)} \text{ for } 2 \leq i \leq r,$$

where

$$\sigma_{i-1}(t) = \sum_{j=1}^{t} \{k_{i-1,1} z_j^{(i-1)} + k_{i-1,0}(z_j^{(i-1)} \oplus 1)\}, \ \sigma_0(t) = t$$

and

$$z_t^{(1)} = x_{t-1}, \ x_0 = 0.$$

Let T_i be the period of i-cascaded generator and

$$K_{i-1} = \sigma_{i-1}(T_{i-1}) \text{ for } 2 \leq i \leq r.$$

Theorem 3. *If $LFSR_1, LFSR_2, \ldots, LFSR_r$ are primitive polynomials of degrees $n_1, n_2, \ldots, n_r$ respectively with $n_1 \mid n_2 \mid \ldots \mid n_r$ and $(K_{i-1}, 2^{n_i} - 1) = 1$ for $2 \leq i \leq r$, then* *a. The period of* $\{z_t^{(r)}\}$ *is*

$$T_r = \prod_{j=1}^{r}(2^{n_j}-1),$$

b. The characteristic polynomial of $\{z_t^{(r)}\}$ *is*

$$f_1(x)\prod_{j=2}^{r} f_j^{(K_{j-1})}(x^{T_{j-1}}),$$

c. The linear complexity of $\{z_t^{(r)}\}$ *is*

$$L_r = \sum_{i=1}^{r} n_i \prod_{j=0}^{i-1}(2^{n_j}-1), \text{ where } n_0 = 1.$$

Proof. The proof of this Theorem is similar to the proof of Theorem 2. Here, we briefly give the proof as follow.

We prove (a) by induction, when $r = 1$, it is just one register and nothing to prove, when $r = 2$, the result is followed from Theorem 1 (a). Now, we assume that it is true for $r = m > 2$, then the period of $\{z_t^{(m)}\}$ is T_m. Consider Liu-Xiao $(m+1)$-cascaded generator, by Lemma 2 (a), the period of clock-controlled register is equal to $T_m.(2^{n_{m+1}}-1)/(K_m, 2^{n_{m+1}}-1)$. Given that $(K_m, 2^{n_{m+1}}-1) = 1$, so, the period of $\{y_{\sigma_m(t)}^{(m+1)}\}$ is T_{m+1}. Furthermore, $z_t^{(m+1)} = z_{t-1}^{(m)} \oplus y_{\sigma_m(t)}^{(m+1)}$, hence, the period of $\{z_t^{(m+1)}\}$ is $LCM\{T_{m+1}, T_m\} = T_{m+1}$ as T_m divides T_{m+1}.

We prove (b) and (c) by induction. When $r = 1$, it is just one register and nothing to prove. When $r = 2$, the result is followed from Theorem 1(b). Now, we assume that the result is true for $r = m > 2$, then the linear complexity of $\{z_t^{(m)}\}$ is L_m. Now, we consider Liu-Xiao $(m+1)$ -cascaded generator. Since the period of Liu-Xiao m-cascaded generator is T_m and $n_1 \mid n_2 \mid \ldots \mid n_{m+1}$, by Lemma 2 (b), the characteristic polynomial of $\{y_{\sigma_m(t)}^{(m+1)}\}$ is $f_{m+1}^{(K_m)}(x^{T_m})$ and irreducible. Hence, the linear complexity of $\{y_{\sigma_m(t)}^{(m+1)}\}$ is $n_{m+1}T_m$. We know that $z_t^{(m+1)} = z_{t-1}^{(m)} \oplus y_{\sigma_m(t)}^{(m+1)}$. Then the characteristic polynomial of $\{z_t^{(m+1)}\}$ is LCM of the characteristic polynomials of $\{z_t^{(m)}\}$ and $\{y_{\sigma_m(t)}^{(m+1)}\}$, that is, $f_{m+1}^{(K_m)}(x^{T_m})f_1(x)\prod_{j=2}^{m} f_j^{(K_{j-1})}(x^{T_{j-1}})$. Thus, the linear complexity of $\{z_t^{(m+1)}\}$ is $L_m + n_{m+1}T_m$ which is equal to L_{m+1}. Therefore, (b) and (c) are true for all $r \geq 1$. ◇

In the above Theorem part (c), if we let $n_1 = n_2 = \ldots = n_r$ ($=n$ say), then the period and linear complexity of this r-cascaded generator are $(2^n-1)^r$ and $n\sum_{j=0}^{r-1}(2^n-1)^j$ respectively, which are the same as that of [7] proved by Liu and Xiao in 1992. Therefore, we obtain a generalised result of Liu and Xiao.

6 Conclusion

In this paper, we obtained new results on the linear complexity of Gollmann-Chambers and Liu-Xiao r-cascaded generators. We also derived their characteristic polynomials. This allows us to have more choices on choosing the parameters $n_1 \mid n_2 \mid \ldots \mid n_r$ such that the sequences of cascaded generators achieve high linear complexity and long period. But, the more general result (for example, no restriction on $n_1 \mid n_2 \mid \ldots \mid n_r$) of the cascaded generators is still not known. Hence it is interesting to further explore or study the properties of such generators.

Acknowledgement The author would like to thank Prof. Harald Niederreiter and the anonymous referees for their helpful comments and suggestions.

References

1. T. Beth and F.C. Piper; The Stop-and-go generator, Advances in Cryptology, EUROCRYPT'84, Lecture Notes in Computer Science, Springer-Verlag, New York, **209** (1985), 88–92.
2. D. Gollmann, Pseudo random properties of cascade connections of clock controlled shift registers, Advances in Cryptology, EUROCRYPT'84, Lecture Notes in Computer Science, Springer-Verlag, New York, **209** (1985), 93–98.
3. D. Gollmann, Linear recursions of cascaded sequences, *Contributions to General Algebra* **3**, (1984), 173–179.
4. D. Gollmann and W.G. Chambers, A cryptanalysis of $\text{Step}_{k,m}$-Cascades, Advances in Cryptology, EUROCRYPT'89, Lecture Notes in Computer Science, Springer-Verlag, New York, **434** (1990), 680–687.
5. C.G. Günther, Alternating step generators controlled by De Bruijn sequences, Advances in Cryptology, EUROCRYPT'87, Lecture Notes in Computer Science, Springer-Verlag, New York, **304** (1988), 5–14.
6. R. Lidl and H. Niederreiter, "Introduction to finite fields and their applications", Cambridge University Press, 1986.
7. X. Liu and G.Z. Xiao, Linear complexity of a type of clock-controlled sequences, *Electronic letters* **28 No 8** (1992), 709–710.
8. B. Smeets, A note on sequences generated by clock controlled shift registers, Advances in Cryptology, EUROCRYPT'85, Lecture Notes in Computer Science, Springer-Verlag, New York, **219** (1986), 142–148.
9. R. Vogel, On the linear complexity of cascaded sequences, Advances in Cryptology, EUROCRYPT'84, Lecture Notes in Computer Science, Springer-Verlag, New York, **209** (1985), 99–109.

Feedback with Carry Shift Registers over $Z/(N)$

Jinzhong Xu[1] and Andrew Klapper[1]

Department of Computer Science,
University of Kentucky
Lexington, KY, 40506-0046
USA.
Email: {jxu,klapper}@cs.uky.edu
Project sponsored by the NSF under grant number NCR-9400762.

Abstract. For any integer $N > 1$, we construct feedback with carry shift registers over $\mathbf{Z}/(N)$ (N-FCSRs). The case when N is a prime, was investigated by Klapper and Goresky [1, 2]. We present the basic properties of N-FCSRs. We solve the register synthesis problem: given a small number of symbols of a sequence over $\mathbf{Z}/(N)$ synthesize the smallest N-FCSR that outputs the sequence.

Keywords: Feedback shift register, pseudo-random sequence generator, linear span, 2-adic span, stream cipher, register synthesis.

1 Introduction

Linear feedback shift registers (or LFSRs) provide a systematic method of analyzing sequences over finite fields [7]. This analysis and the speed of these devices have led to their use in the design of pseudo-random generators [5, 7] for applications such as stream ciphers, error correction, radar, CDMA, and software testing.

Sequences over other alphabets may be useful, but their analysis is generally more difficult. For example, one can consider LFSRs over the ring $\mathbf{Z}/(N)$ where N is a composite integer. However various methods of analysis of LFSRs, such as the Berlekamp-Massey algorithm [4, 7], are difficult to generalize to this setting due to the existence of zero divisors in $\mathbf{Z}/(N)$.

Much of the analysis of LFSRs can be based on the interpretation of a sequence over a ring as a power series in one variable. Klapper and Goresky developed a class of registers called *feedback with carry shift registers* (or FCSRs) together with an extensive analysis based on the interpretation of a sequence over a field $\mathbf{Z}/(p)$, p prime, as a so-called p-adic number [1, 2]. In particular, they used De Weger's rational approximation algorithm [8] to solve the FCSR-synthesis problem: given a short prefix of a sequence, find the minimal size FCSR that generates the sequence.

The purpose of this paper is to extend the FCSR architecture to sequences over $\mathbf{Z}/(N)$, with N an arbitrary integer greater than 1. The resulting generators are called N-FCSRs. We first describe the architecture of N-FCSRs and some of their basic properties.

The main result of the paper is a solution to the N-FCSR synthesis problem: given a prefix of a sequence $A = (a_0, a_1, \cdots)$, $0 \le a_i < N$, find the smallest N-FCSR that outputs A. Let $\lambda(A)$ be the size of this smallest N-FCSR, called the *N-adic span of A*. We describe an algorithm that requires at most $9\lambda(A) + 30$ symbols of A to produce the desired N-FCSR. The time complexity of the algorithm is $O(\lambda(A)(\lambda(A)C + N^{3/2}D\log N))$, where D is the time needed to divide numbers less than N^3 and C is the time needed to multiply numbers less than N.

If $\alpha(A,N) = \sum_{i=0}^{\infty} a_i N^i$, then the register synthesis problem is equivalent to finding the minimal (by a certain measure) integers u, q with $\alpha(A,N) = -u/q$. As it turns out, the approach used by Klapper and Goresky to solve the register synthesis problem when N is prime does not work when N is composite. Rather, we use a modification of the original Berlekamp-Massey algorithm. The general structure is that we maintain an approximation to $-u/q$ at each stage. An adjustment is made if the approximation is wrong for the next symbol. A significant difficulty arises here that does not occur with LFSRs due to the carry in arithmetic with N-adic numbers. To handle this, a more complicated adjustment is made that makes the approximation correct for the next three symbols.

For every composite integer N this gives a new measure of cryptographic security. Any sequence over $\{0, 1, \cdots, N-1\}$ is not secure for use in a stream cipher if its N-adic span is small. For if this is the case, then a known plaintext attack using only a small part of the key sequence will reveal enough information to construct a generator of the sequence. In particular, a binary sequence can be considered to be a sequence over this set in several ways – for example, we can encode $\lfloor \log N \rfloor$ consecutive bits as elements of $\mathbf{Z}/(N)$ or we can simply treat 0 and 1 as elements of $\mathbf{Z}/(N)$. Thus linear span is only one of an infinite family of measures that must be considered when evaluating the cryptographic security of a sequence.

For example, the sequence A over $\mathbf{Z}/(15)$ that is the 15-adic expansion of $-1/3449$ has 15-adic span equal to 3 and period 3448. Thus 57 elements of the sequence suffice to recover a simple generator of the sequence. Elements of $\mathbf{Z}/(15)$ can be encoded using 4 bits. Encoded in binary, A has period 13792, while 218 bits suffice to recover a simple generator. However, it can be shown that the sequence has linear span 3446. We have not done a systematic search for examples of this type, but it is likely that they abound.

2 Register Description

Let $N > 1$ be an integer and $S = \{a : 0 \le a \le N-1\}$. For any integer $r \ge 1$, the state of an N-Feedback with Carry Shift Register (or N-FCSR) of length r consists of r integers $a_0, a_1, \cdots, a_{r-1} \in S$ and an arbitrary integer $M = M_{r-1}$.

The state change function is determined by $r+1$ integers $d, q_1, q_2, \cdots, q_r \in S$ such that $\gcd(d, N) = 1$ and $q_r \neq 0$ as follows:

Step 1: Compute the integer sum $\sigma = M_{r-1} + a_{r-1}q_1 + a_{r-2}q_2 + \cdots + a_0 q_r$.

Step 2: Find $a_r \in S$ and $M_r \in \mathbf{Z}$ such that $\sigma = da_r + M_r N$.

Step 3: Output a_0 and replace the state $(M_{r-1}, a_0, a_1, \cdots, a_{r-1})$ by the new state $(M_r, a_1, a_2, \cdots, a_r)$.

Step 2 is always possible since $\gcd(d, N) = 1$. An N-FCSR is depicted in Figure 1.

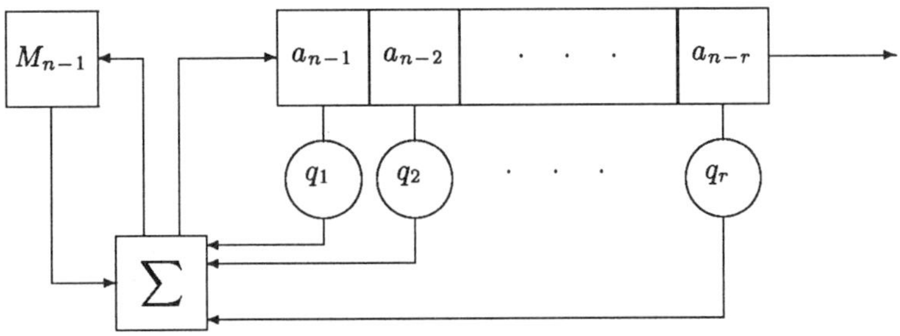

Fig. 1. An N-FCSR.

This device outputs a sequence $\{a_0, a_1, a_2, \cdots,\}$ over S by iterating the state change operation. The integer

$$q = -d + q_1 N + q_2 N^2 + \cdots + q_r N^r$$

is called the *connection number*. We let $q_0 = -d$. The connection number is analogous to the connection polynomial of an LFSR [7].

The output sequence of an N-FCSR is eventually periodic. This follows from the fact that the memory is bounded in any infinite execution. In fact, suppose w is the number of non-zero q_i among $\{q_1, \cdots, q_r\}$. Suppose at some state the register has memory M. If $|M| \leq w(N-1)$, then this remains true at all later states. If $|M| > w(N-1)$, then in the next state the memory decreases at least by one. Thus after a finite number of iterations the absolute value of the memory is bounded by $w(N-1)$. The hardware memory required to implement such an N-FCSR is fully determined by the parameters $N, w \leq r$, and the initial memory M.

The analysis of LFSRs and FCSRs arises by identifying a binary sequence with a power series or a 2-adic number, respectively. Similarly, for any $N > 1$, we can identify a a sequence $A = (a_0, a_1, \cdots)$ of elements of $\mathbf{Z}/(N)$ with an N-adic number, i.e., a formal power series

$$\alpha(A,N)=\sum_{i=0}^{\infty}a_iN^i.$$

We call this the *generating number* of A. The arithmetic of such numbers is similar to that using formal power series, but with carry. See [3] for a treatment of the case when N is prime. The arithmetic in the composite case is essentially the same.

Proposition 1. *A sequence A over $\mathbf{Z}/(N)$ is eventually periodic if and only if $\alpha(A,N)=-u/q$ with $u,q\in\mathbf{Z}$ and $\gcd(q,N)=1$. A is strictly periodic if and only if $0\le u<q$.*

It follows that the generating number of an output sequence from an N-FCSR is a rational number with denominator relatively prime to N.

Theorem 1. *Let $A=(a_0,a_1,\cdots)$ be the output from an N-FCSR with connection number q, $q_0=-d$, and initial memory M_{r-1}. Then*

$$\alpha(A,N)=\frac{u}{q}=\frac{\sum_{n=0}^{r-1}(\sum_{i=0}^{n}q_ia_{n-i})N^n-M_{r-1}N^r}{q}.$$

Moreover, A has period $T=ord_q(N)$ if $\gcd(u,q)=1$. In particular, if q is prime and N is a primitive root in $\mathbf{Z}/(q)^*$, then $T=q-1$.

Thus for any prime number q there is an N and an N-FCSR with connection number q whose output period is $q-1$.

3 The Register Synthesis Problem

The Berlekamp-Massey algorithm solves the LFSR synthesis problem of finding the shortest LFSR that generates a given sequence [4]. The synthesis problem for FCSRs in the case of 2-adic numbers is solved by a modification of De Weger's rational approximation algorithm [2,8], and these results apply to p-adic numbers when p is prime as well. In this section we solve the register synthesis problem for N-FCSRs when N is not necessarily a prime number. The extension of the Berlekamp-Massey algorithm for LFSRs over $\mathbf{Z}/(N)$ with N not a prime was carried out by Reeds and Sloane [6].

Definition 1. *For a sequence $A=\{a_i:i\ge 0\}$ over S, the N*-adic span *of A is the size of the smallest N-FCSR generating A plus the maximum number of bits used for the memory.*

In order to solve the synthesis problem for N-FCSRs, we:

1. Design an efficient algorithm that finds the reduced rational representation of the connection number of the sequence A. The running time and the number of bits required for convergence are polynomial in the N-adic span.

2. From the rational representation $\alpha = u/q$, efficiently construct an N-FCSR that generates the given sequence.

The main goal of this section is to describe and analyze these algorithms. We first consider (1). For any integer x, we define the *index of* x as

$$\mathrm{ind}_N(x) = \begin{cases} t & \text{if } 0 < x = a_0 + a_1N + \cdots + a_tN^t,\ a_i \in S,\ a_t \neq 0, \\ \mathrm{ind}_N(-x) & \text{if } x < 0, \\ -\infty & \text{if } x = 0. \end{cases}$$

Proposition 2. *Let $x, y \in \mathbf{Z}$. We have*

$$\mathrm{ind}_N(xy) \leq \mathrm{ind}_N(x) + \mathrm{ind}_N(y) + 1.$$

In particular, if $s \in S, y > 0$, then $\mathrm{ind}_N(sy) \leq \mathrm{ind}_N(y) + 1$. Also

$$\mathrm{ind}_N(x + y) \leq \max(\mathrm{ind}_N(x), \mathrm{ind}_N(y)) + 1.$$

Proof: If one of x, y is zero, the proposition is true when we take the convention $a + (-\infty) = -\infty$ for any finite integer a. For the product inequality, we assume both x and y are non-zero and positive. Let $s = \mathrm{ind}_N(x)$, $t = \mathrm{ind}_N(y)$. We then have two expansions:

$$x = a_0 + a_1N + \cdots a_sN^s \text{ and } y = b_0 + b_1N + \cdots b_tN^t.$$

It follows that $xy \leq (N^{s+1} - 1)(N^{t+1} - 1) = N^{s+t+2} - N^{s+1} - N^{t+1} + 1$, and then $\mathrm{ind}_N(xy) \leq s + t + 1 = \mathrm{ind}_N(x) + \mathrm{ind}_N(y) + 1$.

For the addition inequality, we have that $|x + y| \leq |x| + |y| \leq (N^{s+1} - 1) + (N^{t+1} - 1)$ and then $\mathrm{ind}_N(x + y) = \mathrm{ind}_N(|x + y|) \leq (s + 1)$ or $(t + 1)$. □

For any pair of integers (h, r) with $r \neq 0$, we define the following function induced by the index function. We always assume the second component in any integer pair to be non-zero unless the contrary is stated explicitly.

Definition 2. *For $N > 1$, $\Phi_N(h, r) = \max\{\mathrm{ind}_N(h), \mathrm{ind}_N(r)\}$.*

By the definition and the properties of the index function, one can show the following.

Proposition 3. *For any pairs (h_1, r_1), (h_2, r_2) and any $k \geq 0$, we have:*

1. $\Phi_N(h_1 + h_2, r_1 + r_2) \leq \max\{\Phi_N(h_1, r_1), \Phi_N(h_2, r_2)\} + 1$.
2. $\Phi_N(h_1r_2 + r_1h_2, r_1r_2) \leq \Phi_N(h_1, r_1) + \Phi_N(h_2, r_2) + 2$.
3. $\Phi_N(N^k(h_1, r_1)) = k + \Phi_N(h_1, r_1)$.

In the original Berlekamp-Massey algorithm, the degree of a polynomial and the maximal degree of a pair of polynomials act as norm functions. In our rational approximation algorithm, we use the index function and the function Φ in a similar way. Since we are doing integer arithmetic, multiplication and addition

```
Rational_Approximation
begin
input A = {a_i : 0 ≤ a_i ≤ k, i ≥ 0}
α ⟵ 1 + N Σ_{i=0}^{k} a_i N^i
(h_0, r_0) ⟵ (0, 1)
(h_1, r_1) ⟵ (a_0 + a_1 N + a_2 N^2, 1 + N^3)
m ⟵ 0
for (i = m + 1 to k − 1)
    if ((h_i − r_i α) ≢ 0 (mod N^{i+1})) {
        if ( ∃s ≠ 0 : (|s| ≤ N^{3/2} + 2) ∧ (N^{i+3} | s(h_i − r_i α)))
            (h_{i+1}, r_{i+1}) ⟵ s(h_i, r_i)
        else {
            Find s, t such that (s, t) ≠ (0, 0), |s|, |t| ≤ N^{3/2} + 2
                and N^{i+3} | s(h_i − r_i α) + tN^{i−m}(h_m − r_m α)
            (h_{i+1}, r_{i+1}) ⟵ s(h_i, r_i) + tN^{i−m}(h_m, r_m)
        }
        if (Φ_N(h_{i+1}, r_{i+1}) > Φ_N(h_i, r_i) and
            Φ_N(h_i, r_i) ≤ i − m + Φ_N(h_m, r_m) and t ≠ 0)
            m ⟵ i
    }
find u*, q* with u*/q* = h_k/r_k and gcd(u*, q*) = 1
output u, q with 1 + N(u/q) = u*/q*
end
```

Fig. 2. Rational Approximation Algorithm.

create carries and thus increase the values of ind and Φ. The rational approximation algorithm appears in Figure 2. The multi-step correction when we update our current approximation to the rational representation of α is designed to force convergence despite the carries resulting from integer arithmetic.

For $N = 3$, the bounds on the integers s, t inside the loop need to change to $[-5, 5]$ (see Lemma 1). To prove correctness and fast convergence of the algorithm, we first ensure that the algorithm makes progress correctly at each index i.

Lemma 1. *The integers s and t exist except when $N = 3$. For $N = 3$, the bounds need to be enlarged to be $[-5, 5]$, with $|s| \neq 5$ or $|t| \neq 5$.*

Proof: For integers x and y, there are unique $w_0, w_1, w_2 \in S$ such that:

$$x(b_i + b_{i+1}N + b_{i+2}N^2) + y(c_m + c_{m+1}N + c_{m+2}N^2) \equiv w_0 + w_1 N + w_2 N^2 \pmod{N^3}.$$

Let $\Psi(x, y) = (w_0, w_1, w_2)$. We need to find suitable integers s, t such that $\Psi(s, t) = (0, 0, 0)$. There are N^3 distinct triples (w_0, w_1, w_2).

Suppose we choose s, t in the range from $[-f(N), f(N)]$. Then there $(2f(N)+1)^2$ pairs to choose. Thus Φ is not one to one if $(2f(N)+1)^2 > N^3$. This holds if

$$f(n) \geq \left\lceil N^{3/2}/2 \right\rceil . \tag{1}$$

By taking the difference between two pairs with the same image, we obtain a pair in the range $[-2f(N), 2f(N)]$ with image $(0,0,0)$. For $N \geq 4$, equation (1) holds if $f(N) = N^2/4$.

For $N = 3$, Ψ is not one to one if s, t are chosen in the range $[-2, 3]$. Thus there is a non-zero pair such that $-5 \leq s, t \leq 5$ and $\Psi(s,t) = (0,0,0)$. Suppose $\Psi(s,t) = (0,0,0)$, but $|s| = |t| = 5$. Since 5 is relatively prime to 3, we can factor it out. □

To find s and t efficiently, we can pre-compute a table. The size of the table is at most N^3 (see the proof of Lemma 1). Alternatively, when N is small, we can simply conduct an exhaustive search. When N is large, the following allows us to speed this up.

Lemma 2. *In the algorithm, s and t can be found in time $O(N^{3/2} D \log N)$ where D is the time required to divide numbers less than N^3.*

Proof: Let $N > 3$. Let $x = x_0 + x_1 N + x_2 N^2 \equiv (h_i - r_i \alpha)/N^i \pmod{N^3}$ and $y = y_0 + y_1 N + y_2 N^2 \equiv (h_m - r_m \alpha)/N^m \pmod{N^3}$ with $x_0 \neq 0, y_0 \neq 0$ and $0 \leq x_i, y_i \leq N-1$. By Lemma 1, there are s, t with $|s|, |t| \leq \lceil N^{3/2} \rceil + 2$ such that $sa + tb \equiv 0 \pmod{N^3}$. We can find them by reducing the search space.

Let $u = \gcd(N^3, x, y)$ and $K = N^3/u$. Then $sx' \equiv -ty' \pmod{K}$, $x = x'u$, and $y = y'u$ for some integers x' and y'. Let $d = \gcd(K, x')$, $x' = x''d$, and $K = K'd$ for some integers x'' and K'. Then $\gcd(d, y') = 1$.

This implies that $d|t$, so that $t = t'd$ and $sx'' = -t'y' \pmod{K'}$ for some integer t'. Since $\gcd(x'', K') = 1$, there is an inverse v of x'' modulo K'. That is, $vx'' + cK' = 1$ for some integer c. This gives us the equation

$$s \equiv -t'y'v \pmod{K'}.$$

Now we need only search for t' such that $|t'| \leq (N^{3/2}+2)/(du)$ and s is congruent modulo K' to an element between $-(N^{3/2}+2)/(du)$ and $(N^{3/2}+2)/(du)$. Since greatest common divisors and quotients can be computed in time $O(D \log N)$, the bound follows. □

We modify α at the start of the algorithm so that initially the pair (h_m, r_m) satisfies (a) $(h_m - \alpha r_m) \equiv 0 \pmod{N^m}$ but not $0 \pmod{N^{m+1}}$ and (b) there is no integer s with $|s| \leq N^2/2$ so that $N^3 | s(a_m + a_{m+1}N + a_{m+2}N^2)$. The algorithm is designed to make both (a) and (b) true for indices i at which we update using both s and t. This guarantees that $s \neq 0, t \neq 0$ whenever they are used, and then $r_i \neq 0$ for each pair (h_i, r_i) in the algorithm.

To make precise statements about the algorithm, it is convenient to classify the the indices.

Definition 3. *Let i be an index.*

1. *If* $h_i - \alpha r_i \equiv 0 \pmod{N^i}$ *but not* $\pmod{N^{i+1}}$, *then* i *is an* updating index.
2. *If* i *is an updating index and* (h_{i+1}, r_{i+1}) *is obtained by multiplying* (h_i, r_i) *by an integer* s, *then* i *is a* type 1 updating index.
3. *If* i *is an updating index and* (h_{i+1}, r_{i+1}) *is obtained as a linear combination of* (h_i, r_i) *and* (h_m, r_m), *then* i *is a* type 2 updating index.
4. *If* $i = 0$ *or* i *is a type 2 updating index and* $\Phi_N(h_i, r_i) \leq i - m + \Phi_N(h_m, r_m)$ *and* $\Phi_N(h_{i+1}, r_{i+1}) > \Phi_N(h_i, r_i)$, *then* i *is a* turning point. *If* i_1 *and* i_2 *are consecutive turning points, then we say* i_2 *is the turning point next to or following* i_1.

Note that an index i is a turning point if it is either the initial index $m = 0$ or it is an index i where m is replaced by i.

We next show that the pair (h_i, r_i) corresponds to a well defined non-zero rational number.

Lemma 3. *Let* i *be a type 2 updating index, and* (s, t) *be the pair used in the combination. Then neither* s *nor* t *is zero.*

Proof: Recall that $N^{i+3} | s(h_i - \alpha r_i) + t(h_m - \alpha r_m)N^{i-m}$. If $s = 0$, then $N^{i+3} | t(h_m - \alpha r_m)N^{i-m}$, and $N^{m+3} | t(h_m - \alpha r_m)$. This is impossible because m is a turning-point. If $t = 0$, then $N^{i+3} | (h_i - \alpha r_i)$. This is also impossible because i is a type 2 updating index. □

Lemma 4. *Let* m *be a turning point. For any index* $i \geq (m+1)$ *before the next turning point,* (h_m, r_m) *and* (h_i, r_i) *are* **Z***-linearly independent. At any updating index* i, $(h_{i+1}, r_{i+1}) \neq (0, 0)$.

Proof: The proof is by induction. Note that at the initial stage, we have $(h_m, r_m) = (0, 1)$ and $(h_{m+1}, r_{m+1}) = (a_m N^m + a_{m+1} N^{m+1} + a_{m+2} N^{m+2}, 1 + N^{m+3})$ with $a_m \neq 0$. This shows that $(h_m, r_m), (h_{m+1}, r_{m+1})$ are **Z**-linearly independent.

Suppose $(h_m, r_m), (h_i, r_i)$ are **Z**-linearly independent and i is an updating index. If i is a type 1 updating index, we have $(h_{i+1}, r_{i+1}) = s(h_i, r_i), s \neq 0$. So $(h_m, r_m), (h_{i+1}, r_{i+1})$ are still **Z**-linearly independent. If i is a type 2 updating index, there are $s \neq 0, t \neq 0$ such that

$$(h_{i+1}, r_{i+1}) = s(h_i, r_i) + tN^{i-m}(h_m, r_m).$$

Suppose there are x, y such that $x(h_{i+1}, r_{i+1}) + y(h_m, r_m) = (0, 0)$. Then

$$xs(h_i, r_i) + (xtN^{i-m} + y)(h_m, r_m) = (0, 0).$$

This implies that $xs = 0, xtN^{i-m} + y = 0$. Since $s \neq 0$, it follows that $x = 0$, so $y = 0$. This shows that $(h_m, r_m), (h_{i+1}, r_{i+1})$ are **Z**-linearly independent. In particular, $(h_{i+1}, r_{i+1}) \neq (0, 0)$. A similar argument shows that if i is a type 2 updating index, then $(h_i, r_i), (h_{i+1}, r_{i+1})$ are **Z**-linearly independent. Since a new turning point is obtained only by type 2 updating, it follows that if i is a turning-point, $(h_i, r_i), (h_{i+1}, r_{i+1})$ are updated into the new $(h_m, r_m), (h_{m+1}, r_{m+1})$, then they are still **Z**-linearly independent. This completes the proof. □

Lemma 5. *At any updating index i, we have* $\mathrm{ind}_N(h_i) < i$.

Proof: The proof is again by induction on i. Note that at the initial stage, $h_m = 0$. Hence $\mathrm{ind}_N(h_m) < m$. Now we suppose it is true for every updating index $k \leq i$. We prove it is true for the next updating index. By the inductive hypothesis we have $\mathrm{ind}_N(h_i) < i$ and $\mathrm{ind}_N(h_m) < m$. Since $h_{i+1} = sh_i$ or $h_{i+1} = sh_i + tN^{i-m}h_m, \mathrm{ind}_N(h_i) < i$, and $\mathrm{ind}_N(N^{i-m}h_m) < i$, it follows that $|h_{i+1}| \leq N^2(N^{i+1} - 1) < N^{i+3}$ and $\mathrm{ind}_N(h_{i+1}) < i + 3$. On the other hand, we at least have $h_{i+1} = h_{i+2} = h_{i+3}$, hence we have $\mathrm{ind}_N(h_{i+3}) < i + 3$. Since the next updating index $j \geq i + 3$ and $h_j = h_{i+1}$, this shows that at the next updating index j, $\mathrm{ind}_N(h_j) = \mathrm{ind}_N(h_{i+3}) < i + 3 \leq j$. □

Theorem 2. *For every i, neither h_i nor r_i is zero.*

Proof: The result is true initially. Since it is true for type 1 updating, we only need to consider the type 2 updating case with $j = i + 1$, i an updating index. We have $(h_{i+1}, r_{i+1}) = s(h_i, r_i) + tN^{i-m}(h_m, r_m)$. Suppose $r_{i+1} = 0$. Then $h_{i+1} = h_{i+1} - r_{i+1}\alpha \equiv 0 \pmod{N^{i+3}}$. If $h_{i+1} \neq 0$, then $\mathrm{ind}_N(h_{i+1}) \geq i + 3$. By Lemma 5, we have $\mathrm{ind}_N(h_i) < i$ and, $\mathrm{ind}_N(N^{i-m}(h_m, r_m)) < (i-m)+m = i$. It follows that $\mathrm{ind}_N(h_{i+1}) \leq i+2 < i+3$. This contradiction shows that $h_{i+1} = 0$. Therefore $(h_{i+1}, r_{i+1}) = (0, 0)$. This contradicts Lemma 4. □

Thus the pair (h_i, r_i) represents a rational number whose N-adic expansion approximates α up to the $(i - 1)$-th term. The algorithm is *convergent at an index i* if $h_i/r_i = \alpha = u/q$. The following ensures that if (h_i, r_i) approximates α up to the i-th position, but $\Phi_N(h_i, r_i)$ is much smaller than i, then the algorithm is convergent at i.

Lemma 6. *If $i > \Phi_N(u, q) + \Phi_N(h_i, r_i) + 2$, then $h_i/r_i = u/q$.*

Proof: We have $h_i/r_i - u/q = bN^i/qr_i$ for some b. If $b \neq 0$, then $\Phi_N(bN^i, qr_i) \geq i$. By Proposition 3,

$$\begin{aligned}\Phi_N(bN^i, qr_i) &= \Phi_N(h_iq - r_iu, r_iq)\\ &\leq \Phi_N(r, q) + \Phi_N(h_i, r_i) + 2.\end{aligned}$$

This is a contradiction, so $b = 0$. □

We want to show that if the algorithm is not convergent at some index, then there is a larger index that is a turning point. Before updating (h_i, r_i) to (h_{i+1}, r_{i+1}) by using s and t, there are two cases that may cause convergence. The first is when the index repeatedly increases with no update. In this case $\Phi_N(h_i, r_i)$ is unchanged, but approximates more and more bits. Lemma 6 says that if the number of these updatings is large enough, the algorithm converges. The second case is when (h_i, r_i) is updated to (h_{i+1}, r_{i+1}) by multiplying by an integer s. Then $\Phi_N(h_{i+1}, r_{i+1}) \leq \Phi_N(h_i, r_i) + 2$. On the other hand, $h_{i+1} - \alpha r_{i+1} = s(h_i - \alpha r_i) \equiv 0 \pmod{N^{i+3}}$. This shows that the Φ-value increases by at most 2, but it approximates at least 3 more bits. Thus if this case repeats enough times without a turning point, the algorithm converges.

Next we show that if the algorithm is not convergent by index i, some later index is a turning point. We need to understand the pattern of updatings between a pair of turning points.

Lemma 7. *Let m be a turning point and i be an updating index before the next turning point. If $\Phi_N(h_i, r_i) \leq i - m + \Phi_N(h_m, r_m)$, then this is true for every $j > i$ before the next turning point.*

Proof: We only need to consider the next updating index j after i. We have (1) $(h_{i+1}, r_{i+1}) = s(h_i, r_i)$ or (2) $(h_{i+1}, r_{i+1}) = s(h_i, r_i) + tN^{i-m}(h_m, r_m)$. Recall that an updating occurs if and only if $h_j - r_j\alpha \equiv 0 \pmod{N^i}$ but $\not\equiv 0 \pmod{N^{i+1}}$. Note that $(h_{i+1}, r_{i+1}) = \cdots = (h_j, r_j)$ and $j - i \geq 3$. In case (1) we have

$$\begin{aligned}\Phi_N(h_j, r_j) &= \Phi_N(h_{i+1}, r_{i+1}) \\ &\leq \Phi_N(h_i, r_i) + 2 \\ &\leq i - m + \Phi_N(h_m, r_m) + 2 \leq (j - m) + \Phi_N(h_m, r_m).\end{aligned}$$

In case (2) we have $\Phi_N(h_{i+1}, r_{i+1}) \leq \Phi_N(h_i, r_i)$ because i is a type 2 updating index, but not a turning point. Therefore

$$\begin{aligned}\Phi_N(h_j, r_j) &= \Phi_N(h_{i+1}, r_{i+1}) \\ &\leq j - m + \Phi_N(h_m, r_m)\end{aligned}$$

remains true by the assumption. □

An index $k > m$ is called *normal* if $\Phi_N(h_k, r_k) \leq (k - m) + \Phi_N(h_m, r_m)$. Thus Lemma 7 shows that once a normal updating is reached, all further updating indices are normal until the next turning point.

Lemma 8. *Let m and m_1 be consecutive turning points, let t be the number of non-normal updatings between m and m_1, and let $m < i \leq m_1$. Then*

$$\Phi_N(h_i, r_i) \leq 2t + \Phi_N(h_{m+1}, r_{m+1}).$$

Proof: The value of $\Phi_N(h_j, r_j)$ only increases when j is a non-normal update point. At a non-normal type 2 update point j, $j - m + \Phi_N(h_m, r_m) < \Phi_N(h_j, r_j)$. Therefore

$$\Phi_N(h_{j+1}, r_{j+1}) \leq \max(\Phi_N(h_j, r_j), j - m + \Phi_N(h_m, r_m)) + 2 \leq \Phi_N(h_j, r_j) + 2$$

and the lemma follows. □

If the difference $i - \Phi_N(h_i, r_i)$ is large enough, then by Lemma 6, the algorithm converges. Thus we want to bound $\Phi_N(h_i, r_i)$. First we bound the increase at each updating.

Lemma 9. *If* $(h_{i+1}, r_{i+1}) = s(h_i, r_i) + tN^{i-m}(h_m, r_m)$ *then*

$$\Phi_N(h_{i+1}, r_{i+1}) \leq \max(\Phi_N(h_i, r_i), i - m + \Phi_N(h_m, r_m)) + 2.$$

Proof: Let $\mu = \max(\Phi_N(h_i, r_i),\ i - m + \Phi_N(h_m, r_m))$, and $h_{i+1} = sh_i + tN^{i-m}h_m$. Note that $\mathrm{ind}_N(h_{i+1}) = \mathrm{ind}_N(|h_{i+1}|)$. When $N > 3$, we have that $|s|, |t| \leq \lfloor N^2/2 \rfloor \leq N^2/2$, and then

$$|h_{i+1}| \leq (N^2/2)(N^{\mu+1} - 1) + (N^2/2)(N^{\mu+1} - 1) < N^{\mu+3} - 1.$$

Hence $\mathrm{ind}_N(z) \leq (\mu + 2)$. Similarly for $r_{i+1} = sr_i + tN^{i-m}r_m$.

When $N = 3$, by Lemma 1, s, t are chosen so that $|s|, |t| \leq 5$, but $|s|$ and $|t|$ are not both 5. It follows that

$$|h_{i+1}| \leq 5(N^{\mu+1} - 1) + 4(N^{\mu+1} - 1) = 9(N^{\mu+1} - 1) < N^{\mu+3} - 1.$$

Similarly for r_{i+1}, and this completes the proof. □

Let m and m_1 be consecutive turning points. We define the increase from m to m_1 as

$$\beta_{m_1} = \Phi_N(h_{m_1}, r_{m_1}) - \Phi_N(h_{m+1}, r_{m+1}).$$

For the initial m, $\beta_m = 0$. Let k_m be the number of turning points less than m.

Lemma 10. *At any turning point* m,

$$\Phi_N(h_{m+1}, r_{m+1}) \leq (m + 3) + 2k_m + \sum_{j \leq m} \beta_j - \Phi_N(h_m, r_m)$$

and

$$2k_m + \sum_{j \leq m} \beta_j \leq 2m/3.$$

Proof: Initially $k_m = 0, \beta_m = 0$. Note that $\Phi_N(h_{m+1}, r_{m+1}) = m + 3$ and $\Phi_N(h_m, r_m) = 0$, hence the lemma is true at the first turning point. Suppose it is true at turning point m and m_1 is the next turning point. Let $m < j_1 < \cdots < j_w = m_1$ be the updatings from m to m_1, and $u_1 = j_1 - m$, $u_i = j_i = j_{i+1} \geq 3$. Then $m_1 = m + u_1 + \cdots + u_w$. Since m_1 is a turning point, there exist s and t such that

$$(h_{m_1+1}, r_{m_1+1}) = s(h_{m_1}, r_{m_1}) + tN^{m_1 - m}(h_m, r_m).$$

By induction, the fact that $-\Phi_N(h_{m+1}, r_{m+1}) = \beta_{m_1} - \Phi_N(h_{m_1}, r_{m_1})$, and by Lemma 9 we have

$$\begin{aligned} &\Phi_N(h_{m_1+1}, r_{m_1+1}) \\ &\quad \leq (m_1 - m) + 2 + \Phi_N(h_m, r_m) \end{aligned}$$

$$\begin{aligned}
&\leq (m_1 - m) + 2 + (m+3) + 2k_m + \sum_{j\leq m} \beta_j - \Phi_N(h_{m+1}, r_{m+1}) \\
&= (m_1+3) + 2(k_m+1) + \sum_{j\leq m} \beta_j + \beta_{m_1} - \Phi_N(h_{m_1}, r_{m_1}) \\
&= (m_1+3) + 2k_{m_1} + \sum_{j\leq m_1} \beta_j - \Phi_N(h_{m_1}, r_{m_1}).
\end{aligned}$$

The second estimate remains. It is true initially. Assume

$$3(k_m + \sum_{j\leq m} \beta_j/2) \leq m.$$

By Lemma 8, $\beta_{m_1} = \Phi_N(h_{m_1}, r_{m_1}) - \Phi_N(h_{m+1}, r_{m+1}) \leq 2(w-1)$, so

$$\begin{aligned}
m_1 &= m + u_1 + \cdots + u_w \\
&\geq 3(k_m + \sum_{j\leq m} \beta_j/2) + u_1 + \cdots + u_w \\
&\geq 3(k_m + \sum_{j\leq m} \beta_j/2) + 3w \\
&\geq 3(k_m + \sum_{j\leq m} \beta_j/2) + 3 + 3\beta_{m_1}/2 \\
&= 3(k_{m_1} + \sum_{j\leq m_1} \beta_j/2).
\end{aligned}$$

Equivalently, we have

$$2k_{m_1} + \sum_{j\leq m_1} \beta_j \leq \frac{2m_1}{3}.$$

which proves the lemma. □

We define λ_m to be the smallest $\Phi_N(h,r)$ among all the pairs (h,r) with $h - \alpha r \equiv 0 \pmod{N^m}$. For an eventually periodic sequence $A = \{a_i \in S\}$ and

$$\alpha = \sum_{i=0}^{\infty} a_i N^i = u/q$$

with u and q relatively prime, we define $\lambda(A) = \Phi_N(u,q)$.

Lemma 11. *At any turning point m,*

$$\Phi_N(h_{m+1}, r_{m+1}) - \lambda_{m+1} \leq 5 + 2k_m + \sum_{j\leq m} \beta_j.$$

Proof: By Lemma 10, we have that

$$\Phi_N(h_{m+1}, r_{m+1}) \leq (m+3) + 2k_m + \sum_{j\leq m} \beta_j - \Phi_N(h_m, r_m).$$

Furthermore, $h/r - h_m/r_m = (hr_m - rh_m)/rr_m$ with $hr_m - rh_m = bN^m \neq 0$ for some b. Thus $\Phi_N(hr_m - rh_m, rr_m) \geq m$. On the other hand, $\Phi_N(hr_m - rh_m, rr_m) \leq \Phi_N(h,r) + \Phi_N(h_m, r_m) + 2$. Consequently, $\lambda_{m+1} = \Phi_N(h,r) \geq (m-2) - \Phi_N(h_m, r_m)$.

It follows that

$$\begin{aligned}\Phi_N(h_{m+1}, r_{m+1}) - \lambda_{m+1} &\leq (m+3) + 2k_m + \sum_{j\leq m} \beta_j - \Phi_N(h_m, r_m) \\ &\quad -(m - 2 - \Phi_N(h_m, r_m)) \\ &= 5 + 2k_m + \sum_{j\leq m} \beta_j.\end{aligned}$$

This proves Lemma 11. □

We are now ready to state the main theorem on the convergence of the algorithm.

Theorem 3. *Let i be any index and $\alpha(A) = u/q$ with u and q relatively prime. Then when $i > 9\lambda(A)+30$, the algorithm is convergent at i. That is, $h_i/r_i = u/q$. Thus u/q can be found by factoring out* $\gcd(h_i, r_i)$.

Proof: Let m be the largest turning point less than or equal to i. First suppose $m > 6(3+\lambda(A))$. Then we have $(h_{m+1}q - r_{m+1}u)/(r_{m+1}q) = (bN^{m+1})/(r_{m+1}q)$ for some integer b. If $b \neq 0$, we have the inequality:

$$\begin{aligned}\Phi_N(h_{m+1}q - r_{m+1}u, r_{m+1}q) &\leq 2 + \Phi_N(h_{m+1}, r_{m+1}) + \Phi_N(u,q) \\ &\leq 2 + 5 + 2k_m + \sum_{j\leq m} \beta_j + \lambda_{m+1} + \lambda(A) \\ &\leq 7 + \frac{2m}{3} + 2.\end{aligned}$$

On the other hand, $\Phi_N(bN^{m+1}, r_{m+1}q) \geq (m+1)$. When $m > 6(3+\lambda(A))$, we have $m+1 \leq 7 + 2m/3 + 2\lambda(A) < m+1$. This contradiction implies $b = 0$, so the algorithm is convergent at m.

Now suppose $m \leq 6(3+\lambda(A))$. Let $i = m+t+1$, so $t+1 > 3\lambda(A)+12$. The number of update points up to i is at most $(t+1)/3$, and each update at a non-turning point j increases $\Phi_N(h_j, r_j)$ by at most 2, so $\Phi_N(h_i, r_i) \leq \Phi_N(h_{m+1}, r_{m+1}) + 2(t+1)/3$. By Lemmas 10 and 11, $\Phi(h_i, r_i) \leq \lambda(A) + 5 + 2m/3 + 2(t+1)/3$.

Suppose $m > 3\lambda(A)+9$. Thus $i = m+t+1 \geq (m+t+1)/3 + (\Phi(h_i, r_i) - \lambda(A) - 5) > 2\lambda(A) + 22/3 + (\Phi(h_i, r_i) - \lambda(A) - 5) \geq \lambda(A) + \Phi(h_i, r_i) + 2$. By Lemma 6 the algorithm is convergent at i.

Finally, suppose $m \leq 3\lambda(A)+9$. Then $t > 6\lambda(A)+21$. Thus $i = m+t+1 \geq (m+t+1)/3 + (\Phi(h_i, r_i) - \lambda(A) - 5) > \lambda(A) + \Phi(h_i, r_i) + 2$. Again, by Lemma 6 the algorithm is convergent at i. □

We can use Theorem 3 and Lemma 2 to estimate the time complexity of the algorithm.

Theorem 4. *The algorithm converges in time $O(\lambda(A)(\lambda(A)C + N^{3/2}D \log N))$, where D is the time needed to divide numbers $< N^3$ and C is the time needed to multiply numbers $< N$.*

For any given eventually periodic sequence A over S, the rational approximation algorithm finds the generating number $\alpha = u/q$ with $\gcd(u, q) = 1$ and $\gcd(N, q) = 1$. Given u and q, it is possible to efficiently construct an N-FCSR with length $r \leq \text{ind}_N(q) + 1$, that generates an eventually periodic sequence over S whose corresponding N-adic number is $\alpha = u/q$. The details are similar to the case when N is prime [2].

References

1. A. Klapper and M. Goresky, 2-adic shift registers, in "Fast Software Encryption, vol. 809, Lecture Notes in Computer Science," Springer Verlag, New York, 1994, 174–178.
2. A. Klapper and M. Goresky, Feedback shift registers, 2-adic span, and combiners with memory, *Journal of Cryptology* **10** (1997), 111–147.
3. N. Koblitz, "*p*-Adic Numbers, *p*-Adic Analysis, and Zeta Functions," Graduate Texts in Mathematics Vol. 58, Springer-Verlag, New York, 1984.
4. J. Massey, Shift-register synthesis and BCH decoding, *IEEE Trans. Info. Th.*, **IT-15(1)** (1969) 122–127.
5. J. Massey and R. Rueppel, Methods of, and Apparatus for , Transforming a Digital Data Sequence into an Encoded Form, *U.S. Patent* **4797922** (1989).
6. J. Reeds and N. Sloane, Shift-register synthesis (modulo *m*), *SIAM J. Comp.*, **14** (1985) 505-513.
7. R. Rueppel, "Analysis and Design of Stream Ciphers," Springer-Verlag, New York, 1986.
8. B.M.M. de Weger, Approximation lattices of *p*-adic numbers, *J. Num. Th.*, **24** (1986) 70–88.

Author Index

Other titles in the DMTCS series:

Combinatorics, Complexity, Logic, Proceedings of DMTCS '96
D S Bridges, C Calude, J Gibbons, S Reeves, I Witten (Eds)
981-3083-14-x

Formal Methods Pacific '97 – Proceedings of FMP '97
L Groves and S Reeves (Eds)
981-3083-31-x

The Limits of Mathematics. A Course on Information Theory and the Limits of Formal Reasoning
G J Chaitin
981-3083-59-x

Unconventional Models of Computation
C S Calude, J Casti and MJ Dineen (Eds)
981-3083-69-7

Quantum Logic
K Svozil
981-4021-07-5

Proceedings of the International Refinement Workshop and Formal Methods Pacific '98
J Grundy, M Schwenke and T Vickers (Eds)
981-4021-16-4

Computing with Biomolecules. Theory and Experiments
G Paun (Ed)
981-4021-05-9

People and Ideas in Theoretical Computer Science
C S Calude (Ed)
981-4021-13-x

Combinatorics, Computation and Logic. Proceedings of DMTCS'99 and CATS'99
C S Calude and M J Dinneed (Eds)
981-4021-56-3

Polynomials. An Algorithmic Approach
M Mignotte and D Stefanescu
981-4021-51-2